Reminiscences of the Anglo-Boer War
Annotated and Illustrated

General Ben Viljoen, Commander of the South African Boer War Exhibit Company, 1904.[1]

Also Including The Transvaal Military Officer Album

Volume 6 of the series, *The Last Horse War*

Series compiled, annotated, illustrated and expanded by Payton Fireman, JD.

ISBN: 978-1-7320124-9-3

Front Cover Photo Credits:.[2]

Back Cover Photo Credits.[3] [4]

Frontispiece Photo Credits:[5]

1 Please see the Bibliography for license links and other information regarding the material in this volume.

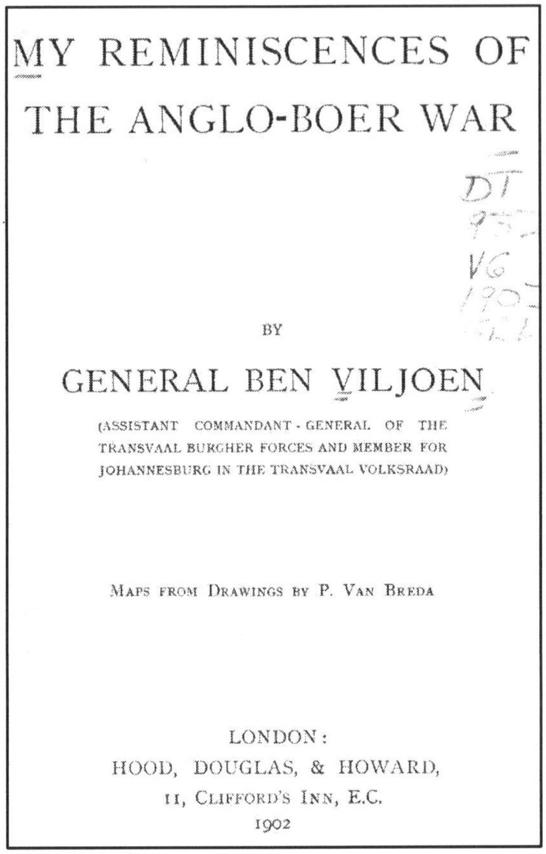

MY REMINISCENCES OF THE ANGLO-BOER WAR

BY

GENERAL BEN VILJOEN

(ASSISTANT COMMANDANT-GENERAL OF THE TRANSVAAL BURGHER FORCES AND MEMBER FOR JOHANNESBURG IN THE TRANSVAAL VOLKSRAAD)

MAPS FROM DRAWINGS BY P. VAN BREDA

LONDON:
HOOD, DOUGLAS, & HOWARD,
11, CLIFFORD'S INN, E.C.
1902

Publisher's Logo.

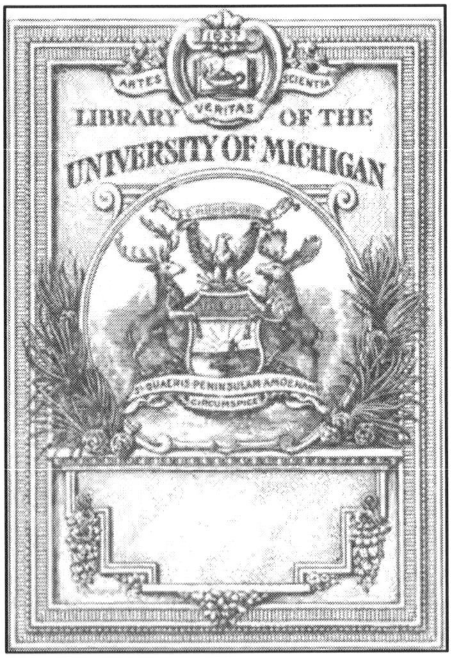

University of Michigan Copy.

Preserved by the University of Michigan. Digitized by Google and the Internet Archive in 2007 with funding from Microsoft Corporation.

Contents

Full list of Chapters

This detailed listing of individual chapters is included as well as the abbreviated one at the beginning of the volume.

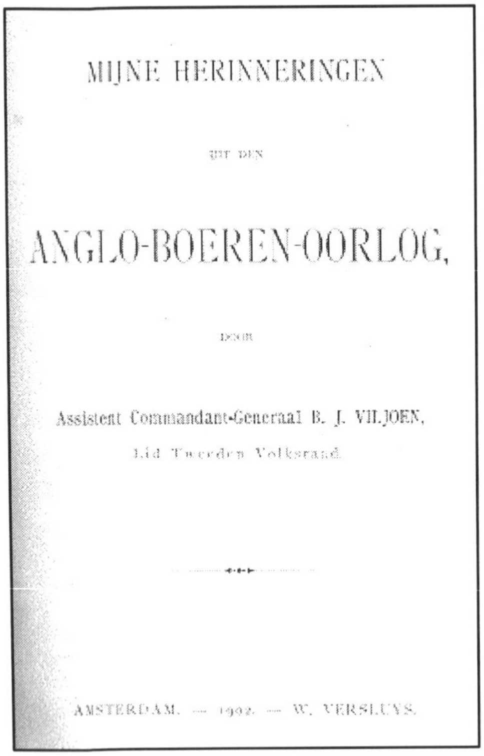
MIJNE HERINNERINGEN

UIT DEN

ANGLO-BOEREN-OORLOG,

DOOR

Assistent Commandant-Generaal B. J. VILJOEN,

Lid Tweeden Volksraad.

AMSTERDAM. — 1902. — W. VERSLUYS.

Front Page of the Original Dutch language Publication.[6]

PF Preface

There are only a few first-person accounts by senior commanders of the Boer armies. So far as can be found, Louis Botha did not write any extensive memoir about his service and command during the War.[2] Neither did Jacobus de la Rey, although his wife, Jacoba did.[3] Commandants Ben Viljoen and Christiaan de Wet are the two most senior military commanders who wrote extensive first person accounts in the immediate aftermath of the conflict. Both memoirists have an engaging style of writing with amazing experiences to relate. They also provide names, dates and other thousands of granular details that, as always, tie in to the other narratives in this series.

Ben and his narrative is so specific about the places and times where events take place that I was able to map the locations with some degree of specificity. The book is not as well annotated as some of the previous volumes in this series but by now it should be apparent that any mentioned place or name can be further identified with a little extra searching. I have been able to include images and thankfully, being able to quickly search for text or names in the previous books and resources I've assembled made the process easier but it was still quite time consuming.

I have also made use of the chronology that I've assembled to reference events. It's slant it very pro-British but the dates and locations are accurate. The British based chronology often phrases defeats as victories but so does our author. This point-counterpoint is interesting and a space saving way to augment the work without taking up a lot of room.[4]

Technical Considerations—As usual after adding all the maps, references and illustrations space in the book came at a premium in order to keep the total page count below around 375 pages; but in the process the Military Officer Album (MOA) has found its rightful home.[5] These are Ben's Brothers in Arms and it is remarkable how many he must have met in the field and again while he was a prisoner on St. Helena. Of the remainder, many are noted as not only staying in the field until the end of the conflict but are also cited as having taken an active part in the peace negotiation assembly of Commandants in May of 1902. You can cross reference with the names listed in the Appendix of De Wet's volume of those who participated in the final deliberations.

This work has 52 chapters. If I inserted a blank page for each chapter it would extend the book by that number. I have made the choice to simply begin a new chapter where it occurs on the page unless it is already very close to the bottom of the page where the previous chapter ends.

I wish I could completely annotate and illustrate this work. By now I know many of the participants who are mentioned by name. Many have previously been annotated and illustrated in prior volumes. There is just not enough space in this book for that to occur.

Text and Spelling: This is a real and intractable problem. The spelling in the text does not match the spelling in the chronology. Neither match the spelling used in the maps and those spellings do not match the ones used in the original image captions. Since the maps are printed in the book one has to use the spelling used in the maps themselves or create yet more confusion.

I have come to the conclusion that there simply was no agreed upon spelling for many of the place names and events. However I must note that it seems that the British were particularly determined to use different spellings than those used by the native Afrikaans authors. Nothing agrees with anything else. There is simply no way to unscramble it all. Fortunately while the spelling varies the names are unique enough that one is able to determine what location is referred to.

I've generally retained the original captions of the photographs and this usually includes the time and place which is, except in rare cases not contemporaneous with the events described. I've tried to provide representative images of the engagement, piece of equipment, or scene of action as best as my research will allow.

[2] There are several biographies of Louis Botha and they are listed in the Series Appendix along with some research on the topic of memoirs. See Discussion: Louis Botha did not publish any extensive first person memoir of his Boer War experiences. https://www.angloboerwar.com/forum/11-research/33431-louis-botha-s-memoirs

[3] See her Dutch work "A Woman's Wanderings and Trials During the Anglo-Boer War" translated by Lucy Hotz, in the bibliography.

[4] Chronology adapted from Creswicke, South Africa and the Transvaal War Vols. 2–7 and With Roberts to the Transvaal by C.N. Robinson.

[5] MOA refers to the Military Officer Album cited in the Bibliography.

Many of the maps that follow Ben's campaigns are from Times' History volumes 4 and 5 as indicated. The best online version is the University of Toronto set of copies.[7] [8] Those files, in color, have more discernible detail than the enhanced black and white versions in this work.[9] I've processed and reprocessed these maps to the limit of my ability and still they are only barely adequate. I've underlined and identified many locations but still the notations are hard to find within the images. I wish they were a better and more easily followed guide for the casual reader but that is not the case. It's too late to take them out and would detract from the work. For scholars who desire to closely follow the campaigns, they are useful in identifying the areas of activity by comparison to the online versions which have better legibility. For serious readers I might suggest having a pencil to circle locations on the maps the reader wants to be able to refer to more easily.

I continue to deprecate italics and quotation marks for obvious quotes. The form and style of my citations has not improved. All uncited images are original to the volume in which they appear.

Ben quotes Dr. Conan Doyle's work, 'The Great Boer War,'[6] often. There must have been a copy of this work on St. Helena where Ben was interred as a prisoner of war. His memory is remarkable, given that he was a prisoner of war and without many if any of his records. It appears he knows the places, names, dates backwards and forwards. Later I realized that given the number of his compatriots who were on St. Helena with him; he had a ready source of other individuals and their recollections to assist in refreshing his own.

A few chapters relate events that are not in chronological order. For instance, Chapter 8 takes place in March through May, 1900; but Chapter 9 goes back to the previous December of 1899.

The word "Kaffir" is an Arabic word meaning one who does not believe in the religion of Mahomet. It was introduced into South Africa by the Portuguese and subsequently applied to the tribes living on the North Eastern side of the Cape Colony.[7] It did not acquire the pejorative meaning that it is currently attached to it until later in the 20th Century.

Glossary

See the Glossary in Volume 1 for many Afrikaans terms.[8] Terms that apply specifically to this word are:

Boschveld: Literally means 'bush field'. It usually refers to a specific part of SA which includes parts of Limpopo and North West Provinces (two of our most northern provinces) but also stretches into Botswana and Zimbabwe, so areas close to the Tropic of Capricorn, sub-tropical and mainly open grassland, with bush.

Hoogeveld: Literally means 'high field' and is in the central / northern area of the plateau of SA. Predominantly open, flat grassland. Historically this area Johannesburg / Pretoria / Free State area as the 'Highveld'. The area closer to the coast, below the escarpment, in the province of Mpumalanga, closer to Mozambique, is referred to as the 'Lowveld.'

Pannetjes: These are small pans. But: Can be a hill, bulge or bump.

Rannjes: This means literally 'a little bit.'

Rantes: This literally means 'Scraps.'

Horse Sickness: This is African Horse Sickness, a highly infectious and often fatal viral disease.[9]

[6] Doyle, Arthur Conan. The Great Boer War. Vol 1 1900. See the Bibliography infra for the full citation.

[7] Handbook of the Boer War. Footnote 4 in chapter 1 'Roundheads of South Africa.'

[8] Many thanks to my friend in SA Wendy Wrinch-Schulz for identifying these somewhat archaic Afrikaans terms that Ben Viljoen uses often in this work.

[9] African Horse Sickness is a disease that is spread by insect vectors. The biological vector of the virus is the Culicoides (midges) species, but this disease can also be transmitted by species of mosquitoes including Culex, Anopheles, and Aedes including A. aegypti, and species of ticks such as Hyalomma and Rhipicephalus.
Horses are the most susceptible host with close to 90% mortality of those affected, followed by mules (50%) and donkeys (10%). African donkeys and zebras very rarely display clinical symptoms, despite high virus titres in blood, and are thought to be the natural reservoir of the virus. AHS manifests itself in four different forms: https://en.wikipedia.org/wiki/African_horse_sickness

Maps

General maps of South Africa have been omitted from this volume, since they have been almost overused in previous volumes in the series. Uncited maps are either original to the book or cited in this map section and the cite is not reproduced when the map or a portion is reused.

Eastern Transvaal RR Maps

Much of Ben's service takes place in proximity to the Pretoria to Delagoa Bay rail system that connects the Transvaal to Portuguese South Africa. Many of these 'stations' will be referred to and mapped in the text.

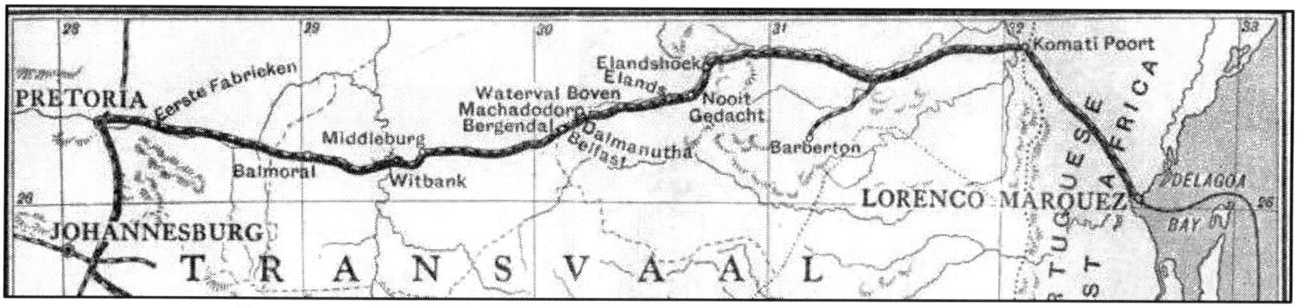

General Map of SA.[10]

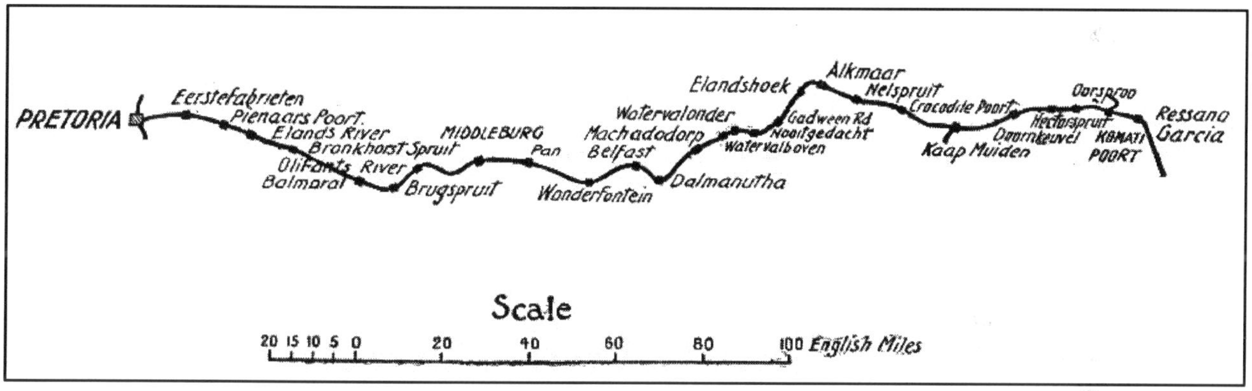

Pretoria to Komati Poort Railway.[11]

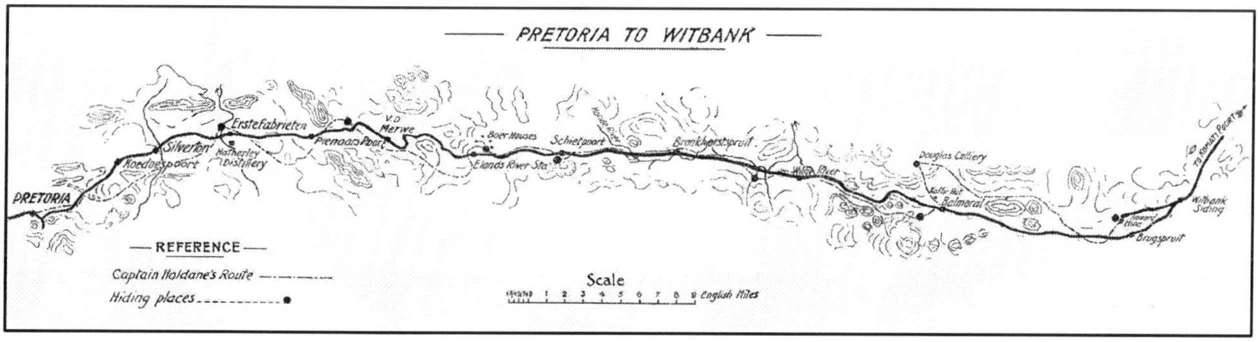

Pretoria to Whitbank.[12]

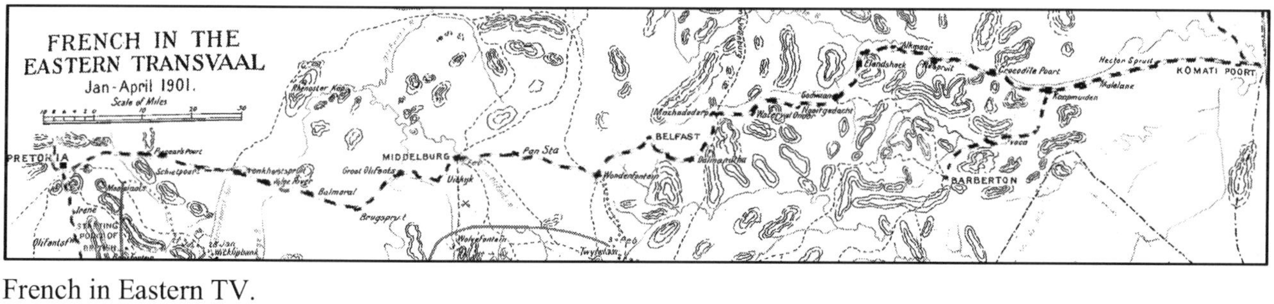

French in Eastern TV.

Davitt Map

This map identifies a large number of locations in North Eastern Transvaal where Viljoen is conducts his operations.

Davitt, Boer Fight for Freedom, 3rd ed. 1902. (Duquesne Scan) South Africa.[13]

Hillegas Map

This map contains details of North Eastern Transvaal as well as some indication of topography.

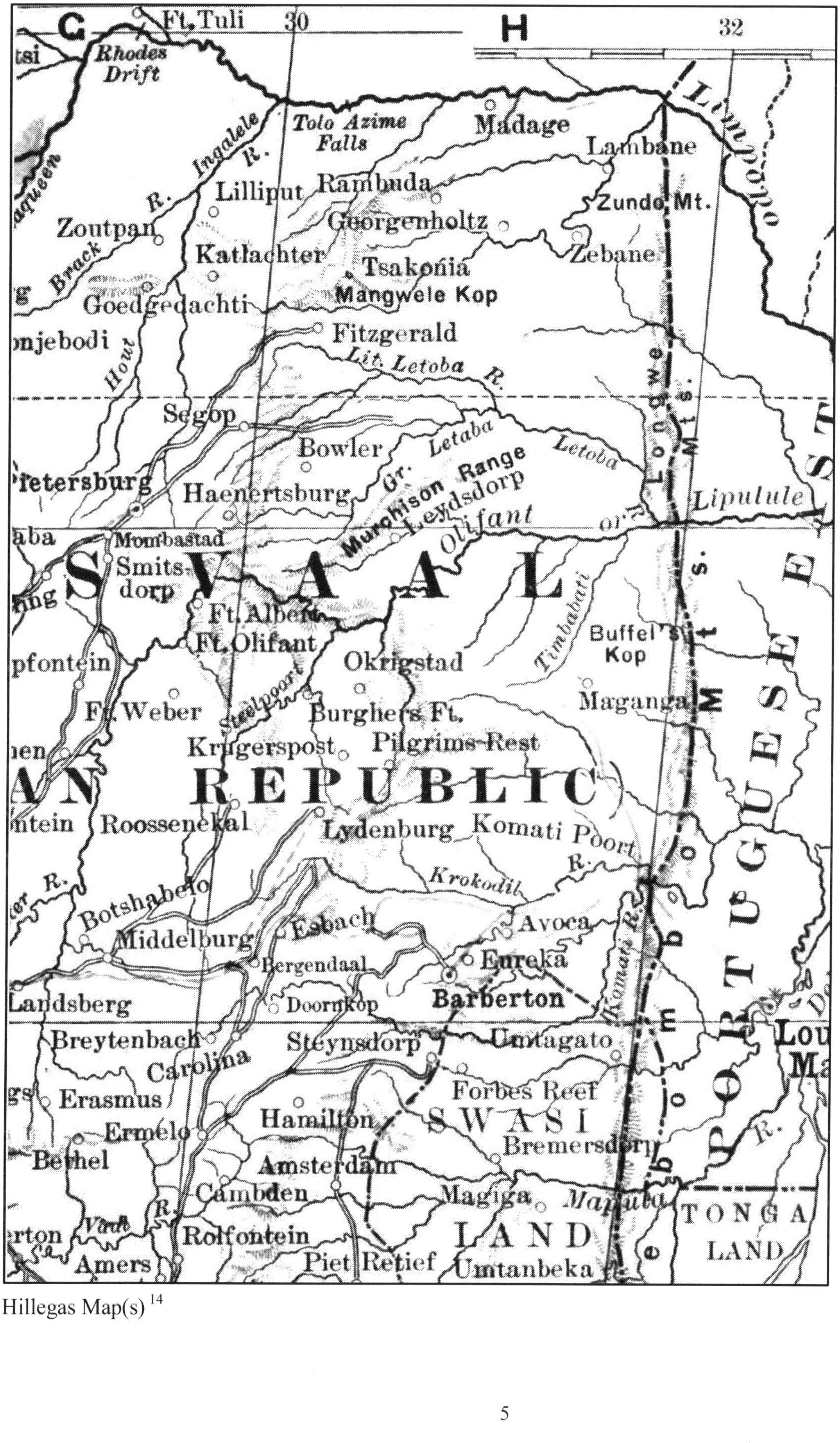

Hillegas Map(s) [14]

Walker Map

This map identifies generalized topographic features and the names of rivers.

Map By: Geo. H. Walker & Co. Boston. Copyright 1899, Library of Congress Maps and Charts # 2877.[15]

Times' History Maps

The following maps are from Times' History volumes 4 and 5 as indicated. The best online version is the University of Toronto set of copies.[16] [17] Those files, in color, have more discernible detail than the enhanced black and white versions in this work.[18] I've processed and reprocessed the maps to the limit of my ability and still they are only barely adequate. I might suggest having a pencil to circle locations on the maps the reader wants to be able to refer to easily.

Transvaal Government Locations

Wanderings of Transvaal Govt.[19] This map is used extensively to follow Ben's activity.

The following are the locations assigned by Leo Amery's, Times History volumes to where the Transvaal Government was located during the conflict.

With volume 4 and 5 being published in 1906 and 1907, it is apparent that the team at the Times, had read both Christiaan De Wet's Book and Ben Viljoen's as well as the entire catalog of Boer War literary production and that their mapping team had incorporated many of those location references in Times' maps. But not all. Trying to locate these places, along with many others, has proved a challenge and I have not been completely successful

Most Times History maps were produced by Spottiswoode & Co. Ltd.[20]

Hectorspruit, 17 Sept. 1900
Pretorius Kop.
Macmac.
Spekboom R. 2 Oct.
Boschhoek, 13-30 Oct.
Draaikraal, 31 Oct.-13 Nov.
Steelpoort R., 14 Nov.
Paardeplaats, 30 Nov. 1900-5 Apr. 01.
Pietersburg, 8-12 Jan. 1901 (3 members of Govt)
Near Pan Station 6 Apr.
Spitz Kop Farm 9 Apr.
Rietspruit, 16 & 23 Apr.
Beginderlyn, 19 Apr.
Vyfhoek, 21-22 Apr.
Vlakfontein, 2 May
Klipbank, 3 May
De Emigratie, 11 May
Mooiplaats, 12-16 May
Spitz Kop, 17 May
Isihlele R., 26 May
Piet Retief, 6 June
St Helena, 8 June
Lang Kloof, 9 June
De Emigratie, 12 June
Van Oudtshoornstroom 14 June
Kranspoort, 15 June
Kalbas Fn, 16 June
Yzervark Fn
Driefontein 18 June
Rietspruit, 19 June
Strehla, 23 June
Kromdraai, 28 June-1 July
Wildebeest Fn 3 July
Kaffir Kraal 4 July
Eerste Geluk, 5 July
Rietfontein

Onverwacht, to 17 July
Zwakfontein, 18 July
Graspan
Vaalbank
Uitkomst, 5 August
Uitgezocht, 6-9 August
Klipbank, 10 August
Compies R., 11-12 August
Schimmelhoek, 13-16 August
Delport, 17-27 August
Athole, 28 Aug.-12 Sept
Maryvale 14-17 Sept.
Bellskop, 20 Sept.
Nerston, 21-23 Sept.
Athole, 24 Sept.-16 Oct.
Mooiplaats, 17-24 Oct.
Klipstapel, 25 Oct.
Kranspoort, 26-30 Oct.
Brakfontein, 31 Oct-3 Nov.
Roodepoort, 4-12 Nov.
Tafel Kop, 13-14 Nov
Kranspan, 14-22 Nov.
Smithfield, 23-25 Nov.
Twyfelaar, 26 Nov.
Windhoek, 18 Dec.
Paardeplaats to 6 Jan 1902.
Windhoek, 7-21 Jan.
Oshoek, 21 Jan.
Grady's Farm, to 9 Feb.
Witpoort, 10 Feb.
Bothasberg, 11 Feb.
Langkloof, 13-17 Feb.
Stroomwater, 17 Feb.
Zwartfontein, 28 Feb.
Stroomwater 29 Feb.-6 Mch.
Witnek, 11 & 14-21 Mch.
12 March
Balmoral, 22 March.

Latitude and Longitude

Following are the latitude and longitude coordinates of as many of the locations as I could locate. Fortunately Google Earth still recognizes the Afrikaans names when searching for South African locations as an alternate to the current location name or "Pin" associated with the place name, which has been changed.

As mentioned in previous volumes the limited variety of nomenclature used by the Dutch immigrants regarding the naming of places leaves something to be desired. The same term or term is used for many similar features and it is hard to sort out. A few more examples for the record:

Modder River (Mud River), Zand River (Sand River), Olifants River (Elephants River) Whitpan (White Pan) Dozens of 'Tafel Kop's (Table Top Kopje) Spitz Kop; and the list goes on from there.

Following are as many locations as I could identify using Google Earth. Some have a higher degree of confidence than others but they all do relate to identified and 'pinned' locations in the program as of 2025. I know where some of the other locations are to within a mile or two but have declined to just put in pins based upon that guesswork. Some have locations I have been completely unable to identify.

If this was a database, like I want it to be, I would just insert hyperlinks and you could go there.

	Wanderings of the Transvaal Government.			
	Location	Date	Latitude	Longitude
1	Hector Spruit	Sept. 17 1900	25°26'23.28"S	31°40'51.83"E
2	Pretorious Kop		25°10'0.00"S	31°16'0.00"E
3	Mac Mac		25° 0'14.99"S	30°48'55.38"E
4	Spekboom R	October 2nd	24°52'7.97"S	30°23'54.41"E
5	Boschhoek	Oct. 13-30	24°33'55.01"S	28°17'59.37"E
6	Draaikraal	10/31-11/13	25°12'59.28"S	30° 3'2.28"E
7	Steelport R	Nov 14	24°49'33.63"S	30° 7'27.58"E
8	Paardeplaats	Nov 30 1900	30° 7'27.58"E	29°42'59.55"E
9	Pietersburg	Jan 8-12 1901	23°53'53.18"S	29°26'59.99"E
10	Pan Station (Approx)	April 6	25°45'2.71"S	29°39'29.44"E
11	Spitz Kop 1	April 9	25° 7'30.55"S	30°46'41.11"E
12	Rietspruit	April 16 & 23	26°13'24.88"S	29° 5'58.07"E
13	Begideerlyn	April 19		
14	Vyfhoek	April 21-22	26°41'16.97"S	27° 6'26.48"E
15	Paardeplaats	April 30		
16	Vlakfontein	May 2	26°22'42.09"S	27°53'3.12"E
17	Klipbank	May 3	26°41'47.24"S	30° 4'47.91"E
18	De Emigratie	May 11	26°32'48.89"S	29°59'19.12"E
19	Moiplaats	May 12-16	25°48'17.01"S	28°24'3.67"E
20	Spitz Kop Farm	May 17		
21	Ishlele River	May 26th		
22	Piet Retief	June 6	27° 0'35.54"S	30°48'15.03"E
23	St. Helena	June 8	26°57'34.23"S	30°23'44.60"E
24	Lang Kloof	Jun 9		
25	De Emigratie	June 12	26°32'48.89"S	29°59'19.12"E
26	Van Oudshoornstroom	June 14	26°30'32.51"S	30° 0'41.86"E
27	Kranspoort	June 15	25°26'11.71"S	29°25'12.25"E
28	Klabasf	June 16		
29	Yzervarkf	June 17	25°36'59.98"S	28°41'0.13"E
30	Rietfontein/Drietfontein	June 18	26°59'12.97"S	30°26'7.02"E
31	Rietfontein (2)		24°35'43.38"S	30°31'52.53"E
32	Rietspruit	June 19	26°13'24.88"S	29° 5'58.07"E
33	Strehla	June 23		
34	Kromdraai	June 28- July 1		
35	Wildebeestf	July 3		
36	Kaffir Kral	July 4		
37	Erstegluk	July 5		
38	Rietfontein (2)		24°35'43.38"S	30°31'52.53"E
39	Onverwacht	To July 17	23°41'0.11"S	27°41'0.18"E
40	Zwakfontein	July 18	25°23'60.00"S	29°10'60.00"E
41	Graspan		27°40'2.22"S	24°52'44.75"E
42	Vaalbank		25° 9'36.88"S	28°50'26.68"E
43	Uitkomst	August 5	27°36'29.60"S	30° 8'1.53"E
44	Uitgezocht	August 6-9	28°27'26.78"S	30° 7'27.43"E
45	Klipbank	August 10	28°33'44.00"S	29°46'29.80"E
46	Compies	August 11-12		
47	Schimmelhoek	August 13-16		
48	Delport	August 17-27	26°14'27.92"S	28°11'55.20"E
49	Athole	Aug 28-Sept 12	26° 6'59.58"S	28° 4'0.87"E
50	Maryvale	Sept. 14-17	26° 9'14.20"S	28° 5'13.30"E
51	Bellskop	Sept 20th	28°31'5.00"S	29°46'25.00"
52	Bellskop 2		26°28'0.00"S	30°47'60.00"E
53	Nerston	Sept 21-23	26°34'10.05"S	30°47'26.58"E
54	Athole (2)	Sept 24-Oct. 16	26°14'27.92"S	28°11'55.20"E
55	Moiplaats	October 17-24		
56	Klipstapel	October 25th	25°37'22.00"S	29° 6'16.00"E
57	Kranspoort	October 26-30		
58	Brakfontein	Oct 31-Nov 3	25°58'59.99"S	28°47'0.00"E
59	Roodepoort	Nov. 4-12	26° 8'50.18"S	27°54'3.83"E
60	Tafel Kop	Nov 13-14	25° 2'9.12"S	29°30'34.56"E
61	Kraspan	Nov. 14-22		
62	Smithfield	Nov 23-25		
63	Twyfelaar	Nov 26		
64	Windhoek	Dec 18	25°25'12.27"S	30° 5'51.43"E
65	Paardeplaats	January 6 1902	25° 2'15.48"S	29°42'59.55"E
66	Windhoek 2	January 7-21	25°25'12.27"S	30° 5'51.43"E
67	Oshoek	Jan 21	26°12'37.04"S	26°12'37.04"S
68	Grady's Fram	To Feb 9, 1902		
69	Witpoort	Feb 10	27°10'43.67"S	26° 7'11.38"E
70	Bothasberg	Feb 11	27°31'23.00"S	29° 6'4.00"E
71	Langkloof	Feb 13-17		
72	Stroomwater	Feb 17		
73	Zwartfontein	Feb 28	26°49'33.78"S	28°10'10.91"E
74	Stroomwater	Feb 29- March 6		
75	Witnek	Feb 11 & 14-21 March		
76	Balmoral	March 22nd	25°52'0.00"S	28°58'0.00"E

The Transvaal

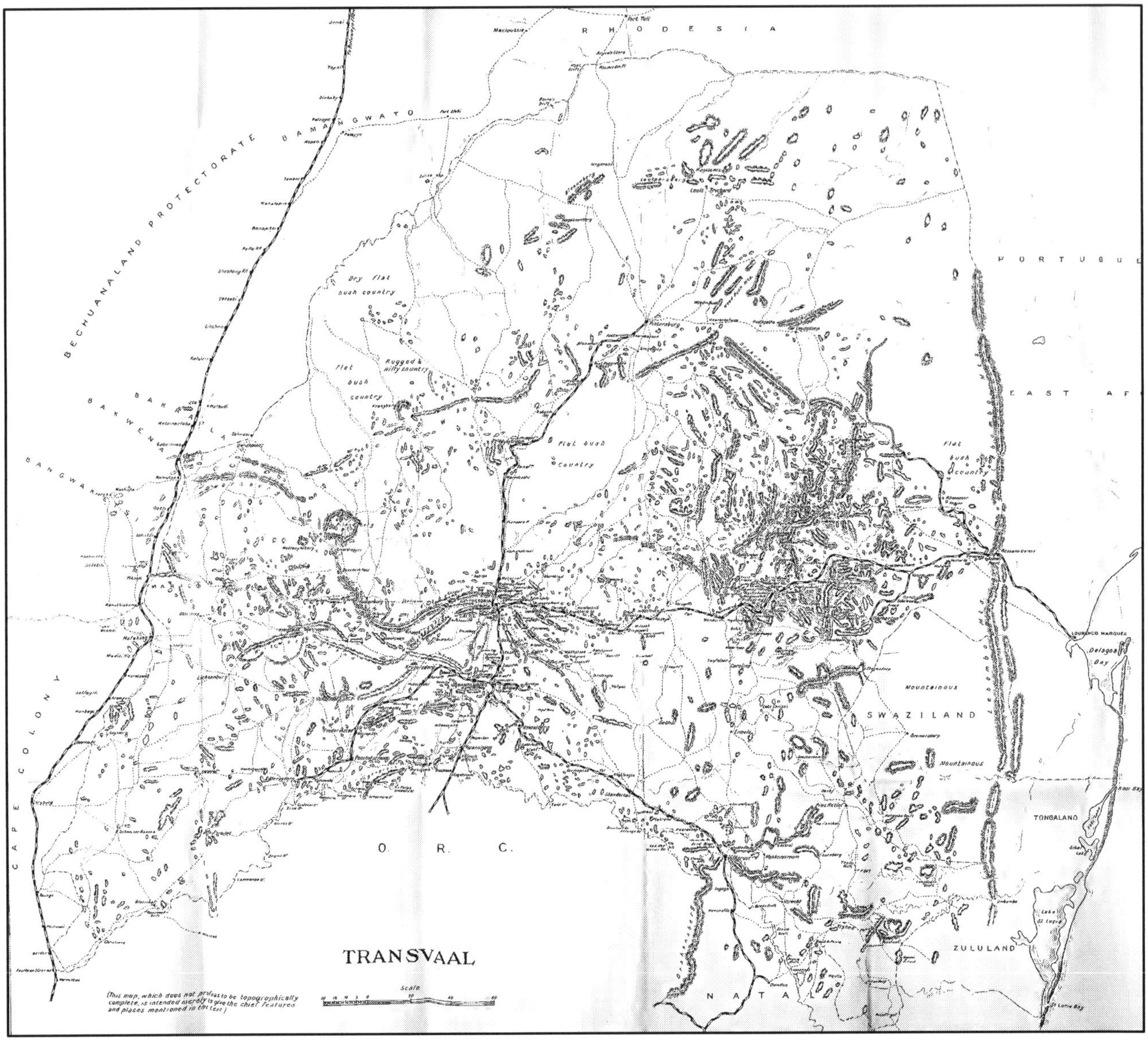

Transvaal Vol 4.[21] PF: Also reproduced in Vol 5. This is the best copy I was able to obtain.

Plummer & Bindon Blood in NE Transvaal

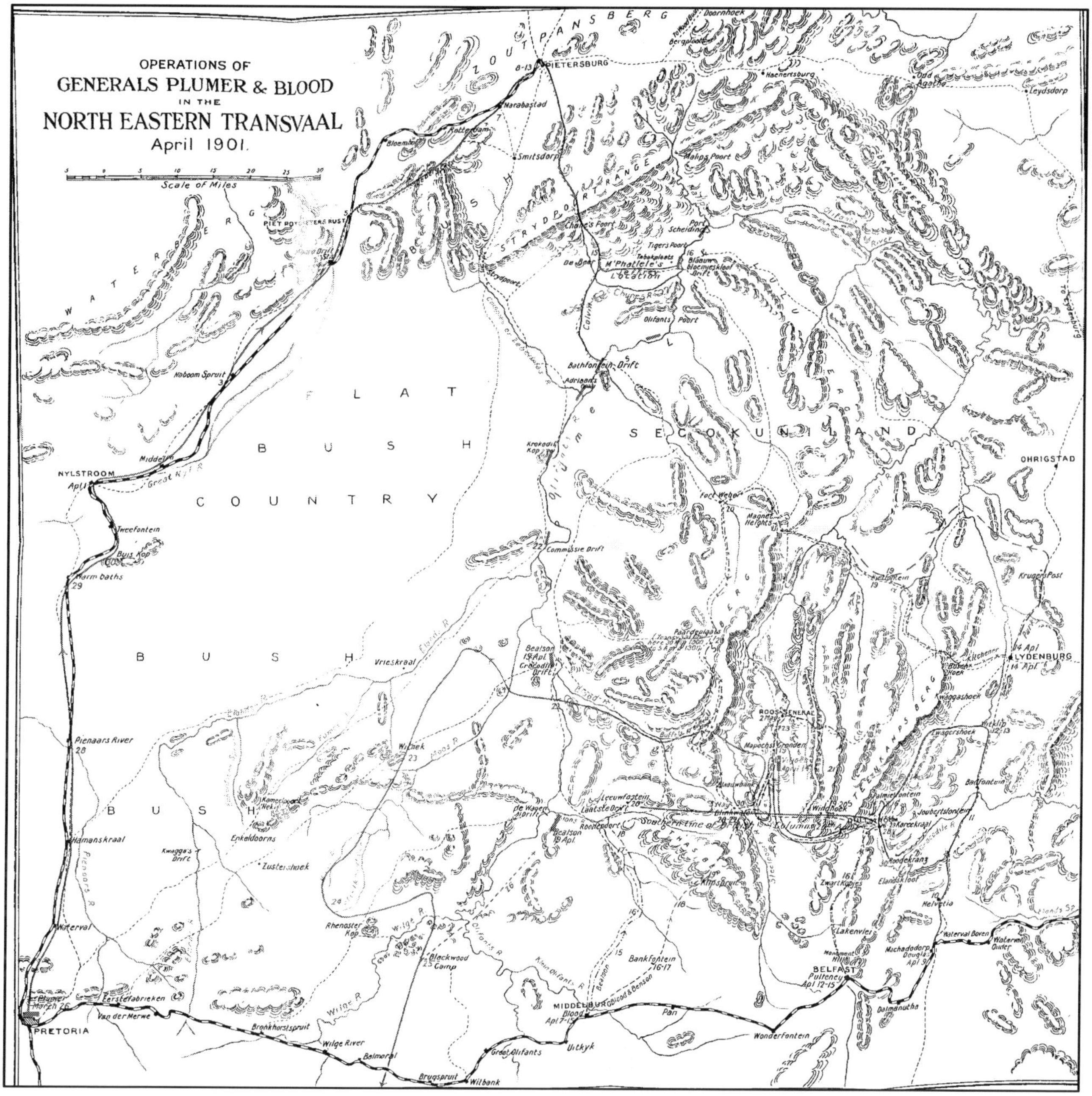

Plummer & Bindon Blood in NE Transvaal.[22]

French in E TV

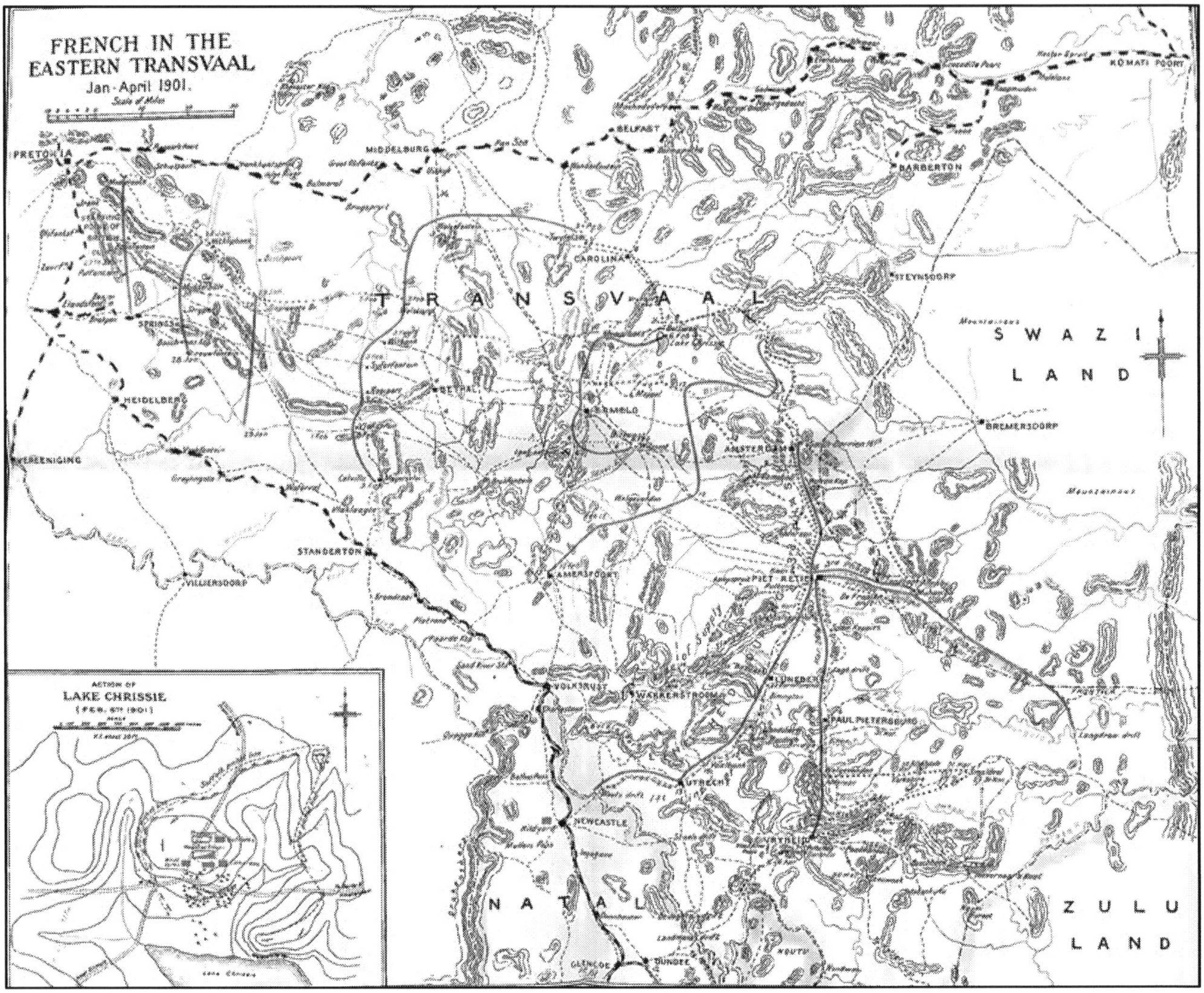

French in E TV.[23]

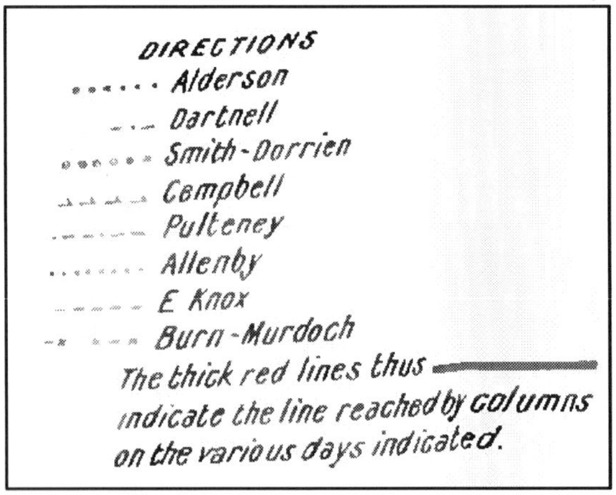

E & N.E. Transvaal

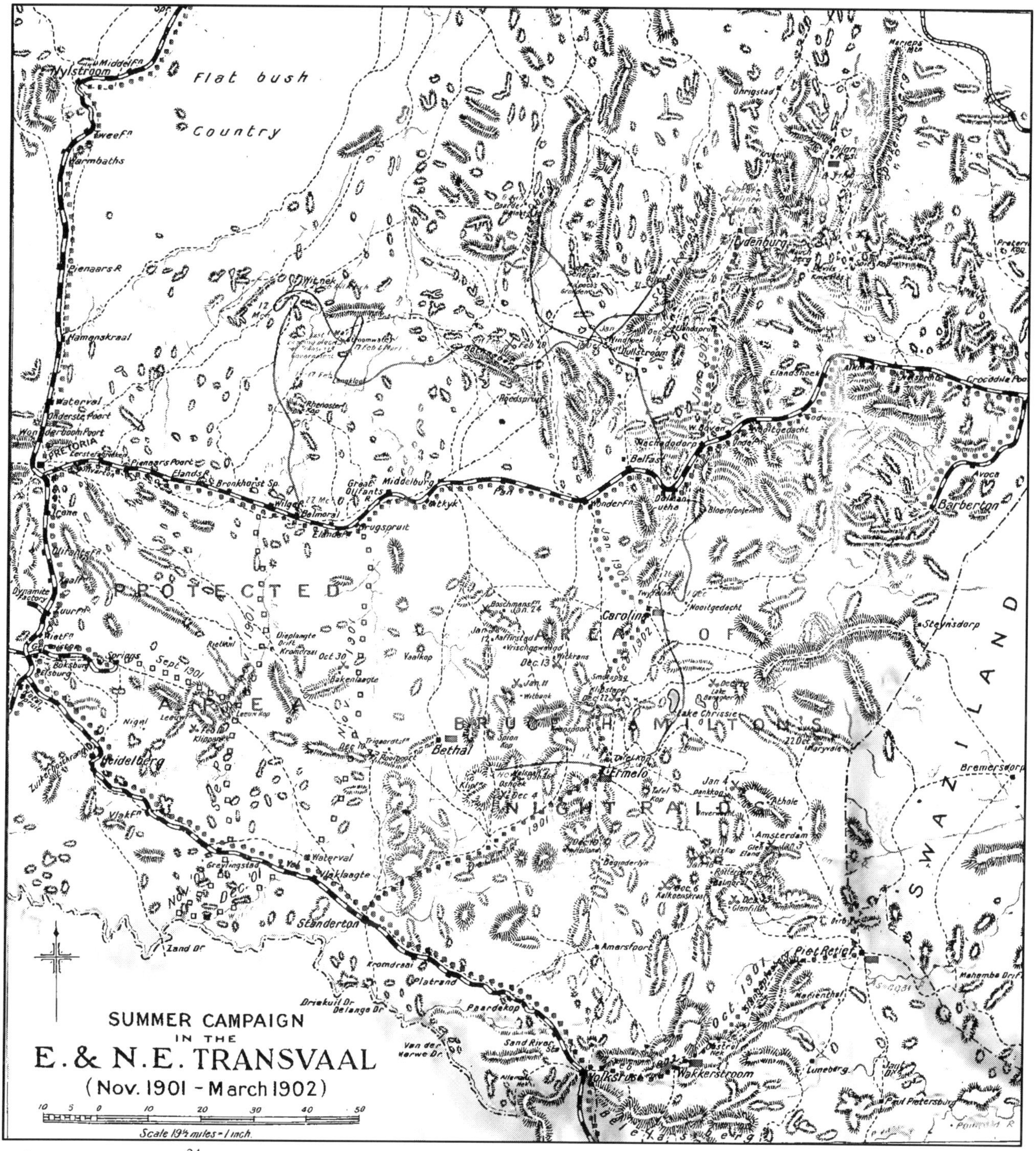

E & N.E. Transvaal.[24]

DIRECTIONS

- Blockhouses
- S. A. C. Posts
- Flight of Transvaal Govt
- × Scene of an action or night-raid.

Botha in Natal

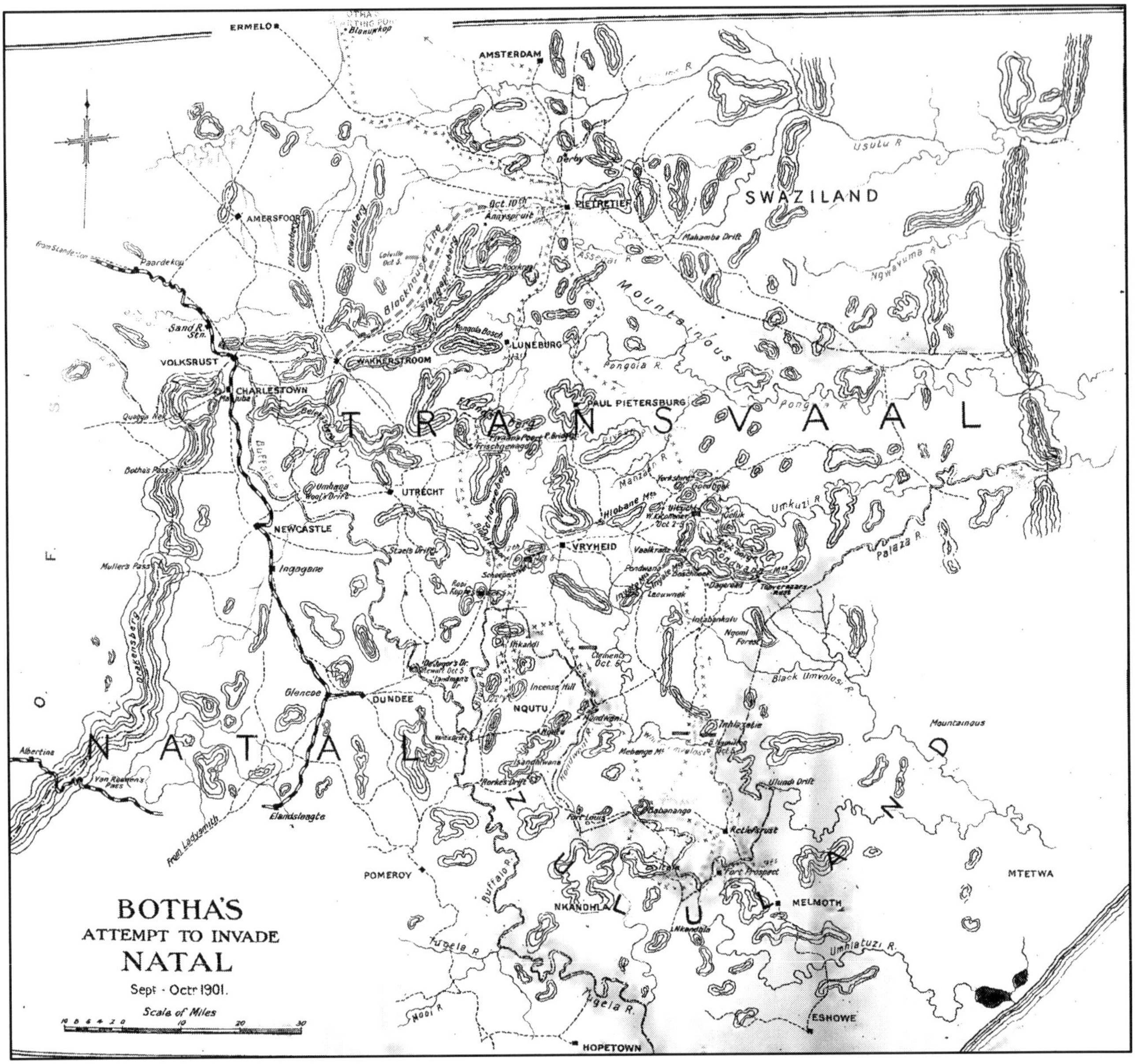

Botha in Natal.[25]

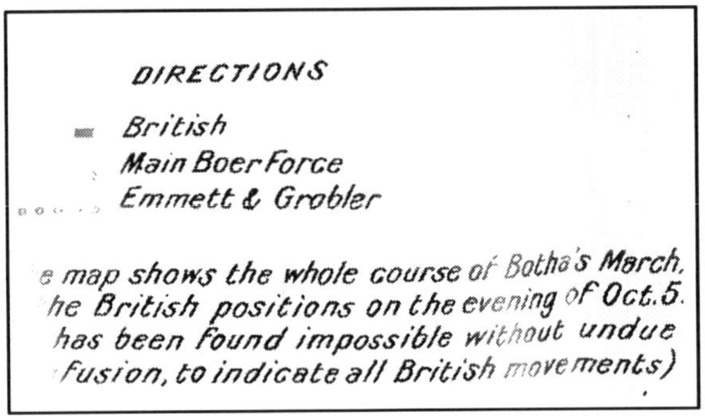

Fort Italia

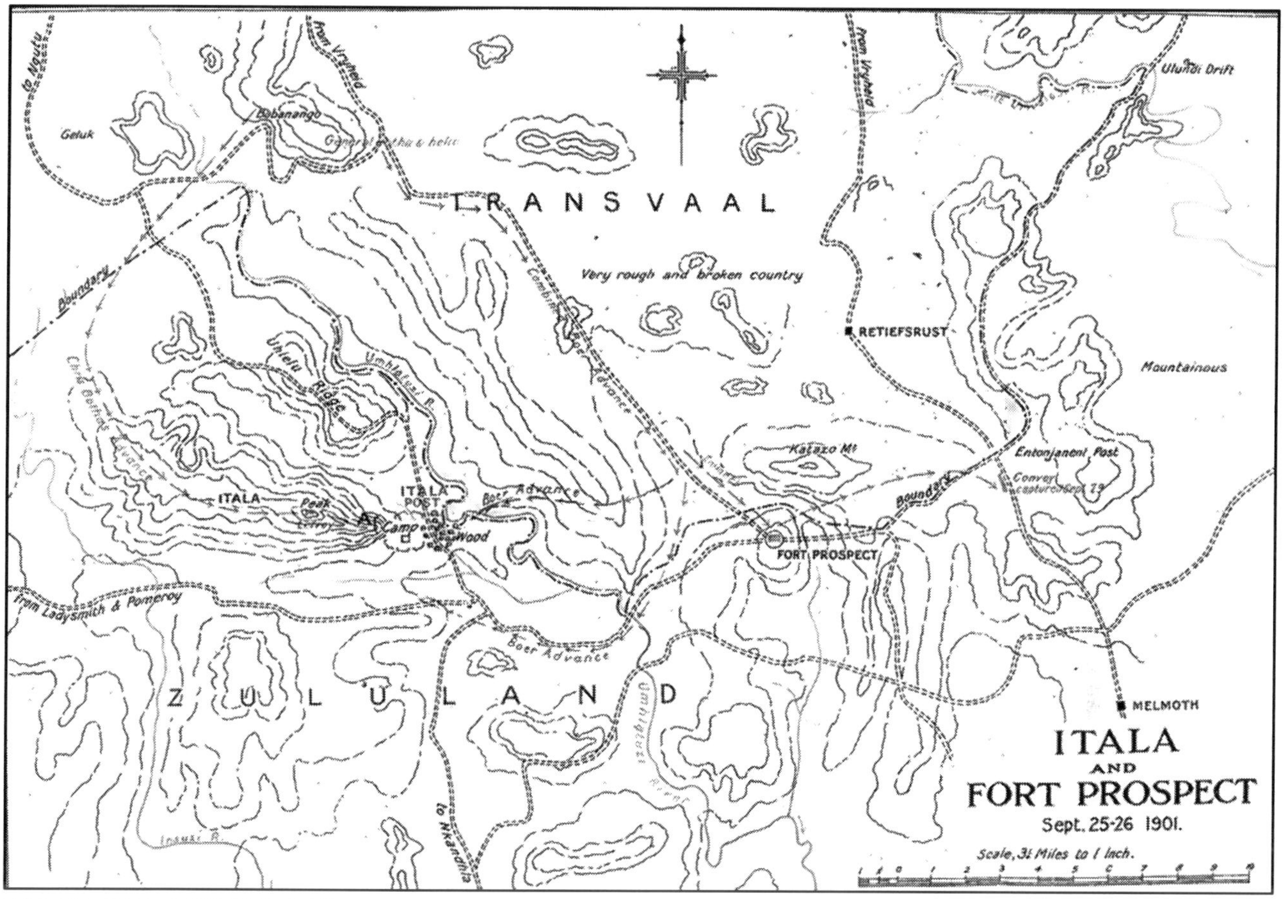

DIRECTIONS
A. Sangared spur, Key to British position

The defences of Itala are only roughly indicated. They are not to scale.

9/26/1901 Attacks on Fort Itala and Fort Prospect. Boers repulsed with very heavy losses at both places. The attempt of Botha and De Wet to invade Natal foiled.

Viljoen and his Adjutants

General Ben Viljoen and some of his Commandants and Veld-cornets.[26] Dr. Nethling [Assumed Standing Back Left] Docks Young [Right]. PF: Ben is to viewer left of his brother Wynand. Wynand J. Viljoen. Picture from Military Officer Album.[10 11]

Only identified as "Commandant Ben Viljoen and staff.'[27] PF: I believe Wynand is far right. The man with the baton is the same in both images.

General Ben Viljoen.[28] Cover photo is a crop of this one.

[10] Commandant Wynand Viljoen, brother of General Ben Viljoen, was born in the Cape Colony on February 29, 1876 and saw the battle at Rensburg Siding, Cape Colony, on February 16, 1900, as a member of the Johannesburg Police Force. At the end of 1900 he became one of its commanders and took the place of General B. Viljoen when he was taken prisoner of war. Coming. Viljoen represented his department at the peace negotiations at the Association. See Military Officer Album infra. Referred to as (MOA) for subsequent references.

[11] PF: Named adjutants in book. My younger brother, Wynand J. Viljoen, who, at the time of writing, is, I believe, still in this position."
[Leading Ben's Commando]
Chapter 11: The next morning I was in a fast train to Natal, accompanied by my adjutant. Rokzak. My other adjutant, Du Preez.
On sending two of my adjutants—Rokzak and Koos Nel—
My cook, Jan Smith, and my messmates were curious to know the reason I did not "wade in," for they always admired my ferocious appetite.
My old coloured groom "Mooiroos," who followed behind leading my horse, evidently thought the same."
My three adjutants, Bester, Redelinghuisen, and J. Viljoen, carefully dressed.
I went into the English line accompanied by my adjutants, [Jacobus] Nel and Bedeluighuis, and my secretary, Lieutenant W. Malan.
Field-Cornet R. D. Young [Probably Docs Young]
[We] set out from Pilgrim's Rest accompanied by Adjutants Nel, Coetzee, Bester, and Potgieter,
My brave and faithful adjutants, Jacobus Nel and L. Jordaan.

My Reminiscences of The Anglo-Boer War

By General Ben Viljoen.
(Assistant Commandant-General of the Transvaal Burgher Forces
And Member for Johannesburg in the Transvaal Volksraad)

General Ben Viljoen and his Secretary (Mr. J. Visser).

Maps from Drawings by P. Van Breda

London: Hood, Douglas, & Howard, 11, Clifford's Inn, E.C. 1902.

Also Including: The Transvaal Military Officer Album

Colonel Brinkman's Preface.

General Ben Viljoen, while engaged on this work, requested me to write a short introduction to it. This request I gladly comply with.

General Viljoen was a prisoner-of-war at Broad bottom Camp, St. Helena, where, after two years' service in South Africa, I was stationed with my regiment. It was at the General's further request that I conveyed this work to Europe for publication.

The qualities which particularly endeared this brave and justly-famous Boer officer to us were his straightforwardness and unostentatious manner, his truthfulness, and the utter absence of affectation that distinguishes him. I am certain that he has written his simple narrative with candour and impartiality, and I feel equally certain, from what I know of him, that this most popular of our late opponents has reviewed the exciting episodes of the War with an honesty, an intelligence, and a humour which many previous publications on the War have lacked.

During his stay at St. Helena I became deeply attached to General Viljoen; and in conclusion I trust that this work, which entailed many hours of labour, will yield him a handsome recompense.

Theodore Brinckman, C.B.

Colonel Commanding, 3rd, The Buffs (East Kent Regt.)

Tarbert, Loch Fyne, Scotland.

September, 1902

The Author to the Reader.

In offering my readers my reminiscences of the late War, I feel that it is necessary to ask their indulgence and to plead extenuating circumstances for many obvious shortcomings.

It should be pointed out that the preparation of this work was attended with many difficulties and disabilities, of which the following were only a few:—

(1) This is my first attempt at writing a book, and as a simple Afrikander I lay no claim to any literary ability.

(2) When captured by the British forces I was deprived of all my notes, and have been compelled to consult and depend largely upon my memory for my facts and data. I would wish to add, however, that the notes and minutiae they took from me referred only to events and incidents covering six months of the War. Twice before my capture, various diaries I had compiled fell into British hands; and on a third occasion, when our camp at Dalmanutha was burned out by a "grass-fire," other notes were destroyed.

(3) I wrote this book while a prisoner-of-war, fettered, as it were, by the strong chains with which a British "parole" is circumscribed. I was, so to say, bound hand and foot, and always made to feel sensibly the humiliating position to which we, as prisoners-of-war on this island, were reduced. Our unhappy lot was rendered unnecessarily unpleasant by the insulting treatment offered us by Colonel Price, who appeared to me an excellent prototype of Napoleon's custodian, Sir Hudson Lowe.[29] [12]

One has only to read Lord Rosebery's[13] work, "The Last Phase of Napoleon," to realise the insults and indignities Sir Hudson Lowe heaped upon a gallant enemy.

We Boers experienced similar treatment from our custodian, Colonel Price, who appeared to be possessed with the very demon of distrust and who conjured up about us the same fantastic and mythical plans of escape as Sir Hudson Lowe attributed to Napoleon. It is to his absurd suspicions about our safe custody that I trace the bitterly offensive regulations enforced on us.

While engaged upon this work, Colonel Price could have pounced down upon me at any moment, and, having discovered the manuscript, would certainly have promptly pronounced the writing of it in conflict with the terms of my "parole."[14]

I have striven as far as possible to refrain from criticism, except when compelled to do so, and to give a coherent story, so that the reader may easily follow the episodes I have sketched. I have also endeavoured to be impartial, or, at least, so impartial as an erring human being can be who has just quitted the bloody battlefields of a bitter struggle.

But the sword is still wet, and the wound is not yet healed.

I would assure my readers that it has not been without hesitation that I launch this work upon the world. There have been many amateur and professional writers who have preceded me in overloading the reading public with what purport to be "true histories" of the War. But having been approached by friends to add my little effort to the ponderous tomes of War literature, I have written down that which I saw with my own eyes, and that which I personally experienced. If seeing is believing, the reader may lend credence to my recital of every incident I have herein recounted.

During the last stages of the struggle, when we were isolated from the outside world, we read in newspapers and other printed matter captured from the British so many romantic and fabulous stories about ourselves, that we were sometimes in doubt whether people in Europe and elsewhere would really believe that we were ordinary human beings and not legendary monsters. On these occasions I read circumstantial reports of my death, and

[12] Sir Hudson Lowe, GCMG KCB (28 July 1769–10 January 1844) was a British Army officer and colonial administrator who served as the governor of Saint Helena from 1816 to 1821. Seeing service during the French Revolutionary and Napoleonic Wars, he is best known for serving as the de facto jailor of Napoleon when he was in exile on Saint Helena. https://en.wikipedia.org/wiki/Hudson_Lowe

[13] Archibald Philip Primrose, 5th Earl of Rosebery, 1st Earl of Midlothian (7 May 1847–21 May 1929) was a British Liberal Party politician who served as Prime Minister of the United Kingdom from March 1894 to June 1895. https://en.wikipedia.org/wiki/Archibald_Primrose,_5th_Earl_of_Rosebery

[14] Ben Arrived on St. Helena on February 25th, 1902. The War ended on June 1st. On June 26th the first Boer Pow's left St Helena for Cape Town and the last departed on October 21st.

once a long, and by no means flattering, obituary (extending over several columns of a newspaper) in which I was compared to Garibaldi, "Jack the Ripper," and Aguinaldo.

On another occasion I learned from British newspapers of my capture, conviction, and execution in the Cape Colony for wearing the insignia of the Red Cross. I read that I had been brought before a military court at De Aar and sentenced to be shot, and what was worse, the sentence was duly confirmed and carried out. A very lurid picture was drawn of the execution. Bound to a chair, and placed near my open grave, I had met my doom with "rare stoicism and fortitude." "At last," concluded my amiable biographer, "this scoundrel, robber, and guerilla leader, Viljoen, has been safely removed, and will trouble the British Army no longer." I also learned with mingled feelings of amazement and pride that, being imprisoned at Mafeking at the commencement of hostilities, General Baden-Powell had kindly exchanged me for Lady Sarah Wilson.[15]

To be honest, none of the above-mentioned reports were strictly accurate. I can assure the reader that I was never killed in action or executed at De Aar, I was never in Mafeking or any other prison in my life (save here at St. Helena), nor was I in the Cape Colony during the War. I never masqueraded with a Red Cross, and I was never exchanged for Lady Sarah Wilson. Her ladyship's friends would have found me a very poor exchange.

It is also quite inaccurate and unfair to describe me as a "thief" and "a scoundrel." It was, indeed, not an heroic thing to do, seeing that the chivalrous gentlemen of the South African Press who employed the epithets were safely beyond my view and reach, and I had no chance of correcting their quite erroneous impressions. I could neither refute nor defend myself against their infamous libels, and for the rest, my friend "Mr. Atkins"[16] kept us all exceedingly busy.

That which is left of Ben Viljoen after the several "coups de grace" in the field and the tragic execution at De Aar, still "pans" out at a fairly robust young person—quite an ordinary young fellow, indeed, thirty-four years of age, of middle height and build. Somewhere in the Marais Quartier of Paris—where the French Huguenots came from—there was an ancestral Viljoen from whom I am descended. In the War just concluded I played no great part of my own seeking. I met many compatriots who were better soldiers than myself; but on occasions I was happily of some small service to my Cause and to my people.

The chapters I append are, like myself, simple in form. If I have become notorious it is not my fault; it is the fault of the newspaper paragraphist, the snap-shooter, and the autograph fiend; and in these pages I have endeavoured, as far as possible, to leave the stage to more prominent actors, merely offering myself as guide to the many battlefields on which we have waged our unhappy struggle.

I shall not disappoint the reader by promising him sensational or thrilling episodes. He will find none such in these pages; he will find only a naked and unembellished story.

Ben J. Viljoen.

(Assistant Commandant-General of the Republican Forces.)

St. Helena,

June, 1902

[15] Sarah Wilson was a war correspondent for the Daily Mail and a cousin of Winston Churchill, she was captured and exchanged for an accused horse thief. See: Sarah Wilson, South African Memories Social, Warlike & Sporting from Diaries Written at the Time chapters IX and X. https://en.wikipedia.org/wiki/Sarah_Wilson_(war_correspondent)

[16] 'Tommy Adkins' is the nickname of a British Soldier, (often just Tommy).

Chapter 1. The War Clouds Gather.

In 1895 the political clouds gathered thickly and grew threatening. They were unmistakable in their portent. War was meant, and we heard the martial thunder rumbling over our heads.

The storm broke in the shape of an invasion from Rhodesia on our Western frontiers, a raid planned by soldiers of a friendly power.

However one may endeavour to argue the chief cause of the South African war to other issues, it remains an irrefutable fact that the Jameson Raid was primarily responsible for the hostilities which eventually took place between Great Britain and the Boer Republics.

Mr. Rhodes, the sponsor and deus ex machina[17] of the Raid, could not agree with Mr. Paul Kruger, and had failed in his efforts to establish friendly relations with him. Mr. Kruger, quite as stubborn and ambitious as Mr. Rhodes, placed no faith in the latter's amiable proposals, and the result was that fierce hatred was engendered between the two Gideons, a racial rancour spreading to fanatical lengths.

Dr. Jameson's stupid raid is now a matter of history; but from that fateful New Year's Day of 1896 we Boers date the terrible trials and sufferings to which our poor country has been exposed. To that mischievous incident, indeed, we directly trace the struggle now terminated.

This invasion, which was synchronous with an armed rebellion at Johannesburg, was followed by the arrest and imprisonment of the so-called gold magnates of the Witwatersrand. Whether these exceedingly wealthy but extremely degenerate sons of Albion and Germania deserved the death sentence pronounced upon their leaders at Pretoria for high treason it is not for me to judge.

I do recall, however, what an appeal for mercy there went up, how piteously the Transvaal Government was petitioned and supplicated, and finally moved "to forgive and forget." The same faction who now press so obdurately for "no mercy" upon the Colonial Afrikanders who joined us, then supplicated all the Boer gods for forgiveness.[18]

Meantime the Republic was plagued by the rinderpest scourge,[19] which wrought untold havoc throughout the country.

Cattle Dead from Rinderpest in South Africa, 1896.[30]

Shooting Rinderpest Oxen.[31]

[17] Deus ex machina 'God from the machine') is a plot device whereby a seemingly unsolvable problem in a story is suddenly or abruptly resolved by an unexpected and unlikely occurrence. Its function is generally to resolve an otherwise irresolvable plot situation, to surprise the audience, to bring the tale to a happy ending, or act as a comedic device.

[18] The Jameson Raid was a botched raid against the Transvaal carried out by British colonial administrator Leander Jameson, who was employed by Cecil Rhodes. It involved 500 British South Africa Company police and launched from Rhodesia over the New Year weekend of 1895–96. The raid was intended to trigger an uprising by the primarily British expatriate workers (known as Uitlanders) in the Transvaal, but it failed. https://en.wikipedia.org/wiki/Jameson_Raid

Later, Jameson became Prime Minister of the Cape Colony (1904–08) and one of the founders of the Union of South Africa. On his death in 1917, he was buried next to Cecil Rhodes and the 34 BSAC soldiers of the Shangani Patrol (killed in 1893 in the First Matabele War) in the Matobos Hills, near Bulawayo.

[19] Rinderpest was an infectious viral disease of cattle, domestic water buffalo, and many other species of even-toed ungulates, including gaurs, buffaloes, large antelope, deer, giraffes, wildebeests, and warthogs. Death rates during outbreaks were usually extremely high, approaching 100% in immunologically naïve populations. On 28 June 2011, the UN Food and Agriculture Organization officially recognized global freedom from the deadly cattle virus. On this day, the FAO Conference, the highest body of the UN agency, adopted a resolution declaring the eradication of rinderpest. https://en.wikipedia.org/wiki/Rinderpest

This scourge was preceded by the dynamite disaster at Vrededorp (near Johannesburg) and the railway disaster at Glencoe in Natal. It was succeeded by a smallpox epidemic, which, in spite of medical efforts, grew from sporadic to epidemic and visited all classes of the Rand, exacting victims wherever it travelled.

During the same period difficulties occurred in Swaziland[20] necessitating the despatch of a strong commando to the disaffected district and the maintenance of a garrison at Bremersdorp. The following year hostilities were commenced against the Magato tribe[21] in the north of the Republic.

After an expensive expedition, lasting six months, the rebellion was quelled. There was little doubt that the administration of unfaithful native commissioners was in part responsible for the difficulties, but there is less doubt that external influences also contributed to the rebellion. This is not the time, however, to tear open old wounds.

Mr. Rhodes has disappeared from the stage for ever; he died [March 26th, 1902] as he had lived. His relentless enemy Mr. Kruger, who was pulling the strings at the other end, is still alive. Perhaps the old man may be spared to see the end of the bloody drama; it was undoubtedly he and Mr. Rhodes who played the leading parts in the prologue.

Which of these two "Big Men" took the greatest share in bringing about the Disaster which has drenched South Africa with blood and draped it in mourning, it would be improper for me at this period to suggest. Mr. Rhodes has been summoned before a Higher Tribunal; Mr. Kruger has still to come up for judgment before the people whose fate, and very existence as a nation, are, at the time of writing, wavering in the balance.

We have been at one another's throats, and for this we have to thank our "statesmen." It is to be hoped that our leaders of the future will attach more value to human lives, and that Boer and Briton will be enabled to live amicably side by side.

A calm and statesmanlike government by men free from ambition and racial rancour, by men of unblemished reputation, will be the only means of pacifying South Africa and keeping South Africa pacified.

Chapter 2. And the War Storm Breaks.

It was during a desultory discussion of an ordinary sessions of the Second Volksraad, in which I represented Johannesburg, that one day in September, 1899—to be precise, the afternoon of the 28th—the messenger of the House came to me with a note, and whispered, "A message from General Joubert, Sir; it is urgent, and the General says it requires your immediate attention."

The Volksraad building, Pretoria.[32]

[20] Chief Makgoba had several clashes with invading white Afrikaner farmers (Trekboers) who occupied land belonging to indigenous people in and around Makgobaskloof. As the Transvaal administration began annexing land and imposing taxes on the indigenous populations, Chief Makgoba instructed his people to rebel against the annexation of land and the imposition of taxes. On June 3, 1895, the administration launched a full-scale war against Chief Makgoba and other local chiefs. He was defeated. His followers say he committed suicide to avoid capture. Campaign against Chief Makgoba (Magoeba) 1895. https://en.wikipedia.org/wiki/Chief_Makgoba And: https://en.wikipedia.org/wiki/List_of_battles_in_South_Africa

[21] See The campaign against Bantu chief M'pefu (Mpefu) of the Venda people, 1898. https://en.wikipedia.org/wiki/Venda_people https://en.wikipedia.org/wiki/List_of_battles_in_South_Africa

Commando with Johannesburg in the background.[33]

Johannesburg Volunteer Infantry.[34]

I broke the seal of the envelope with some trepidation. I guessed its contents, and a few of my colleagues in the Chamber hung over me almost speechless with excitement, whispering curiously, "Jong, is dit fout?"—"Is this correct. Is it war?"

Everybody knew, of course, that we were in for a supreme crisis, that the relations between Great Britain and our Republic were strained to the bursting point, that bitter diplomatic notes had been exchanged between the governments of the two countries for months past, and that a collision, an armed collision, was sooner or later inevitable.

Being "Fighting-Commandant" of the Witwatersrand goldfields, and, therefore, an officer of the Transvaal army, my movements on that day excited great interest among my colleagues in the Chamber. After reading General Joubert's note I said, as calmly as possible: "Yes, the die is cast; I am leaving for the Natal frontier. Good-bye. I must now quit the house. Who knows, perhaps for ever!"

General Joubert's mandate was couched as follows:—

"You are hereby ordered to proceed with the Johannesburg commando to Volksrust to-morrow, Friday evening, at 8 o'clock. Your field cornets have already received instructions to commandeer the required number of burghers and the necessary horses, waggons, and equipment. Instructions have also been given for the necessary railway conveyances to be held ready. Further instructions will reach you."

Previous to my departure next morning I made a hurried call at Commandant-General Joubert's offices. The ante-chamber leading to the Generalissimo's "sanctum-sanctorum" was crowded with brilliantly-uniformed officers of our State Artillery, and it was only by dint of using my elbows very vigorously that I gained admission to my chief-in-command.

The old General seemed to feel keenly the gravity of the situation. He looked careworn and troubled: "Good-morning, Commandant," he said; "aren't you away yet?"

I explained that I was on my way to the railway station, but I thought before I left I'd like to see him about one or two things.

"Well, go on, what is it?" General Joubert enquired, petulantly.

"I want to know, General Joubert," I said, "whether England has declared war against us, or whether we are taking the lead. And another thing, what sort of general have I to report myself to at Volksrust?"

The old warrior, without looking up or immediately answering me, drew various cryptic and hieroglyphic pothooks and figures on the paper before him. Then he suddenly lifted his eyes and pierced me with a look, at which I quailed and trembled.

He said very slowly: "Look here; there is as yet no declaration of war, and hostilities have not yet commenced. You and my other officers should understand that very clearly, because possibly the differences between ourselves and Great Britain may still be settled. We are only going to occupy our frontiers because England's attitude is extremely provocative, and if England see that we are fully prepared and that we do not fear her threats, she will perhaps be wise in time and reconsider the situation. We also want to place ourselves in a position to prevent and quell a repetition of the Jameson Raid with more force than we exerted in 1896."

An hour afterwards I was on board a train travelling to Johannesburg in the company of General Piet Cronje and his faithful wife. General Cronje told me that he was proceeding to the western districts of the Republic to take up the command of the Potchefstroom and Lichtenburg burghers. His instructions, he said, were to protect the Western frontier.

I left General Cronje at Johannesburg on the 29th September, 1899, and never saw him again until I met him at St. Helena nearly two and a half years afterwards, on the 25th March, 1902. When I last saw him we greeted each other as free men, as free and independent legislators and officers of a free Republic. We fought for our rights to live as a nation.

Now I meet the veteran Cronje a broken old man, captive like myself, far away from our homes and our country.

Then and Now!

Then we went abroad free and freedom-loving men, burning with patriotism. Our wives and our women-folk watched us go; full of sorrow and anxiety, but satisfied that we were going abroad in our country's cause.

And Now!

Two promising and prosperous Republics wrecked, their fair homesteads destroyed, their people in mourning, and thousands of innocent women and children the victims of a cruel war.

There is scarcely an Afrikander family without an unhealable wound. Everywhere the traces of the bloody struggle; and, alas, most poignant and distressing fact of all, burghers who fought side by side with us in the earlier stages of the struggle are now to be found in the ranks of the enemy.

These wretched men, ignoring their solemn duty, left their companions in the lurch without sense of shame or respect for the braves who fell fighting for their land and people.

Oh, day of judgment! The Afrikander nation will yet avenge your treachery.

Johannesburg Commando.[35]

Burgomaster Schutte of Johannesburg at the departure of the Burghers.[36]

Chapter 3. The Invasion of Natal.

After taking leave of my friend Cronje at Johannesburg Station, my first duty was to visit my various field cornets. About four o'clock that afternoon I found my commando was as nearly ready as could be expected. When I say ready, I mean ready on paper only, as later experience showed. My three field cornets were required to equip 900 mounted men with waggons and provisions, and of course they had carte blanche to commandeer. Only fully enfranchised burghers of the South African Republic were liable to be commandeered, and in Johannesburg town there was an extraordinary conglomeration of cosmopolitans amenable to this gentle process of enlistment.

It would take up too much time to adequately describe the excitement of Johannesburg on this memorable day. Thousands of Uitlanders were flying from their homes, contenting themselves, in their hurry to get away, to stand in Kaffir or coal trucks and to expose themselves cheerfully to the fierce sun, and other elements. The streets were palpitating with burghers ready to proceed to the frontier that night, and with refugees speeding to the stations. Everybody was in a state of intense feeling. One was half-hearted, another cheerful, and a third thirsting for blood, while many of my men were under the influence of alcohol.

Johannesburg Uitlanders, fleeing from the expected war, fighting for seats in open trucks.[37] Crowding in Cars to get away from Johannesburg on First War Scare.[38]

When it was known that I had arrived in the town my room in the North Western Hotel was besieged. I was approached by all sorts of people pleading exemption from commando duty. One Boer said he knew that his solemn duty was to fight for his country and his freedom, but he would rather decline. Another declared that he could not desert his family; while yet another came forward with a story that of his four horses, three had been commandeered, and that these horses were his only means of subsistence. A fourth complained that his waggons and mules had been clandestinely (although officially) removed.

Many malingerers suddenly discovered acute symptoms of heart disease and brought easily-obtained doctor's certificates, assuring me that tragic consequences would attend their exposure in the field. Ladies came to me pleading exemption for their husbands, sisters for brothers, mothers for sons, all offering plausible reasons why their loved ones should be exempted from commando duty.

It was very difficult to deal with all these clamorous visitors. I was much in the position of King Solomon, though lacking his wisdom. But I would venture to say that his ancient majesty himself would have been perplexed had he been in my place. It is necessary that the reader should know that the main part of the population was composed of all nationalities and lacked every element of Boer discipline.

On the evening of the 29th of September, I left with the Johannesburg commando in two trains. Two-thirds of my men had no personal acquaintance with me, and at the departure there was some difficulty because of this. One burgher came into my private compartment uninvited. He evidently forgot his proper place, and when I suggested to him that the compartment was private and reserved for officers, he told me to go to the devil, and I was compelled to remove him somewhat precipitately from the carriage. This same man was afterwards one of my most trustworthy scouts.

The following afternoon we reached Standerton, where I received telegraphic instructions from General Joubert to join my commando to that of Captain [A.. H.] Schiel, who was in charge of the German Corps, and to place myself under the supreme command of Jan Kock, a member of the Executive Council, who had been appointed a general by the Government.

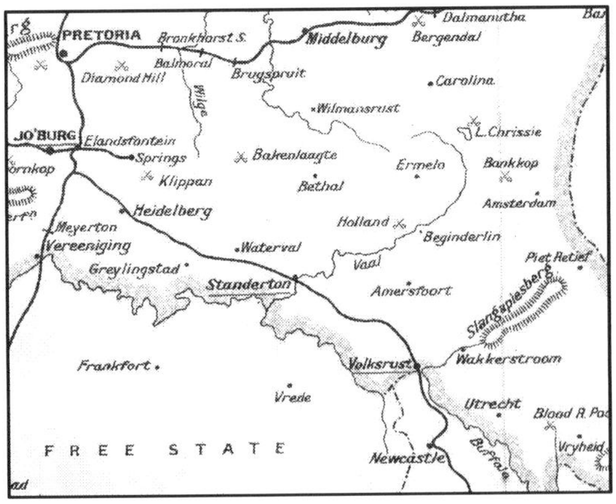

Standerton is 1/2 way between Johannesburg and Volksrust. A view of Standerton.[39]

We soon discovered that quite one-third of the horses we had taken with us were untrained for the serious business of fighting, and also that many of the new burghers of foreign nationality had not the slightest idea how to ride.

Our first parade, or "Wapenschouwing" gave food for much hilarity. Here one saw horses waltzing and jumping, while over there a rider was biting the sand, and towards evening the doctors had several patients. It may be stated that although not perfectly equipped in the matter of ambulances, we had three physicians with us, Doctors Visser, Marais, and Shaw. Our spiritual welfare was being looked after by the Reverends Nel and Martins, but not for long, as both these gentlemen quickly found that commando life was unpleasant and left us spiritually to ourselves, even as the European Powers left us politically. But I venture to state that no member of my commando really felt acutely the loss of the theological gentlemen who primarily accompanied us.

On the following day General Kock and a large staff arrived at the laager, and, together with the German Corps, we trekked to Paardakop and Klip River, in the Orange Free State, where we were to occupy Botha's Pass. My convoy comprised about a hundred carts, mostly drawn by mules, and it was amusing to see the variety of provisions my worthy field-cornets had gathered together. There were three full waggons of lime-juice and other unnecessary articles which I caused to be unloaded at the first halting-place to make room for more serviceable provisions.

Botha's Pass south of Volksrust, north of Newcastle.[40] Loaded ox wagon.[41]

It should be mentioned that of my three field-cornets only one, the late Piet Joubert of Jeppestown, actually accompanied my commando. The others sent substitutes, perhaps because they did not like to expose themselves to the change of air. We rested some days at the Klip River, in the Orange Free State, and from thence I was sent with a small escort of burghers by our General to Harrismith to meet a number of Free State officers. After travelling two days I came upon Chief Free State Commandant [Marthinus] Prinsloo, who afterwards deserted, and other officers. The object of my mission was to organise communications with these officers. On the 11th of October, having returned to my commando, we received a report that our Government had despatched the Ultimatum to England, and that the time specified for the reply to that document had elapsed. Hostilities had begun.

We received orders to invade Natal, and crossed the frontier that very evening. I, with a patrol of 50 men, had not crossed the frontier very far when one of my scouts rode up with the report that a large British force was in sight on the other side of the River Ingogo. I said to myself at the time: “If this be true the British have rushed up fairly quickly, and the fat will be in the fire very soon.”

We then broke into scattered formation and carefully proceeded into Natal. After much reconnoitring and concealment, however, we soon discovered that the “large English force” was only a herd of cattle belonging to friendly Boers, and that the camp consisted of two tents occupied by some Englishmen and Kaffirs who were mending a defective bridge.

Crossing the border into Natal. Tent camp in background.[42]

“Men accompanied by their native servants.”[43] [22]

We also came across a cart drawn by four bullocks belonging to a Natal farmer, and I believe this was the first plunder we captured in Natal. The Englishman, who said he knew nothing about any war, received a pass to proceed with his servants to the English lines, and he left with the admonition to in future read the newspapers and learn when war was imminent. Next day our entire commando was well into Natal.

The continuous rain and cold of the Drakenbergen rendered our first experience of veldt life, if not unbearable, very discouraging. We numbered a fairly large commando, as Commandant J. Lombard,[23] commanding the Hollander corps, had also joined us. Close by Newcastle we encountered a large number of commandos, and a general council of war was held under the presidency of Commandant General Joubert. It was here decided that Generals Lukas Meyer[24] and Dijl Erasmus[25] should take Dundee, which an English garrison held, while our commandos under General Kock[26] were instructed to occupy the Biggarburg Pass. Preceded by scouts we wound our way in that direction, leaving all our unnecessary baggage in the shape of provisions and ammunition waggons at Newcastle.

General Lucas Meyer General Louis Botha and Commandant D. J. E. Erasmus.[44]

[22] Also known as an Aachterrijder, or achterryder. A mounted groom or other attendant; a military servant, or batman. agterryer - definition of agterryer in A Dictionary of South African English - DSAE

[23] Johannes Petrus la Grange Lombard. See Military Officer Album (MOA).

[24] Lucas Meyer MOA.

[25] General Daniel J. E. Erasmus. See MOA. And: https://en.wikipedia.org/wiki/Daniel_Jacobus_Elardus_Erasmus

[26] J. H M. Kock. MOA.

One of my acting field-cornets and the field-cornets of the German commando, prompted by goodness knows what, pressed forward south, actually reaching the railway station at Elandslaagte. A goods train was just steaming into the station, and it was captured by these foolhardy young Moltkes. I was much dissatisfied with this action, and sent a messenger ordering them to retire after having destroyed the railway. On the same night I received instructions from General Kock to proceed with two hundred men and a cannon to Elandslaagte, and I also learned that Captain Schiel[27] and his German Corps had left in the same direction.

Imagine, we had gone further than had actually been decided at the council of war, and we pressed forward still further without any attempt being made to keep in touch with the other commandos on our left and right. Seeing the inexpediency of this move, I went to the General in command and expressed my objections to it. But General Kock was firmly decided on the point, and said, "Go along, my boy." We reached Elandslaagte at midnight; it was raining very heavily. After scrambling for positions in the darkness, although I had already sufficiently seen that the lie of the land suggested no strategic operations, we retired to rest. Two days later occurred the fateful battle.

The Capture of the Train at Elandslaagte.

Chapter 4. Defeated at Elandslaagte.

In the grey dawn of the 21st of October a number of scouts I had despatched overnight in the direction of Ladysmith returned with the tidings that "the khakis were coming." "Where are they, and how many are there of them?" I asked. "Commandant," the chief scout replied, "I don't know much about these things, but I should think that the English number quite a thousand mounted men, and they have guns, and they have already passed Modderspruit." To us amateur soldiers this report was by no means reassuring, and I confess I hoped fervently that the English might stay away for some little time longer.

It was at sunrise that the first shot I heard in this war was fired. Presently the men we dreaded were visible on the ridges of hills south of the little red railway station at Elandslaagte. Some of my men hailed the coming fight with delight; others, more experienced in the art of war, turned deadly pale. That is how the Boers felt in their first battle. The awkward way in which many of my men sought cover, demonstrated at once how inexperienced in warfare we youngsters were. We started with our guns and tried a little experimental shooting.

[27] A. H. Schiel MOA Captured at Ellandslagte. "Colonel A. H. Schiel was till the end of the war a prisoner at St. Helena."

Staats Artillery at start of conflict. All arms & accoutrements are pristine, as is the Long Tom itself.[45]
Long Tom in charge of Boers En. Route to Modder River.[46] PF: This must have been taken at the same time.

Walker's Maritzburg Artillery with Hotchkiss Gun.[47]

Loading Drill.[48]

Firing Drill.[49]

French 75 mm field Gun (Canon de 75 modèle 1897).[28]

Staats Artillerie with French 75mm guns.[50]

[28]"In 1896 the Transvaal imported six QF field guns from Schneider & Co to form the "1ste Rijdende Battery" (First Mounted Battery) or more commonly, the "French Battery." A letter in the Pretoria archives gives their factory numbers as 3244 to 3249. Ironically in the Transvaal the guns soon became known as the "French Krupps"! Canon de 75 modèle 1897 - Wikipedia

"Adjutant Roos and Gunners with a 15-Pounder [*sic*] Creusot Gun on the Tugela."[51] Lieutenant and Adjutant Carl Roos.[52]

The second and third shots appeared to be effective; at any rate, as far as we could judge, they seemed to disturb the equanimity of the advancing troops. I saw an ammunition cart deprived of its team and generally smashed.

The British guns appeared to be of very small calibre indeed. Certainly they failed to reach us, and all the harm they did was to send a shell through a Boer ambulance within the range of fire. This shot was, I afterwards ascertained, purely accidental. When the British found that we too, strange to say, had guns, and, what is more, knew how to use them, they retired towards Ladysmith. But this was merely a ruse; they had gone back to fetch more.

Netherlands Ambulance at Modderspruit.[53]

Still, though it was a ruse, we were cleverly deceived by it, and while we were off-saddling and preparing the mid-day meal they were arranging a new and more formidable attack. From the Modderspruit siding they were pouring troops brought down by rail, and although we had a splendid chance of shelling the newcomers from the high kopje we occupied, General Kock, who was in supreme command of our corps, for some reason which has never been explained, refused to permit us to fire upon them. I went to General Kock and pleaded with him, but he was adamant. This was a bitter disappointment to me, but I consoled myself with the thought that the

This gun was far ahead of its time and a pioneer of quick firing artillery. Not only did it use single-piece brass cased charges, it was also equipped with two buffer cylinders. The gun was designed to fire shrapnel at long ranges with a flat trajectory and a high muzzle velocity. This strained the carriage and buffers, which often were in need of repairs. The gun is easily recognisable by its unusual low carriage and interrupted screw breechblock." From: Boer Artillery Trophies in Britain, Canada and India MC Heunis, O.V.S.A.C. Study No. 5, May-Jun 2003. 75mm Creusot QF (Matèriel de Campagne de 75 m/m a tir Rapide Système Schneider-Canet) Credit Neville Constantine. See Anglo Boer War - Artillery and Ammunition - Page 15 - Boer War Forum

General was much older than myself, and had been fighting since he was a baby. I therefore presumed he knew better. Possibly if we younger commanders had had more authority in the earlier stages of the war, and had had less to deal with arrogant and stupid old men, we should have reached Durban and Cape Town.

I must here again confess that none of my men displayed any of the martial determination with which they had so buoyantly proceeded from Johannesburg. To put it bluntly, some of them were "footing" it and the English cavalry, taking advantage of this, were rapidly outflanking them. The British tactics were plain enough. General French had placed his infantry in the centre with three field batteries (fifteen pounders), while his cavalry, with Maxims, encompassed our right and left. He was forming a crescent, with the obvious purpose of turning our position with his right and left wing. When charging at the close of the attack the cavalry, which consisted mainly of lancers, were on both our flanks, and completely prevented our retreat. It was not easy to estimate the number of our assailant's forces. Judging roughly, I calculated they numbered between 5,000 and 6,000, while we were 800 all told, and our artillery consisted merely of two Nordenfeldt guns with shell, and no grape shot.

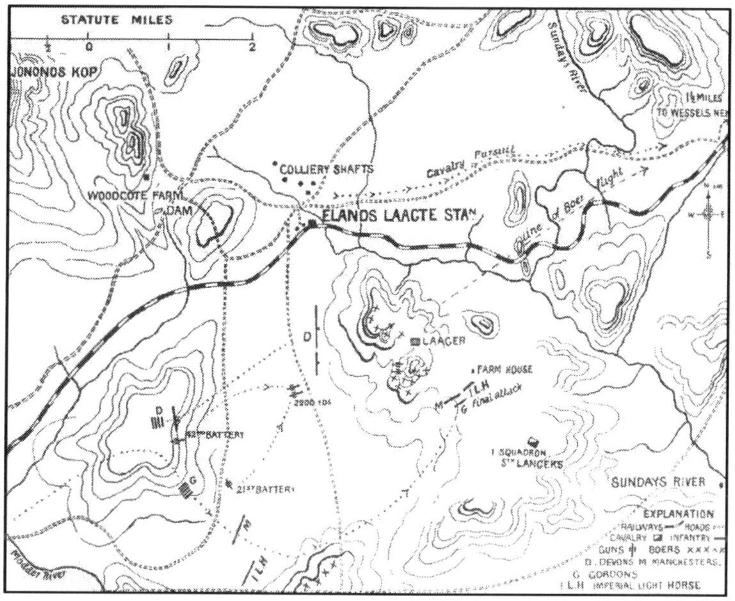

Elandslaagate Battle Area.[54]

General Sir John French, K.C.B.[55 29]

The British certainly meant business that day. It was the baptismal fire of the Imperial Light Horse,[30] a corps principally composed of Johannesburgers, who were politically and racially our bitter enemies. And what was more unfortunate, our guns were so much exposed that they were soon silenced. For a long time we did our best to keep our opponents at bay, but they came in crushing numbers, and speedily dead and maimed burghers covered the veldt. Then the Gordon Highlanders and the other infantry detachments commenced to storm our positions. We got them well within the range of our rifle fire, and made our presence felt; but they kept pushing on with splendid determination and indomitable pluck, though their ranks were being decimated before our very eyes.

This was the first, as it was the last time in the War that I heard a British band playing to cheer attacking "Tommies." I believe it used to be a British war custom to rouse martial instincts with lively music, but

[29] John Denton Pinkstone French, 1st Earl of Ypres, (28 September 1852–22 May 1925) Re: Elandslaagate; French was ordered to investigate reports that the Boers had taken Elandslaagte. Taking with him the 5th Lancers, six squadrons of Natal Carbineers and Natal Mounted Rifles, a battery of field artillery and a brigade of infantry under Colonel Ian Hamilton, he made contact with the Boers at 13:00 that day. White was initially cautious but on 21 October, having learned of Symons' victory at Talana the previous day, he permitted French to attack. Concerned at French's lack of experience at commanding infantry, White initially proposed that his chief of staff Maj-Gen Archibald Hunter take command, but Hunter advised that French should be left in command. White himself came merely to observe. French enjoyed numerical superiority of around 3:1.

French's most important role was as Commander-in-Chief of the British Expeditionary Force (BEF) for the first year and a half of the First World War. https://en.wikipedia.org/wiki/John_French,_1st_Earl_of_Ypres

[30] The Imperial Light Horse was raised by the British in Johannesburg on 21 September 1899 for service in the Second Boer War. Its initial strength was 444 officers and men. It was informally known as the "Reformers Regiment" as many of its officers served on the Reform Committee, or more commonly the Uitlander Regiment by the Transvaal Government and the Boer Commandos. https://en.wikipedia.org/wiki/Johannesburg_Light_Horse_Regiment

something must have gone wrong with the works in this War, there must have occurred a rift in the lute, for ever after this first battle of Elandslaagte the British abandoned flags, banners, and bands and other quite unnecessary furniture.

About half an hour before sunset, the enemy had come up close to our positions and on all sides a terrible battle raged. To keep them back was now completely out of the question. They had forced their way between a kloof, and while rushing up with my men towards them, my rifle was smashed by a bullet. A wounded burgher handed me his and I joined Field-Cornet Peter Joubert who, with seven other burghers, was defending the kloof. We poured a heavy fire into the British, but they were not to be shaken off. Again and again they rushed up in irresistible strength, gallantly encouraged by their brave officers. Poor Field-Cornet Joubert perished at this point.

When the sun had set and the awful scene was enveloped in darkness there was a dreadful spectacle of maimed Germans, Hollanders, Frenchmen, Irishmen, Americans, and Boers lying on the veldt. The groans of the wounded were heartrending; the dead could no longer speak. Another charge, and the British, encouraged by their success, had taken our last position, guns and all. My only resource now was to flee, and the battle of Elandslaagte was a thing of the past.

General Kock (Center) and his staff.[56]

Kock and staff identified.[57] [31]

Elandslaagte and train.[58]

Gordon Highlanders advancing at Elandslaagte.[59]

Just as the main Boer line began to wave white flags of surrender and the British ceased fire, General Kock led a counterattack, dressed in a top hat and his Sunday best. His party drove back the British in confusion, but they were rallied by the infantry commander, Ian Hamilton, and Kock and his assault force were killed or severely wounded. Kock was captured and died a week later while a prisoner in Ladysmith.

[31] "General Kock And His Staff: The Heroes Of Elandslaagte, 21 October 1899. J. H. M. Kock, the gallant old Transvaaler, Member of the Executive Raad, died at Ladysmith, on the 31st of October 1899, of inflammation of the lungs contracted on the field of battle, where he was wounded and lay long in the pouring rain before the English ambulance found him. His illness was increased by sorrow at his defeat. His son, Judge Kock, fell into the hands of the enemy, but was exchanged by after his father's death. General White Count Zeppelin was killed after firing off all his ammunition and eventually defending himself to the last against the charging Lancers with his whip. Colonel A. H. Schiel was till the end of the war a prisoner at St. Helena. Albedyll succeeded him in his command. With the exception of Potgieter and the lieutenant of the Transvaal Artillery, who did not take part in the battle, the others are all members of the German Corps, of which Dr. Elsburger stood at the head of the ambulance." Heroes of the Boer War. Review of Reviews, 1903.

After this last counterattack the remaining Boers mounted their horses and began to retreat. A squadron of lancers and another of dragoons were able to get in amongst them with lances and sabers, inflicting severe casualties.

Imperial Light Horse.[60]

Charge of the 5th Lancers.[61]

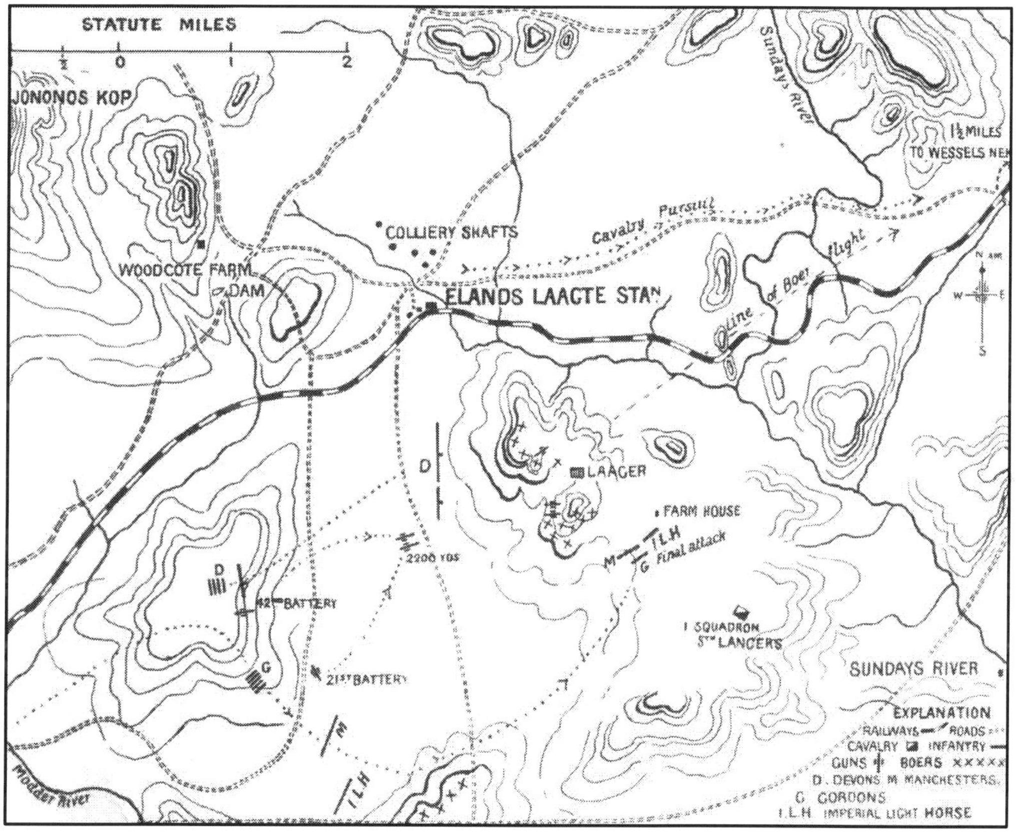

Battle of Elandslaagte.[62] Note colliery shafts and pursuit of cavalry as well as Sundays River.

Landscape of Elandslaagte.[63] "Elandslaagte. The flat ground where the 5th Lancers charged the retreating Boers. [Note Coal mine shaft and works. Northern Natal is underlain by coal seams which is onc rcason the British were prepared to defend it rather than retreating to the plains of Natal.][64]

Searching for wounded at night.[65]

Removing wounded after the Battle of Elandslaagte.[66]

Carrying the wounded from the battlefield of Elandslaagte.[67] [68] "Mrs. Otto Krantz, the wife of a hunter. Accompanied her husband to Natal and fought by his side at Elandslaagte, also in the battles on the Tugela, and after wars in the Free State. She is by no means the only Boer woman who has carried a rifle in the war, and used it."[69] And.[70]

The Hollander Corps leaving Pretoria for the front.[32]

Dutch Corps Monument.[71] [33]

[32] Departure of The Hollander Corps from Pretoria 6 October 1899. "There was not a foreigner in South Africa, with the exception of the British, who did not take side with the Boers when war had once become inevitable. The Hollanders were the first to form themselves into a Corps, and they rendered invaluable services during the war." Image: Heroes of the Boer War. Review of Reviews.

[33] The Dutch Corps Monument is a provincial heritage site in Kliprivier in the KwaZulu-Natal district of South Africa. It was designed by Gerard Moerdijk. [a] The monument was vandalized in 2014. In 1982, it was described in the Government Gazette as
This memorial, which was designed by G. Moerdyk of Pretoria, was built by the firm Barker and Nel of Volksrust and was erected by the citizens of The Netherlands in memory of members of the Dutch Corps who died during the battle of Elandslaagte on 21 October 1899.
Dutch Corps Monument - Wikipedia

"The German (Hollander) Corps. Before the Battle the of Elandslaagte (October 21st, 1899). The corps lost many killed and wounded in this battle."[72 34]

Casualties In Action	Killed	Wounded		Missing and Prisoners			Forces Engaged		
	Officers	Men	Officers	Men	Officers	Men	Total	British	Boer	UK Loss
Elandslaagte	5	50	30	169	0	4	258	3,500	2,000	7.4%

Boer Forces: 46 killed 105 wounded, 181 missing or captured.

Chapter 5 Pursued by the Lancers.

Another last look at the bloody scene. It was very hard to have to beat an ignominious retreat, but it was harder still to have to go without being able to attend to one's wounded comrades, who were piteously crying aloud for help. To have to leave them in the hands of the enemy was exceedingly distressing to me. But there was no other course open, and fleeing, I hoped I might "live to fight another day." I got away, accompanied by Fourie and my Kaffir servant [Mooiroos]. "Let us go," I said, "perhaps we shall be able to fall in with some more burghers round here and have another shot at them."

Behind us the British lancers were shouting "Stop, stop, halt you —— Boers!" They fired briskly at us, but our little ponies responded gamely to the spur and, aided by the darkness, we rode on safely. Still the lancers did not abandon the chase, and followed us for a long distance. From time to time we could hear the.pitiful cries and entreaties of burghers who were being "finished off," but we could see nothing. My man and I had fleet horses in good condition, those of the pursuing lancers were big and clumsy.

My adjutant, Piet Fourie, however, was not so fortunate as myself. He was overtaken and made a prisoner. Revolvers were being promiscuously fired at us, and at times the distance between us and our pursuers grew smaller. We could plainly hear them shouting "Stop, or I'll shoot you," or "Halt, you damned Boer, or I'll run my lance through your blessed body."

We really had no time to take much notice of these pretty compliments. It was a race for life and freedom. Looking round furtively once more I could distinguish my pursuers; I could see their long assegais;[35] I could

[34] The Battle of Elandslaagte (10/21/1899) was also notable for being the first and last battle of the volunteer Hollanderkorps. The Hollanderkorps was a group of ca. 150 Dutch volunteers which had been established a mere month earlier. During the battle the Hollanderkorps suffered 9 fatalities, including Herman Coster, along with fellow officer Cars Geerts de Jonge and seven soldiers: P.J. van den Broek, H. van Cittert, J.A. Lepeltak Kieft, Jan Moora, J.Th. Rummeling, M. Schaink, and F.W. Wagner. A further 35 of the Hollanders were taken prisoner by the British. Among the prisoners was Willem Frederik Mondriaan (brother of the Dutch artist Piet Mondrian) who was wounded in the battle. Although he was able to crawl away from the battlefield he was taken prisoner by the British shortly after. He was later sent to Saint Helena as a prisoner of war, returning to South Africa in 1903. Cornelis Vincent 'Cor' van Gogh, the brother of the Dutch artist Vincent van Gogh, also fought in the battle where he was wounded and taken prisoner. He died shortly after. The Boer generals were deeply unhappy with the Hollanderkorps' performance, and it was disbanded after the battle though several hundred Dutch volunteers continued to fight in Boer regiments. The names of the deceased Hollanders, including Coster, were inscribed at a monument at the location of the battle. The monument was destroyed by vandals in 2014. Battle of Elandslaagte - Wikipedia

hear the snorting of their unwieldy horses, the clattering of their swords. These unpleasant combinations were enough to strike terror into the heart of any ordinary man.

Everything now depended upon the fleetness and staying power of my sturdy little Boer pony, Blesman. He remained my faithful friend long after he had got me out of this scrape; he was shot, poor little chap, the day when they made me a prisoner. Poor Blesman, to you I owe my life! Blesman was plainly in league against all that was British; from the first he displayed Anglophobia of a most acute character. He has served me in good stead, and now lies buried, faithful little heart, in a Lydenburg ditch.

In my retreat Sunday River had to be crossed. It was deep, but deep or not, we had to get through it. We were going at such a pace that we nearly tumbled down the banks. The precipice must have been very steep; all I remember is finding myself in the water with Blesman by my side. The poor chap had got stuck with his four legs in the drift sand. I managed to liberate him, and after a lot of scrambling and struggling and wading through the four foot stream, I got to the other side. On the opposite bank the British were still firing. I therefore decided to lie low in the water, hoping to delude them into thinking I was killed or drowned. My stratagem was successful. I heard one of my pursuers say, "We've finished him," and with a few more pyrotechnic farewells they retraced their steps towards Ladysmith.

On the other side, however, more horsemen came in pursuit. Unquestionably the British, fired by their splendid success, were following up their victory with great vigour, and again I was compelled to hide in the long grass into which my native servant, with Ethiopian instinct, had already crept. While I was travelling along on foot my man had rescued my horse from the muddy banks of the river.

When all was said and done I had escaped with a good wetting. Now for Newcastle. I had still my rifle, revolver, and cartridges left to me; my field-glass I had lost, probably in the river. Water there was plenty, but food I had none. The track to Newcastle to a stranger, such as I was in that part of the country, was difficult to discover. To add to my perplexities I did not know what had happened at Dundee, where I had been told a strong British garrison was in occupation.[36] Therefore, in straying in that direction I ran the risk of being captured.

Finally, however, I came upon a kaffir kraal. I was curtly hailed in the kaffir language, and upon my asking my swarthy friends to show me the road, half a dozen natives, armed with assegais, appeared on the scene. I clasped my revolver, as their attitude seemed suspicious. After they had inspected me closely, one of the elders of the community said: "You is one of dem Boers vat runs avay? We look on and you got dum dum to-day. Now we hold you, we take you English magistrate near Ladysmith."

But I know my kaffir, and I sized up this black Englishman instantly. "The fact is," I said, "I'm trekking with a commando of 500 men, and we are doing a bit of scouting round your kraal. If you will show me the way to the Biggersbergen I will give you 5s. on account." My amiable and dusky friend insisted on 7s. 6d., but after I had intimated that if he did not accept 5s. I should certainly burn his entire outfit, slaughter all his women and kill all his cattle, he acquiesced. A young Zulu was deputed as my guide, but I had to use my fists and make pretty play with my revolver, and generally hint at a sudden death, or he would have left me in the lurch. He muttered to himself for some time, and suddenly terminated his soliloquy by turning on his heels and disappearing in the darkness.

The light of a lantern presently showed a railway station, which I rightly guessed to be Waschbank. Here two Englishmen, probably railway officials, came up to me, accompanied by my treacherous guide. The latter had obviously been good enough to warn the officials at the station of my approach, but luckily they were unarmed. One of them said, "You've lost your way, it appears," to which I replied, "Oh, no, indeed; I'm on the right track I think." "But," he persisted, "you won't find any of your people here now; you've been cut to pieces at Elandslaagte and Lukas Meyer's and Erasmus's forces round Dundee have been crushed. You had better come along with me to Ladysmith. I promise you decent treatment." I took care not to get in between them, and, remaining at a little distance, said, revolver in hand, "Thanks very much, it's awfully good of you. I have no business to transact in Ladysmith for the moment and will now continue my journey. Good-night." "No, no, no, wait a minute," returned the man who had spoken first, "you know you can't pass here." "We shall see about that," I said. They rushed upon me, but ere they could overpower me I had levelled my revolver. The first

[35] Spears.

[36] He is referring to the Battle of Talana Hill that took place on October 20th. https://en.wikipedia.org/wiki/Battle_of_Talana_Hill

speaker tried to disarm me, but I shook him off and shot him. He fell, and as far I know, or could see, was not fatally wounded. The other man, thinking discretion the better part of valour, disappeared in the darkness, and my unfaithful guide had edged away as soon as he saw the glint of my gun.

My adventures on that terrible night were, however, not to end with this mild diversion. About an hour after daybreak, I came upon a barn upon which the legend "Post Office Savings Bank" was inscribed. A big Newfoundland dog lay on the threshold, and although he wagged his tail in a not unfriendly manner, he did not seem disposed to take any special notice of me. There was a passage between the barn and some stables at the back and I went down to prospect the latter. What luck if there had been a horse for me there! Of course I should only have wanted to borrow it, but there was a big iron padlock on the door, though inside the stables I heard the movements of an animal. A horse meant to me just then considerably more than three kingdoms to King Richard. For the first time in my life I did some delicate burglary and housebreaking to boot. But the English declare that all is fair in love and war, and they ought to know.

I discovered an iron bar, which enabled me to wrench off the lock from the stable door, and, having got so far with my burglarious performance, I entered cautiously, and I may say nervously. Creeping up to the manger I fumbled about till I caught hold of a strap to which the animal was tied, cut the strap through and led the horse away. I was wondering why it went so slowly and that I had almost to drag the poor creature along. Once outside I found to my utter disgust that my spoil was a venerable and decrepit donkey. Disappointed and disheartened, I abandoned my booty, leaving that ancient mule brooding meditatively outside the stable door and clearly wondering why he had been selected for a midnight excursion. But there was no time to explain or apologise, and as the mule clearly could not carry me as fast as my own legs, I left him to his meditations.

At dawn, when the first rays of the sun lit up the Biggersbergen in all their grotesque beauty, I realised for the first time where I was, and found that I was considerably more than 12 miles from Elandslaagte, the fateful scene of yesterday. Tired out, half-starved and as disconsolate as the donkey in the stable, I sat myself on an anthill.

Using an ant hill as a makeshift stove.[73]

Ant mounds of "compass termites" oriented north/south.[74]

For 24 hours I had been foodless, and was now quite exhausted. I fell into a reverie; all the past day's adventures passed graphically before my eyes as in a kaleidoscope; all the horrors and carnage of the battle, the misery of my maimed comrades, who only yesterday had answered the battle-cry full of vigour and youth, the pathos of the dead who, cut down in the prime of their life and buoyant health, lay yonder on the veldt, far away from wives and daughters and friends for ever more.

While in a brown study on this anthill, 30 men on horseback suddenly dashed up towards me from the direction of Elandslaagte. I threw myself flat on my face, seeking the anthill as cover, prepared to sell my life dearly should they prove to be Englishmen. As soon as they observed me they halted, and sent one of their number up to me. Evidently they knew not whether I was friend or foe, for they reconnoitred my prostrate form behind the anthill with great circumspection and caution; but I speedily recognised comrades-in-arms. I think the long tail which is peculiar to the Basuto pony enabled me to identify them as such, and one friend, who was their outpost, brought me a reserve horse, and what was even better, had extracted from his saddle-bag a tin of welcome bully beef to stay my gnawing hunger. But they brought sad tidings, these good friends. Slain on the battlefield lay Assistant-Commandant J. C. Bodenstein and Major Hall, of the Johannesburg Town Council, two of my bravest officers, whose loss I still regret.

We rode on slowly, and all along the road we fell in with groups of burghers. There was no question that our ranks were demoralised and heartsick. Commandant-General Joubert had made Dannhauser Station his headquarters and thither we wended our way. But though we approached our general with hearts weighed down with sorrow, so strange and complex a character is the Boers', that by the time we reached him we had gathered together 120 stragglers, and had recovered our spirits and our courage. I enjoyed a most refreshing rest on an unoccupied farm and sent a messenger to Joubert asking him for an appointment for the following morning to hand in my report of the ill-fated battle. The messenger, however, brought back a verbal answer that the General was exceedingly angry and had sent no reply. On retiring that night I found my left leg injured in several places by splinters of shell and stone. My garments had to be soaked in water to remove them, but after I had carefully cleaned my wounds they very soon healed.[37]

The next morning I waited on the Commandant-General. He received me very coldly, and before I could venture a word said reproachfully: "Why didn't you obey orders and stop this side of the Biggarsbergen, as the Council of War decided you should do?" He followed up the reproach with a series of questions: "Where's your general?" "How many men have you lost?" "How many English have you killed?" I said deferentially: "Well, General, you know I am not to be bullied like this. You know you placed me in a subordinate position under the command of General Kock, and now you lay all the blame for yesterday's disaster on my shoulders. However, I am sorry to say General Kock is wounded and in British hands. I don't know how many men we have lost; I suppose about 30 or 40 killed and approximately 100 wounded. The British must have lost considerably more, but I am not making any estimate."

General Joubert with His Son-In-Law, Malan, and His Secretary, Bracht, in His Tent Before Ladysmith.[75]

General Joubert and Staff at Modder Spruit.[76]

The grey-bearded generalissimo cooled a little and spoke more kindly, although he gave me to understand he did not think much of the Johannesburg commando. I replied that they had been fighting very pluckily, and that by retiring they hoped to retrieve their fortunes some other day. "H'm," returned the General, "some of your burghers have made so masterly a retreat that they have already got to Newcastle, and I have just wired Field-Cornet Pienaar, who is in charge, that I should suggest to him to wait a little there, as I propose sending him some railway carriages to enable him to retreat still further.

As for those Germans and Hollanders with you, they may go to Johannesburg; I won't have them here any more."

"General," I protested, "this is not quite fair. These people have volunteered to fight for, and with us; we cannot blame them in this matter. It is most unfortunate that Elandslaagte should have been lost, but as far as I can see there was no help for it." The old General appeared lost in thought; he seemed to take but little notice of what I

[37] Born in 1869 he is 30 years old.

said. Finally he looked up and fixed his small glittering eyes upon me as if he wished to read my most inmost thoughts.

"Yes," he said, "I know all about that. At Dundee things have gone just as badly.

Lukas Meyer made a feeble attack, and Erasmus left him in the lurch. The two were to charge simultaneously, but Erasmus failed him at a critical moment, which means a loss of 130 men killed and wounded, and Lukas Meyer in retreat across the Buffalo River. And now Elandslaagte on the top of all! All this owing to the disobedience and negligence of my chief officers."

The old man spoke in this strain for some time, until I grew tired and left. But just as I was on the point of proceeding from his tent, he said: "Look here, Commandant, reorganise your commando as quickly as you can, and report to me as soon as you are ready." He also gave me permission to incorporate in the reorganised commando various Hollander and German stragglers who were loafing round about, although he seemed to entertain an irradicable prejudice against the Dutch and German corps.

The Commandant of the Hollander corps, Volksraad Member Lombard, came out of the battle unscathed; his captain, Mr. B. J. Verselewel de Witt Hamer, had been made a prisoner; the Commandant of the German corps, Captain A. Schiel, fell wounded into British hands, while among the officers who were killed in action I should mention Dr. H. J. Coster,[38] the bravest Hollander the Transvaal ever saw, the most brilliant member of the Pretoria Bar, who laid down his life because in a stupid moment Kruger had taunted him and his compatriots with cowardice.

Hollander Corps at Pretoria.[77]

Funeral of General Koch.[78]

[38] Dr. J. H. Coster. (MOA)

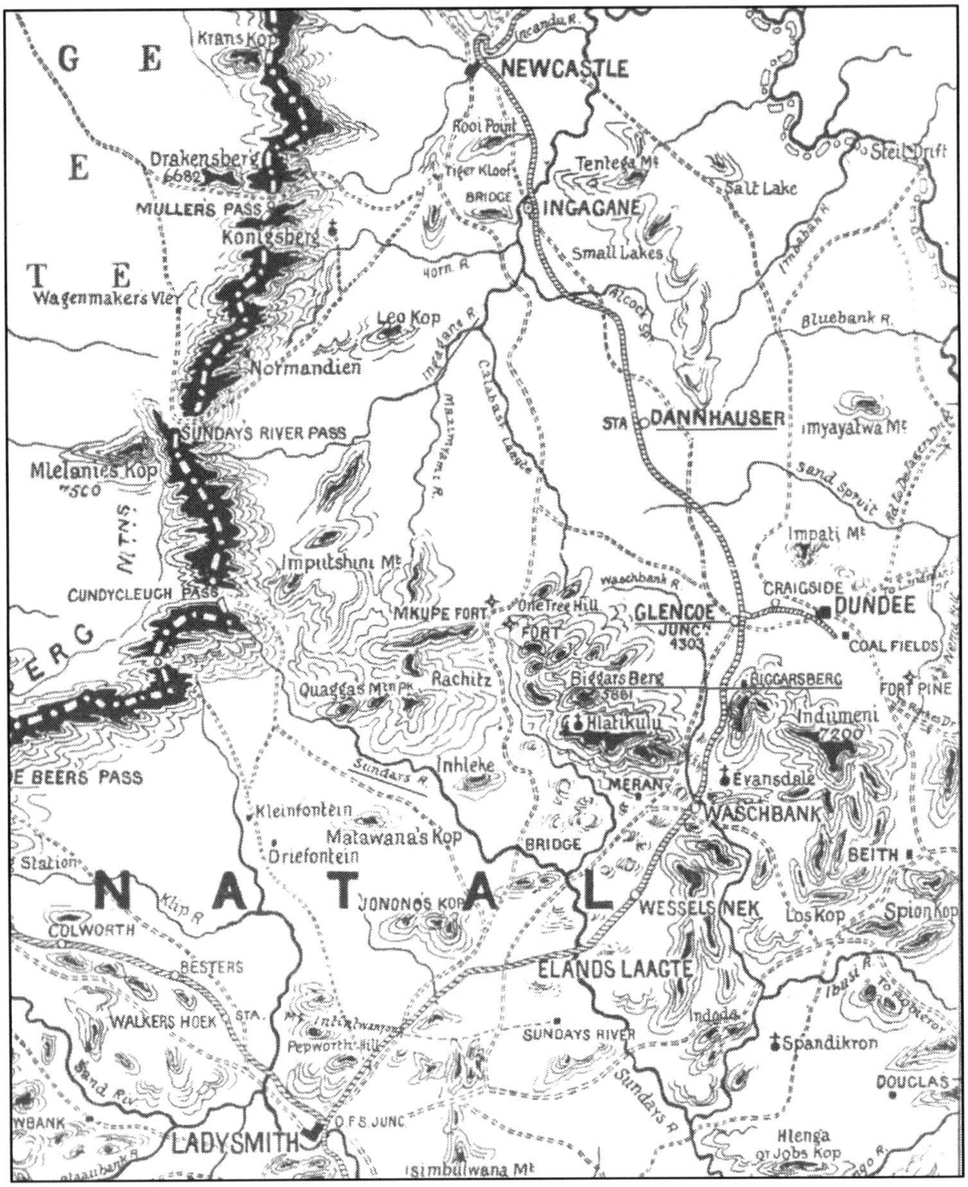

Northern Natal Detail.[79] Note the 'Biggars Berg' range of hills near Glencoe. Also Dannhauser station to the north. Also the Sundays River and the northern Igagane river flowing north past Newcastle.

Chapter 6. Risking Joubert's Anger.

After the above unpleasant but fairly successful interview with our Commander-in-Chief, I left the men I had gathered round me in charge of a field-cornet, and proceeded by train to Newcastle to collect the scattered remnants of my burghers, and to obtain mules and waggons for my convoy. For, as I have previously stated, it was at Newcastle we had left all our commissariat-waggons and draught cattle under a strong escort. On arrival I summoned the burghers together, and addressing them in a few words, pointed out that we should, so soon as possible, resume the march, in order to reach the fighting line without delay, and there retrieve the pride and honour of our commando.

"Our beloved country," I said, "as well as our dead, wounded and missing comrades, require us not to lose courage at this first reverse, but to continue the righteous struggle even against overwhelming odds," and so on, in this strain.

I honestly cannot understand why we should have been charged with cowardice at the battle of Elandslaagte, although many of us seemed to apprehend that this would be the case. We had made a good fight of it, but overwhelmed by an organised force of disciplined men, eight or ten times our number, we had been vanquished, and the British were the first to admit that we had manfully and honourably defended our positions.

To put a wrong construction on our defeat was a libel on all who had bravely fought the fight, and I resented it. There are such things as the fortunes of war, and as only one side can win, it cannot always be the same. However, I soon discovered that a small number of our burghers did not seem inclined to join in the prolongation of the struggle. To have forced them to rejoin us would have served no purpose, so I thought the best policy would be to send them home on furlough until they had recovered their spirits and their courage. No doubt the scorn and derision to which they would be subjected by their wives and sisters would soon induce them to take up arms again and to fulfil the duties their country required. I therefore requested those who had neither the courage nor the inclination to return to the front to fall out, and about thirty men fell back, bowing their heads in shame. They were jeered at and chaffed by their fellows, the majority of whom had elected to proceed. But the shock of Elandslaagte had been too much for the weaker brethren, who seemed deaf to every argument, and only wanted to go home. I gave each of these a pass to proceed by rail to Johannesburg, which read as follows:—

"Permit.................................... to go to Johannesburg on account of cowardice, at Government's expense."

They put the permit in their pockets without suspecting its contents, and departed with their kit to the station to catch the first available train.

The reader will now have formed an idea of the disastrous moral effect of this defeat, and the subsequent difficulty of getting a commando up to its original fighting strength. But in spite of this I am proud to say that by far the greater number of the Johannesburgers were gathered round me and prepared to march to meet the enemy once more.

My trap and all its contents had been captured by the enemy at Elandslaagte, and I found it necessary to obtain new outfits, &c., at Newcastle. This was no easy matter, as some of the storekeepers had moved the greater part of their goods to a safer place, while some commandos had appropriated most of the remainder. What was left had been commandeered by Mr. J. Moodie, a favourite of General Joubert, who was posing there as Resident Justice of the Peace; and he did not feel inclined to let any of these goods out of his possession. By alternately buying and looting, or in other words stealing, I managed to get an outfit by the next morning, and at break of day we left for Dannhauser Station, arriving there the same evening without further noteworthy incident.

Next day, when the Johannesburg corps turned out, we numbered 485 mounted men, all fully equipped. On arrival at Glencoe Station I received a telegram from General Joubert informing me that he had defeated the enemy at Nicholson's Nek near Ladysmith that day (October 30, 1899) taking 1,300 prisoners, who would arrive at Glencoe the following morning. He desired me to conduct them to Pretoria under a strong escort. What a flattering order! To conduct prisoners-of-war, taken by other burghers! Were we then fit for nothing but police duty?

Receiving British Prisoners at Pretoria.[80]

British Prisoners leaving Pretoria for Waterval.[81]

However, orders have to be obeyed, so I sent one of my officers with 40 men to take the prisoners to Pretoria, and reported to the Commandant-General by telegram that his order had been executed, also asking for instructions as to where I was to proceed with my commando. The reply I received was as follows:—

"Pitch your camp near Dundee, and maintain law and order in the Province, also aid the Justice of the Peace in forwarding captured goods, ammunition, provisions, etc., to Pretoria, and see that you are not attacked a second time."

This was more than flesh and blood could bear; more than a "white man" could stand. It was not less than a personal insult, which I deeply resented. Evidently my chief had resolved to keep us in the background; he would not trust our commando in the fighting line. In short, he would not keep his word and give us another chance to recoup our losses.

I had, however, made up my mind, and ordered the commando to march to Ladysmith. If the General would not have me at the front I should cease to be an officer. And, although I had no friends of influence who could help me I resolved to take the bull by the horns, and leave the rest to fate.

On the 1st November, 1899, we reached the main army near Ladysmith, and I went at once to tell General Joubert in person that my men wanted to fight, and not to play policemen in the rear of the army. Having given the order to dismount I proceeded to Joubert's tent, walked in with as much boldness as I could muster, and saluted the General, who was fortunately alone. I at once opened my case, telling him how unfair it was to keep us in the rear, and that the burghers were loudly protesting against such treatment. This plea was generally used throughout the campaign when an officer required something to be granted him. At first the old General was very wrathful. He said I had disobeyed his orders and that he had a mind to have me shot for breach of discipline. However, after much storming in his fine bass voice, he grew calmer, and in stentorian tones ordered me for the time being to join General Schalk Burger, who was operating near Lombard's Kop in the siege of Ladysmith.

General Schalk Burger.[82] [39] General Schalk Burger's commando.[83] Appears to be Schalk on left by the wagon.

That same evening I arrived there with my commando and reported myself to Lieut-General Burger. One of his adjutants, Mr. Joachim Fourie,[40] who distinguished himself afterwards on repeated occasions and was killed in action near his house in the Carolina district, showed me a place to laager in. We pitched our tents on the same spot where a few days before Generals White and French had been defeated, and there awaited developments.

At this place the British, during the battle of Nicholson's Nek, had hidden a large quantity of rifle and gun ammunition in a hole in the ground, covering it up with grass, which gave it the appearance of a heap of rubbish. One of the burghers who feared this would be injurious to the health of our men in camp, set the grass on fire, and this soon penetrated to the ammunition. A tremendous explosion occurred, and it seemed as if there were a real battle in progress. From all sides burghers dashed up on horseback to learn where the fighting was taking place.

General Joubert sent an adjutant to enquire whether the Johannesburgers were now killing each other for a change, and why I could not keep my men under better control. I asked this gentleman to be kind enough to see for himself what was taking place, and to tell the Commandant-General that I could manage well enough to keep my men in order, but could not be aware of the exact spot where the enemy had chosen to hide their ammunition.

Meanwhile, it became daily more evident to me how greatly Joubert depreciated my commando, and that we would have to behave very well and fight very bravely to regain his favour. Other commandos also seemed to have no better opinion, and spoke of us as the laager which had to run at Elandslaagte, forgetting how even General Meyer's huge commando had been obliged to retreat in the greatest confusion at Dundee. If all the

[39] See MOA And: https://en.wikipedia.org/wiki/Schalk_Willem_Burger

[40] See MOA.

details of this Dundee engagement were published it would be discovered that it was a Boer disaster only second to that of Elandslaagte.

We were now, however, at any rate at the front. I sent out my outposts and fixed my positions, which were very far from good; but I decided to make no complaints. We had resolved to do our very best to vindicate our honour, and to prove that our accusers had no reason to call us either cowards or good-for-nothings.

Chapter 7. The Boer General's Superstitions.

A few days after we had arrived before Ladysmith we joined an expedition to reconnoitre the British entrenchments, and my commando was ordered near some forts on the north-westerly side of the town. Both small and large artillery were being fired from each side. We approached within 800 paces of a fort; it was broad daylight and the enemy could therefore see us distinctly, knew the exact range, and received us with a perfect hailstorm of fire. Our only chance was to seek cover behind kopjes and in ditches, for on any Boer showing his head the bullets whistled round his ears. Here two of my burghers were severely wounded, and we had some considerable trouble to get them through the firing line to our ambulance. At last, late in the afternoon, came the order to retire, and we retired after having achieved nothing.

I fail to this day to see the use of this reconnoitring, but at Ladysmith everything was equally mysterious and perplexing. It was perhaps that my knowledge of military matters was too limited to understand the subtle manoeuvres of those days. But I have made up my mind not to criticise our leader's military strategy, though I must say at this juncture that the whole siege of Ladysmith and the manner in which the besieged garrison was ineffectually pounded at with our big guns for several months, seem to me an unfathomable mystery, which, owing to Joubert's untimely death, will never be explained satisfactorily. But I venture to describe Joubert's policy outside Ladysmith as stupid and primitive, and in another chapter I shall again refer to it.

After another fortnight or so, we were ordered away to guard another position to the **south-west of Ladysmith**, as the Free State commando under Commandant Nel, and, unless I am mistaken, under Field-Cornet Christian de Wet (afterwards the world-famous chief Commander of the Orange Free State, and of whom all Afrikanders are justly proud), had to go to Cape Colony.[41]

Here I was under the command of Dijl [Daniel] Erasmus, who was then General and a favourite of General Joubert. We had plenty of work given us. Trenches had to be dug and forts had to be constructed and remodelled. At this time an expedition ventured to Estcourt, under General Louis Botha, who replaced General [Lucas] Meyer, sent home on sick leave. My commando joined the expedition under Field-Cornet J. Kock, who afterwards caused me a lot of trouble.

I can say but little of this expedition to Estcourt, save that the Commander-in-Chief accompanied it. But for his being with us, I am convinced that General Botha would have pushed on at least as far as Pietermaritzburg, for the English were at that time quite unable to stop our progress. But after we got to Estcourt, practically unopposed, Joubert, though our burghers had been victorious in battle after battle, ordered us to retreat. The only explanation General Joubert ever vouchsafed about the recall of this expedition was that in a heavy thunderstorm which had been raging for two nights near Estcourt, two Boers had been struck by lightning, which, according to his doctrine, was an infallible sign from the Almighty that the commandos were to proceed no further.

It seems incredible that in these enlightened days we should find such a man in command of an army; it is, nevertheless, a fact that the loss of two burghers induced our Commandant-General to recall victorious commandos who were carrying all before them. The English at Pietermaritzburg, and even at Durban, were trembling lest we should push forward to the coast, knowing full well that in no wise could they have arrested our progress. And what an improvement in our position this would have meant! As it was, our retirement encouraged the British to push forward their fighting line so far as Chieveley Station, near the Tugela river, and the commandos had to take up a position in the "randjes,"[42] on the westerly banks of the Tugela.

[41] This is correct, see De Wet's account in Three Years' War.

[42] Randje translates literally to 'Brink' or 'Edge' in Afrikaans and Dutch.

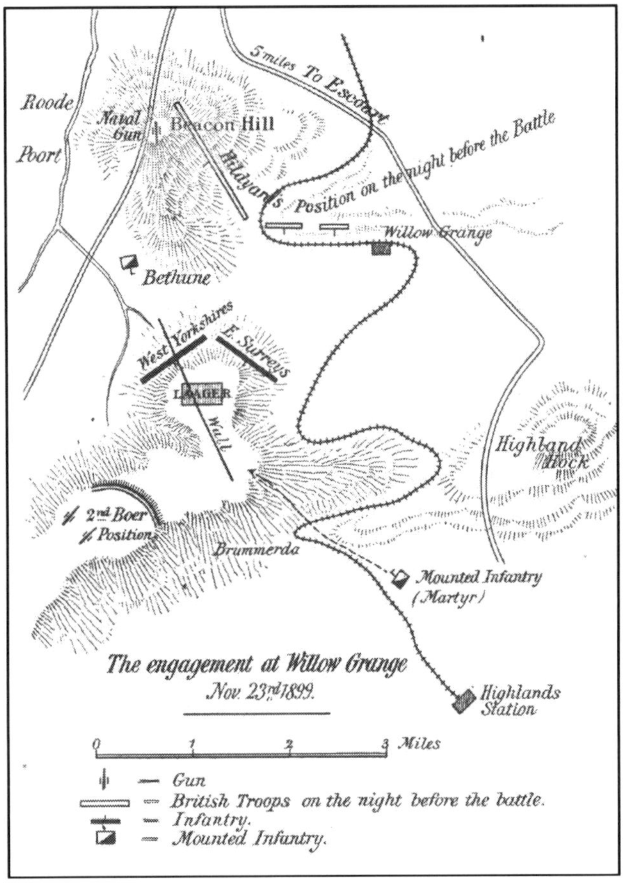

Willow Grange.[84 43]

Along the Tugela—Coming Suddenly Upon an English Outpost.

Boer War: bringing in the wounded the Willow Grange fight. Process print.[85]

[43] The Battle of Willow Grange took place on 23 November 1899 during the Second War of Independence at Willow Grange near Estcourt. Here the Boer Forces under the leadership of General Piet Joubert won a victory over the British Forces. The battle was fought in bad weather; the lightning struck one Civilian dead and two others unconscious, and also struck to death six horses. One British soldier was also struck by lightning. https://af.wikipedia.org/wiki/Slag_van_Willow_Grange

Chapter 8. The "Great Powers" to Intervene.

During the retreat of our army to the frontier of the Transvaal Republic nothing of importance occurred. Here again confusion reigned supreme, and none of the commandos were over-anxious to form rearguards. Our Hollander Railway Company made a point of placing a respectful distance between her rolling-stock and the enemy, and, anxious to lose as few carriages as possible, raised innumerable difficulties when asked to transport our men, provisions and ammunition. Our generals had meantime proceeded to Laing's Nek by rail to seek new positions, and there was no one to maintain order and discipline.

About 150 Natal Afrikanders who had joined our commandos when these under the late General Joubert occupied the districts about Newcastle and Ladysmith, now found themselves in an awkward position. They elected to come with us, accompanied by their families and live stock, and they offered a most heartrending spectacle. Long rows of carts and wagons wended their way wearily along the road to Laing's Nek. Women in tears, with their children and infants in arms, cast reproachful glances at us as being the cause of their misery. Others occupied themselves more usefully in driving their cattle. Altogether it was a scene the like of which I hope never to see again.

Supply wagon being pulled out of mud by troops and oxen.[86]

The Natal kaffirs now had an opportunity of displaying their hatred towards the Boers. As soon as we had left a farm and its male inhabitants had gone, they swooped down on the place and wrought havoc and ruin, plundering and looting to their utmost carrying capacity. Some even assaulted women and children, and the most awful atrocities were committed. I attach more blame to the whites who encouraged these plundering bands, especially some of the Imperial [Light Horse, colonial] troops and Natal men in military service. Not understanding the bestial nature of the kaffirs, they used them to help carry out their work of destruction, and although they gave them no actual orders to molest the people, they took no proper steps of preventing this.

When our commando passed through Newcastle, we found the place almost entirely deserted, excepting for a few British subjects who had taken an oath of neutrality to the Boers.

"Panorama of Newcastle, now invested by the Boers."[87]

I regret to have to state that during our retreat a number of irresponsible persons set fire to the Government buildings in that town. It is said that an Italian officer burned a public hall on no reasonable pretext; certainly he never received orders to that effect. As may be expected of an invading army, some of our burgher patrols and other isolated bodies of troops looted and destroyed a number of houses which had been temporarily deserted. But with the exception of these few cases, I can state that no outrages were committed by us in Natal, and no property was needlessly destroyed.

On our arrival at Laing's Nek a Council of War was immediately held to decide our future plans.

We now found ourselves once more on the old battlefields of 1880 and 1881, where Boer and Briton had met 20 years before to decide by trial of arms who should be master of the S. A. Republic. Traces of that desperate struggle were still plainly visible, and the historic height of Majuba stood there, an isolated sentinel, recalling to us the battle in which the unfortunate Colley lost both the day and his life.[44]

Arrival of a commando at Majuba Hill.[88]

Majuba Hill from the south.[89]

Officers Graves. Majuba in the distance..[90]

General Colley, killed at Majuba.[91]

I was told off to take up a position in the Nek where the wagon-road runs to the east across the railway-tunnel, and here we made preparations for digging trenches and placing our guns. Soon after we had completed our entrenchments we once more saw the enemy. They were lying at Schuinshoogte on the Ingogo, and had sent a mounted corps with two guns to the Nek. Although we had no idea of the enemy's strength, we were fully prepared to meet the attack; the Pretoria, Lydenburg and other laagers were posted to the left on the summit of Majuba Hill, and other commandos held good positions on the east. But the enemy evidently thought that we had fled all the way back to Pretoria, and not expecting to find the Nek occupied, advanced quite unconcerned. We fired a few volleys at them, which caused them to halt in considerable surprise, and, replying with a little artillery fire, they quickly returned to Schuinshoogte. We had, however, to be on our guard both day and night. It was bitterly cold at the time and a strong easterly wind was blowing.

[44] The Battle of Majuba Hill on 27 February 1881 was the final and decisive battle of the First Boer War that was a resounding victory for the Boers. The British Major General Sir George Pomeroy Colley occupied the summit of the hill on the night of 26–27 February 1881. The summit was then stormed by the Boer forces using effective use of cover and superior marksmanship. A total of 285 British were killed, captured or wounded. The revelation that professionally trained soldiers were defeated by young farmboys led by a smattering of older soldiers proved to be a major blow to Britain's prestige and negotiating position in the treaty that ended the war. https://en.wikipedia.org/wiki/Battle_of_Majuba_Hill And: https://en.wikipedia.org/wiki/George_Pomeroy_Colley

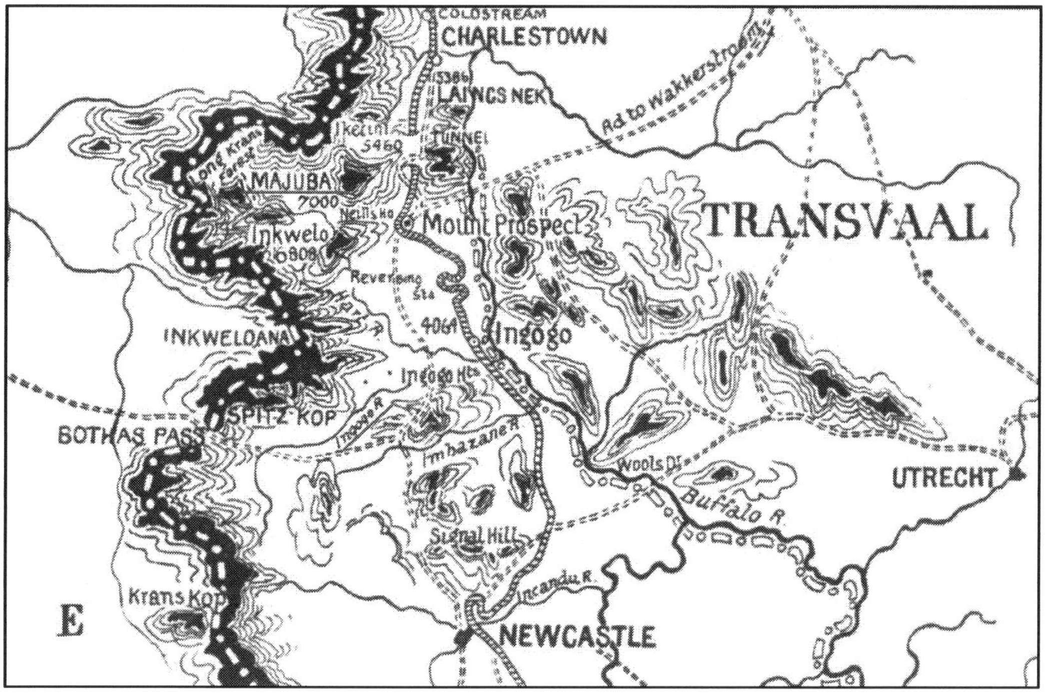

Note Majuba, Laings Nek and Ingogo River. PF: Subequently I will generally omit 'Note' and just refer to locations identified

Drakensberg Railway Switchbacks.[92]

Next day something occurred which afforded a change to the monotony of our situation, namely, the arrival from Pretoria of Mr. John Lombaard, member of the First Volksraad for Bethel. He asked permission to address us and informed us that we need only hold out another fortnight, for news from Europe had reached them to the effect that the Great Powers had decided to put an end to the War. This communication emanating from such a semi-official source was believed by a certain number of our men, but I think it did very little to brighten up the spirits of the majority, or arouse them from the lethargy into which they seemed to have fallen. A fortnight passed, and a month, without us hearing anything further of this expected intervention, and I have never been able to discover on whose authority and by whose orders Mr. Lombaard made to us that remarkable communication.

Meantime, General Buller did not seem at all anxious to attack us, perhaps fearing a repetition of the "accidents" on the Tugela; or possibly he thought that our position was too strong. For some reason, therefore, Laing's Nek was never attacked, and Buller afterwards, having made a huge "detour," broke through Botha's Pass. Meanwhile, Lord Roberts and his forces were marching without opposition through the Orange Free State, and I was ordered to proceed to Vereeniging with my commando. We left Laing's Nek on the 19th of May, and proceeded to the Free State frontier by rail.

Tunnel Near Laing's Nek Wrecked by the Boers. South View.[93] Tunnel Near Laing's Nek Wrecked by the Boers. North View.[94] Laing's nek Tunnel Reopened.[95]

Transvaal commando train.[96]

Chapter 9. Colenso and Spion Kop Fights.

Eight days after my commando had been stationed in my new position under General Erasmus, I received instructions to march to Potgietersdrift, on the Upper Tugela, near Spion Kop, and there to put myself at Andries Cronje's[45] disposal. This gentleman was then a general in the Orange Free State Army, and although a very venerable looking person, was not very successful as a commander. Up to the 14th of December, 1899, no noteworthy incident took place, and nothing was done but a little desultory scouting along the Tugela, and the digging of trenches.

At last came the welcome order summoning us to action; and we were bidden to march on Colenso Heights with 200 men to fill up the ranks, as a fight was imminent. We left under General Cronje and arrived the next morning at daybreak, and a few hours after began the battle now known to the world as the Battle of Colenso (15th December, 1899).

[45] Pieter Arnoldus Cronje See MOA and https://en.wikipedia.org/wiki/Piet_Cronj%C3%A9

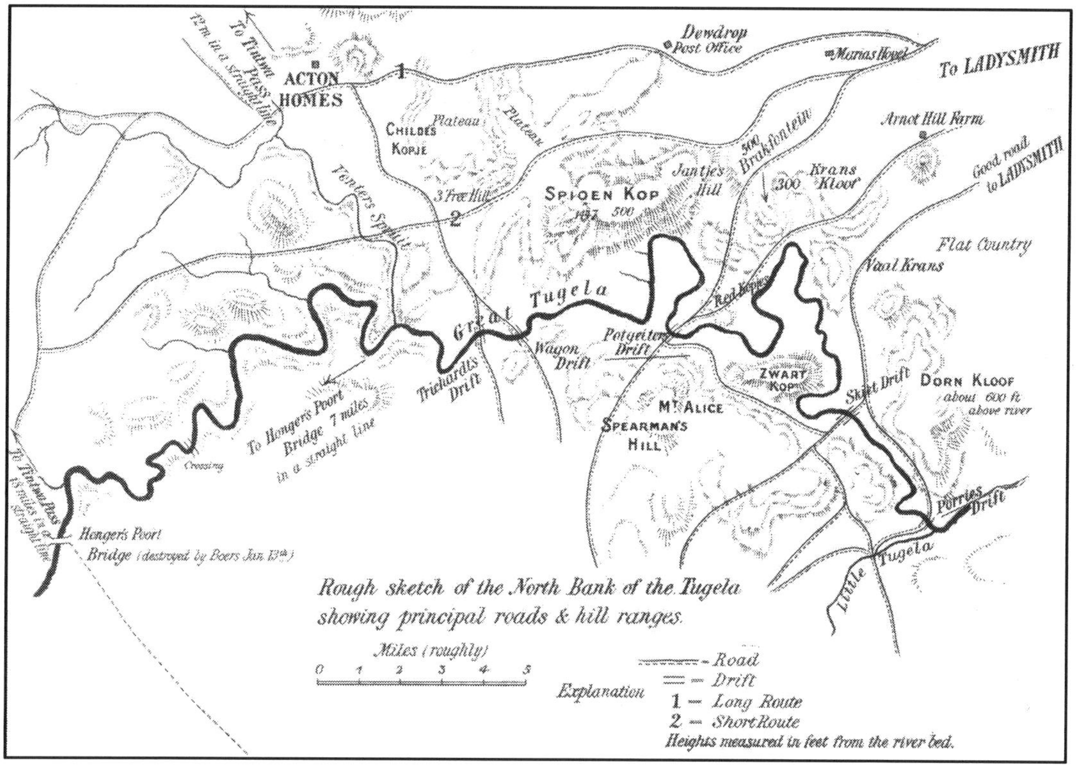

Spion Area of Operations. North Bank of the Tugela.[97] Note Potgeiters Drift and then Porries Drift at the junction of the Little Tugela with the larger stream.

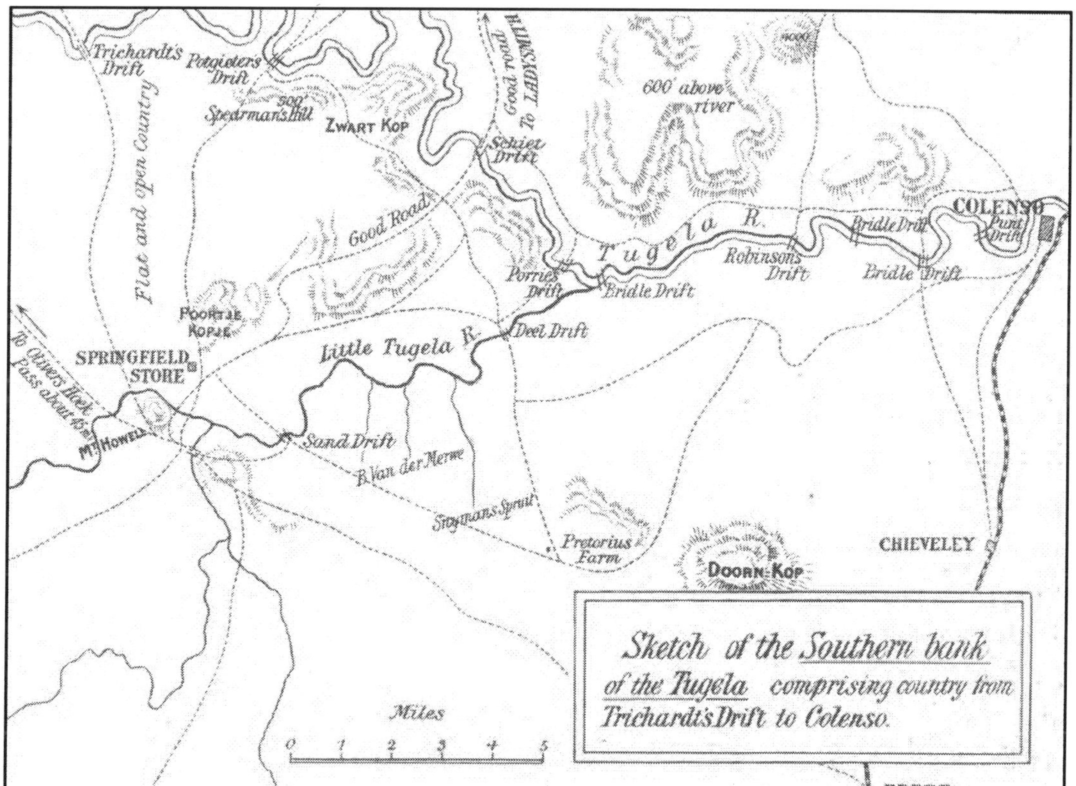

South Bank of the Tugela.[98] Note Potgeiters drift at top and Colenso on right.

I afterwards heard that the commandos under General Cronje were to cross the river and attack the enemy's left flank. This did not happen, as the greatest confusion prevailed owing to the various contradictory orders given by the generals. For instance, I myself received four contradictory orders from four generals within the space of ten minutes. I, however, took the initiative in moving my men up to the river to attempt the capture of a battery of guns on the enemy's left flank which had been left unprotected, as was the case with the ten guns which fell into our hands later in the day.

I had approached within 1,400 paces of the enemy, and my burghers were following close behind me when an adjutant from General Botha (accompanied by a gentleman named C. Fourie,[46] who was then also parading as a general) galloped up to us and ordered us at once to join the Ermelo commando, which was said to be too weak to resist the attacks of the enemy.

We hurried thither as quickly as we could round the rear of the fighting line, where we were obliged to off-saddle and walk up to the position of the Ermelo burghers. This was no easy task; the battle was now in full swing, and the enemy's shells were bursting in dozens around us, and in the burning sun we had to run some miles.

When we arrived at our destination Mr. Fourie (the pseudo general) and his adjutant could nowhere be found. As to the Ermelo burghers, they said they were quite comfortable, and had asked for no assistance.

Not a single shell had reached them, for a clump of aloe trees stood a hundred yards away, which the English presumably had taken for Boers, judging by the terrific bombardment these trees were being subjected to.

By this time the attack was repulsed, and General Buller was in full retreat to Chieveley, though our commando had been unable to take an active part in the fighting, at which we were greatly disappointed. It is much to be regretted that the retreat of the enemy was not followed up at once. Had this been done, the campaign in Natal would have taken an entirely different aspect, and very probably would have been attended by a more favourable conclusion. I consider myself far from a prophet, but this I know; and if we had then and on subsequent occasions followed up our successes, the result of the Campaign would have been far more satisfactory to us.

After I had assisted in bringing away through the river the guns we had taken, and seen to other matters which required my immediate attention, I was ordered to remain with the Ermelo commando at Colenso, near Toomdrift, and to await there further instructions.

Boers with the captured guns after the battle.[47] Captured British field gun at Viljoen's Drift.[99] Captured British field gun at Viljoen's Drift.[100] This gun, one of those captured at Colenso, was being sent down to be used around Bloemfontein. The coloured gentlemen are mine bots, going home to Basutoland.

Captured British Cannon.[101]

[46] Christian E. Fourie MOA.

[47] Apparently the breech blocks had not been removed and the guns were fully functional. Since the Boers still controlled the railroad between Colenso and the outskirts of besieged Ladysmith they were able to use it to transport the weapons. Image source: britishbattles.com

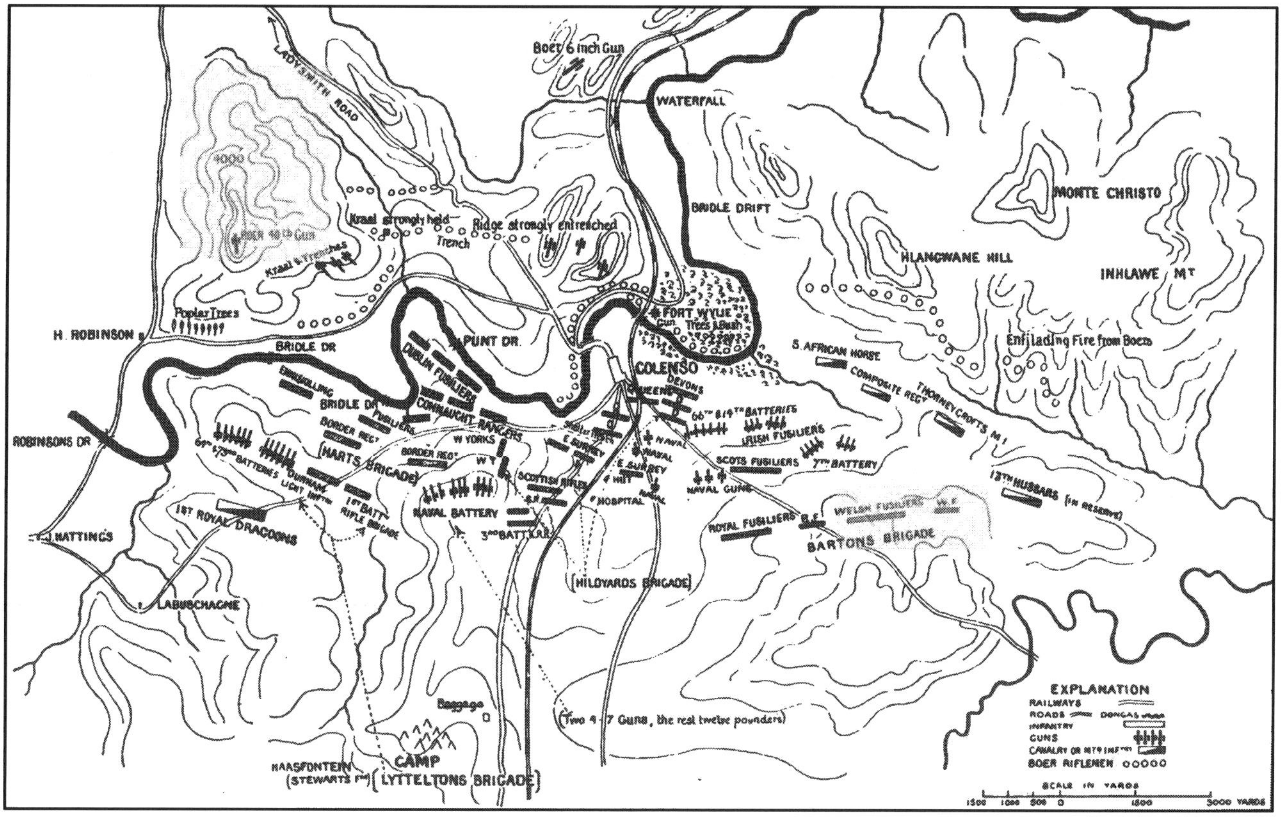

Battle of Colenso.[102]

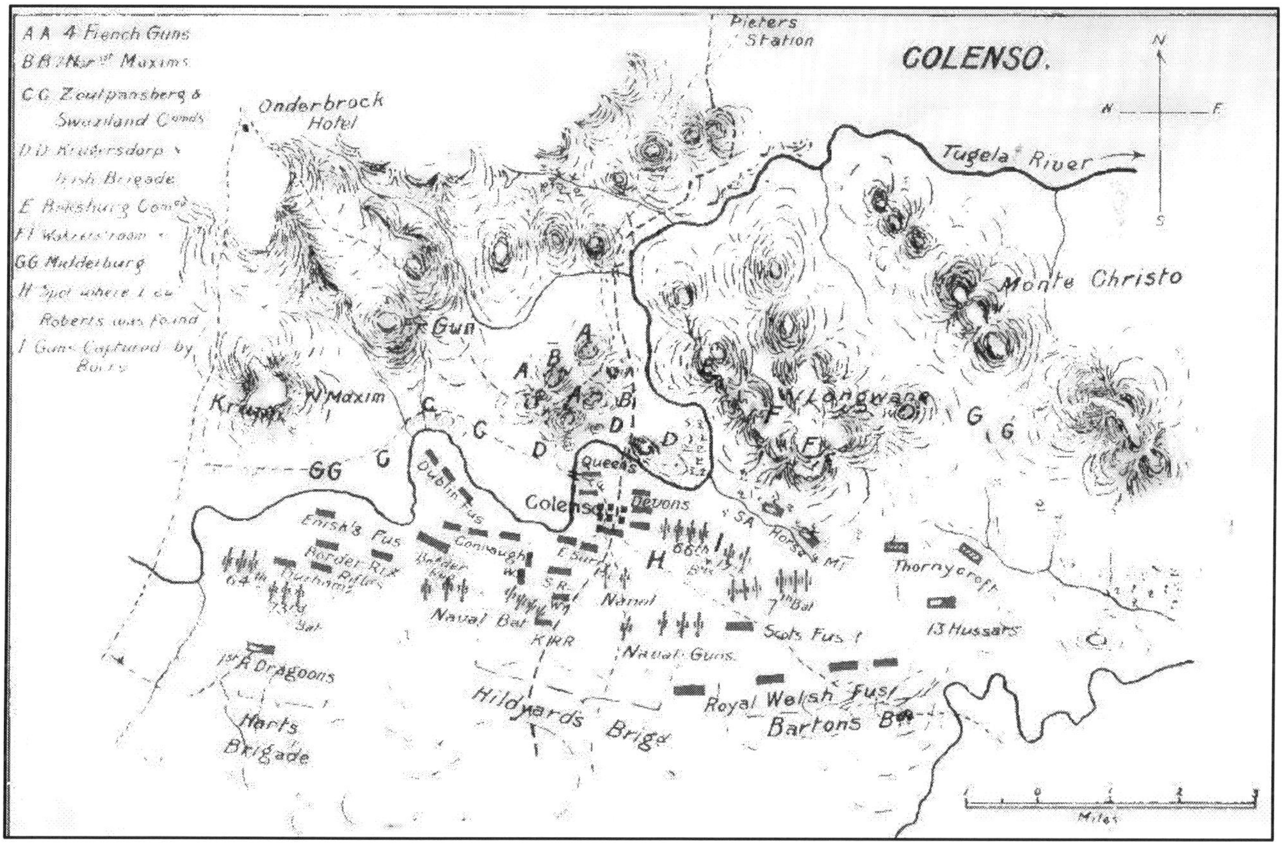

With apologies for the quality. This is the best that could be obtained from the various available editions of Viljoen's book.

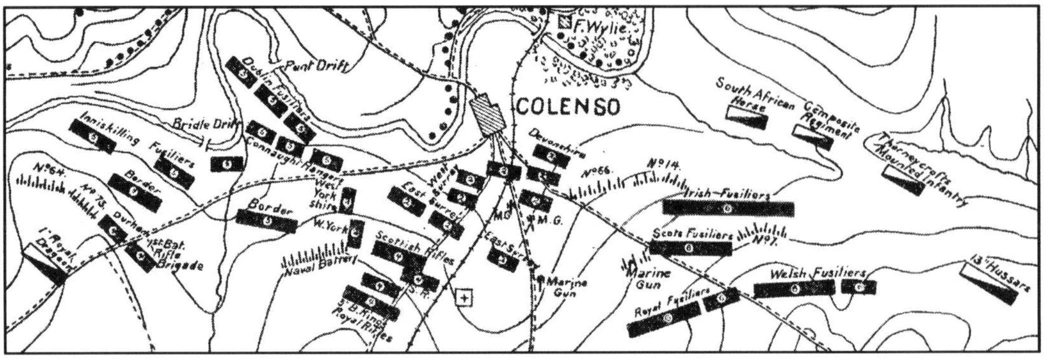

Battle of Colenso.[103] Battle Map of Colenso.[48] British: 143 killed, 755 wounded, 240 missing, 10 guns captured. Boers: 8 killed, 30 wounded.

A few weeks of inactivity followed, the English sending us each day a few samples of their shells from their 4·7 Naval guns. Unfortunately, our guns were of much smaller calibre, and we could send them no suitable reply. As a rule we would lie in the trenches, and a burgher would be on the look-out. So soon as he saw the flash of an English gun, he would cry out; "There's a shell," and we then sought cover, so that the enemy seldom succeeded in harming us.

One day one of these big shells fell amongst a group of fourteen burghers who were at dinner. The shell struck a sharp rock, which it splintered into fragments, and was emitting its yellow lyddite; but, fortunately, the fuse refused to burn, and the shell did not explode, so we had a narrow escape that day from a small catastrophe.

My laager had been at Potgietersdrift all this time, and for the time being we were deprived of our tents. We were not sorry, therefore, when we were ordered to leave Colenso and to return to our camp.

A few days after we were told off to take up a position at the junction of the Little and the Big Tugela, between Spion Kop and Colenso. Here we celebrated our first Christmas in the field; our friends at Johannesburg had sent us a quantity of presents by means of a friend, Attorney Raaff, comprising cakes, cigars, cigarettes, tobacco and other luxuries. Along this part of the Tugela we found a fair quantity of vegetables, and poultry, and as their respective owners had fled we were unable to pay for what we had. We were obliged, therefore, to "borrow" all these things for the banquet befitting to the occasion.

But General Buller had not quite finished with us yet. He marched on Spion Kop, but with the exception of a feint attack nothing of importance happened then. One day I went across the river with a patrol to discover what the enemy was doing, when we suddenly came across nine English spies, who fled as soon as they saw us. We galloped after them, trying to cut them off from the main body, which was at a little distance away from us, and would no doubt have overtaken them, but, riding at a breakneck speed over a mountain ridge, we found ourselves suddenly confronted with a strong English mounted corps, apparently engaged in drilling. We were only 500 paces away from them, and we jumped off our horses, and opened fire. But there were only a dozen of us, and the enemy soon began sending us a few shells, and prepared to attack us with their whole force. About a hundred mounted men, with horses in the best of condition, set off to pursue us.

We were obliged to ride back by the same path we had come by, which was fortunate for us, as we knew the way and could ride through crevices and dongas without any hesitation. In this way we soon gave our pursuers the slip.

Buller's forces seemed at first to have the intention of forcing their way through near Potgietersdrift, and they took possession of all the "randts" on their side of the river, causing us to strengthen the position on our side. We thus had to shift our commando again to Potgietersdrift, where we soon had the enemy's Naval guns playing on our positions. This continued day and night for a whole week.

It seemed as if General Buller were determined to annihilate all the Boers with his lyddite shells, so as to enable the soldiers to walk at their leisure to the release of Ladysmith. Certainly we suffered considerably from lyddite fumes.

[48] The Battle of Colenso was the third and final battle fought during the Black Week of the Second Boer War. It was fought between British and Boer forces from the independent South African Republic and Orange Free State in and around Colenso, Natal, South Africa on 15 December 1899. Inadequate preparation, lack of reconnaissance and uninspired leadership led to a British defeat. Battle of Colenso - Wikipedia

The British next made a feint attack near Potgietersdrift, advancing with a great clamour till they had come within 2,000 paces of us, where they occupied various "randts" and kopjes, always under cover of their artillery. Once they came a little too close to our positions, and we suddenly opened fire on them. The result was that their ambulance waggons were seen to become very busy driving backwards and forwards.

This "feint," however, was only made in order to divert our attention, while Buller was concentrating his troops and guns on Spion Kop. The ruse succeeded to a large extent, and on the 21st January the memorable battle of Spion Kop (near the Upper Tugela) began.

General Warren, who, I believe, was in command here, had ordered another "feint" attack from the extreme right wing. General Cronje and the Free Staters had taken up a position at Spion Kop, assisted by the commandos of General Erasmus and Schalk Burger.

The fight lasted the whole of that day and the next, and became more and more fierce. Luckily General Botha appeared on the scene in time, and re-arranged matters so well and with so much energy that the enemy found itself well employed, and was kept in check at all points.

I had been ordered to defend the position at Potgietersdrift, but the fighting round Spion Kop became so serious that I was obliged to send up a field cornet with his men as a reinforcement, which was soon followed by a second contingent, making altogether 200 Johannesburgers in the fight, of whom nine were killed and 18 wounded. The enemy had reached the top of the "kop" on the evening of the second day of the fight, not, however, without having sustained considerable losses. At this juncture one of our generals felt so disheartened that he sent away his carts, and himself left the battlefield.

But General Botha kept his ground like a man, surrounded by the faithful little band who had already borne the brunt of this important battle. And one can imagine our delight when next morning we found that the English had retreated, leaving that immense battlefield, strewn with hundreds of dead and wounded, in our hands.

"What made them leave so suddenly last night," was the question we asked each other then, and which remains unanswered to this day.

General Warren has stated that the cause of his departure was the want of water, but I can hardly credit that statement, as water could be obtained all the way to the top of Spion Kop; and even had it been wanting it is not likely that after a sacrifice of 1,200 to 1,300 lives the position would have been abandoned on this account alone. Our victory was undoubtedly a fluke.

Major Wright [Or possibly Major Preston] demanding possession of the wounded.[49] Searching for casualties after the Battle of Spion Kop.[104] British: 243 killed 1,250 wounded. Boer: 68 killed 267 wounded.[50]

[49]"The Boers at first refused to allow the British volunteer ambulance men to carry off the wounded. Major Wright, who had been wounded at Elandslaagte and now serving under the Red Cross, persisted in his demand, claiming his right under the Geneva Convention, and finally carried his point in a personal interview with Commandant Botha." Image Source: With the Flag to Pretoria, H. W. Wilson Vol 1 1900 By R.B.M. Paxton, After a sketch by Ernest Prater.

British troops retreating across the Tugela River after the Battle of Spion Kop.[105] Note the ambulance wagon in the middle of the river with the red cross. Ambulance Corps crossing the Tugela.[106]

"Wounded at Spion Kop." "Stretcher-bearers crossing the pontoon bridge at Trichard's [drift]."[51]

"The British trench in 'the murderous acre' on Spion Kop after the battle."[52] Spion Kop in 2005.[53]

[50] PF: There was an informal truce on the 24th, for the removal of the wounded and burial of the dead. Over the next two days the Boer forces could have made life very difficult for the retiring forces, with shelling or even counterattacking. They chose not to do so.

[51] With Roberts to the Transvaal. C.N. Robinson [1900] Navy and Army.

[52] This picture appears to be taken before internment. The soldiers appear to have been laid to rest. Image: File: Spion Kop Grave 1 and also: File:Spioenkop1.jpg - Wikipedia

[53] File:Spioenkoploopgraf3.JPG - Wikipedia Spion Kop 2005: This photo shows a section of the British graves at the site of the Battle of Spioenkop. Many of the fallen soldiers were buried in the trenches where they died. These graves therefore give an indication of where the trenches were located at the time of the battle. Author: Renier Maritz. See Also: Spion Kop Battlefield Memorials - Wikipedia

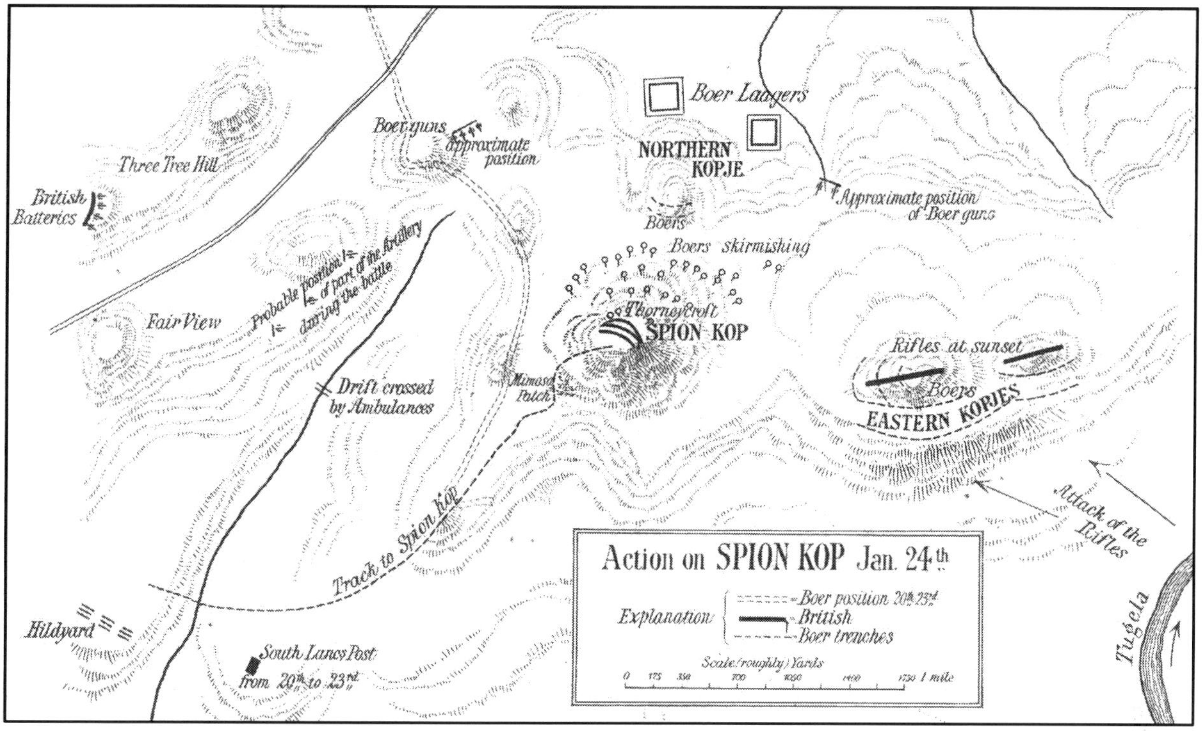

Action on Spion Kop.[107] Viljoen's forces would have been on the Eastern Kopjes in support.

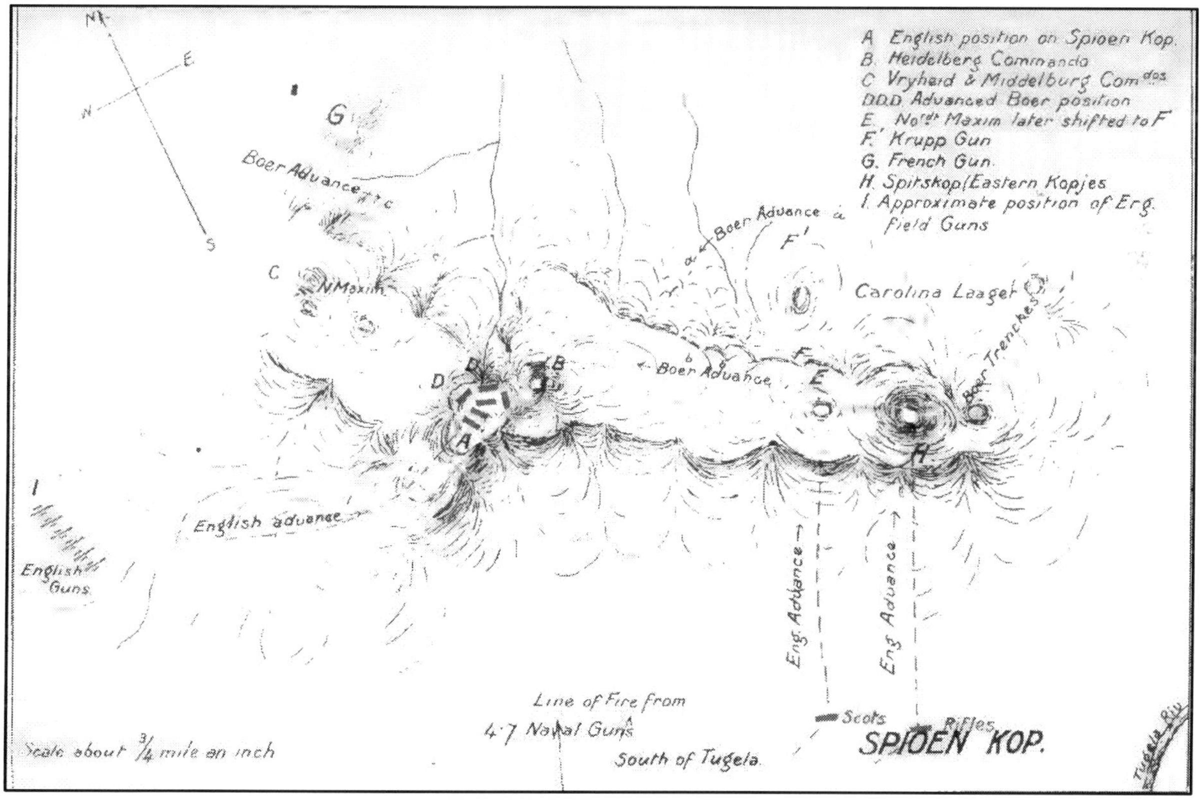

The best available image from the various editions of Viljoen's book.

Chapter 10. The Battle of Vaal Krantz.

Soon after his defeat at Spion Kop, General Buller, moved by the earnest entreaties for help from Ladysmith, and pressed by Lord Roberts, attempted a third time to break through our lines. This time my position had to bear the onslaught of his whole forces. For some days it had been clear to me what the enemy intended to do, but I wired in vain to the Commander-in-Chief to send me reinforcements, and I was left to defend a front, one and a half miles in length, with about 400 men. After many requests I at last moved General Joubert to send me one of the guns known as "Long Toms," which was placed at the rear of our position, and enabled us to command the Vaalkrantz, or, as we called it, "Pontdrift" kopjes. But instead of the required reinforcements, the Commander sent a telegram to General Meyer to Colenso, telling him to come and speak to me, and to put some heart into me, for it seemed, he said, "as if I had lost faith."

Boer 'Long Tom' Gun in use at Vaal Krantz February 5th, 1900.[108] Note shell in loader's hands and bag of powder on the shoulder of assistant. Long Tom.[54] showing breech and Powder bag used along with ramrod/swab and water bucket.[109] This may be the same crew but in a different location.

General [Lucas] Meyer came, and I explained to him how matters stood, and that I should not be able to check the enormous attacking force with my commando alone. The British were at this time only 7,000 paces away from us. The required assistance, however, never came, although I told the General that a faith strong enough to move Majuba Hill would be of no avail without a sufficient number of men.

Early in the morning of the 5th February, 1900, my position was heavily bombarded, and before the sun had risen four of my burghers had been put hors de combat.[55] The enemy had placed their naval guns on the outskirts of the wood known as "Zwartkop" so as to be able to command our position from an elevation of about 400 feet. I happened to be on the right flank with ninety-five burghers and a pom-pom; my assistant, Commandant Jaapie du Preez, commanding the left flank. The assailants threw two pontoon bridges across the river and troops kept pouring over from 10 o'clock in the morning. The whole of the guns' fire was now concentrated on my position; and although we answered with a well-directed fire, they charged time after time.

British Pontoon Bridge over the Tugela River.[110]

Piquet crossing a Pontoon Bridge.

[54] The 155 mm Creusot Long Tom was a French siege gun (artillery piece) manufactured by Schneider et Cie in Le Creusot, France and used by the Boers in the Second Boer War as field guns. Four guns, along with 4,000 common shells, 4,000 shrapnel shells and 800 case shot were purchased by the Transvaal in 1897.

[55] Out of the fight.

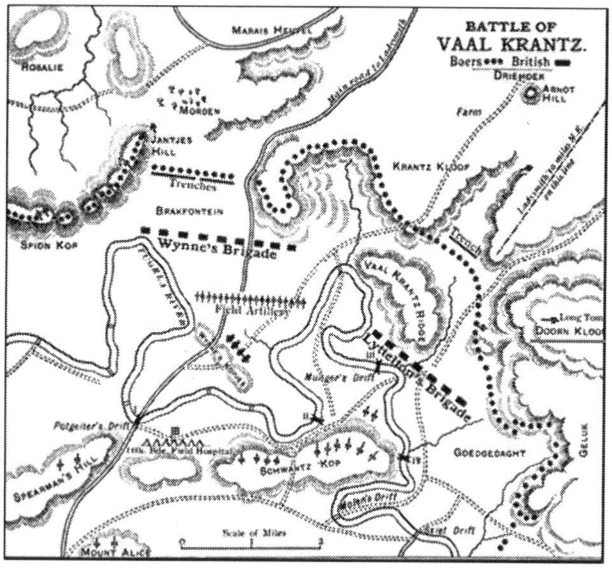

Vaal Krantz viewed from across the Tugela River.[111] "Map of the action at Vaal Krantz."[112] Note Long Tom on Dorn Kloof.

Panorama of the Action of the Val Kranz, at 2 P.M. on February 5th.[113]

1st Durham Light Infantry moving through the gun line to attack Val Krantz.[56]

[56]"This photograph shows the Durhams in the act of passing one of the field guns. It was taken under fire, and is a quite typical view of a modern battlefield–the infantry advancing in open order, the gunners taking what shelter their gun affords. Even the insignificant size of the figures, especially of the trees in the middle distance, lost as they are in the expanse of landscape is quite characteristic. Except for the noise, and the occasional dropping of a wounded man, a spectator would hardly know than anything serious was going on." Image source: With the Flag to Pretoria, H. W. Wilson Vol 1 1900 and britishbattles.com Photo by Lieut. Salmond. P.315.

Rene' Bull Sketch: "A false position: Vaal Krantz, which we found untenable on February 8th" [1900].[114]

The number of my fighting men was rapidly diminishing. I may say this was the heaviest bombardment I witnessed during the whole of the campaign. It seemed to me as if all the guns of the British army were being fired at us.

Their big lyddite guns sent over huge shells, which mowed down all the trees on the kopje, while about fifty field pieces were incessantly barking away from a shorter range. Conan Doyle, in his book, "The Great Boer War," states that the British had concentrated no less than seventy-three guns on that kopje.

British Artillery on Zwart Kop [Swartz Kop] firing on Val Krantz: February 5th, 1900[115] One of the naval guns on Zwart Kop bombarding Vaal Krantz.[116]

Naval Guns on Zwart kop.[117] Cameronians Maxim Gun on Zwart Kop during Spion Kop Battle.[118]

In vain I implored the nearest Generals for reinforcements and requested our artillery in Heaven's name to aim at the enemy's guns. At last, however, "Long Tom" commenced operations, but the artillerymen in charge had omitted to put the powder in a safe place and it was soon struck by a lyddite shell which set the whole of it on fire. This compelled us to send to the head laager near Ladysmith for a fresh supply of powder.

On looking about me to see how my burghers were getting on I found that many around me had been killed and others were wounded. The clothes of the latter were burnt and they cried out for help in great agony.

Our pom-pom had long since been silenced by the enemy, and thirty of my burghers had been put out of the fight. The enemy's infantry was advancing nearer and nearer and there was not much time left to think. I knelt down behind a kopje, along with some of the men, and we kept firing away at 400 paces, but although we sent a good many to eternal rest, the fire of the few burghers who were left was too weak to stem the onslaught of those overwhelming numbers.

A lyddite shell suddenly burst over our very heads. Four burghers with me were blown to pieces and my rifle was smashed. It seemed to me as if a huge cauldron of boiling fat had burst over us and for some minutes I must have lost consciousness. A mouthful of brandy and water (which I always carried with me) was given me and restored me somewhat, and when I opened my eyes I saw the enemy climbing the kopje on three sides of us, some of them only a hundred paces away from me.

Vaal Krantz Summit.[119]

The Durham Light Infantry Capturing Vaal Krantz.[120]

I ordered my men to fall back and took charge of the pom-pom, and we then retired under a heavy rifle and gun fire. Some English writers have made much ado about the way in which our pom-pom was saved, but it was nothing out of the ordinary. Of the 95 burghers with me 29 had been killed, 24 wounded.

West Yorkshires on Val Krantz. February 5th, 1900.[121]

British troops waiting to attack at the Battle of Val Krantz.[57]

[57] "This photograph was taken under fire during the battle of Vaal Krantz, and shows the Brakfontien position in the distance. The men in the foreground are awaiting the order to advance." Image source: With the Flag to Pretoria, H. W. Wilson Vol 1 1900 and britishbattles.com

When I had a few minutes rest I felt a piercing pain in my head, and the blood began to pour from my nose and ears.

We had taken up another position at 1,700 paces, and fired our pom-pom at the enemy, who now occupied our position of a few minutes before. Our other guns were being fired as well, which gave the British an exciting quarter of an hour. On the right and left of the positions taken by them our burghers were still in possession of the "randten"; to the right Jaapie du Preez, with the loss of only four wounded, kept his ground with the rest of my commando.

The next morning the fight was renewed, and our "Long Tom" now took the lead in the cannon-concert, and seemed to make himself very unpleasant to the enemy.

The whole day was mainly a battle of big guns. My headache grew unbearable, and I was very feverish. General Botha had meanwhile arrived with reinforcements, and towards evening things took a better turn.

But I was temporarily done for, and again lost consciousness, and was taken to the ambulance. Dr. Shaw did his best, I hear, for me; but I was unconscious for several days, and when I revived the doctor told me I had a slight fracture of the skull caused by the bursting of a shell. The injuries, however, could not have been very serious for ten days after I was able to leave my bed. I then heard that the night I had been taken to the hospital, the British had once more been forced to retire across the Tugela, and early in the morning of the 7th of February our burghers were again in possession of the kopje "Vaalkrantz," round which such a fierce fight had waged and for the possession of which so much blood had been spilled.

So far as I could gather from the English official reports they lost about 400 men, while our dead and wounded numbered only sixty-two.

Casualties In Action	Killed		Wounded		Missing and Prisoners			Forces Engaged		
	Officers	Men	Officers	Men	Officers	Men	Total	British	Boer	UK Loss
Vaal Krantz	2	23	18	326	0	5	374	15,000	10,000	2.5%

Taking into consideration the determination with which General Buller had attacked us, and how dearly he had paid for this third abortive attempt, the retreat of his troops remains as much of a mystery to me as that at Spion Kop.

Our "Long Tom" was a decided success, and had proved itself to be exceedingly useful.

The Battle of "Vaalkrantz" kopje was to me and to the Johannesburg commando undoubtedly the most important and the fiercest fight in this war, and although one point in our positions was taken, I think that on the whole I may be proud of our defence. About two-thirds of its defenders were killed or wounded before the enemy took that spot, and all who afterwards visited the kopje where our struggle had taken place had to admit that unmistakable evidence showed it to be one of the hottest fights of the Natal campaign.

All the trees were torn up or smashed by shells, great blocks of rock had been splintered and were stained yellow by the lyddite; mutilated bodies were lying everywhere—Briton and Boer side by side; for during the short time "Vaalkrantz" had been in their possession the English had not had an opportunity of burying the bodies of friends or foe.

I think I may quote a few paragraphs of what Dr. Doyle says in his book about this engagement:—

> "The artillery-fire (the "Zwartkop" guns and other batteries) was then hurriedly aimed at the isolated "Vaalkrantz" (the real object of the attack), and had a terrific effect. It is doubtful whether ever before a position has been exposed to such an awful bombardment. The weight of the ammunition fired by some of the cannon was greater than that of an entire German battery during the Franco-Prussian war."

Royal Navy 4.7 inch guns with spotting telescope in foreground at Vaal Krantz.[122] Photo by Rene' Bull, showing one of H.M.S. Terrible's 4.7-inch Naval Guns being hauled into position during the battle of Colenso.[58]

The 5-inch Howitzer or Siege Gun.[123] 37th R.A. howitzer battery going to Maddox Hill to throw lyddite into the Boer Laager.[124]

Dr. Doyle, however, is not sure about the effect of these powerful guns, for he says:—

> "Although the rims of the kopje were being pounded by lyddite and other bombs it is doubtful whether this terrific fire did much damage among the enemy, as seven English officers and 70 men were lying dead on the kopje against only a few Boers, who were found to have been wounded."

Of the pom-pom, which I succeeded in saving from the enemy's hands, the same writer says:—

> "It was during this attack that something happened of a more picturesque and romantic nature than is usually the case in modern warfare; here it was not a question of combatants and guns being invisible or the destruction of a great mass of people. In this case it concerns a Boer gun, cut off by the British troops, which all of a sudden came out of its hiding-place and scampered away like a frightened hare from his lair. It fled from the danger as fast as the mules' legs would take it, nearly overturning, and jolting and knocking against the rocks, while the driver bent forward as far as he could to protect himself from the shower of bullets which were whistling round his ears in all directions. British shells to the right of him, shells to the left of him bursting and spluttering, lyddite shrapnel fuming and fizzing and making the splinters fly. But over the "randtje" the gun disappeared, and in a few minutes after it was in position again, and dealing death and destruction amongst the British assailants."

[58] Due to a severe bout of malarial fever, René Bull was forced to return from the seat of war in early April 1900. Within two weeks of landing at Kingstown, Ireland, he embarked upon an extensive lecture tour of the British Isles, his first venue being the Gaiety Theatre, Dublin (on 23rd April). Reports indicate that his lecture included 200 lantern slides. Text and Image Credit Neville Constantine: https://www.angloboerwar.com/forum/21-uniforms-badges-and-equipment/31231-artillery-and-ammunition?start=234

Boer Pom-Pom, Or Vickers-Maxim Gun. This gun was in action at Alleman's Nek and elsewhere.[125] Note the belt feed ammunition held by the man behind the gun. Long Tom.[126] showing breech and Powder bag used along with ramrod/swab and water bucket.[127]

While I was under treatment in Dr. Shaw's ambulance I was honoured by a visit from General Joubert, who came to compliment me on what he called the splendid defence of Vaalkrantz, and to express his regret at the heavy loss sustained by our commando. I heard from Dr. Shaw that after the battle the groans and cries of the wounded burghers could be heard in the immediate neighbourhood of the English outposts. Some burghers volunteered to go, under cover of the darkness, to see if they could save these wounded men. They cautiously crept up to the foot of the kopjes, from where they could plainly see the English sentinels, and a little further down found in a ditch two of our wounded, named Brand and Liebenberg; the first had an arm and a leg smashed, the latter had a bullet in his thigh.

One can imagine what a terrible plight they were in after laying there for two nights and a day, exposed to the night's severe cold and the day's scorching sun. Their wounds were already decomposing, and the odour was most objectionable.

The two unfortunate men were at once carried to the laager and attended to with greatest care. Poor Liebenberg died of his wounds soon after. Brand, the youngest son of the late President Brand, of the Orange Free State, soon recovered, if I remember rightly.

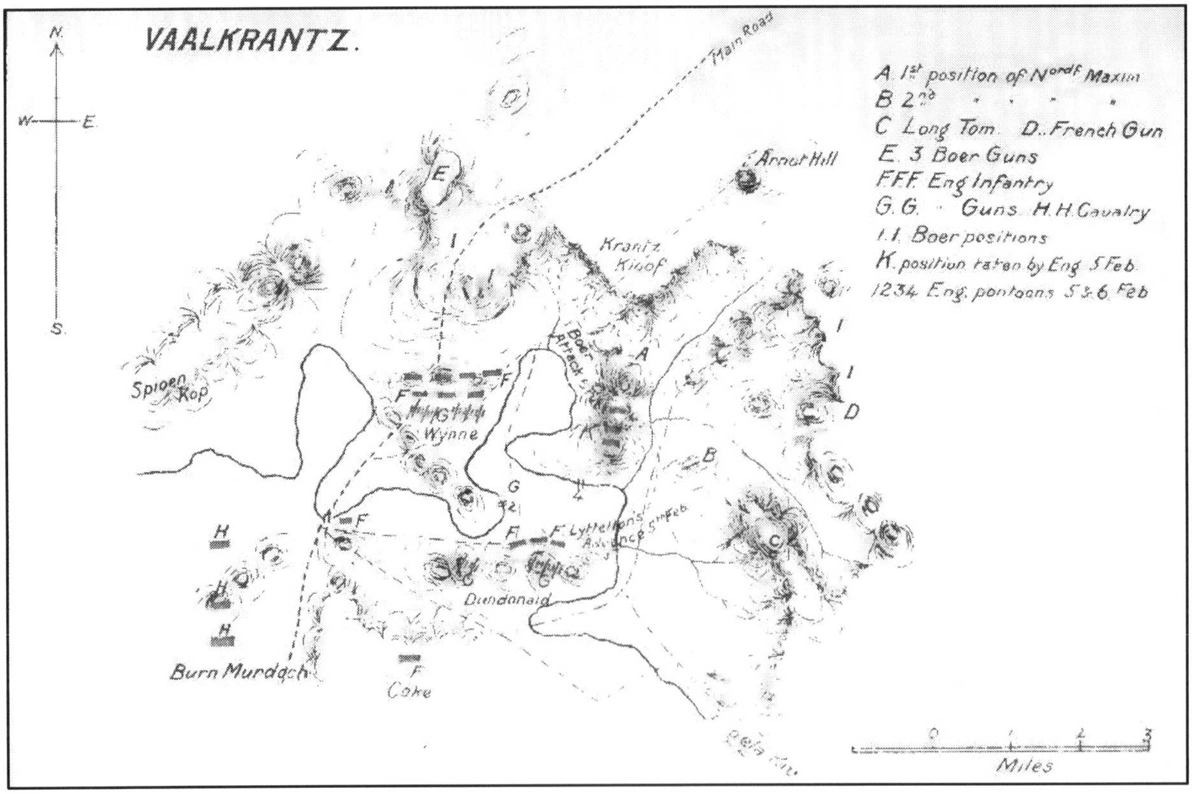

Original Map from Ben's Book.

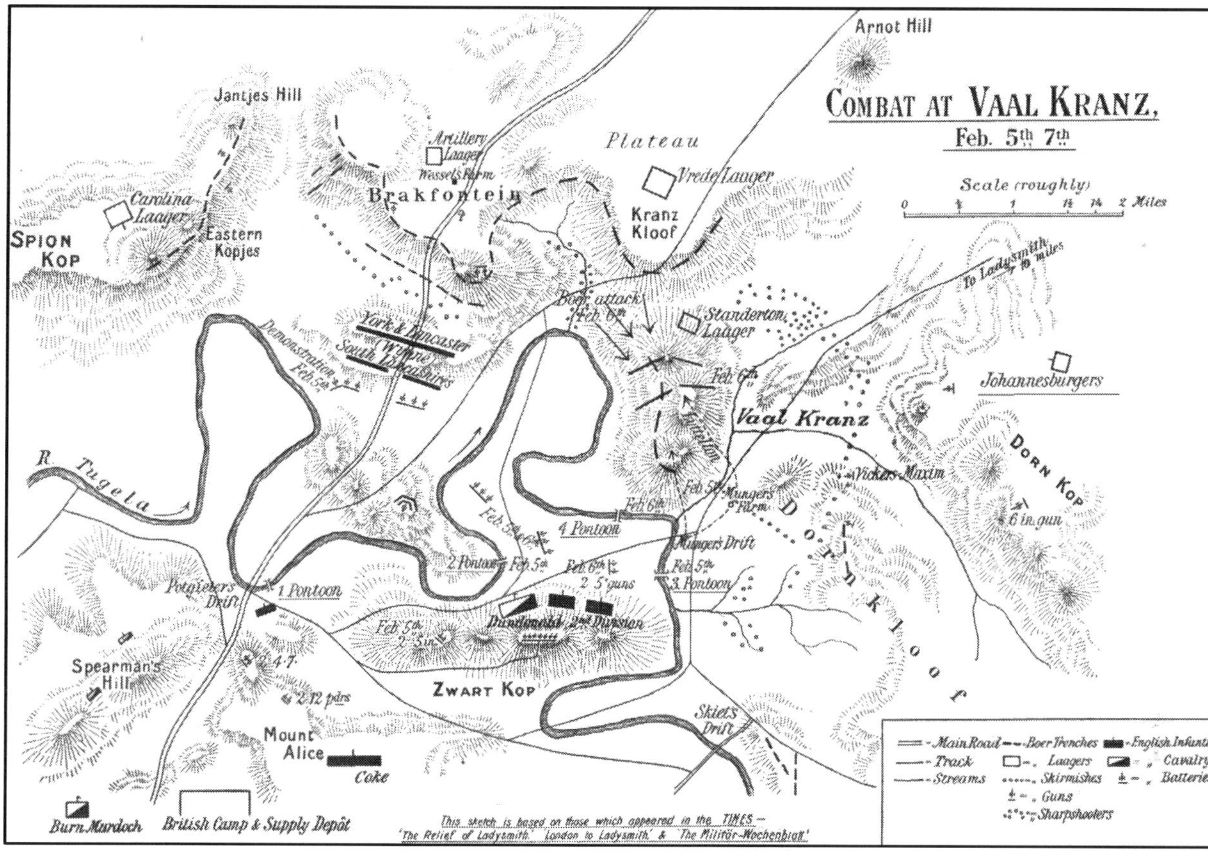

Combat at Vaal Krantz.[128] Note Johannesburgers on Dorn Kop.

At the risk of incurring the displeasure of a great number of people by adding the following statement to my description of the battle of Vaalkrantz, I feel bound to state that Commandant-General Joubert, after our successes at Colenso, Spion Kop, and Vaalkrantz, asked the two State Presidents, Kruger and Steyn, to consider the urgency of making peace overtures to the English Government. He pointed out that the Republics had no doubt reached the summit of their glory in the War. The proposal read as follows:

That the Republican troops should at once evacuate British territory, compensation to be given for the damage to property, etc., inflicted by our commandos, against which the British Government was to guarantee that the Republics should be spared from any further incursions or attacks from British troops, and to waive its claim of Suzerainty; and that the British Government should undertake not to interfere with the internal affairs and legal procedure of the two Republics, and grant general amnesty to the colonial rebels.

Commander-in-Chief Joubert defended these proposals by pointing out that England was at that moment in difficulties, and had suffered repeated serious defeats. The opportunity should be taken, urged the General.

He was supported by several officers, but other Boer leaders contended that Natal, originally Boer territory, should never again be ceded to the enemy. As we heard nothing more of these proposals, I suppose the two State Presidents rejected them.

March 1900 Offer and Reply

The battle of Vaal Krantz took place on February 5th-7th, 1900.

On March 5th, 1900 the Presidents of both Republics wrote directly to Robert Cecil[59] (The Prime Minister of the United Kingdom and known as the Marquess of Salisbury) in an attempt to negotiate an end to the conflict on honorable terms.[60] [61]

During the previous month the following events had transpired:

2/16/1900 Capture of Cingolo Hill by Redvers Buller's force. Roberts occupied Jacobsdal.

2/18/1900 Fighting at Paardeberg, where Cronje was being surrounded. Buller Captures Monte Cristo.

2/19/1900 Capture of Hlangwane by the Fusilier Brigade. Bombardment of Cronje's position began.

2/27/1900 Cronje, with 44 commandants and other officers of all grades, and over 3500 men, surrendered unconditionally to Roberts. Redvers Buller's force captured the Boer position at Pieters. This action opened the road to Ladysmith. Boers retired north to Ladysmith.

2/28/1900 Relief of Ladysmith after 120 day's investment.

3/5/1900 General Gatacre occupied Stormberg without opposition. Ovetures of Peace made by Kruger and Steyn to British Govt.

3/7/1900 Roberts dispersed Boers near Poplar Grove. General Gatacre reached Burghersdorp.

Kruger and Steyn's Letter to Lord Salisbury

From the States-President of the South African Republic and the Orange Free State

Bloemfontein, 5th March, 1900.

To His Excellency Lord Salisbury, London.

The blood and tears of the thousands who have suffered through this war, and the prospect of all the moral and material ruin which now threatens South Africa, render it necessary for both parties carrying on the war to ask themselves calmly, and in the faith of the Trinity, for what they are fighting and if the aims of both justify all this horrible misery and devastation.

On this account, and with an eye to the assertion of several English Statesmen that the war was begun and carried on with the determined end to undermine Her Majesty's authority in South Africa, and to establish in the whole of South Africa a Government independent of Her Majesty's Government, we consider it our duty to declare that this War was only commenced as a measure of defence and for the purpose of obtaining a guarantee for the threatened independence of the South African Republic, and was only continued in order to ensure the indisputable independence of both Republics as Sovereign International States, and to obtain the assurance that the subjects of Her Majesty who have taken part with us in the war will not suffer the least hurt either in their lives or their possessions.

On these conditions alone we demand, as in the past, to see peace restored in South Africa, and an end made to the wrong that now exists there.

But if Her Majesty's Government has decided upon destroying the independence of the Republic, nothing remains to us and our people but to persist to the bitter end on the road now taken, notwithstanding the overpowering might of the British Empire, trusting that God, who has lit the inextinguishable fire of the love of liberty in our hearts, and in the hearts of our fathers, will not abandon us, but will fulfil His work in us, and in our descendants.

[59] Robert Arthur Talbot Gascoyne-Cecil, 3rd Marquess of Salisbury (3 February 1830–22 August 1903), known as Lord Salisbury, was a British statesman and Conservative politician who served as Prime Minister of the United Kingdom three times for a total of over thirteen years. He was also Foreign Secretary before and during most of his tenure. He avoided international alignments or alliances, maintaining the policy of "splendid isolation." Robert Gascoyne-Cecil, 3rd Marquess of Salisbury - Wikipedia

[60] CDR Note: 'This correspondence will be found in Chapter 30.' [PF: It is in the Appendix with the correspondence. Chapter 30 is Steyn's Letter to Kitchener. It is reproduced here as well for its relevance to the story and the struggle.]

[61] In Dr. Conan Doyle's work, Cause and Conduct, these two communications are reproduced as well. The differences in language are notable, since each purports to be a full transcription.

We hesitated to lay this declaration earlier before Your Excellency, because we were afraid that as long as the advantage was on our side, and our Army had in their occupation positions of defence far into the British Colonies, such a declaration would have hurt the feelings of the English nation; but now that the prestige of the British Empire may be considered to be restored, through the capture of one of our armies, and we are compelled by this to sacrifice other positions which our armies occupied, this difficulty is removed, and we can no longer hesitate to tell you, in the face of the whole civilized world, why we are fighting, and on what conditions we are prepared to make peace.

Reply

From Lord Salisbury

To their Excellencies the States-Presidents of the South African Republic and Orange Free State.

London, 11th March, 1900.

I have the honour to acknowledge the receipt of your Honour's cable, dated 5th March, from Bloemfontein, of which the purport is principally whether Her Majesty's Government will acknowledge the indisputable independence of the South African Republic and Orange Free State and treat them as Sovereign International States, and will offer to conclude the war on these conditions.

In the beginning of October of this year, there was peace between the Queen and the two Republics, under the Convention which then held good. There was a discussion carried on during a few months between Her Majesty's Government and the South African Republic, of which the purport was the amendment of very serious grievances under which English inhabitants suffered in the South African Republic.

In the course of these negotiations, the South African Republic obtained the knowledge that Her Majesty's Government had made considerable preparations for war, and had taken steps to provide the necessary reinforcements for the English garrisons at Cape Colony and Natal. No inroad on the rights guaranteed by the Conventions had, until then, taken place on the English side.

Suddenly the South African Republic, after having two days previously issued an insulting ultimatum, declared War on Her Majesty; and the Orange Free State, with which there had been no disagreement, took a similar step.

Thereupon an inroad was made into Her Majesty's territory by the two Republics; three towns within the British frontier were besieged, a great part of the two Colonies was over-run, with great destruction of property and life, and the Republics claimed the right to treat the inhabitants of Her Majesty's territory as if this territory had been annexed by one of these States.

The Transvaal having these actions in view, had for years stored up, on an enormous scale, military provisions, which could only have been destined for use against England.

Your Excellencies made some remarks of a negative nature concerning the aim for which these preparations were made. I do not consider it necessary to discuss the question which you have thus raised, but the consequences of the preparations, made in great secrecy, have been that the British Empire has found itself forced to repel an inroad which has brought on a costly war, and caused the loss of thousands of valuable lives.

This great misfortune has been the punishment that Great Britain has had to undergo during the last few years for having suffered the two Republics to exist.

Keeping in sight the use which the two Republics have made of the position presented to them, and the misfortunes which their unprovoked attacks on Her Majesty's territory have brought, Her Majesty's Government can only reply to Your Honour's telegram by saying that they are not prepared to acknowledge the independence either of the South African Republic, or of the Orange Free State.

Chapter 11. The Turn of the Tide.

After the English forces had retreated from Vaalkrantz across the Tugela, a patrol of my commando under my faithful adjutant, J. Du Preez, who had taken my place for the time being, succeeded in surprising a troop of fifty Lancers, of the 17th regiment, I believe, near Zwartkop, east of the Tugela, and making them prisoners after a short skirmish. Among these men, who were afterwards sent to Pretoria, was a certain Lieutenant Thurlington. It was a strange sight to see our patrol coming back with their victims, each Boer brandishing a captured lance.

Being still in the hospital in feeble health without any prospect of a speedy recovery, I took the doctor's advice and went home to Rondepoort, near Krugersdorp, where my family was staying at the time, and there, thanks to the careful treatment of my kind doctor and the tender care of my wife I soon recovered my strength.

Ben Viljoen and his family: Helena Beatrix Els (1871–1950) and Wynand Eugene Viljoen (1891–1978).[129]

On the 25th of February I received a communication from my commando to the effect that General Buller had once more concentrated his forces on Colenso and that heavy fighting was going on. The same evening I also had a telegram from President Kruger, urging me to rejoin my commando so soon as health would allow, for affairs seemed to have taken a critical turn. The enemy appeared to mean business this time, and our commando had already been compelled to evacuate some very important positions, one of which was Pieter's Heights.

Then the news came from Cape Colony that General Piet Cronje had been surrounded at Paardeberg, and that as he stubbornly refused to abandon his convoy and retreat, he would soon be compelled by a superior force to surrender.

The next morning I was in a fast train to Natal, accompanied by my faithful adjutant. Rokzak. My other adjutant, Du Preez, had meantime been ordered to take a reinforcement of 150 men to Pieter's Heights, and was soon engaged in a desperate struggle in the locality situated between the Krugersdorpers' and the Middleburgers' positions.

The situation was generally considered very serious when I arrived near the head laager at Modderspruit late in the evening of the 27th of February, unaware of the unfavourable turn things had taken during the day at Paardeberg, in the Cape Colony, and on the Tugela. We rode on that night to my laager at Potgietersdrift, but having to go by a roundabout way it took us till early next morning before we reached our destination. The first thing I saw on my arrival was a cart containing ten wounded men, who had just been brought in from the fighting line, all yellow with lyddite.

Field-cornet P. van der Byl, who came fresh from the fight near Pieter's Heights, told me that these burghers had been wounded there. I asked them what had happened and how matters stood. "Ah, Commandant," he replied, "things are in a very bad way! Commandant Du Preez and myself were called to Pieter's Heights three days ago, as the enemy wanted to force their way through. We were in a very awkward position, the enemy storming us again and again; but we held our own, and fired on the soldiers at 50 paces.

The English, however, directed an uninterrupted gun fire at our commandos, and wrought great havoc. Early Sunday morning the other side asked for a truce to enable them to bury their dead who were lying too close to our positions to be got at during the fighting. Many of their wounded were lying there as well, and the air was rent during 24 hours with their agonised groans, which were awful to hear. We, therefore, granted an armistice till 6 o'clock in the evening." (This curiously coincided in time with Lord Roberts' refusal to General Piet Cronje at Paardeberg to bury his dead).

"The enemy," continued the field-cornet, "broke through several positions, and while we were being fired at by the troops which were advancing on us, we were attacked on our left flank and in the rear. Assistant-Commandant Du Preez, and Field-Cornet Mostert, were both severely wounded, but are now in safe hands. Besides these, 42 of our burghers were killed, wounded, or taken prisoners; we could only bring 16 of our wounded with us. The Krugersdorpers, too, have suffered severely. The enemy has pushed through, and I suppose my burghers are now taking up a position in the "randten" near Onderbroekspruit."

Here was a nice state of things! When I had left my commando 15 days previously, we had had heavy losses in the battle of Vaal krantz, and now again my burghers had been badly cut up. We had lost over 100 men in one month.

View of Pieters Hill from south of the Tugela River showing lines of British infantry moving up the hill towards the Boer trenches on the crest.[130]

A 4.7 Inch Gun on Pieters Hill.[131]

But there was no time to lose in lamenting over these matters, for I had just received information that General P. Cronje had been taken prisoner with 4,000 men. The next report was to the effect that the enemy was breaking through near Onderbroekspruit, and that some burghers were retiring past Ladysmith. I was still in telegraphic communication with the head laager, and at once wired to the Commandant-General for instructions. The answer was:—

"Send your carts back to Modderspruit (our headquarters) and hold the position with your mounted commandos."

The position indicated was on the Upper Tugela, on a line with Colenso. My laager was about 20 miles away from the head laager; the enemy had passed through Onderbroekspruit, and was pushing on with all possible speed to relieve Ladysmith, so that I now stood in an oblique line with the enemy's rear. I sent out my carts to the south-west, going round Ladysmith in the direction of Modderspruit. One of my scouts reported to me that the Free State commandos which had been besieging Ladysmith to the south, had all gone in the direction of Van Reenen's Pass; another brought the information that the enemy had been seen to approach the village, and that a great force of cavalry was making straight for us.

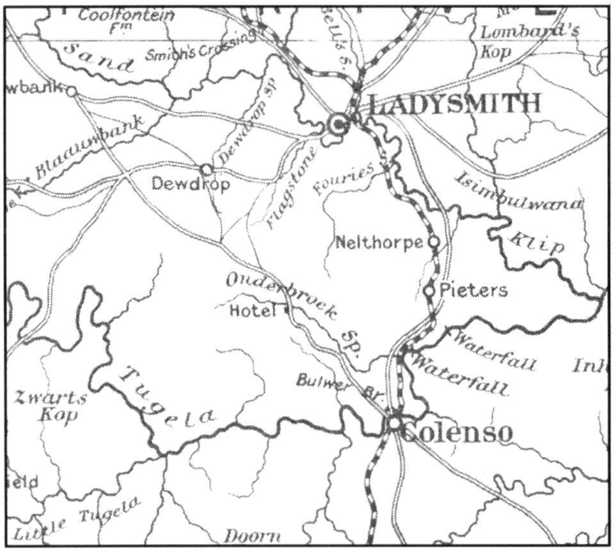

Ladysmith and the Battlefields on the Tugela.[132] Note Onderbruek Spruit. Ben is at Potigers Drift at the junction of the Little Tugela and the main channel. Second Map, note Van Reenen's Pass through the Drakensberg.

The Boer commando guarding Van Reenan's Pass.[133]

General Joubert's instructions were therefore inexplicable to me, and if I had carried them out I would probably have been cut off by the enemy. My burghers were also getting restless, and asked me why, while all the other commandos were retiring, we did not move.

The Drakensburg Mountains between the Natal and Transvaal borders.[134] "Van Reenen's Pass is a cut in the Drakensburg Mountains between the Orange Free State and Natal. This shows you the kind of country our troops may have to negotiate."[135]

An example of the Great Escarpment. Giants Castle in the Drakensberg Range.[136]

Cronje's surrender had had a most disheartening effect on them; there was, in fact, quite a panic among them. I mounted a high kopje from which I could see the whole Orange Free State army, followed by a long line of quite 500 carts and a lot of cattle, in full retreat, and enveloped in great clouds of red dust. To the right of Ladysmith I also noticed a similar melancholy procession. On turning round, I saw the English in vast numbers approaching very cautiously, so slowly, in fact, that it would take some time before they could reach us. Another and great force was rushing up behind them, also in the direction of Ladysmith.

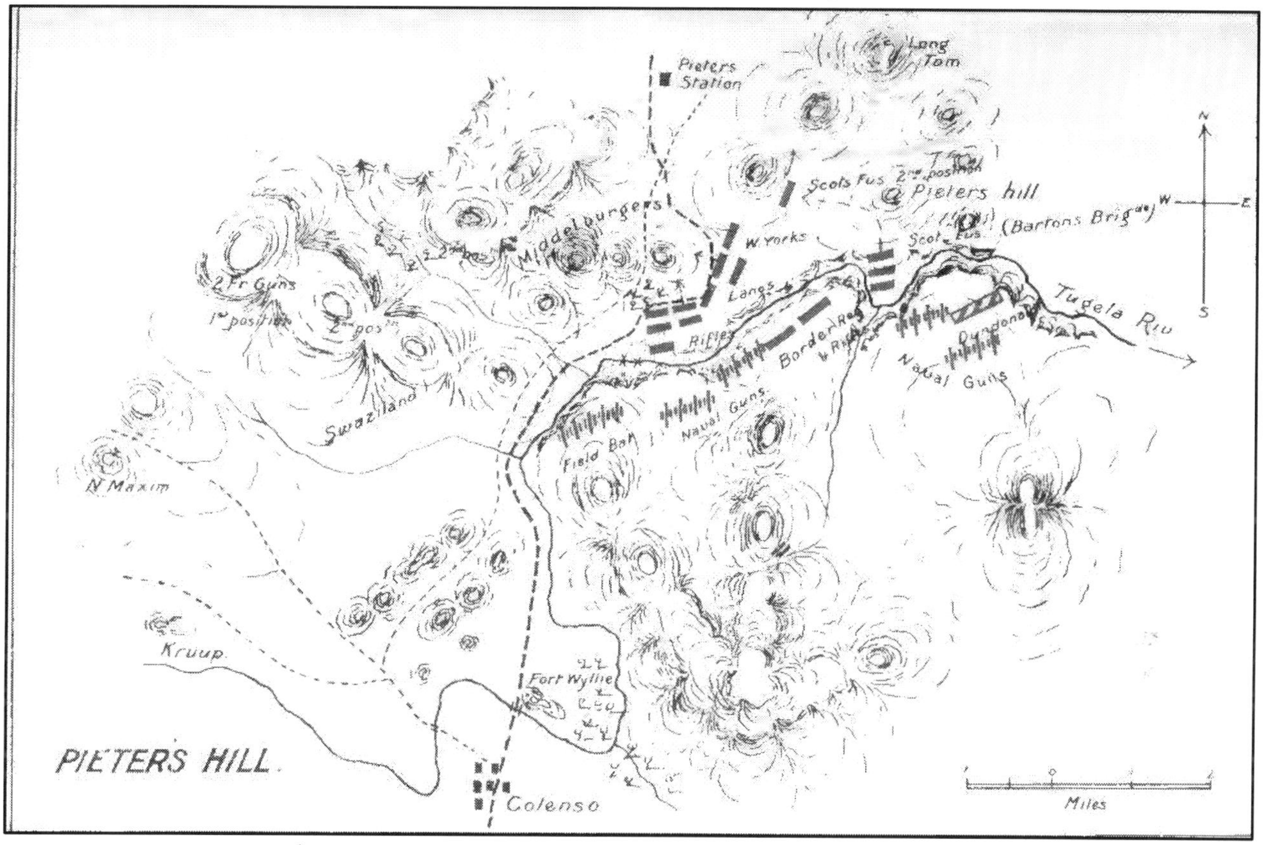

Pieter's Hill, February 27th, 1900.

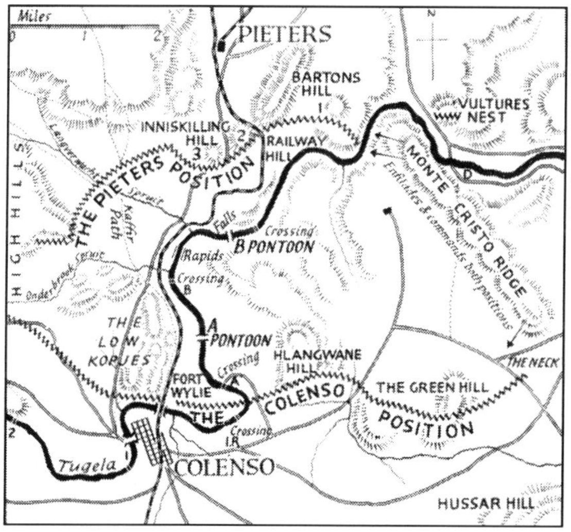

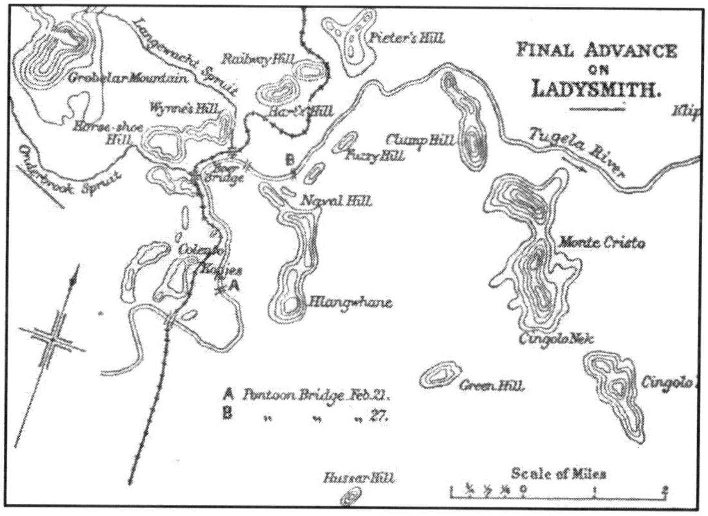

Portion of Relief of Ladysmith.[137]

Final Advance on Ladysmith.[138] Note Onderbrook Spruit.

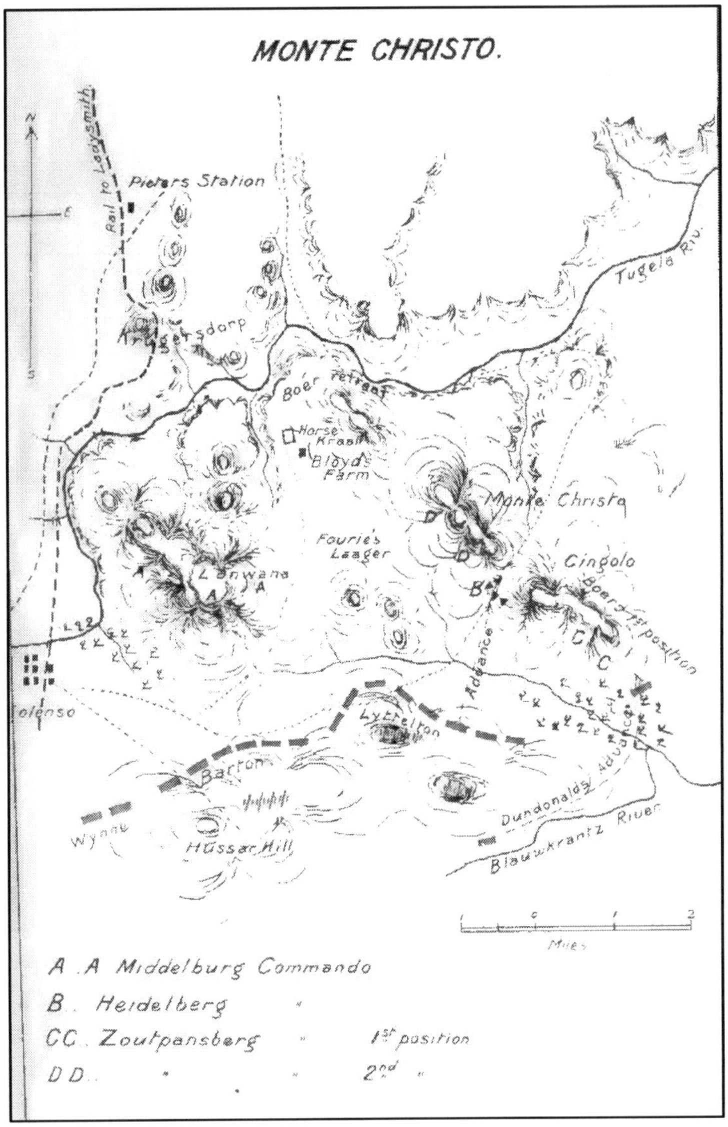

This map of Monte Christo is included but not described in Viljoen's book.[62] [63]

[62] The Battle of Tugela (or Thukela) Heights, also known as the Battle of Pieters Hill, Battle of the Pieters, or the Battle of the Tugela River, consisted of a series of military actions lasting from 14 February through to 27 February 1900 in which General Sir Redvers Buller's British army forced Louis Botha's Boer army to lift the Siege of Ladysmith during the Second Boer War. https://en.wikipedia.org/wiki/Battle_of_the_Tugela_Heights

[63] Battle of the Tugela Heights Chronology:
2/16/1900 Capture of Cingolo Hill by Redvers Buller's force. Roberts occupied Jacobsdal. Flight of Cronje's force and occupation of Majersfontein by the Guards. General Kelly-Kenny, in pursuit of Cronje retiring east with 10,000 men on Bloemfontein, captured 78 waggons with stores, 2 waggons with Mauser rifles, and 8 waggons with shell belonging to Cronje's column.
2/17/1900 Cronje's force overtaken and surrounded at Paardeberg. General Brabant engaged the enemy near Dordrecht.
2/18/1900. Capture of Monte Cristo. General Lyttelton's Division, drove the Boers across the river.
2/19/1900 Capture of Hlangwane by the Fusilier Brigade. The Boers evacuated the hill, and left a large camp behind them. Bombardment of Cronje's position began. Boer reinforcements driven back. Cronje asked for armistice, but Kitchener demanded his surrender; Cronje refused, and was then bombarded heavily. Reoccupation of Dordrecht. General Brabant entered the town.
2/20/1900 General Hart occupied Colenso. Roberts defeated Boer reinforcements at Paardeberg.
2/21/1900 General Wynne's Lancashire Brigade and other portions of the 5th Division crossed the Tugela at Colenso.
2/22/1900 Two Batallions of Norcott's Brigade and Lancashire move forward against Boer low kopjes 'Wynnes Hill' and Horseshoe Hill. 300 British killed or wounded.
2/23/1900 The Boers' position at Grobler's Kloof attacked.
2/23/1900 General Hart leads the Inniskilling Fusiliers, Connaught Rangers, Dublin Fusiliers and Imperial Light Infantry against Pieters Position and 'Inniskilling Hill'. Fighting continues throughout the night.
2/24/1900 Stalemate before Inniskilling Hill. Wounded remain in 'no man's land' between trenches.
2/26/1900 Finding the passage of the river near Colenso commanded by strong entrenchments, Redvers Buller sent his guns and baggage back to the south side of the Tugela, and found a new crossing.

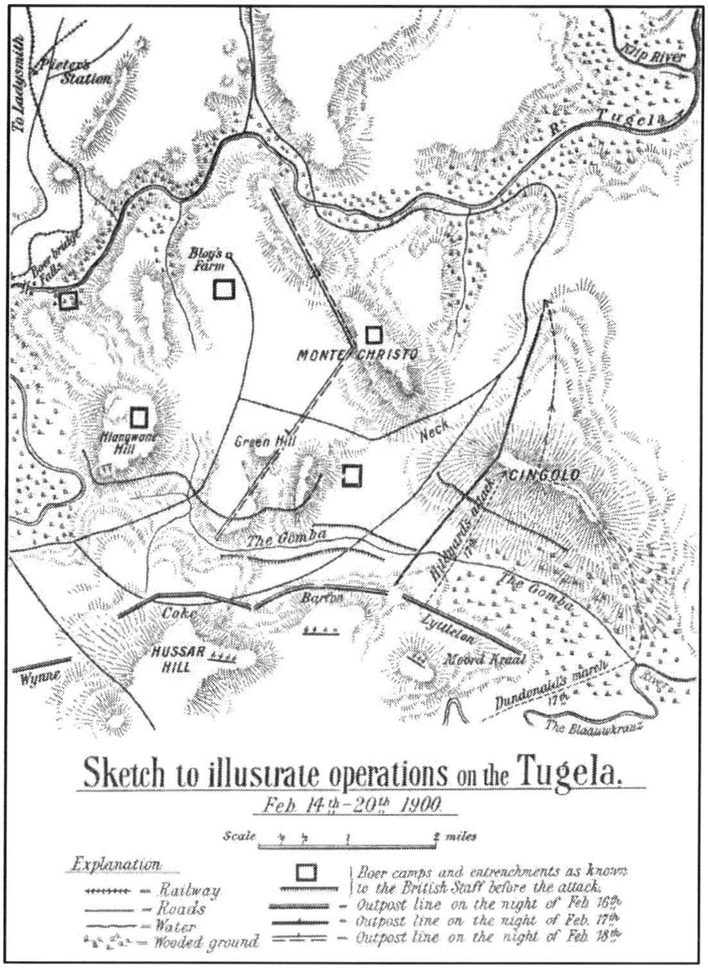

Operations Feb 14th to 20th.[139]

It must have been a race for the Distinguished Service Order or the Victoria Cross to be won by the one who was first to enter Ladysmith. We knew that the British infantry, aided by the artillery, had paved the way for relief, and I noticed the Irish Fusiliers on this occasion, as always, in the van. But Lord Dundonald rushed in and was proclaimed the hero of the occasion.

Before concluding this chapter I should like to refer to a few incidents which happened during the Siege of Ladysmith. It is unnecessary to give a detailed description of the destruction of "Long Tom" at Lombards kop or the blowing up of another gun west of Ladysmith, belonging to the Pretoria Commando. The other side have written enough about this, and made enough capital out of them; and many a D.S.O. and V.C. has been awarded on account of them.

The Boer "Long Tom" in retreat from Ladysmith.[140]

2/27/1900 (on anniversary of Majuba, 1881) Cronje, with 44 commandants and other officers of all grades, and over 3500 men, surrendered unconditionally to Roberts. Redvers Buller's force captured the Boer position at Pieters. This action opened the road to Ladysmith. Boers retired north to Ladysmith.

Alas, I can put forward nothing to lessen our dishonour. As regards the "Long Tom" which was blown up, this was a piece of pure treachery, and a shocking piece of neglect, Commandant Weilbach, who ought to have defended this gun with the whole of his Heidelberg Commando, was unfaithful to his charge. The Heidelbergers, however, under a better officer, subsequently proved themselves excellent soldiers.

A certain Major Erasmus was also to blame. He was continually under the influence of some beverage which could not be described as "aqua pura"; and we, therefore, expected little from him. But although the planning and the execution of the scheme to blow up "Long Tom" was a clever piece of work, the British wasted time and opportunity amusing themselves in cutting out on the gun the letters "R.A." (Royal Artillery), and the effect of the explosion was only to injure part of the barrel. After a little operation in the workshops of the Netherlands South African Railway Company at Pretoria under the direction of Mr. Uggla, our gun-doctor, "Long Tom's" mouth was healed and he could spit fire again as well as before. As to the blowing up of the howitzer shortly after, I will say the incident reflected no credit on General Erasmus, as he ought to have been warned by what happened near Lombards kop, and to have taken proper precautions not to give a group of starving and suffering soldiers an opportunity of penetrating his lines and advancing right up to his guns.

"Long Tom in hospital: A Boer gun under repair at Pretoria."[141] PF: I is fairly certain that Mr. Uggla is in the picture, not certain who it is. One of four Boer 'Long Tom' canons, damaged at Siege of Ladysmith, on its way to the Siege of Kimberley after it was repaired in Pretoria.[142] PF: Possibly because the range of the gun was somewhat shorter than previously and not suitable for field army operations.

Both incidents will be an ugly blot on the history of this war, and I am sorry to say the two Boer officers have never received condign punishment. They should, at any rate, have been called before the Commandant-General to explain their conduct.

The storming of Platrand (Caesar's Camp), south-east of Ladysmith, on the 6th of January, 1900, also turned out badly for many reasons. The attack was not properly conducted owing to a jealousy amongst some of the generals, and there was not proper co-operation.[64]

The burghers who took part in the assault and captured several forts did some splendid work, which they might well be proud of, but they were not seconded as they should have been. The enemy knew that if they lost Platrand, Ladysmith would have to surrender; they therefore defended every inch of ground, with the result that our men were finally compelled to give way. And, for our pains, we sustained an enormous loss in men, which did not improve in any way the broken spirit of our burghers.

[64] In the early hours of 6 January 1900, Boer storming parties under General C. J. de Villiers began climbing Wagon Hill and Caesar's Camp. They were spotted and engaged by British working parties who were emplacing some guns. The Boers captured the edge of Wagon Hill, but could not advance further. Half an hour later, they also captured part of the Caesar's Camp position, but again could not advance beyond the forward edge of the hill. British counter-attacks also failed.
At noon, de Villiers made another attack on Wagon Hill. Some of the exhausted defenders panicked and fled, but Hamilton led reserves to the spot and recaptured some empty gun pits. Although Hamilton suggested that the Boers would withdraw at night, Sir George White demanded that fresh reserves were to retake the entire position immediately. The Devonshire Regiment suffered heavy casualties but only drove the Boers back as far as the crest line, and a stalemate once again ensued.
Late in the afternoon, a terrific rainstorm broke, and the Boers withdrew under cover of it.
The British suffered 175 killed and 249 wounded. 52 dead Boers were left in the British positions, but their total casualties were not recorded. https://en.wikipedia.org/wiki/Siege_of_Ladysmith#Battle_of_Wagon_Hill_(or_Platrand)

Chapter 12. The Great Boer Retreat.

There was clearly no help for it, we had to retreat. I gave orders to saddle up and to follow the example of the other commandos, reporting the fact to the Commandant-General. An answer came—not from Modderspruit this time, but from the station beyond Elandslaagte—that a general retreat had been ordered, most of the commandos having already passed Ladysmith, and that General Joubert had gone in advance to Glencoe. At dusk I left the Tugela positions which we had so successfully held for a considerable time, where we had arrested the enemy from marching to the relief of Ladysmith, and where so many comrades had sacrificed their lives for their country and their people.

General Joubert's camp at Glencoe.[143]

It was a sad sight to see the commandos retreating in utter chaos and disorder in all directions. I asked many officers what instructions they had received, but nobody seemed to know what the orders actually were; their only idea seemed to be to get away as quickly as possible.

Finally, at 9 o'clock in the evening we reached Klip River, where a strange scene was taking place. The banks were crowded with hundreds of mounted men, carts and cattle mingled in utter confusion amongst the guns, all awaiting their turn to cross. With an infinite amount of trouble the carts were all got over one at a time. After a few minutes' rest I decided on consulting my officers, that we should cross the river with our men by another drift further up the stream, our example being followed by a number of other commandos.

Crossing a Drift.[144]

A freight wagon of the Transvaal.[145]

I should point out here that in retreating we were going to the left, and therefore in perilous proximity to Ladysmith. The commandos which had been investing the town were all gone; and Buller's troops had already reached it from the eastern side, and there was really nothing to prevent the enemy from turning our rear, which had perforce to pass Ladysmith on its way from the Tugela. When we had finally got through the drift late that evening, a rumour reached us that the British were in possession of Modderspruit, and so far as that road was concerned, our retreat was effectually cut off.

Shortly before the War, however, the English had made a new road which followed the course of the Klip River up to the Drakensbergen, and then led through the Biggarsbergen to Newcastle. This road was, I believe, made for military purposes; but it was very useful to us, and our wagons were safely got away by it.

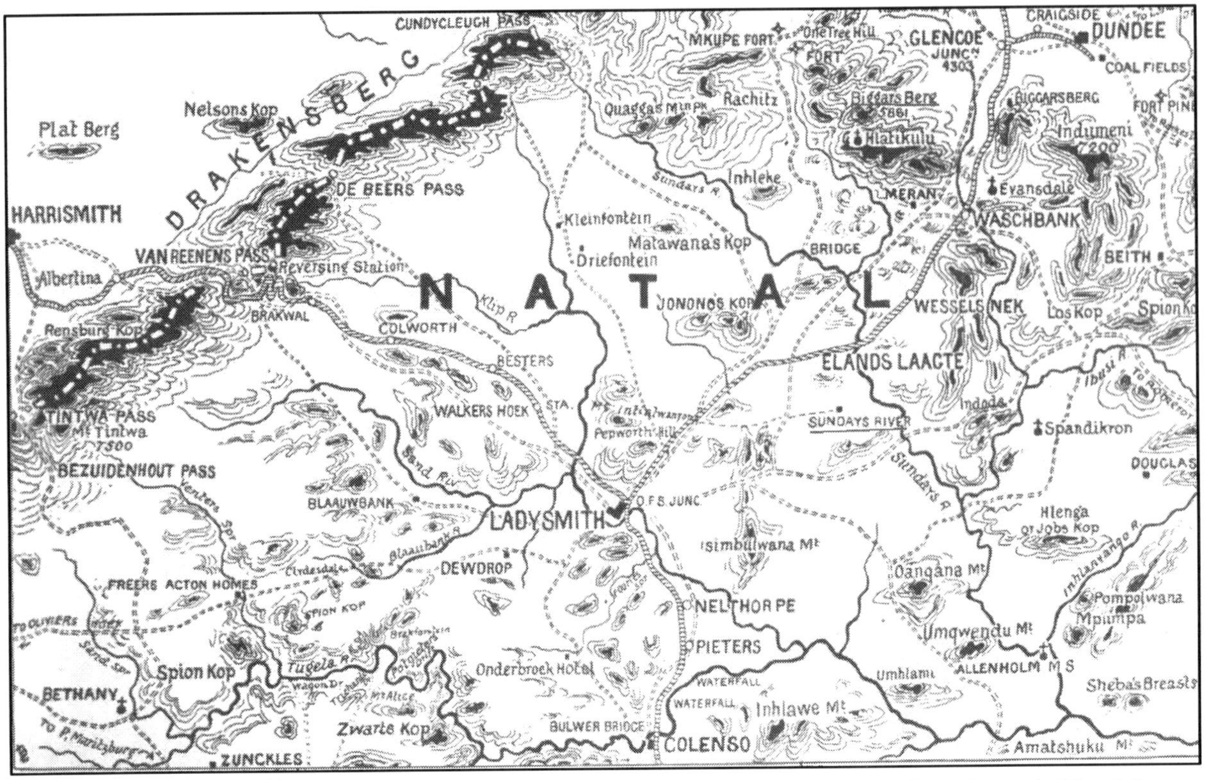

Northern Natal Detail.[146] Note the Klip River above Ladysmith and Sundays River through Elandslaagte as well as Biggarsberg near the top.

Commandant D. Joubert, of the Carolina Commando, then sent a message asking for reinforcements for the Pretoria laager, situated to the north-west of Ladysmith. It was a dark night and the rain was pouring down in torrents, which rendered it very difficult to get the necessary burghers together for this purpose.

I managed, however, to induce a sufficient number of men to come together, and we rode back; but on nearing the Pretoria Laager, I found to my dismay that there were only 22 of us left. What was to be done? This handful of men was of very little use; yet to return would have been cowardly, and besides, in the meantime our laager would have gone on, and would now be several hours' riding ahead of us. I sent some burghers in advance to see what was happening to the Pretoria Laager. It seemed strange to me that the place should still be in the hands of our men, seeing that all the other commandos had long since retired. After waiting fully an hour, our scouts came back with the information that the laager was full of English soldiers, and that they had been able to hear them quarrelling about the booty left behind by the burghers.

It was now two o'clock in the morning. Our Pretoria comrades were apparently safe, and considerably relieved we decided to ride to Elandslaagte which my men would by that time have surely reached. Our carts were sooner or later bound to arrive there, inasmuch as they were in charge of a field-cornet known to us as one of our best "retreat officers." I think it was splendid policy under the circumstances to appoint such a gentleman to such a task; I felt sure that the enemy would never overtake him and capture his carts.

Men of the Pretoria Commando.[147] Camp of the Pretoria Commando. The historical Majuba Hill is in the background.[148]

We followed the main road, which was fortunately not held by the enemy, as had been reported to us. On the way we encountered several carts and waggons which had been cast away by the owners for fear of being caught up by the pursuing troops. Of course the rumour that this road was in possession of the English was false, but it increased the panic among the burghers. Not only carts had been left behind, but, as we found in places, sacks of flour, tins of coffee, mattresses and other jettison, thrown out of the carts to lighten their burden.

Mules with Ammunition Waggons Crossing a Spruit.[149] A big horse train crossing the Wilge river. A wagon is overturned in the bed of the river.[150]

On nearing Elandslaagte we caught up the rear of the fleeing commandos. Here we learned that Generals Botha and Meyer were still behind us with their commandos, near Lombardsdorp. We off-saddled, exhausted and half starving. Luckily, some of the provisions of our commissariat, which had been stored here during the Ladysmith investment, had not been carried away. But, to our disgust, we found that the Commissariat-Commissioner had set fire to the whole of it, so we had to appease our hunger by picking half-burned potatoes out of a fire.

At 7 o'clock next morning General Botha and his men arrived at Elandslaagte and off-saddled in hopes of getting something to eat. They were also doomed to disappointment. Such wanton destruction of God's bounty was loudly condemned, and had Mr. Pretorius, the Commissioner of Stores, not been discreet enough to make himself scarce, he would no doubt have been subjected to a severe "sjamboking." Later in the day a council of war was held, and it was decided that we should all stay there for the day, in order to stop the enemy if they should pursue us. Meantime we would allow the convoys an opportunity of getting to the other side of the Sunday River.

The British must have been so overjoyed at the relief of Ladysmith that Generals Buller and White did not think it necessary to pursue us, at any rate for some time, a consideration for which we were profoundly grateful. Methinks General Buller must have felt that he had paid a big price for the relief of Ladysmith, for it must have cost him many more lives than he had relieved. But in that place were a few Jingos (Natal Jingos) who had to be released, I suppose, at any costs.

My burghers and I had neither cooking utensils nor food, and were anxious to push forward and find our convoys; for we had not as yet learned to live without carts and commissariat. At dusk the generals—I have no idea who they were—ordered us to hold the "randjes" south of the Sunday River till the following day, and that no burghers were to cross the river. This order did not seem to please the majority, but the Generals had put a guard near the bridge, with instructions to shoot any burghers and their horses should they try to get to the other side; so they had perforce, to remain where they were. Now I had only 22 men under my command, and I did not think these would make an appreciable difference to our fighting force, so I said to myself: "To-night we shall have a little game with the generals for once."

We rode towards the bridge, and of course the guard there threatened to fire on us if we did not go back immediately. My adjutant, however, rode up and said: "Stand back, you ——! This is Commandant Viljoen, who has been ordered to hurry up a patrol at ——" (mentioning some place a few miles away) "which is in imminent danger of being captured."

The guards, quite satisfied, stepped back and favoured us with a military salute as we rode by. When we had been riding a little way I heard someone ask them what "people" they were who had passed over the bridge, and I caught the words: "Now you will see that they will all want to cross."

I do not contend I was quite right in acting in this insubordinate manner, but we strongly objected to being put under the guard of other commandos by some one irresponsible general. I went on that night till we reached the Biggarsbergen, and next day sent out scouts in the direction of the Drakensbergen to inquire for the scattered remains of my commando. The mountains were covered with cattle from the laagers about Glencoe Station. The Boers there were cooking food, shoeing their horses, or repairing their clothes; in fact, they were very comfortable and very busy. They remarked: "There are many more burghers yonder with the General; we are quite sure of that." ... "The Commandant-General is near Glencoe and will stop the retreating men."

Boer Camp at Elandslaagte.[151]

In short, as was continually happening in the War, everything was left to chance and the Almighty. Luckily General Botha had deemed it his duty to form a rearguard and cover our retreat; otherwise the English would have captured a large number of laagers, and many burghers whose horses were done up. But, whereas we had too little discipline, the English had evidently too much. It is not for me to say why General Buller did not have us followed up; but it seems that the British lost a splendid chance.

Some days went by without anything of note happening. My scouts returned on the third day and reported that my commando and its laager had safely got through, and could be expected the next day. Meanwhile I had procured some provisions at Glencoe, and for the time being we had nothing to complain about.

I was very much amused next day to receive by despatch-rider a copy of a telegram from Glencoe sent by General Joubert to General Prinsloo at Harrismith (Orange Free State) asking for information regarding several missing commandos and officers, amongst whom my name appeared, while the telegram also contained the startling news that my commando had been reported cut up at Klip River and that I had been killed in action! This was the second time that I was killed, but one eventually gets used to that sort of thing.

I sent, by the despatch-rider, this reply:—

"I and my commando are very much alive!" Adding: "Tell the General we want four slaughter oxen."

The following day I received orders to attend a council of war which was to be held at Glencoe Station. The principal object of this gathering was to discuss further plans of operation, to decide as to where our next positions were to be taken, and where the new fighting line would be formed.

We all met at the appointed time in a big unoccupied hall near Glencoe Station, where General Joubert opened the last council that he was to conduct in this world. Over 50 officers were present and the interest was very keen for several reasons. In the first place we all desired some official information about the fate of General Cronje and his burghers at Paardeburg, and in the second place some expected to hear something definite about the intervention of which so much had been said and written of late. In fact many thought that Russia, France, Germany or the United States of America would surely intervene so soon as the fortunes of war began to turn against us. My personal opinion was stated just before the war at a public meeting, held in Johannesburg, where I said: "If we are driven to war we must not rely for deliverance on foreign powers, but on God and the Mauser."

Some officers thought we ought to retire to our frontiers as far as Laing's Nek, and it was generally believed that this proposal would be adopted. According to our custom General Joubert opened the council with an address, in which he described the situation in its details. It was evident that our Commandant-General was very low-spirited and melancholy, and was suffering greatly from that painful internal complaint which was so soon to put an end to his career.

No less than eleven assisting commandants and fighting generals were present, and yet not one could say who was next in command to General Joubert. I spoke to some friends about the irregularities which occurred during our retreat from Ladysmith: how all the generals were absent except Botha and Meyer, while the latter was on far from good terms with General Joubert since the unfortunate attack on Platrand. This was undoubtedly due to the want of co-operation on the part of the various generals, and I resolved if possible, to bring our army into a closer union. I therefore proposed a motion:—

"That all the generals be asked to resign, with the exception of one assistant commandant-general and one fighting general."

Commandant Engelbrecht had promised to second my proposal, but when it was read out his courage failed him. The motion, moreover, was not very well received, and when it was put to the vote I found that I stood alone, even my seconder having forsaken me. As soon as an opportunity presented itself I asked General Joubert who was to be second in command. My question was not answered directly, but egged on by my colleagues, I asked whether General Botha would be next in command. To this he replied: "Yes, that is what I understand—."

And if I am not mistaken, this was the first announcement of the important fact that Botha was to lead us in future.

Much more was said and much arranged; some of the commandos were to go to Cape Colony and attempt to check the progress of Lord Roberts, who was marching steadily north after Cronje's surrender. Finally each officer had some position assigned to him in the mountain-chain we call the Biggarsbergen. I was placed under General Meyer at Vantondersnek, near Pomeroy, and we left at once for our destination. From this place a pass leads through the Biggarsbergen, about 18 miles from Glencoe Station.

Image: General Joubert Opening a Council of War with Prayer.

Chapter 13. Driven from the Biggarsbergen.

We spent the next few weeks in entrenching and fortifying our new positions. General Botha had left with some men for the Orange Free State which Lord Roberts, having relieved Kimberley, was marching through. General Joubert died about this time at Pretoria, having been twenty-one years Commandant-General of the South African Republic.[65] He was without doubt one of the most prominent figures in the South African drama.

Body of General Joubert Being Received at Pretoria.[152]

Funeral of Commandant-General Piet Joubert, in Pretoria. [153] Funeral of General Joubert—The Casket.[154]

General Botha now took up the chief command and soon proved himself to be worthy of holding the reins. He enjoyed the confidence and esteem of our whole army, a very important advantage under our trying circumstances.

Assisted by De Wet he was soon engaged in organizing the commandos in the Orange Free State, and in attempting to make some sort of a stand against the British, who were now marching through the country in overwhelming numbers. In this Republic the burghers had been under the command of the aged General Prinsloo, who now, however, had become so downhearted that the supreme command was taken from him and given to General De Wet. Prinsloo surrendered soon after, in doing which he did his people his greatest service; it was, however, unfortunate that he should have succeeded in leading with him 900 burghers into the hands of the enemy.

[65] General Joubert. General Piet Joubert died on March 27th, 1900 from injuries suffered in a fall from his horse the previous November. Sir George White, the defender of Ladysmith, summed up Jouberts character when he called him "a soldier and a gentleman, and a brave and honourable opponent".[Chisholm, Hugh, ed. (1911). "Joubert, Petrus Jacobus". Encyclopædia Britannica. Vol. 15 (11th ed.). Cambridge University Press. p. 522. And See: https://en.wikipedia.org/wiki/Piet_Joubert

General Marthinus J. Prinsloo.[155] [66] Surrender of Prinsloo's Boer forces in Brandwater basin, 1900.[156]

In the Biggarsbergen we had nothing to do but to sleep and eat and drink. On two separate occasions, however, we were ordered to join others in attacking the enemy's camp at Elandslaagte. This was done with much ado, but I would rather say nothing about the way in which the attacks were directed. It suffices to say that both failed miserably, and we were forced to retire considerably quicker than we had come.

Our generals, meantime, were very busy issuing innumerable circulars to the different commandos. It is impossible for me to remember the contents of all these curious manifestos, but one read as follows:—

"A roll-call of all burghers is to be taken daily; weekly reports are to be sent to headquarters of each separate commando, and the minimum number of burghers making up a field-cornetship is therein to be stated. Every 15 men forming a field-cornetship are to be under a corporal; and these corporals are to hold a roll-call every day, and to send in weekly detailed reports of their men to the Field-Cornet and Commandant, who in his turn must report to the General."

Another lengthy circular had full instructions and regulations for the granting of "leave" to burghers, an intricate arrangement which gave officers a considerable amount of trouble. The scheme was known as the "furlough system," and was an effort to introduce a show of organisation into the weighty matter of granting leave of absence. It failed, however, completely to have its desired effect. It provided that one-tenth of each commando should be granted furlough for a fortnight, and then return to allow another tenth part to go in its turn. In a case of sick leave, a doctor's certificate was required, which had to bear the counter-signature of the field-cornet; its possessor was then allowed to go home instead of to the hospital.

Further, a percentage of the farmers were allowed from time to time to go home and attend to pressing matters of their farms, such as harvesting, shearing sheep, etc. Men were chosen by the farmers to go and attend to matters not only for themselves but for other farmers in their districts as well. The net result of all this was that when everybody who could on some pretext or other obtain furlough had done so, about a third of each commando was missing. My burghers who were mostly men from the [Johannesburg area and] Witwatersrand Goldfields, could of course obtain no leave for farming purposes; and great dissatisfaction prevailed. I was inundated with complaints about their unfair treatment in this respect and only settled matters with considerable trouble.

I agree that this matter had to be regulated somehow, and I do not blame the authorities for their inability to cope with the difficulty. It seemed a great pity, however, that the commandos should be weakened so much and

[66] Prinsloo Marthinus General Born in 1838, he was serving as the Commandant of the Winburg Commando in 1867. He was a man of wealth and served as an elder of the church. At the start of the Boer War, he commanded Free State commandos in Natal. In the attack on Caesar's Camp on 6 Jan 1900, he was accused of indecision. He commanded the defensive line on the Tugela after the battle of Spion Kop. When Ladysmith was relieved and Buller pushed north, Prinsloo commanded the Free State commandos at Van Reenen's Pass. Now aged 62, he was seen as too old for combat and retired to his farm. He re-engaged and operated in the Free State. On 30 July 1900 he was captured near Fouriesburg in the Brandwater Basin with over 4,000 men, 2,800 cattle, 4,000 sheep, 5,000 horses and 2 million rounds. He died in 1903. See Also: Marthinus Prinsloo - Wikipedia

that the fighting spirit should be destroyed in this fashion. Of course it was our first big war and our arrangements were naturally of a very primitive character.

It was the beginning of May before our friends the enemy at Ladysmith and Elandslaagte began to show some signs of activity. We discovered unmistakable signs that some big forward movement was in progress, but we could not discover on which point the attack was to be directed. Buller and his men were marching on the road along Vantondersnek, and I scented heavy fighting for us again. I gathered a strong patrol and started out to reconnoitre the position. We found that the enemy had pitched their camp past Waschbank in great force, and were sending out detachments in an easterly direction. From this I concluded that they did not propose going through Vantondersnek, but that they intended to attack our left flank at Helpmakaar. This seemed to me, at any rate, to be General Buller's safest plan.

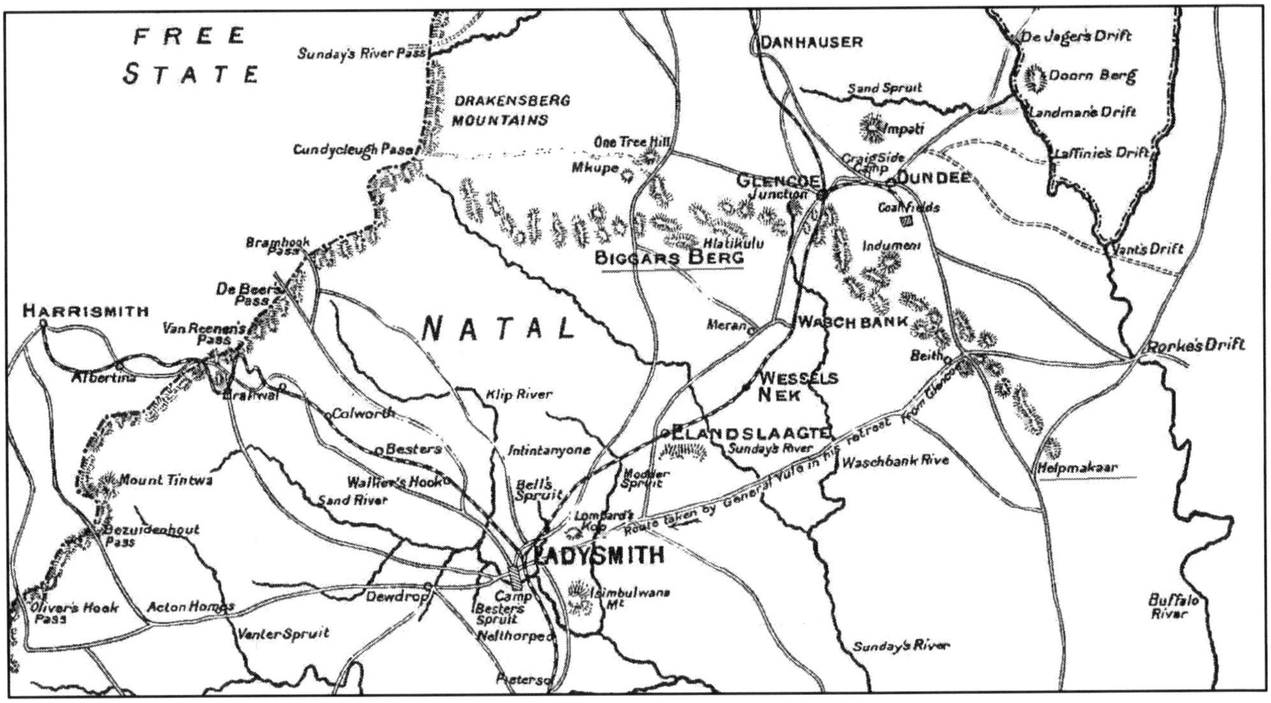

Northern Natal.[157] Note Helpmakaar on right at the east end of the Biggarsberg chain

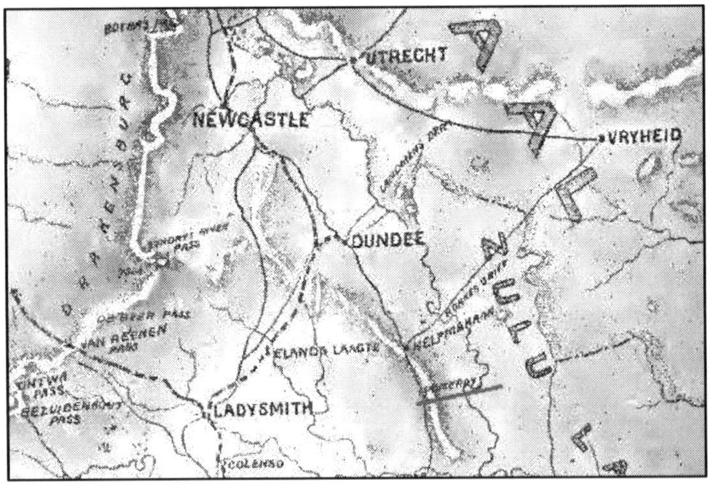

Birch Map of Natal.[158] Pomeroy, east of Ladysmith and just south of Helpmakaar. Portion of Buller's advance through Natal.[159]
Pomeroy at bottom with Helpmakaar just above.

Helpmakaar was east of my position; it is a little village elbowed in a pass in the Biggarsbergen. By taking this point one could hold the key to our entire extended line of defence, as was subsequently only too clearly shown. I pointed this out to some of our generals, but a commandant's opinion did not weigh much just then; nor was any notice taken of a similar warning from Commandant Christian Botha, who held a position close to mine with the Swaziland burghers.

We had repeated skirmishes with the English outposts during our scouting expeditions, and on one occasion we suddenly encountered a score of men of the South African Light Horse.[67] [68]

We noticed them in a "donk" (a hollow place) thickly covered with trees and bushes, but not before we were right amongst them. It appears they mistook us for Englishmen, while we thought at first they were members of Colonel Blake's Irish Brigade. Many of them shook hands with us, and a burgher named Vivian Cogell asked them in Dutch: "How are you, boys?"

To which an Englishman, who understood a little Dutch, answered: "Oh, all right; where do you come from?"

Vivian replied: "From Viljoen's commando; we are scouting."

Then the Englishman discovered who we were, but Vivian gave the man no time for reflection. Riding up to him, he asked: "What regiment do you belong to?"

"To the South African Light Horse," answered the Englishman.

"Hands up!" retorted Vivian, and the English-Afrikander threw down his gun and put up his hands.

"Hands up! Hands up!" was the cry now universally heard, and although a few escaped, the majority were disarmed and made prisoners. It had been made a rule that when a burgher captured a British soldier he should be allowed to conduct him to Pretoria, where he could then obtain a few days' leave to visit his family. This did much to encourage our burghers to make prisoners, although many lost their lives in attempting to do so.

The next day, General Buller marched on Helpmakaar, passing close to our position. We fired a few shots from our Creusot gun, and had several light skirmishes. The enemy, however, concentrated the fire of a few batteries on us, and our guns were soon silenced.

General L. Meyer had arrived with some reinforcements close to Helpmakaar, but the position had never been strengthened, and the sole defending force consisted of the Piet Retief burghers, known as the "Piet Retreaters," together with a small German corps. The result was easy to predict. The attack was made, and we lost the position without seriously attempting to defend it. Buller was now, therefore, in possession of the key to the Boer position in Natal, a position which we had occupied for two months—and could therefore, have fortified to perfection—and whose strategic importance should have been known in its smallest details. I think our generals, who had a sufficient force at their disposal, of which the mobility has become world-famed, should have been able to prevent such a fiasco as our occupation of the splendid line of defence in the Biggarsbergen turned out to be.

Here, for the first time in the war, General Buller utilised his success, and followed up our men as they were retreating on Dundee. He descended by the main waggon track from Helpmakaar, and drove the commandos like sheep before him. I myself was obliged to move away in hot haste and join the general retreat. Once or twice our men attempted to make a stand, but with little success.[69]

When we reached Dundee the enemy gradually slackened off pursuit, and at dark we were clear of them. Satisfied with their previous day's success, and sadly hampered by their enormous convoys, the English now allowed us to move on at our leisure.[70]

[67] The South African Light Horse regiment of the British Army were raised in Cape Colony in 1899 and disbanded in 1907. The commanding officer tasked with raising the regiment was Major (locally a Lieutenant Colonel) the Honourable Julian Byng (10th Hussars) who would go on to rise to the rank of Field Marshal) South African Light Horse - Wikipedia

[68] South African Light Horse Article: https://www.angloboerwar.com/?option=com_content&view=article&id=467 1,727 Records.

[69] This is the same route that Yule took when he retreated from Dundee towards Ladysmith after the battle of Talana Hill.

[70] 5/13/1900 General Buller advanced towards the Biggarsberg.
5/14/1900 Occupation of Dundee.
5/15/1900 Occupation of Glencoe.
5/15/1900 Christiana occupied.
5/17/1900 Colonel Mahon, at the head of the relief force, entered Mafeking.
5/18/1900 Occupation of Newcastle.
5/20/1900 Colonel Bethune's Mounted Infantry ambushed near Vryheid. In Natal.
5/29/1900 Battle of Doornkop: Boers defeated.
5/30/1900 Occupation of Utrecht by General Hildyard. Charles Warren defeated the enemy near Douglas.

Chapter 14. Dispirited And Demoralised.

Our first intention was to proceed to Vereeniging, there to join General Botha's forces. At Klip River Station, that preceding Vereeniging, I was ordered, however, to leave my carts behind and proceed with my men to Vaalbank, as the enemy were advancing with forced marches, and had compelled all the other commandos to fall back on Vereeniging.

Mounted Infantry Crossing the Vaal River.[160] Thursday, May 24, 1900 British Army entered the Transvaal, crossing the Vaal near Parys, unopposed.

First camp inside the Transvaal.[161]

On our way we met groups of retreating burghers, each of whom gave us a different version of the position. Some said that the enemy had already swept past Vereeniging, others that they could not now be stopped until they reached Johannesburg. Further on, we had the good fortune to encounter General Botha and his staff.

The General ordered me to take up a position at the Gatsrand, near the Nek at Pharaohsfontein, as the British, having split their forces up into two parts, would send one portion to cross the Vaal River at Lindeque's Drift, whilst the other detachments would follow the railway past Vereeniging. Generals Lemmer[71] and Grobler[72] were already posted at the Gatsrand to obstruct the enemy's progress.

[71] H. R. Lemmer. See MOA

[72] Probably F. A. Grobler. There are four senior commanders of that name in MOA.

I asked General Botha how we stood. He sighed, and answered: "If only the burghers would fight we could stop them easily enough; but I cannot get a single burgher to start fighting. I hope their running mood will soon change into a fighting mood. You keep your spirits up, and let us do our duty."

"All right, General," I answered, and we shook hands heartily.

We rode on through the evening and at midnight halted at a farm to give our horses rest and fodder. The owner of the farm was absent on duty, and his family had been left behind. On our approach the women-folk, mistaking us for Englishmen, were terrified out of their wits. Remembering the atrocities and horrors committed in Natal on the advance of the Imperial troops, they awaited the coming of the English with the greatest terror. On the approach of the enemy many women and children forsook their homes and wandered about in caves and woods for days, exposed to every privation and inclemency of the weather, and to the attacks of wandering bands of plundering kaffirs.

Southern Transvaal.[162] Note the Gatsrand west of Vereeniging. This is the only map that shows that feature.

Mrs. van der Merwe, whom we met here, was exceedingly kind to us, and gave us plenty of fodder for our horses. We purchased some sheep, and slaughtered them and enjoyed a good meal before sunrise; and each one of us bore away a good-sized piece of mutton as provisions for the future.

Our scouts, whom we had despatched overnight, informed us that Generals Lemmer and Grobler had taken up their stand to the right of Pharaohsfontein in the Gatsrand, and that the English were approaching in enormous force.

By nine in the morning we had taken up our positions, and at noon the enemy came in sight. Our commando had been considerably reduced, as many burghers, finding themselves near their homes, had applied for twenty-four hours' leave, which had been granted in order to allow them to arrange matters before the advance of the English on their farms made it impossible. A few also had deserted for the time being, unable to resist the temptation of visiting their families in the neighbourhood.

Some old burghers approached us and hailed us with the usual "Morning, boys! Which commando do you belong to?"

"Viljoen's."

"We would like to see your Commandant," they answered.

Presenting myself, I asked: "Who are you, and where do you come from, and where are you going to?"

They answered: "We are scouts of General Lemmer and we came to see who is holding this position."

"But surely General Lemmer knows that I am here?"

"Very probably," they replied, "but we wanted to know for ourselves; we thought we might find some of our friends amongst you. You come from Natal, don't you?"

"Yes," I answered sadly. "We have come to reinforce the others, but I fear we can be of little use. It seems to me that it will be here as it was in Natal; all running and no fighting."

"Alas!" they said, "the Free Staters will not remain in one position, and we must admit the Transvaalers are also very disheartened. However, if the British once cross our frontiers you will find that the burghers will fight to the bitter end."

Consoled by this pretty promise we made up our minds to do our best, but our outposts presently brought word that the British were bearing to the right and nearing General Grobler's position, and had passed round that of General Lemmer. Whilst they attacked General Grobler's we attacked their flank, but we could not do much damage, as we were without guns. Soon after the enemy directed a heavy artillery fire on us, to which we, being on flat ground, found ourselves dangerously exposed.

Towards evening the enemy were in possession of General Grobler's position, and were passing over the Gatsrand, leaving us behind. I ordered my commando to fall back on Klipriversberg, while I rode away with some adjutants to attempt to put myself in communication with the other commandos.

The night was dark and cloudy, which rendered it somewhat difficult for us to move about in safety. We occasionally fell into ditches and trenches, and had much trouble with barbed wire. However, we finally fell in with General Lemmer's rearguard, who informed us that the enemy, after having overcome the feeble resistance of General Grobler, had proceeded north, and all the burghers were retreating in haste before them.

We rode on past the enemy to find General Grobler and what his plans were. We rode quite close to the English camp, as we knew that they seldom posted sentries far from their tents. On this occasion, however, they had placed a guard in an old "klipkraal," for them a prodigious distance from their camp, and a "Tommy" hailed us from the darkness.—

"Halt, who goes there?"

I replied "Friend," whereupon the guileless soldier answered:

"Pass, friend, all's well."

I had my doubts, however. He might be a Boer outpost anxious to ascertain if we were Englishmen. Afraid to ride into ambush of my own men, I called out in Dutch:

"Whose men are you?"

The Tommy lost his temper at being kept awake so long and retorted testily, "I can't understand your beastly Dutch; come here and be recognized." But we did not wait for identification, and I rode off shouting back "Thanks, my compliments to General French, and tell him that his outposts are asleep."

This was too much for the "Tommy" and his friends, who answered with a volley of rifle fire, which was taken up by the whole line of British outposts. No harm was done, however, and we soon rode out of range. I gave up looking for General Grobler, and on the following morning rejoined my men at Klipriversberg.

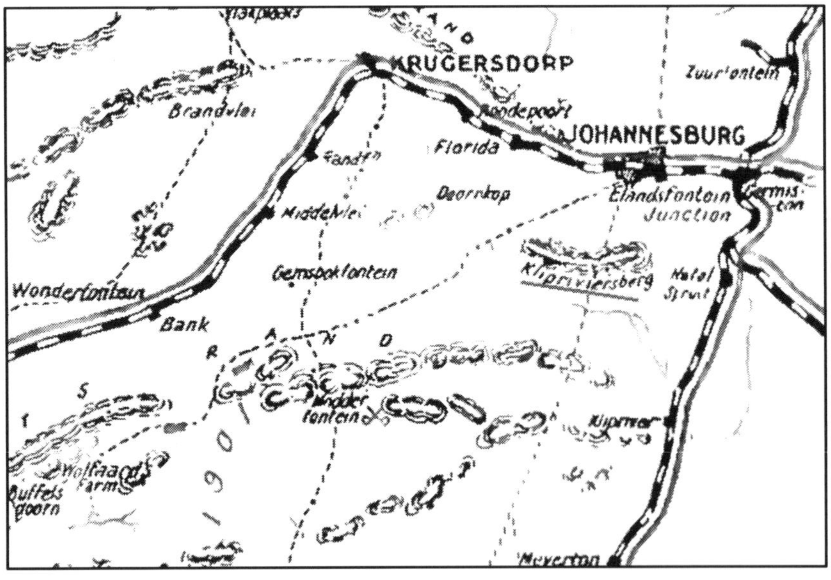

Portion of Western Transvaal. Note Klipriversberg just south of Johannesburg.[163]

It was by no means easy to find out the exact position of affairs. Our scouts reported that the enemy's left wing, having broken through General Grobler's position, were now marching along Van Wijk's Rust. I could, however, obtain no definite information regarding the right wing, nor could I discover the General under whose orders I was to place myself. General Lemmer, moreover, was suffering from an acute disease of the kidneys, which had compelled him to hand over his command to Commandant Gravett,[73] who had proved himself an excellent officer.

General Grobler had lost the majority of his men, or what was more likely the case, they had lost him. He declared that he was unaware of General Botha's or Mr. Kruger's plans, and that it was absurd to keep running away, but he clearly did not feel equal to any more fighting, although he had not the moral courage to openly say so. From this point this gentleman did no further service to his country, and was shortly afterwards dismissed. The reader will now gather an idea of the enormous change which had come over our troops. Six months before they had been cheerful and gay, confident of the ultimate success of their cause; now they were downhearted and in the lowest of spirits. I must admit that in this our officers were no exception.

Those were dark days for us. Now began the real fighting, and this under the most difficult and distressing circumstances; and I think that if our leaders could have had a glimpse of the difficulties and hardships that were before us, they would not have had the courage to proceed any further in the struggle.

"Group of Boer Commandants on the Klip River, Near Johannesburg. This picture was taken just before the great battle on 28 May 1900, when all the forces united under Lord Roberts were needed to capture the positions held by 2,000 Boers. Lord Roberts succeeded, but only after he had encountered the most determined resistance. His losses were very heavy. Johannesburg was occupied by him on the 31st of May."[164]

Early next morning (the 29th May, 1900) we reached Klipspruit, and found there several other commandos placed in extended order all the way up to Doornkop.

Amongst them was that of General De la Rey, who had come from the Western frontier of our Republic, and that of General Snyman, whom I regard as the real defender and reliever of Mafeking, for he was afraid to attack a garrison of 1,000 men with twice that number of burghers.

Before having had time to properly fortify our position we were attacked on the right flank by General French's cavalry, while the left flank had to resist a strong opposing force of cavalry. Both attacks were successfully repulsed, as well as a third in the centre of our fighting line.

The British now marched on Doornkop, their real object of attack being our extreme right wing, but they made a feint on our left. Our line of defence was very extended and weakened by the removal of a body of men who had been sent to Natal Spruit to stop the other body of the enemy from forcing its way along the railway line and cutting off our retreat to Pretoria.

[73] G. H. Gravett. See MOA.

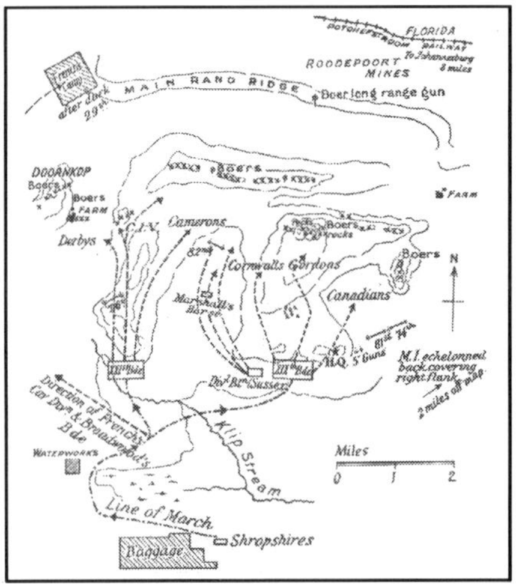

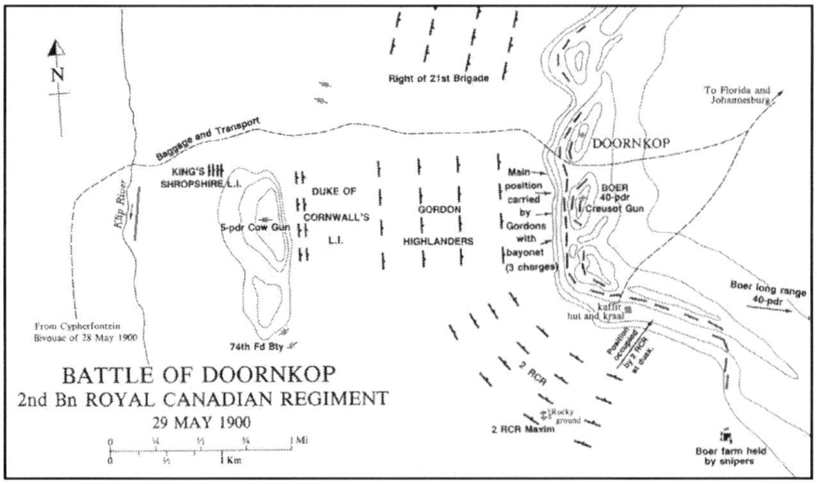

Actions before Johannesburg. May 29th 1900.[165] Note 'Klip Stream.' Battle of Doornkop:[166] And.[74]

The battle lasted till sunset, and was especially fierce on our right, where the Krugersdorpers stood. Early in the evening our right wing had to yield to an overwhelming force, and during the night all the commandos had to fall back. My commando, which should have consisted of about 450 men, only numbered 65 during this engagement; our losses were two men killed. I was also slightly wounded in the thigh by a piece of shell, but I had no time to attend such matters, as we had to retire in haste, and the wound soon healed.

The next day our forces were again in full retreat to Pretoria, where I understood we were to make a desperate stand. About seven o'clock we passed through Fordsburg, a suburb of Johannesburg.

We had been warned not to enter Johannesburg, as Dr. Krause, who had taken from me the command of the town, had already surrendered it to Lord Roberts, who might shell it if he found commandos were there. Our larger commissariat had proceeded to Pretoria, but we wanted several articles of food, and strange to say the commissariat official at Johannesburg would not give us anything for fear of incurring Lord Roberts' displeasure!

I was very angry; the enemy were not actually in possession of the town, and I therefore should have been consulted in the matter; but these irresponsible officials even refused to grant us the necessaries of life!

At this time there was a strong movement on foot to blow up the principal mines about Johannesburg, and an irresponsible young person named Antonie Kock had placed himself at the head of a confederacy with this object in view. But thanks to the explicit orders of General L. Botha, which were faithfully carried out by Dr. Krause, Kock's plan was fortunately frustrated, and I fully agree with Botha that it would have been most impolitic to have allowed this destruction. I often wished afterwards, however, that the British military authorities had shown as much consideration for our property.

We had to have food in any case, and as the official hesitated to supply us we helped ourselves from the Government Stores, and proceeded to the capital. The roads to Pretoria were crowded with men, guns, and vehicles of every description, and despondency and despair were plainly visible on every human face.

Image: A Surprise,—Coyell Meeting the Imperial Light Horse.

[74] By late May 1900, the British army was nearing Johannesburg but across the army's route lay the formidable Klipriversberg Range. In its midst was the position of Doornkop, which the Boers had occupied in force. As part of the general attack, the Canadian infantry moved forward over a low ridge and up a long slope, under fire the whole way. The Boers had set fire to the brush and for part of the distance the troops had to run through flames, while the smoke made navigation and control difficult. By clever use of the ground, aided by the aggressive support of the battalion machine gun section, the Canadians captured their objective with losses of only seven men wounded. https://en.wikipedia.org/wiki/Battle_of_Doornkop

Chapter 15. Occupation of Pretoria.

The enemy naturally profited by our confusion to pursue us more closely than before. The prospect before us was a sad one, and we asked ourselves, "What is to be the end of all this, and what is to become of our poor people? Shall we be able to prolong the struggle, and for how long?"

But no prolongation of the struggle appeared to have entered into our enemy's minds, who evidently thought that the War had now come upon its last stage, and they were as elated as we were downhearted. They made certain that the Boer was completely vanquished, and his resistance effectually put an end to. At this juncture Conan Doyle, after pointing out what glorious liberty and progress would fall to the Boers' lot under the British flag, wrote:—

"When that is learned it may happen that they will come to date a happier life and a wider liberty from that 5th of June which saw the symbol of their nation pass for ever from the ensigns of the world."

Thus, not only did Lord Roberts announce to the world that "the War was now practically over," but Conan Doyle did not hesitate to say the same in more eloquent style.

How England utterly under-estimated the determination of the Boers, subsequent events have plainly proved. It is equally plain that we ourselves did not know the strength of our resolution, when one takes into account the pessimism and despair that weighed us down in those dark days; and as the Union Jack was flying over our Government buildings we might have exclaimed:— "England, we do not know our strength, but you know it still less!"

Nearly all the commandos were now in the neighbourhood of Pretoria, General Botha forming a rearguard, and we determined to defend the capital as well as we could. But at this juncture some Boer officer was said to have received a communication from the Government, informing us that they had decided not to defend the town. A cyclist was taking this communication round to the different commandos, but the Commandant-General did not seem to be aware of it, and we tried in vain to find him so as to discover what his plans were.

The greatest confusion naturally prevailed, and as all the generals gave different orders, no one knew what was going to be done. I believe General Botha intended to concentrate the troops round Pretoria, and there offer some sort of resistance to the triumphant forces of the enemy, and we had all understood that the capital would be defended to the last; but this communication altered the position considerably. Shortly afterwards all the Boer officers met at Irene Estate, near Pretoria, in a council of war, and were there informed that the Government had already forsaken the town, leaving a few "feather-bed patriots" to formally surrender the town to the English.

I thought this decision of easy surrender ridiculous and inexplicable, and many officers joined me in loud condemnation of it. I do not remember exactly all that happened at the time, but I know a telegram arrived from the Commandant-General saying that a crowd had broken open the Commissariat Buildings in Pretoria and were looting them. An adjutant was sent into Pretoria to spread an alarm that the English were entering the town, and this had the effect of driving all the looters out of it. Some of my own men were engaged in these predatory operations, and I did not see them again until three days after.

The English approached Pretoria very cautiously, and directed some big naval guns on our forts built round the town, to which we replied for some time with our guns from the "randten," south-west of the town; but our officers were unable to offer any organised resistance, and thus on the 5th of June, 1900, the capital of the South African Republic fell with little ado into the enemy's hands.

Bloemfontein, the capital of the Orange Free State, had months before suffered the same fate, and thousands of Free Staters had surrendered to the English as they marched from Bloemfontein to the Transvaal. Happily, however, in the Free State President Steyn and General De Wet were still wide awake and Lord Roberts very soon discovered that his long lines of communication were a source of great trouble and anxiety to him. The commandos, meanwhile, were reorganised; the buried Mausers and ammunition were once more resurrected, and soon it became clear that the Orange Free State was far from conquered.

Public Square in Pretoria before the British Occupation.[167] PF: This must have been taken by Davis during the conflict when it was being used as a military camp. Boer Retreat from Pretoria[168]

British Troops just before raising the Union Jack.[169] Hoisting the British flag on the Raadzaal, Pretoria.[170]

Colonials "Marching Past" at the formal Occupation of Pretoria, June 5th 1900.[171] Lord Roberts and staff cheering the Queen--Raising Union Jack on occupation of Pretoria, June 5th, S.A.[172]

The fall of Pretoria, indeed, was but a sham victory for the enemy. A number of officials of the Government remained behind there and surrendered, together with a number of burghers, amongst these faint-hearted brethren being even members of the Volksraad and men who had played a prominent part in the Republic's history; while to the everlasting shame of them and their race, a number of other Boers entered at once into the English service and henceforth used their rifles to shoot at and maim their own fellow-countrymen.[75]

[75] Their particular antagonists were the Imperial Light Horse and the National Scouts.

Chapter 16. Battle of Donkerhoek ("Diamond Hill").

Our first and best positions were now obviously the kopjes which stretched from Donkerhoek past Waterval and Wonderboompoort. This chain of mountains runs for about 12 miles E. and N.E. of Pretoria, and our positions here would cut off all the roads of any importance to Pietersburg, Middelburg, as well as the Delagoa Bay railway. We therefore posted ourselves along this range, General De la Rey forming the right flank, some of our other fighting generals occupying the centre, whilst Commandant-General Botha himself took command of the left flank.

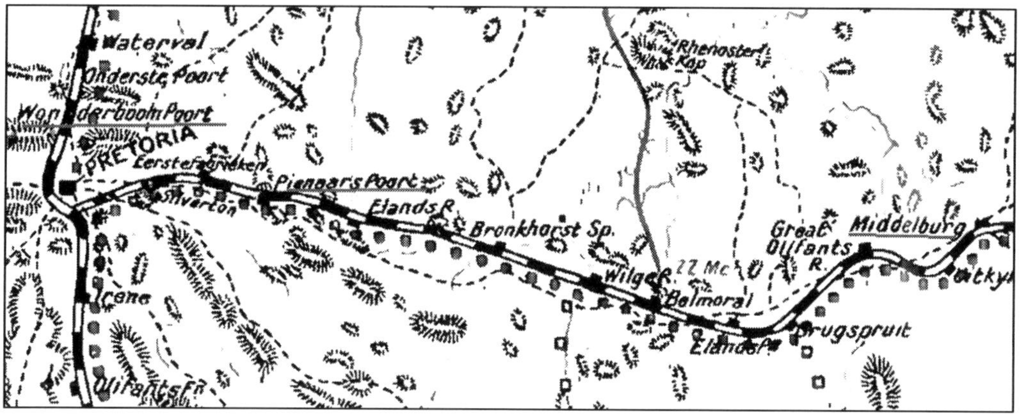

East and N.E. Transvaal.[173] Note Wonderboom Poort just above Pretoria and Pianaars Poort (gate) east of Pretoria. Middelburg, far right. The Boer forces had evacuated Pretoria but had not 'retreated' significant distance. Roberts was now static in an urban area with a functional and hostile army on the outskirts.

On the 11th of June, 1900, Lord Roberts approached with a force of 28,000 to 30,000 men and about 100 guns, in order, as the official despatches had it, "to clear the Boers from the neighbourhood of Pretoria." Their right and left flanks were composed of cavalry,[76] whilst the centre was formed of infantry regiments; their big guns were placed in good positions and their field pieces were evenly distributed amongst the different army divisions.

Towards sunset they began booming away at our whole 13 miles of defence. Our artillery answered their fire from all points with excellent results, and when night fell the enemy retired a little with considerable losses.

The battle was renewed again next day, the enemy attempting to turn our right with a strong flanking movement, but was completely repulsed. Meanwhile I at Donkerpoort proper had the privilege of being left unmolested for several hours. The object of this soon became apparent. A little cart drawn by two horses and bearing a white flag came down the road from Pretoria. From it descended two persons, Messrs. Koos Smit, our Railway Commissioner and Mr. J. F. de Beer, Chief Inspector of Offices, both high officials of the South African Republic. I called out to them from a distance.

"Halt, you cannot pass. What do you want?"

Smit said, "I want to see Botha and President Kruger. Dr. Scholtz is also with us. We are sent by Lord Roberts."

I answered Mr. Smit that traitors were not admitted on our premises, and that he would have to stay where he was. Turning to some burghers who were standing near I gave instructions that the fellows were to be detained.

Mr. Smit now began to "sing small," and turning deadly pale, asked in a tremulous voice if there were any chance of seeing Botha.

"Your request," I replied, "will be forwarded." Which was done.

An hour passed before General Botha sent word that he was coming. Meanwhile the battle continued raging fiercely, and a good many lyddite bombs were straying our way. The "white-flaggists" appeared to be very anxious to know if the General would be long in coming, and if their flag could not be hoisted in a more conspicuous place. The burghers guarding them pointed out, however, that the bombs came from their own British friends.

After a while General Botha rode up. He offered a far from cordial welcome to the deputation.

[76] Ian Hamilton to the British Right and John French on the left.

Dr. Scholtz produced a piece of paper and said Lord Roberts had sent him to enquire why Botha insisted on more unnecessary bloodshed, and why he did not come in to make peace, and that sort of thing.

Botha asked if Scholtz held an authoritative letter or document from the English general, to which the Doctor replied in the negative.

Smit now suggested that he should be allowed to see Mr. Kruger, but Botha declared, with considerable emphasis, "Look here, your conduct is nothing less than execrable,[77] and I shall not allow you to see Mr. Kruger. You are a couple of contemptible scoundrels, and as for Dr. Scholtz, his certificate looks rather dubious. You will go back and give the following message to Lord Roberts:—

"That this is not the first time messages of this description are sent to me in an unofficial manner; that these overtures have also sometimes been made in an insulting form, but always equally unofficially. I have to express my surprise at such tactics on the part of a man in Lord Roberts' position. His Lordship may think that our country is lost to us, but I shall do my duty towards it all the same. They can shoot me for it or imprison me, or banish me, but my principles and my character they cannot assail."

One could plainly see that the conscience-stricken messengers winced under the reproach. Not another word was said, and the noble trio turned on their heels and took their white flag back to Pretoria.

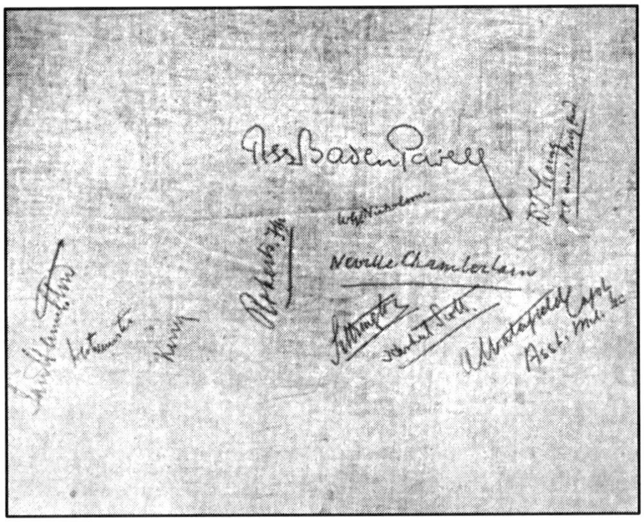

Cape Cart, Requisitioned by Lord Roberts for Peace Negotiations with Louis Botha, Driven by the Military Secretary, Captain Waterfield, to the Boer Lines.[174] What is most instructive is how the flag of truce was authenticated. It is enscribed with the following signatures for authenticity: Roberts (Vertical in Center), Baden Powell, Roberts' Secretary Neville Chamberlain. Kerry, Nicholson, Scott, and others. Flag of Truce.[175]

PF: These images are from a different request for parley. See Ian's March & Century of Wrong for extended discussion of the Laing's Nek peace negotiations between Buller and Christian Botha

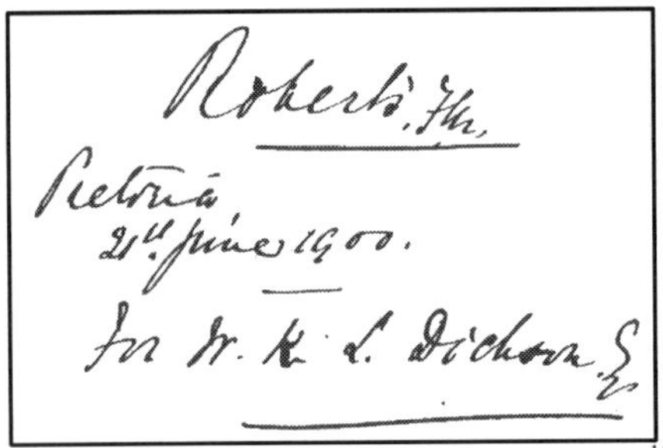

Roberts' Autograph for comparison.[176]

Whether Botha was right in allowing these "hands-uppers" to return, is a question I do not care to discuss, but many burghers had their own opinion about it. Still, if they had been detained by us and shot for high treason, what would not have been said by those who did not hesitate to send our own unfaithful burghers to us to induce us to surrender.

[77] Of the poorest quality.

I cannot say whether Lord Roberts was personally responsible for the sending of these messengers, but that such action was extremely improper no one can deny. It was a specially stupendous piece of impudence on the part of these men, J. S. Smit and J. F. de Beer, burghers both, and highly placed officials of the S. A. Republic. They had thrown down their arms and sworn allegiance to an enemy, thereby committing high treason in the fullest sense of the word. They now came through the fighting lines of their former comrades to ascertain from the commanders of the republican army why the whole nation did not follow their example, why they would not surrender their liberty and very existence as a people and commit the most despicable act known to mankind.

"Pretoria was in British hands!" As if, forsooth, the existence of our nationality began and ended in Pretoria! Pretoria was after all only a village where "patriots" of the Smit and de Beer stamp had for years been fattening on State funds, and, having filled their pockets by means of questionable practices, had helped to damage the reputation of a young and virile nation.

Not only had they enjoyed the spoils of high office in the State Service offices, to which a fabulous remuneration was attached, but they belonged to the Boer aristocracy, members of honourable families whose high birth and qualities had secured for them preference over thousands of other men and the unlimited confidence of the Head of State. Little wonder these gentlemen regarded the fall of Pretoria as the end of the war!

The battle continued the whole day; it was fiercest on our left flank, where General French and his cavalry charged the positions of the Ermelo and Bethel burghers again and again, each time to be repulsed with heavy losses. Once the lancers attacked so valiantly that a hand-to-hand fight ensued. The commandant of the Bethel burghers afterwards told me that during the charge his kaffir servant got among the lancers and called upon them to "Hands up!" The unsophisticated native had heard so much about "hands up," and "hands-uppers," that he thought the entire English language consisted of those two simple words, and when one lancer shouted to him "Hands up," he echoed "Hands up." The British cavalryman thrust his lance through the nigger's arm, still shouting "Hands up," the black man retreating, also vociferously shrieking "Hands up, boss; hands up!"

When his master asked him why he had shouted "Hands up" so persistently though he was running away, he answered: "Ah, boss, me hear every day people say, 'Hands up;' now me think this means kaffir 'Soebat' (to beg). I thought it mean, 'Leave off, please,' but the more I shouted 'Hands up' English boss prod me with his assegai all the same."

On our right General De la Rey had an equally awkward position; the British here also made several determined attempts to turn his flank, but were repulsed each time. Once during an attack on our right, their convoy came so close to our position that our artillery and our Mausers were enabled to pour such a fire into them that the mules drawing the carts careered about the veldt at random, and the greatest confusion ensued. British mules were "pro-Boer" throughout the War. The ground, however, was not favourable for our operations, and we failed to avail ourselves of the general chaos.

Towards the evening of the second day General Tobias Smuts[78] made an unpardonable blunder in falling back with his commandos. There was no necessity for the retreat; but it served to show the British that there was a weak point in our armoury. Indeed, the following day the attack in force was made upon this point. The British had meantime continued pouring in reinforcements, men as well as guns.

About two o clock in the afternoon Smuts applied urgently for reinforcements, and I was ordered by the Commandant-General to go to his position. A ride of a mile and a half brought us near Smuts; our horses were put behind a "randje," the enemy's bullets and shells meantime flying over their heads without doing much harm. We then hurried up on foot to the fighting line, but before we could reach the position General Smuts and his burghers had left it. At first I was rather in the dark as to what it all meant until we discovered that the British had won Smuts' position, and from it were firing upon us. We fell down flat behind the nearest "klips" and returned the fire, but were at a disadvantage, since the British were above us.

[78] Tobias Smuts, see MOA.

I never heard where General Smuts and his burghers finally got to. On our left we had Commandant Kemp[79] with the Krugersdorpers; on the right Field-Cornet Koen Brits. The British tried alternately to get through between one of my neighbours and myself, but we succeeded, notwithstanding their fierce onslaught, in turning them back each time. All we could do, however, was to hold our own till dark. Then orders were given to "inspan" all our carts and other conveyances as the commandos would all have to retire.

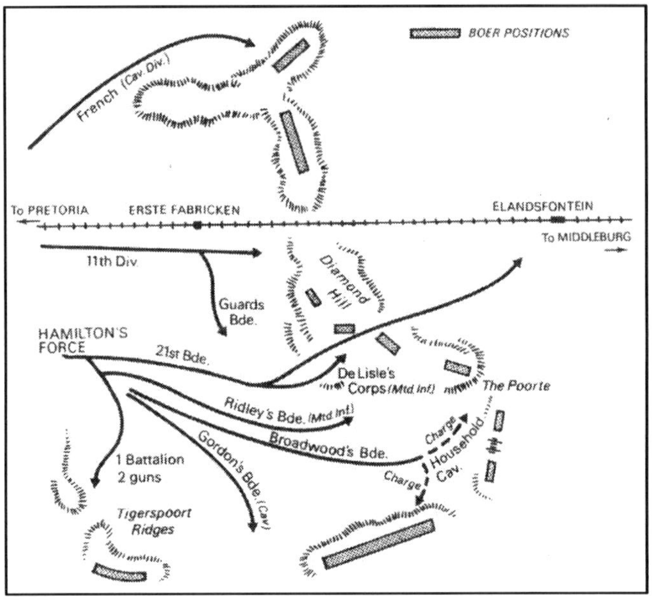

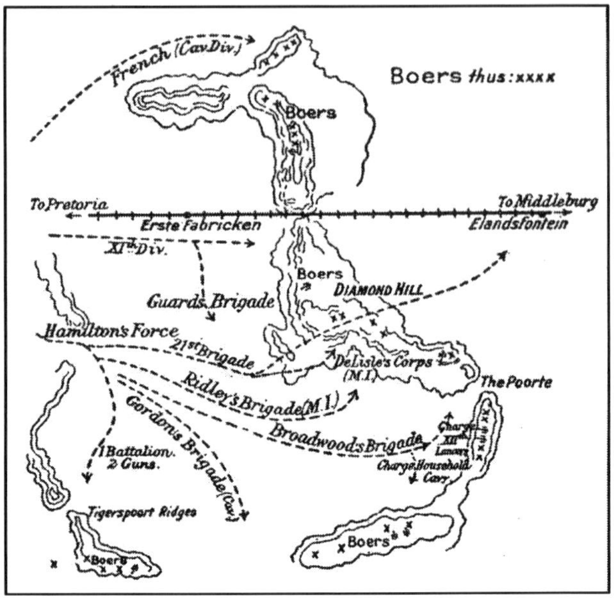

Plan of The Operations for action of Diamond Hill 11–12 June, 1900.[177] Note Tigerspoorte Ridges lower left.
Plan of The Operations of 11th And 12th June, 1900. Scale, Diagram is about 16 miles square. (4*4)[178]

I do not know the extent of the British losses in that engagement. My friend Conan Doyle wisely says nothing about them, but we knew they had suffered very severely indeed. Our losses were not heavy; but we had to regret the death of brave Field-Cornet Roelf Jansen and some other plucky burghers. Dr. Doyle, referring to the engagement, says:

"'The two days' prolonged struggle (Diamond Hill) showed that there was still plenty of fight in the burghers. Lord Roberts had not routed them," etc.

Thus ended the battle of Donkerhoek, and next day our commandos were falling back to the north.

		Killed		Wounded		Missing and Prisoners		
Date	Action	Officers	Men	Officers	Men	Officers	Men	Total
June	Diamond Hill	8	6	16	128	1	3	162

Casualty figures for the action from, Doyle, Arthur Conan. The Great Boer War.[80]

[79] Jan Kemp, see MOA.

[80] Doyle, Arthur Conan. The Great Boer War: Seventh Impression 1900 (September) London, Smith, Elder & Co. ludes with "Halt at Pretoria" Viljoen is dissembling or is using a prior edition that does not have the casualty figures. Figures from Pages 535 & 536 https://archive.org/details/greatboerwar00doyl/page/n7/mode/2up?q=Boer+War

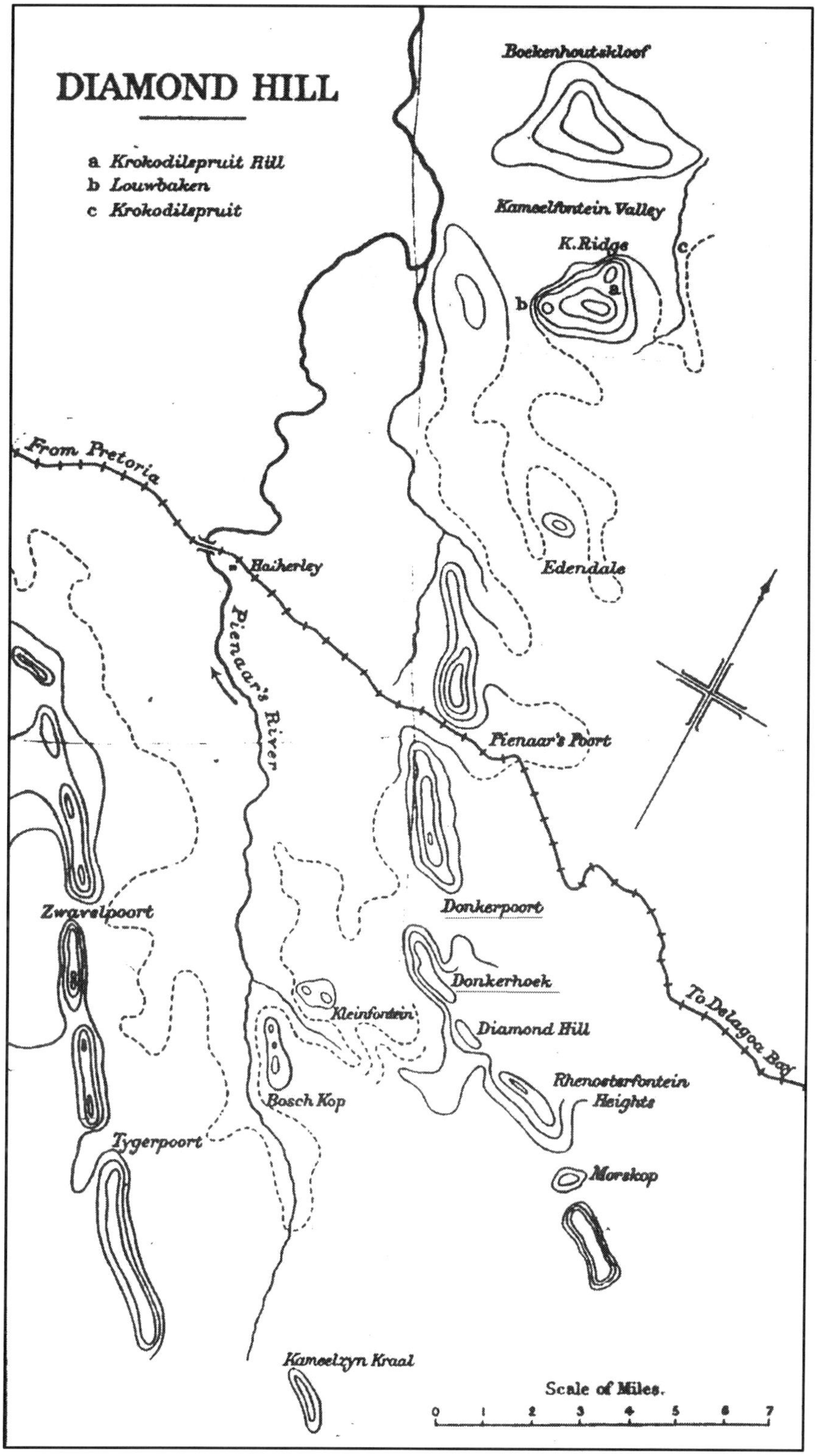

Diamond Hill Area.[179] Note Donkerhoek below Donkerspoort.[81]

[81] Weakened by the long march to Pretoria and the loss of horses and sick men, the British force mustered only 14,000, a third of whom were mounted on wobbly horses. The morning of 12 June with artillery fire from guns escorted to forward positions by a squadron of New South Wales Mounted Rifles allowed a Regular infantry advance that captured Diamond Hill. A counterattack was planned by Botha, supported with fire from Rhenosterfontein Hill. The regular Mounted Infantry from De Lisle's corps advanced to a farm, where two rapid firing pom-poms were positioned, supported by the Western Australian Mounted Infantry of Hatherley Moor. The hill was attacked by the New South Wales Mounted Rifles, who trotted across the plain in extended order, then increased to a gallop under Boer fire before they dismounted at the base of the hill. The mounted rifles advanced up the hill and charged the Boer defenders, forcing the latter to retreat. They held the hill despite Boer artillery fire, which forced Botha to call off the counterattack, as British artillery fire from the hill carried the potential to confusion with the Boer retreat. https://en.wikipedia.org/wiki/Battle_of_Diamond_Hill

Chapter 17. I Become a General.

In our retreat northwards the English did not pursue us. They contented themselves by fortifying the position we had evacuated between Donkerhoek and Wonderboompoort. Meantime our commandos proceeded along the Delagoa Bay Railway until we reached Balmoral Station, while other little divisions of ours were at Rhenosterkop, north of Bronkhorst Spruit.

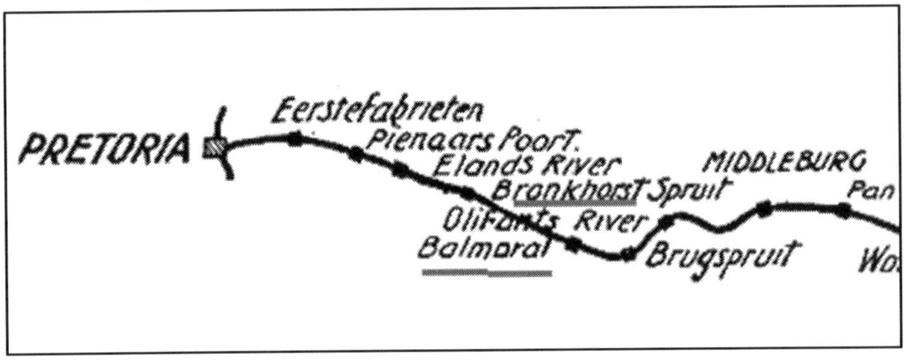

I may state that this general retreat knocked the spirit out of some of our weaker brethren. Hundreds of Boers rode into Pretoria with the white flag suspended from their Mauser barrels. In Pretoria there were many prominent burghers who had readily accepted the new conditions, and these were employed by the British to induce other Boers within reach, by manner of all sorts of specious promises, to lay down their arms. Many more western district Boers quietly returned to their homes. Luckily, the Boer loves his Mauser too well to part with it, except on compulsion, and although the majority of these western Boers handed in their weapons, some retained them.

They retained their weapons by burying them, pacifying the confiding British officer in charge of the district by handing in rusty and obsolete Martini-Henris or a venerable blunderbuss which nobody had used since ancestral Boer shot lions with it in the mediaeval days of the first great trek. The buried Mausers came in very useful afterwards.

Boers Surrendering at the government buildings, Pretoria"[180]

About this time General Buller entered the Republic from the Natal side, and marched with his force through the southern districts of Wakkerstroom, Standerton, and Ermelo. Hundreds of burghers remained on their farms and handed their weapons to the British. In some districts, for instance, at Standerton, the commandant and two out of his three field-cornets surrendered.

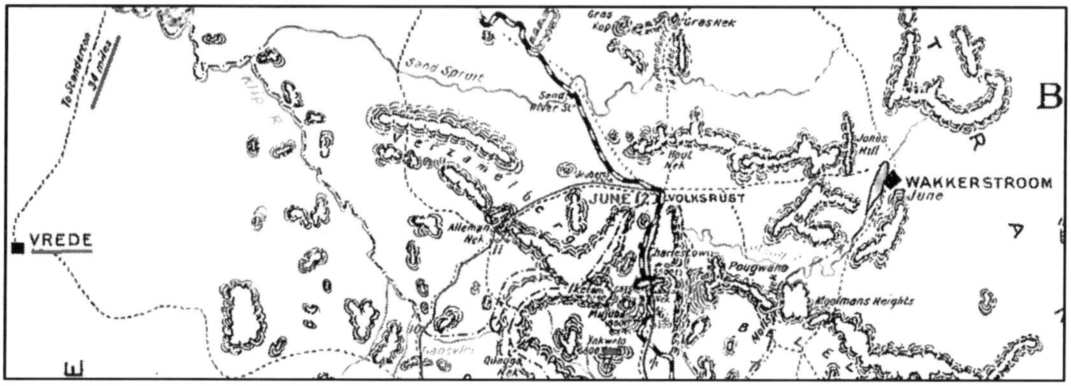

Buller in Natal. Note left side "To Standerton 34 miles" from Vrede.

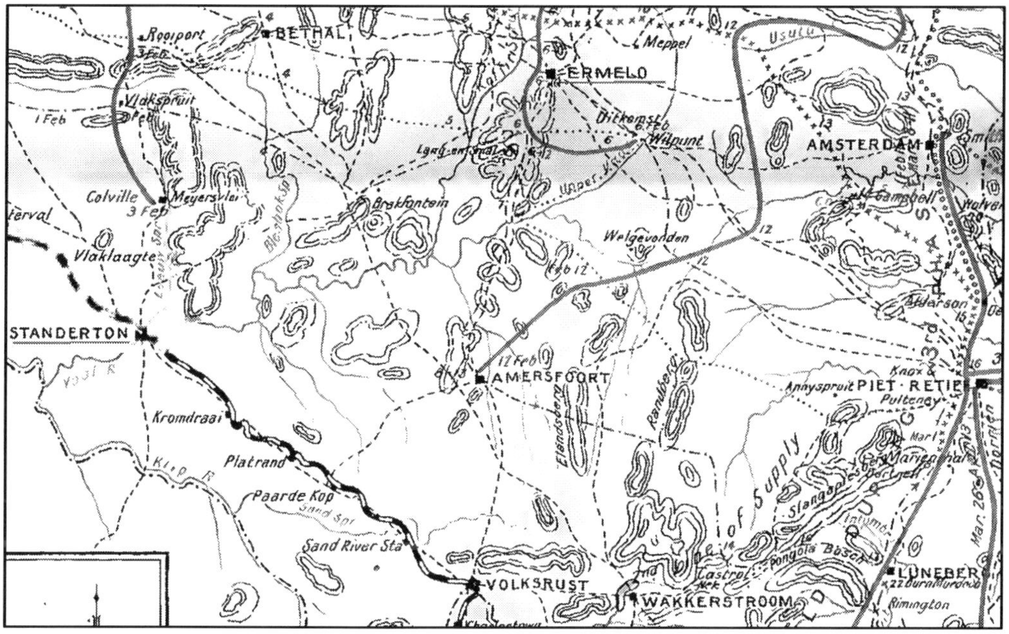

Standerton (Left) and Ermelo (top).

Thus, not only were some commandos without officers, but others entirely disappeared from our army. Still, at the psychological moment a Joshua would appear, and save the situation, as, for instance, in the Standerton district, where Assistant-Field-Cornet Brits led a forlorn hope[82] and saved a whole commando from extinction. The greatest mischief was done by many of our landdrosts, who, after having surrendered, sent out communications to officers and burghers exhorting them to come in.

The majority of our Boer officers, however, remained faithful to their vow, though since the country was partly occupied by the British it was difficult to get in touch with the Commandant-General or the Government, and the general demoralisation prevented many officers from asserting their authority.

Generals Sarel Oosthuizen[83] and H. L. Lemmer,[84] both now deceased, were sent to the north of Pretoria, to collect the burghers from the western districts, and to generally rehabilitate their commandos. They were followed by Assistant-Commandant General J. H. De la Rey and State Attorney Smuts (our legal adviser). It was at this point, indeed, that the supreme command of the western districts was assumed by General De la Rey, who, on his way to the north, attacked and defeated an English garrison at Selatsnek.

The "reorganisation" of our depleted commandos proceeded very well; about 95 per cent. of the fighting Boers rejoined, and speedily the commandos in the western districts had grown to about 7,000 men.

[82] A forlorn hope is a band of soldiers or other combatants chosen to take the vanguard in a military operation, such as a suicidal assault through the breach of a defended position, or the first men to climb a scaling ladder against a defended fortification, or a rearguard, to be expended to save a retreating army, where the risk of casualties is high. Such men were volunteers motivated by the promise of reward or promotion, or men under punishment offered pardon for their offenses, if they survived. https://en.wikipedia.org/wiki/Forlorn_hope

[83] Sarel François Oosthuizen. See MOA and: https://en.wikipedia.org/wiki/Sarel_Oosthuizen

[84] Hermanus Richard Lemmer. See MOA.

But just a few weeks after his arrival in the West Krugersdorp district, poor, plucky Sarel Oosthuizen was severely wounded in the battle of Dwarsvlei, and died of his wounds some time after.[85]

General H. Lemmer, a promising soldier, whom we could ill spare, was killed soon after while storming Lichtenburg under General De la Rey, an engagement in which we did not succeed. We had much trouble in replacing these two brave generals, whose names will live for all time in the history of the Boer Republics.

It is hardly necessary to dwell on the splendid work done by Assistant-Commandant-General De la Rey in the western districts. Commandant-General Botha was also hard worked at this stage, and was severely taxed reorganising his commandos and filling up the lamentable vacancies caused by the deaths of Lemmer and Oosthuizen.

I have already pointed out that General De la Rey had taken with him the remainder of the burghers from the western districts. The following commandos were now left to us:—Krugersdorp and Germiston, respectively, under the then Commandants J. Kemp and C. Gravett, and the Johannesburg police,[181] with some smaller commandos under the four fighting generals, Douthwaith, Snyman (of Mafeking fame), Liebenberg,[86] and Du Toit.[87] The last four generals were "sent home" and their burghers with those of Krugersdorp, Germiston, Johannesburg, Boksburg and the Mounted Police, were placed under my command, while I myself was promoted to the rank of General. I had now under me 1,200 men, all told—a very fair force.

The famous Johannesburg Police, Passing Through Johannesburg.[182 88]

[85] 7/11/1900 The battle of Dwarsvlei in the district of Krugersdorp. The Boer forces led by Gen. Sarel Oosthuizen (known as the Red Bull of Krugersdorp) attacked the English on Oosthuizen's farm. https://af.wikipedia.org/wiki/Slag_van_Dwarsvlei

[86] Probably Petrus Johannes Liebenberg. See MOA.

[87] General Sarel P. du Toit. See MOA.

[88] "The Johannesburg Police (Van Dam's Commando) [Possibly Gerard Mari Johan van Dam, See MOA] "Off to the Front," 26 October 1899. No commando behaved more gallantly In the war than the Johannesburg Police, which, with the other police and the State Artillery, formed the standing army of the country. Originally the Johannesburg Police numbered over a thousand men. After the Battle of Dalmanutha (23—27 August 1900) their strength was reduced to less than 200." Heroes of the Boer War. Review of Reviews, 1903. PF: These men are the much maligned Zarps in the British narratives. They fought bravely at Dalmanutha.

General Hendrik Synman.[183] General Snyman and Commandant Botha. (Captured at Rustenberg in June.)[89] [90]

I can hardly describe my feelings on hearing of my promotion to such a responsible position. For the first time during the War I felt a sort of trepidation. I had all sorts of misgivings; how should I be able to properly guard the interests of such a great commando? Had I a right to do so? Would the burghers be satisfied? It was all very well to say that they would have to be satisfied, but if they had shown signs of dissatisfaction I should have felt bound to resign. I am not in the habit of blinking at facts; they are stern things. What was to become of me if I had to tender my resignation? I was eager and rash, like most young officers, for although the prospects of our cause were not brilliant and our army had suffered some serious reverses, I still had implicit faith in the future, and above all, in the justice of the cause for which we were fighting. And I knew, moreover, that the burghers we now had left with us were determined and firm.

There was only one way open to me: to take the bull by the horns. I thought it my duty to go the round of all the commandos, call the burghers together, tell them I had been appointed, ask them their opinion on the appointment, and give them some particulars of the new organisation.

I went to the Krugersdorp Commando[91] first. All went well, and the burghers comprising the force received me very cordially. There was a lot of questioning and explanations; one of the commandants was so moved by my address that he requested those who were present to conclude the meeting by singing Psalm 134, verse 3,[92] after which he exhorted his fellow burghers in an impassioned speech to be obedient and determined.

The worst of it was that he asked me to wind up by offering a prayer. I felt as if I would gladly have welcomed the earth opening beneath me. I had never been in such a predicament before. To refuse, to have pleaded exoneration from this solemn duty, would have been fatal, for a Boer general is expected, amongst other things, to conduct all proceedings of a religious character. And not only Boer generals are required to do this thing, but all subordinate officers, and an officer who cannot offer a suitable prayer generally receives a hint that he is not worthy of his position. In these matters the burghers are backed up by the parsons.

There was, therefore, no help for it; I felt like a stranger in Jerusalem, and resolved to mumble a bit of a prayer as well as I could. I need not say it was short, but I doubt very much whether it was appropriate, for all sorts of thoughts passed through my head, and I felt as if all the bees in this world were buzzing about my ears. Of

[89] Possibly Commandant J.D.L. Botha. See MOA. Image from: Hillegas, Boers in War, 1900.

[90] 6/14/1900 Rustenburg occupied by General Baden Powell. Botha's rearguard surprised and "thoroughly routed" by General Ian Hamilton's Mounted Infantry. Position on Zand River attacked by 800 Boers with three guns. Enemy driven off by General Knox.

[91] See: https://en.wikipedia.org/wiki/Krugersdorp_Commando And Frederik Jacobus Potgieter in MOA and https://en.wikipedia.org/wiki/Frederik_Jacobus_Potgieter

[92] The Book of Common Prayer translation of the psalm consists of four verses:
Behold now, praise the Lord: all ye servants of the Lord;
Ye that by night stand in the house of the Lord: even in the courts of the house of our God.
Lift up your hands in the sanctuary: and praise the Lord.
The Lord that made heaven and earth: give thee blessing out of Zion.
https://en.wikipedia.org/wiki/Psalm_134

course I had to shut my eyes; I knew that. But I had, moreover, to screw them up, for I knew that everybody was watching me. I closed my eyes very tightly, and presently there came a welcome "Amen."

My old commando was now obliged to find a new commandant and I had to take leave of them in that capacity. I was pleased to find the officers and men were sorry to lose me as their commandant, but they said they were proud of the distinction that had been conferred upon me. Commandant F. Pienaar, who took my place, had soon to resign on account of some rather serious irregularities. My younger brother, W. J. Viljoen, who, at the time of writing,[93] is, I believe, still in this position, replaced him.

At the end of June my commandos marched from Balmoral to near Donkerhoek in order to get in touch with the British. Only a few outpost skirmishes took place.

My burghers captured half a score of Australians near Van der Merwe Station, and three days afterwards three Johannesburgers were surprised near Pienaarspoort. As far as our information went the Donkerhoek Kopjes were in possession of General Pole-Carew, and on our left General Hutton,[94] with a strong mounted force, was operating near Zwavelpoort and Tigerspoort. We had some sharp fighting with this force for a couple of days, and had to call in reinforcements from the Middelburg and Boksburg commandos.

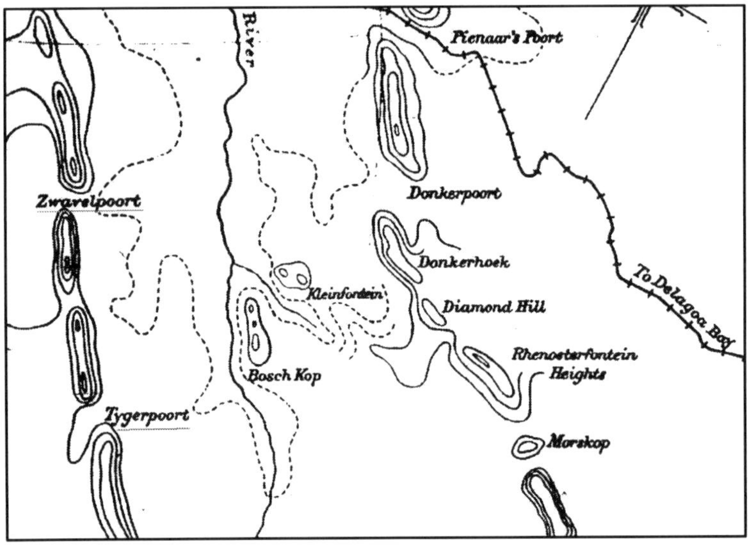

Tygerspoort and Zwavelpoort, left side, along with other familiar features.

Wynand Viljoen.[95]

The fighting line by this time had widely extended and was at least sixty miles in length; on my right I had General Daniel Erasmus with the Pretoria commando, and farther still to the right, nearer the Pietersburg railway, the Waterberg and Zoutpansberg commandos were positioned.[96]

General Pole-Carew[97] tried to rush us several times with his cavalry, but had to retire each time. Commandant-General Botha finally directed us to attack General Hutton's position, and I realised what this involved. It would

[93] PF: From Ben's internment on St. Helena in early 1902.

[94] Lieutenant-General Sir Edward Thomas Henry Hutton, (6 December 1848–4 August 1923) was a British military commander, who pioneered the use of mounted infantry in the British Army and later commanded the Canadian Militia and the Australian Army.
When the Second Boer War with South Africa was on the horizon, Hutton lobbied Canada to participate. Without informing Canada's Prime Minister, Sir Wilfrid Laurier, Hutton published mobilization plans in the Canadian Military Gazette. Laurier's government then received a letter of gratuity from Britain for its decision to aid them in the Boer region. Furious, Laurier announced Canada's refusal to send any contingents. In the end, political pressure was too great, and Canada played a role in the war.
In late March, 1900, he took up the post as Commander of the 1st Mounted Infantry Brigade, a formation made up of Canadian, Australian, and New Zealand troops. This force was actively engaged both during Lord Roberts's advance from Bloemfontein and after the fall of Pretoria. When the brigade was broken up in November 1900, he returned to the United Kingdom.
https://en.wikipedia.org/wiki/Edward_Hutton_(British_Army_officer)

[95] Commandant Wynand Viljoen, brother of General Ben Viljoen, was born in the Cape Colony on February 29, 1876 and saw the battle at Rensburg Siding, Cape Colony, on February 16, 1900, as a member of the Johannesburg Police Force. At the end of 1900 he became one of its commanders and took the place of General B. Viljoen when he was taken prisoner of war. Coming. Viljoen represented his department at the peace negotiations at the Association. See MOA.

[96] 7/11/1900 The Battle of Onderstepoort took place during which a Boer force led by General Freek [Frederick] Grobler defeated a British force. https://af.wikipedia.org/wiki/Slag_van_Onderstepoort

be the first fight I had to direct as a fighting general. Much would depend on the issue, and I fully understood that my influence with, and my prestige among, the burghers in the future was absolutely at stake.

General Hutton's main force was encamped in a "donk" at the very top of the randt, almost equidistant from Tigerspoort, Zwavelpoort and Bapsfontein. Encircling his laager was another chain of "randten" entirely occupied and fortified, and we soon realised what a large and entrenched stretch of ground it was.

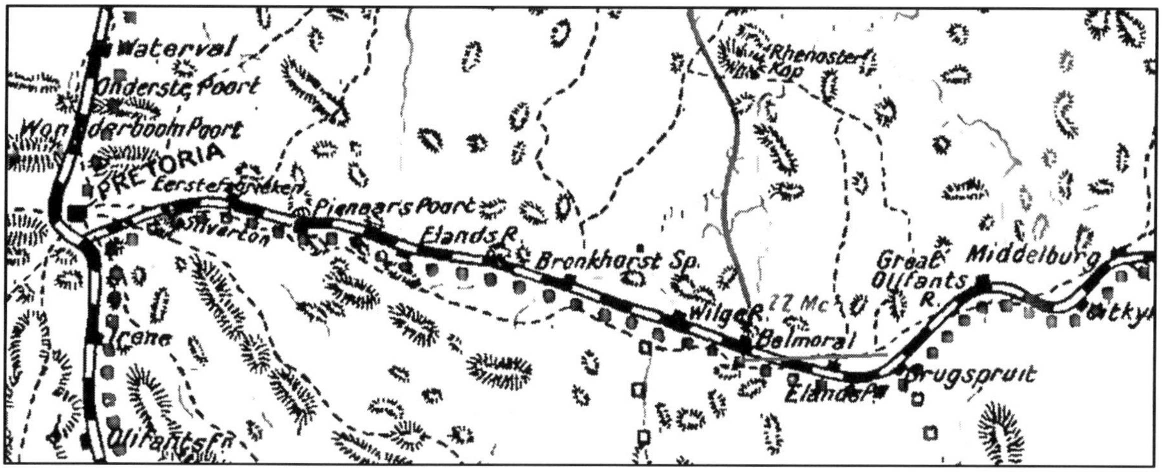

Balmoral is just East of Wilge River station.

The Commandant-General, [Louis Botha] accompanied by the French, Dutch, American and Russian attachés, would follow the attack from a high point and keep in touch with me by means of a heliograph, thus enabling Botha to keep well posted about the course of the battle, and to send instructions if required.

During the night of the 13th of July we marched in the following order: On the right were the Johannesburg and Germiston commandos; in the centre the Krugersdorp and the Johannesburg Police; and on the left the Boksburg and Middelburg commandos. At daybreak I ordered a general storming of the enemy's entrenchments. I placed a Krupp gun and a Creusot on the left flank, another Krupp and some pom-poms to the right, while I had an English 15-pounder (an Armstrong) mounted in the centre. Several positions were taken by storm with little or no fighting. It was my right flank which met with the only stubborn resistance from a strongly fortified point occupied by a company of Australians.

Soon after this position was in our possession, and we had taken 32 prisoners, with a captain and a lieutenant. When Commandant Gravett had taken the first trenches we were stubbornly opposed in a position defended by the Irish Fusiliers, who were fighting with great determination. Our burghers charged right into the trenches; and a hand-to-hand combat ensued. The butt-ends of the guns were freely used, and lumps of rock were thrown about. We made a few prisoners and took a pom-pom, which, to my deep regret, on reinforcements with guns coming up to the enemy, we had to abandon, with a loss of five men. Meanwhile, the Krugersdorpers and Johannesburg Police had succeeded in occupying other positions and making several prisoners, while half a dozen dead and wounded were left on the field.

The ground was so exposed that my left wing could not storm the enemy's main force, especially as his outposts had noticed our march before sunrise and had brought up a battery of guns, and in this flat field a charge would have cost too many lives.

We landed several shells into the enemy's laager, and if we had been able to get nearer he would certainly have been compelled to run.

[97] Reginald Pole-Carew, KCB, CVO (1 May 1849–19 September 1924) was a British Army officer He served as a staff officer with Lord Roberts in the Second Anglo-Afghan War in 1878 and again served with him during the Second Boer War of 1899–1902. He was promoted to the substantive rank of colonel on 27 November 1899, and at the same time appointed in command of the 9th Brigade with the local rank of major-general. As such he was in command of the brigade during the Battle of Modder River on 28 November 1899. In February 1900 he was appointed in command of the Guards Brigade, shortly before the Relief of Kimberley. He later commanded the 11th Division of the South Africa Field Force https://en.wikipedia.org/wiki/Reginald_Pole-Carew_(British_Army_officer)

CAPT. REICHMAN (United States of America). COL. GURKO (Russia). MR. FISCHER (Orange Free State). LIEUT. DUVAL (France).

MILITARY *ATTACHÉS* AND THEIR ADJUTANT, MR. FISCHER. It was difficult for the *attachés* to keep up with the Boer Commandoes, because of the latter's mobility; dangerous too, as was proved by the loss of the Netherlands *attaché*, Lieut. Nix, who was mortally wounded at Sanna's Post (1 April 1900). Bnt they were always eager to be present at the interesting Boer incursions and enthusiastic in their praises of Boer tactics.

Foreign military attaches. Captain Carl Reichman, U.S.A. is on the left.[184 185]

General de La Rey and Colonel Guorko, Russian military attaché.[186]

When darkness supervened we retired to our base with a loss of two killed and seven wounded; whereas 45 prisoners and 20 horses with saddles and accoutrements were evidence that we had inflicted a severe loss upon the enemy. So far as I know, the Commandant-General was satisfied with my work. On the day after the fight I met an attaché. He spoke in French, of which language I know nothing. My Gallic friend then tried to get on in English, and congratulated me in the following terms with the result of the fight: "I congratuly very much you, le Général; we think you good man of war." It was the first time I had bulked in anyone's opinion as largely as a battleship; but I suppose his intentions were good enough.[98 99]

Battle of Witpoort, South African War Memorial (Halifax), Nova Scotia. [187]

[98] 7/16/1900 Determined attacks by Boers on left flank of British posts in the Pretoria district. Enemy driven off with loss.

[99] The battle opened at 06.45 with heavy shelling on Major Munn's troops. Renowned Boer fighter Roland Schikkerling and his comrades went to the north of the New Zealanders' ridge and then charged under heavy rifle fire. The New Zealanders surrendered one of the three hills they occupied, the Boers were in possession of the higher hill which commanded the lower middle hill. Schikkerling captured a captain and twenty New Zealanders and some continued right over the ridge and captured a number of horses. (Boer Willem Morkel du Toit died in the charge.)
The Canadians mounted a counter-attack. Colonel Edwin Alderson sent two squadrons of the Canadians to assist 'the Irish on the kopje which had been vacated by the New Zealanders' and, with the fire from the guns, the position was regained. Lieutenants Borden and John Edgar Burch of 'B' Squadron led a counter-attack. They were successful, but at the cost of their lives. Boer marksmen less than 200 yards distant shot them as they stood up to lead the rifles forward. Lord Roberts reported to the War Office that Borden and Burch "were killed while gallantly leading their men in a counter attack upon the enemy's flank at a critical juncture of his assault upon our position." At 14.00, British officer Edward Hutton moved all of his available troops onto the desperately regained positions and, by sundown, the battle of Witpoort had ended. His losses were seven killed, with 30 wounded. Two officers and 22 soldiers had been taken prisoner. The Canadian losses were heavy and included Lt Borden, the son of the Canadian Minister of Defence. He is buried at the Braamfontein Cemetery, not far from the grave of Willem Morkel du Toit.
Another famous Canadian casualty was Trooper Lorne Mulloy who was blinded in both eyes and awarded the Distinguished Conduct Medal. See also: https://en.wikipedia.org/wiki/Harold_Lothrop_Borden

A few days afterwards Lord Roberts sent a hundred women and children down the line to Van der Merwe Station, despite Botha's vehement protests.

It fell to my lot to receive these unfortunates, and to send them on by rail to Barberton, where they could find a home. I shall not go into a question which is still sub judice;[100] nor is it my present purpose to discuss the fairness and unfairness of the war methods employed against us. I leave that to abler men. I shall only add that these waifs were in a pitiful position, as they had been driven from their homes and stripped of pretty nearly everything they possessed.

"The above picture represents a Train-load of the Wives of Boers who are still in Arms against us, and who have been Deported from the Towns we Occupy and sent by us to Rejoin their Belligerent Husbands. The Women are, apparently, as Hostile as the Men, and can Display their Rancour without Fear. It will be noticed that one has hoisted the Transvaal Boer Flag."[188]

Towards the end of July [General Frederick] Carrington[101] marched his force to Rustenburg, and thence past Wonderboompoort, while another force proceeded from Olifantsfontein in the direction of Witbank Station. We were, therefore, threatened on both sides and obliged to fall back on Machadodorp.[102]

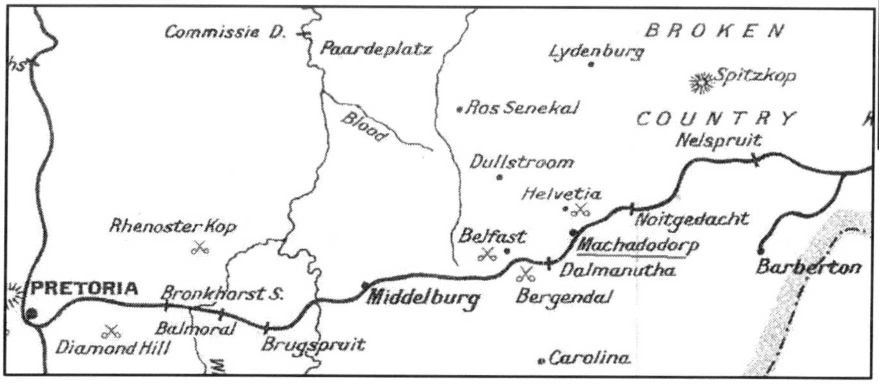

Machadodorp and engagements identified.

Image: The Fight with General Hutton at Olifantsfontein.

[100] Under consideration; being presently considered, litigated, or adjudicated.

[101] Major-General Sir Frederick Carrington, KCB, KCMG (23 August 1844–22 March 1913) was a British Army officer best known for his role in the Second Matabele War. https://en.wikipedia.org/wiki/Frederick_Carrington

[102] 7/21/1900 Advance begun from Pretoria east, along Delagoa Bay Railway. A supply train, with 100 Welsh Fusiliers, captured near Honing Spruit.

7/25/1900 Roberts's force reached Balmoral on the way to Middelburg. French's Cavalry and Hutton's Mounted Infantry put Boers to flight six miles south of Balmoral. General French crosses Oliphant's River.

Chapter 18. Our Camp Burned Out.

The beginning of August saw my commandos falling back on Machadodorp. Those of Erasmus and Grobler remained where they were for the time being, until the latter was discharged for some reason or other and replaced by Attorney Beyers.[103] General [Daniel J. E.] Erasmus suffered rather worse, for he was deprived of his rank as a general and reduced to the level of a commandant on account of want of activity.

Our retreat to Machadodorp was very much like previous experiences of the kind; we were continually expecting to be cut off from the railway by flanking movements and this we had to prevent because we had placed one of our big guns on the rails in an armour-clad railway carriage. The enemy took care to keep out of rifle range, and the big gun was an element of strength we could ill afford to lose. Besides, our Government were now moving about on the railway line near Machadodorp, and we had to check the enemy at all hazards from stealing a march on us. Both at Witbank Station and near Middelburg and Pan Stations we had skirmishes, but not important enough to describe in detail.

After several unsuccessful attempts, the Boer Artillery at last managed to fire the big gun without a platform. It was tedious work, however, as "Long Tom" was exceedingly heavy, and it usually took twenty men to serve it. The mouth was raised from the "kastion" by means of a pulley, and the former taken away; then and not till then could the gunner properly get the range. The carriage vacuum sucking apparatus had to be well fixed in hard ground to prevent recoil. The enemy repeatedly sent a mounted squad to try and take this gun, and then there was hard fighting.

Long Tom on rail carriage.[189]

Boer Camp with Creusot.[190]

One day while we were manoeuvring with the "Long Tom," the veldt burst into flames, and the wind swept them along in our direction like lightning. Near the gun were some loads of shells and gunpowder, and we had to set all hands at work to save them. While we were doing this the enemy fired two pom-poms at us from about 3,000 yards, vastly to our inconvenience.

Various types of ammunition used in the war.[191]

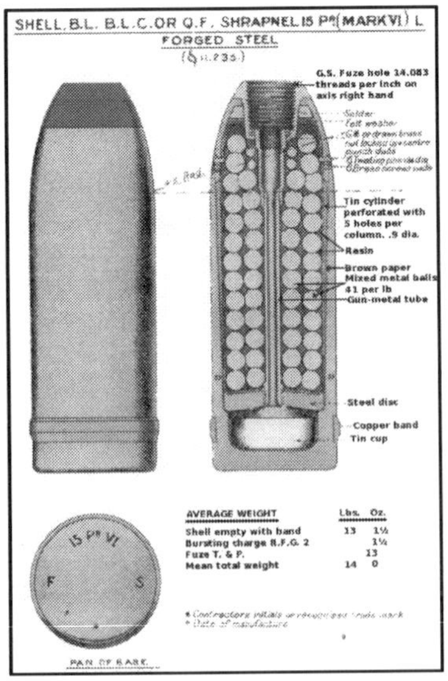

Diagram of Construction of 15 lb. Shrapnel shell.[192]

[103] Christiaan F. Beyers. See MOA and https://en.wikipedia.org/wiki/Christiaan_Frederik_Beyers

As my commando formed a sort of centre for the remainder, Commandant-General Botha was, as a rule, in our immediate neighbourhood, which made my task much easier, our generalissimo taking the command in person on several occasions, if required, and assisting in every possible way.

The enemy pursued us right up to Wonderfontein Station (the first station south-west of Belfast), about 15 miles from Dalmanutha or Bergendal, and waited there for Buller's army to arrive from the Natal frontier.

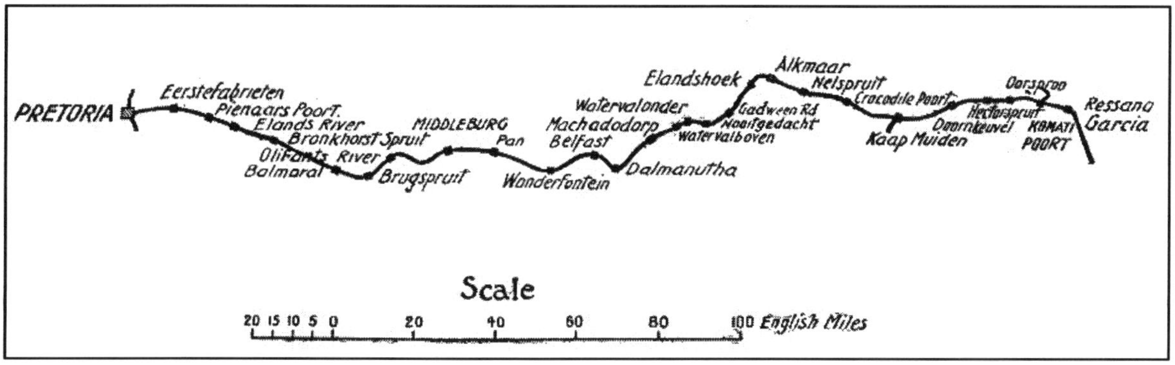

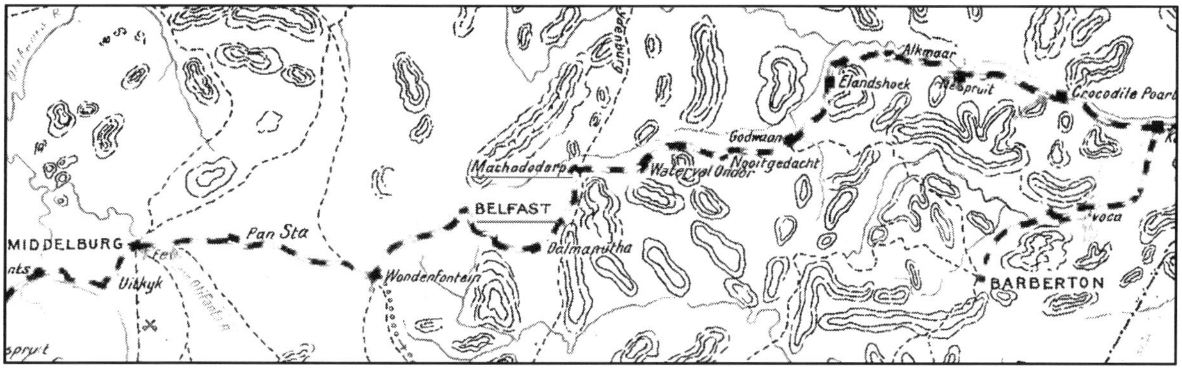

We occupied the "randten" between Belfast and Machadodorp, and waited events. While we were resting there Lord Roberts sent us 250 families from Pretoria and Johannesburg in open trucks, notwithstanding the bitterly cold weather and the continual gusts of wind and snow. One can picture to oneself the deplorable condition we found these women and children in.

But, with all this misery, we still found them full of enthusiasm, especially when the trucks in which they had to be sent on down the line were covered with Transvaal and Free State flags. They sang our National Anthem as if they had not a care in the world.

Many burghers found their families amongst these exiles, and some heartrending scenes were witnessed. Luckily the railway to Barberton was still in our possession, and at Belfast the families were taken over from the British authorities, to be sent to Barberton direct. While this was being done near Belfast under my direction, the unpleasant news came that our camp was entirely destroyed by a grass fire.

The Commandant-General and myself had set up our camp near Dalmanutha Station. It consisted of twelve tents and six carts. This was Botha's headquarters, as well as of his staff and mine. When we came to the spot that night we found everything burned save the iron tyres of the waggon wheels, so that the clothes we had on were all we had left us. All my notes had perished, as well as other documents of value. I was thus deprived of the few indispensable things which had remained to me, for at Elandslaagte my "kit" had also fallen into the hands of the British.

Louis Botha and staff at Belfast, August, 1900. Botha stands in the centre of the group [with a Mauser C96 pistol on his left hip.] Next [to] him stands his adjutant, handing a letter to the Rev. P. McQueen, for the Boston Globe."[193]

The grass had been set on fire by a kaffir to the windward of the camp. The wind had turned everything into a sea of fire in less than no time, and the attempts at stamping out the flames had been of no avail. One man gave us a cart, another a tent; and the harbour at Delagoa Bay being still open (although the Portuguese had become far from friendly towards us after the recent British victories) we managed to get the more urgent things we wanted. Within a few days we had established a sort of small camp near to headquarters.

We had plenty to do at this time—building fortresses and digging trenches for the guns. This of course ought to have been done when we were still at Donkerhoek by officers the Commandant-General had sent to Machadodorp for the purpose. We had made forts for our "Long Toms," which were so well hidden from view behind a rand that the enemy had not discovered them, although a tunnel would have been necessary in order to enable us to use them in shelling the enemy. We were therefore obliged to set to work again, and the old trenches were abandoned. The holes may surprise our posterity, by the way, as a display of the splendid architectural abilities of their ancestors.

Long Tom Pass and Monument.[194] [104]

[104] Long Tom Pass is a mountain pass on the Great Escarpment situated in the Mpumalanga province, on the R37 regional route between Lydenburg and Sabie (South Africa). https://en.wikipedia.org/wiki/Long_Tom_Pass

Chapter 19. Battle of Bergendal (Machadodorp) [Dalmanutha or Belfast].

Let us pass on to the 21st of August, 1900.[105] Buller's army had by this time effected a junction with that of Lord Roberts' between Wonderfontein and Komati River. The commandos under Generals Piet Viljoen[106] and Joachim Fourie had now joined us, and taken up a position on our left, from Rooikraal to Komati Bridge. The enemy's numbers were estimated at 60,000, with about 130 guns, including twelve 4·7 naval guns, in addition to the necessary Maxims.

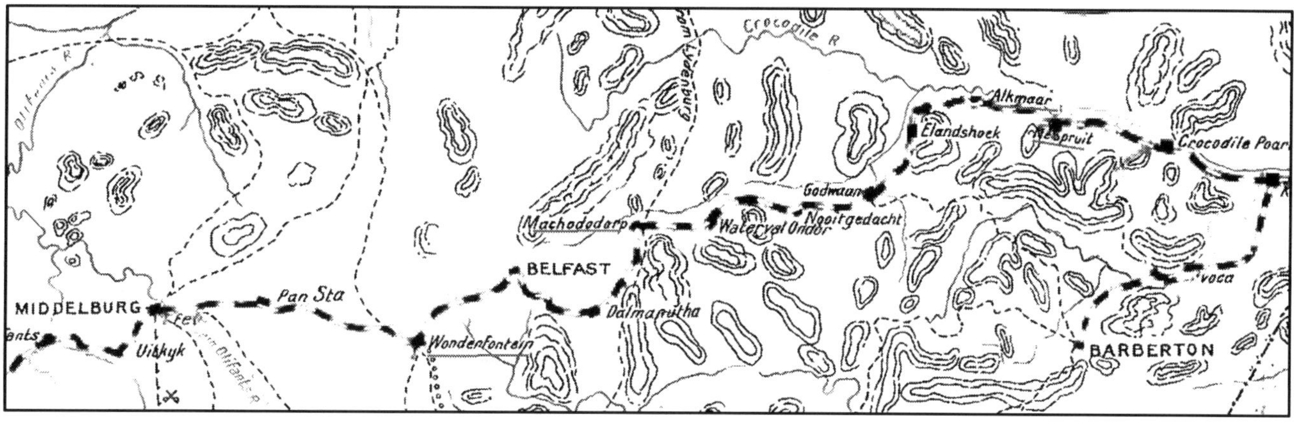

Wonderfontein west of Belfast, Machadodorp to the north and the retreat to Nelspruit east of Alkmaar.

We had about 4,000 men at the most with six Maxims and about thirteen guns of various sizes. Our extreme left was first attacked by the enemy while they took possession of Belfast and Monument Hill, a little eastward, thereby threatening the whole of our fighting lines. My commandos were stationed to the right and left of the railway and partly round Monument Hill. Fighting had been going on at intervals all day long, between my burghers and the enemy's outposts. The fighting on our left wing lasted till late in the afternoon, when the enemy was repulsed with heavy losses; while a company of infantry which had pushed on too far during the fighting, through some misunderstanding or something of that sort, were cut off and captured by the Bethel burghers.

The attack was renewed the next morning, [22nd] several positions being assailed in turn, while an uninterrupted gunfire was kept up. General Duller[107] was commanding the enemy's right flank and General French was in charge of the left. We were able to resist all attacks and the battle went on for six days without a decisive result. The enemy had tried to break through nearly every weak point in our fighting line and found out that the key to all our positions existed in a prominent "randje" to the right of the railway. This point was being defended by our brave Johannesburg police, while on the right were the Krugersdorpers and Johannesburgers and to the left the burghers from Germiston. Thus we had another "Spion Kop" fight for six long days. The Boers held their ground with determination, and many charges were repulsed by the burghers with great bravery. But the English were not to be discouraged by the loss of many valiant soldiers and any failure to dislodge the Boers from the "klip-kopjes." They were admirably resolute; but then they were backed up by a superior force of soldiers and artillery.

On the morning of the 27th of August the enemy were obviously bent on concentrating their main force on this "randje." There were naval guns shelling it from different directions, while batteries of field-pieces pounded away incessantly. The "randje" was enveloped by a cloud of smoke and dust. The British Infantry charged under cover of the guns, but the Police and burghers made a brave resistance. The booming of cannon went on without

[105] 7/26/1900 Philip de Wet, younger brother of Christian de Wet, surrendered at Kroonstad. General Hunter occupied Fouriesburg. General MacDonald, after fighting a rearguard action, blocked Naauwpoort Nek.
7/27/1900 Occupation of Middelburg by advance guard of Roberts without opposition.
7/30/1900 Surrender of Generals Prinsloo, A. J. Villiers, and Crowther, and 4000 Boers to General Hunter.
8/4/1900 Surrender of Harrismith to General MacDonald.
8/10/1900 Discovery of the plot at Pretoria to kidnap Roberts and the British officers. Pursuit of De Wet continued.
8/16/1900 Eland's River garrison relieved. See: https://en.wikipedia.org/wiki/Battle_of_Elands_River_(1900)
8/24/1900 Roberts left for the front in the Eastern Transvaal to operate against General Botha.
8/25/1900 Lieutenant Hans Cordua shot in Pretoria for his participation in the plot against Roberts.
[106] See MOA.
[107] A nickname for General Buller.

intermission, and the storming was repeated by regiment upon regiment. Our gallant Lieutenant Pohlman was killed in this action, and Commandant Philip Oosthuizen was wounded while fighting manfully against overwhelming odds at the head of his burghers. An hour before sunset the position fell into the hands of the enemy. Our loss was heavy—two officers, 18 men killed or wounded, and 20 missing.[108]

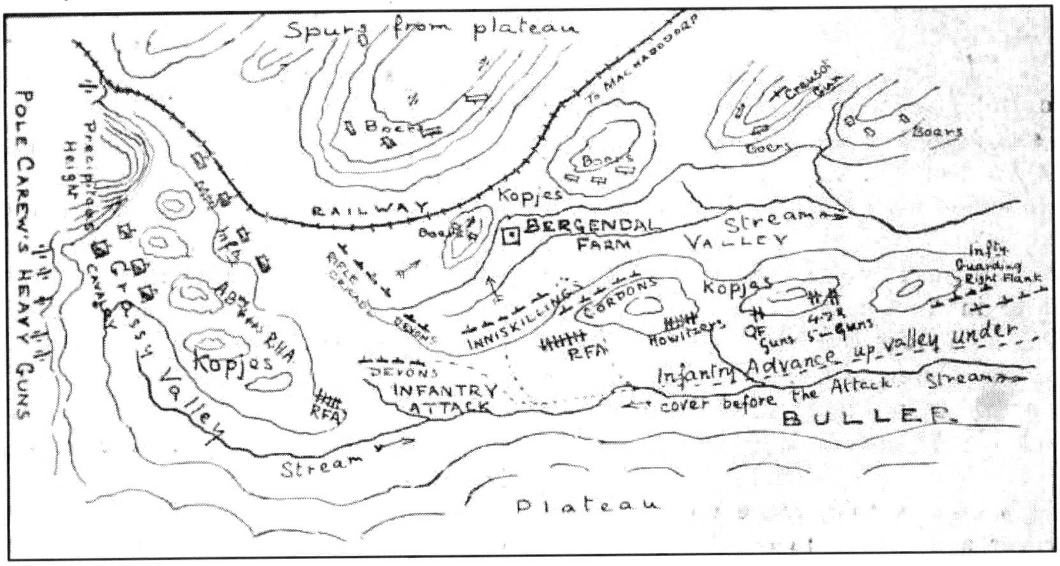

Map of the battle of Bergendal.[195 109]

Thus ended one of the fiercest fights of the war. With the exception of the battle of Vaalkrantz (on the Tugela) our commandos had been exposed to the heaviest and most persistent bombardment they had yet experienced. It was by directing an uninterrupted rifle fire from all sides on the lost "randje" that we kept the enemy employed and prevented them from pushing on any farther that evening.

At last came the final order for all to retire via Machadodorp.

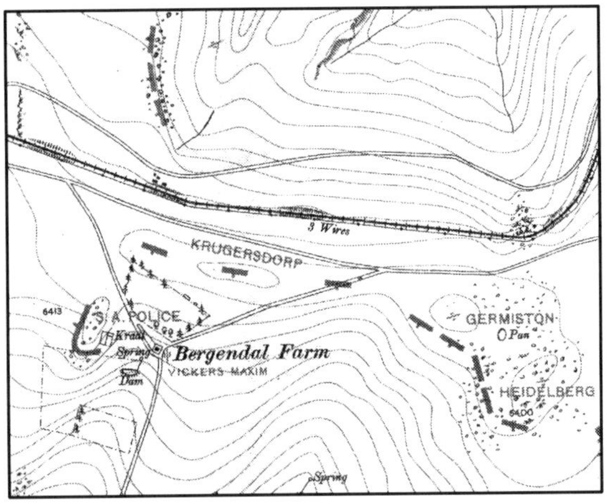

Position of Boers during the battle of Bergendal on 27 August 1900.[196] Bergendal Monuments.[197 110] UK 19,000 men, 82 guns. Boer Forces 7,000 men, 20 guns. Casualties and losses. UK 385 Boers 78.

[108] On August 27th, Buller himself went scouting early that morning. The Boers' right flank did not actually end at the railway line, but it did swing sharply clockwise. It thus formed a salient angle. The result was that their position on the farm Bergendal (occupied by the ZARPs) stuck out like the tip of one's nose or a protuberance. On both sides of this position the Boers' sangars were facing away from the tip, with the result that those occupying them could not assist those on the tip in the case of a frontal attack. Then there was some high ground some 3,000 yards south-west of the salient angle and that was an ideal place for the artillery he wanted to use.
About 36 guns were placed on the high ground and the bombardment of the ZARPs position started at 11:00. After three hours the Rifle Brigade and the Inniskilling Fusiliers charged and burst right through the Boer lines. When the fighting stopped they found that the position had been defended by the Zuid-Afrikaanse Republiek Politie. The British found 14 dead bodies and captured nineteen prisoners. The rest of the wounded were taken away by the Boers and some even survived. British casualties for the day were: officers, 3 dead and 7 wounded; men, 100 wounded or missing and 12 dead.

[109] The Battle of Bergendal (also known as the Battle of Belfast or Battle of Dalmanutha) took place from August 21st to the 27th, 1900. The battle was the last set-piece battle of the war, although the war was still to last another two years. It was also the last time that the Boers' four 155 mm Creusot Long Tom guns were used in the same battle. https://en.wikipedia.org/wiki/Battle_of_Bergendal

Chapter 20. Two Thousand British Prisoners Released.

After the battle of Bergendal there was another retreat. Our Government, which had fled from Machadodorp to Waterval Station, had now reached Nelspruit, three stations further down the line, still "attended," shall I say, by a group of Boer officials and members of the Volksraad, who preferred the shelter of Mr. Kruger's fugitive skirts to any active fighting.

President Kruger's private car used as the capital after evacuation of Pretoria.[198]

There were also hovering about this party half a dozen Hebraic persons of extremely questionable character, one of whom had secured a contract for smuggling in clothes from Delagoa Bay; and another one to supply coffee and sugar to the commandos. As a rule, some official or other made a nice little commission out of these transactions, and many burghers and officers expressed their displeasure and disgust at these matters; but so it was, and so it remained. That same night we marched from Machadodorp [North] to Helvetia, where we halted while a commando was appointed to guard the railway at Waterval Boven.

The next morning a big cloud of dust arose. "De Engelse kom" (the English are coming) was the cry. And come they did, in overwhelming numbers. We fired our cannon at their advance guard, which had already passed Machadodorp: but the British main force stayed there for the day, and a little outpost skirmishing of no consequence occurred.

A portion of the British forces appeared to go from Belfast via Dullstroom to Lydenburg, these operations being only feebly resisted. Our commandos were now parcelled out by the Commandant-General, who followed a path over the Crocodile River bridge with his own section, which was pursued by a strong force of Buller's.

I was ordered to go down the mountain in charge of a number of Helvetia burghers to try and reach the railway, which I was to defend at all hazards. General Smuts, with the remnant of our men went further south towards the road leading to Barberton. Early the next morning we were attacked and again obliged to fall back. That night we stayed at Nooitgedacht.

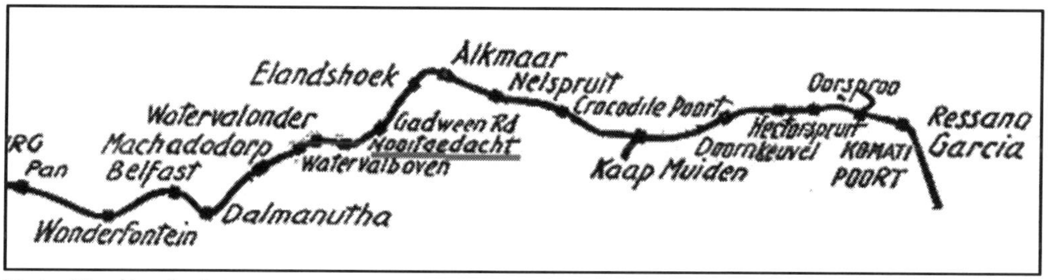

Note Nooitgedacht.

[110] A memorial was erected in 1935 in memory of the twelve Zarps who died in the Battle of Berg-en-dal on 27 Augustus 1900. A much larger second memorial, in honour of all Boers, including the Zarps, who had died in the Eastern Transvaal, was unveiled on 29 August 1970 and a small stone cairn was erected in memory of the 25 officers and men of the 2nd Battalion the Rifle Brigade who died in the battle.

Waterval-Boven, circa 1901. Armoured train during Anglo-Boer War at railway station.[199] Cog Wheel Railway[200] from Waterval Boven to Waterval Onder the line drops at a gradient so steep that the engines used a cog-wheel rack railroad.[201]

Remains of Boer big gun, Waterval

Remains of Boer big gun, Watervaal.[202]

The Boer position at and near Nooitgedacht was unique. Here was a great camp in which 2,000 English prisoners-of-war were confined, but in the confusion the majority of their Boer guards had fled to Nelspruit. I found only 15 burghers armed with Martini-Henry rifles left to look after 2,000 prisoners. Save for "Tommy" being such a helpless individual when he has nobody to give him orders and to think for him, these 2,000 men might have become a great source of danger to us had they had the sense to disarm their fifteen custodians (and what was there to prevent them doing so?) and to destroy the railway, they would have been able not only to have deprived my commando of provisions and ammunition, but also to have captured a "Long Tom." There was, moreover, a large quantity of victuals, rifles, and ammunition lying about the station, of which nobody appeared to take any notice. Of the crowd of officials who stuck so very faithfully to the fugitive Government there was not one who took the trouble to look after these stores and munitions.

On arrival I telegraphed to the Government to enquire what was to be done with the British prisoners-of-war. The answer was: "You had better let them be where they are until the enemy force you to evacuate, when you will leave them plenty of food."

This meant that there would be more D.S.O's or V.C's handed out, for the first "Tommies" to arrive at the prisoners' camp would be hailed as deliverers, and half of them would be certain of distinctions.

I was also extremely dissatisfied with the way the prisoners had been lodged, and so would any officer in our fighting line have been had he seen their condition and accommodation. But those who have never been in a fight and who had only performed the "heroic" duty of guarding prisoners-of-war, did not know what humanity meant to an enemy who had fallen into their hands.

So what was I to do?

To disobey the Government's orders was impossible. I accordingly resolved to notify the prisoners that, "for military reasons," it would be impossible to keep them in confinement any longer.

The next morning I mustered them outside the camp, and they were told that they had ceased to be prisoners-of-war, at which they seemed to be very much amazed. I was obliged to go and speak formally to some of them; they could scarcely credit that they were free men and could go back to their own people.

It was really pleasant to hear them cheer, and to see how pleased they were. A great crowd of them positively mobbed me to shake hands with them, crying, "Thank you, sir; God bless you, sir." One of their senior officers was ordered to take charge of them, while a white-flag message was sent to General Pole-Carew to send for these fine fellows restored to freedom, and to despatch an ambulance for the sick and wounded. My messenger, however, did not succeed in delivering the letter, as the scouts of the British advance-guard were exceedingly drunk, and shot at him; so that the prisoners-of-war had to go out and introduce themselves. I believe they were compelled to overpower their own scouts.

Ten days afterwards an English doctor and a lieutenant of the 17th Lancers came to us, bringing a mule laden with medical appliances and food. The English medico, Dr. Ailward, succeeded, moreover, in getting through our lines without my express permission.

Next morning I accompanied an ambulance train to transport the wounded British to the charge of the British agent at Delagoa Bay.

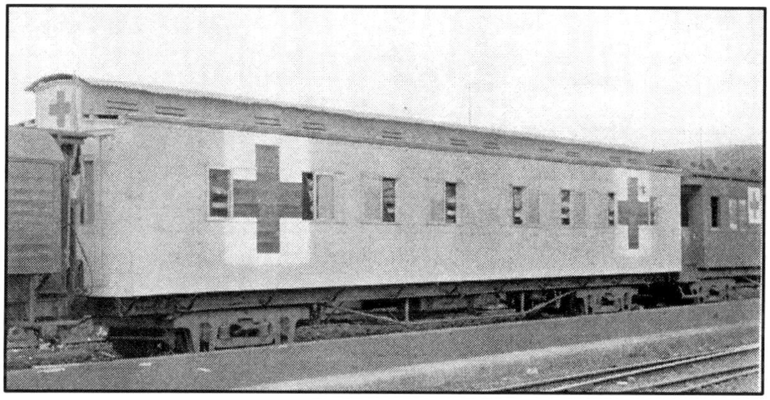

Boer Red Cross Car.[203] Interior of a Boer hospital train.[204]

Outside Nooitgedacht I found four military doctors with a field ambulance.

"Does this officer belong to the Red Cross?" I asked.

"No," was the answer, "he is only with us quite unofficially as a sympathetic friend."

"I regret," said I, "that I cannot allow this thing; you have come through our lines without my permission; this officer no doubt is a spy."

I wired at once for instructions, which, when received read: "That as a protest against the action of the English officers who stopped three of our ambulances, and since this officer has passed through our lines without permission, you are to stop the ambulance and dispatch the doctors and their staff, as well as the wounded to Lourenco Marques."

The doctors were very angry and protested vehemently against the order, which, however, was irrevocable. And thus the whole party, including the Lancers' doctor, were sent to Lourenco Marques that very day. The nearest English General was informed of the whole incident, and he sent a very unpleasant message the next day, of which I remember the following phrases:—

"The action which you have taken in this matter is contrary to the rules of civilised warfare, and will alter entirely the conditions upon which the War was carried on up to the present," etc.

After I had sent my first note we found, on inspection, some Lee-Metford cartridges and an unexploded bomb in the ambulance vans. This fact alone would have justified the retention of the ambulance.

This was intimated again in our reply to General Pole-Carew, and I wrote, inter alia: "Re the threat contained in your letter of the ... I may say I am sorry to find such a remark coming from your side, and I can assure you that whatever may happen my Government, commandants, and burghers are firmly resolved to continue the War on our side in the same civilised and humane manner as it has hitherto been conducted."

This was the end of our correspondence in regard to this subject, and nothing further happened, save that the English very shortly afterwards recovered five out of the eight ambulances we had retained.

Ambulance presented by the Women of Queensland for South African Service, 1900.[205] Ambulance Wagon.[206]

Chapter 21. A Government in Flight.

About this time President Steyn arrived from the Orange Free State and had joined President Kruger, and the plan of campaign for the future was schemed. It was also decided that Mr. Schalk Burger should assume the acting Presidentship, [of the Transvaal] since Mr. Kruger's advanced age and feeble health did not permit his risking the hardships attendant on a warlike life on the veldt.

It was decided Mr. Kruger should go to Europe and Messrs. Steyn and Burger should move about with their respective commandos. They were younger men and the railway, would soon have to be abandoned.

General S. W. Burger.[111]

General Schalk W. Burger.[207]

[111] See MOA, and: https://en.wikipedia.org/wiki/Schalk_Willem_Burger Following the end of the war, Burger visited Europe, including The Hague to raise funds for relief. In 1905, Burger became a founding a member of the Het Volk Party and secured an uncontested seat on the Legislative Council. Burger was a delegate of the Transvaal to the National Convention in 1908 which lead to the forming of the Union of South Africa in 1910. After the Union of South Africa was established, Burger became a member of the South African Party under General Louis Botha and rose to Chairman of the Transvaal branch of the South African Party. In 1911, Burger became a member of the Transvaal Executive Committee and following his election to the Senate in 1913, Burger continued to serve as a senator until his death on 5 December 1918.

We spent the first weeks of September at Godwan River and Nooitgedacht Station, near the Delagoa Bay railway, and had a fairly quiet time of it. General Buller had meanwhile pushed on with his forces via Lydenburg in the direction of Spitskop and the Sabi, on which General Botha had been compelled to concentrate himself after falling back, fighting steadily, while General French threatened Barberton.

I had expected Pole-Carew to force me off the railway line along which we held some rather strong positions, and I intended to offer a stout resistance. But the English general left me severely alone, went over Dwaalheuvel by an abandoned wagon-track, and crossed the plateau of the mountains, probably to try and cut us off through the pass near Duivelskantoor. I tried hard, with the aid of 150 burghers, to thwart his plans and we had some fighting. But the locality was against us, and the enemy with their great force of infantry and with the help of their guns forced us to retire.

About the 11th of September I was ordered to fall back along the railway, via Duivelskantoor and Nelspruit Station, since General Buller was threatening Nelspruit in the direction of Spitskop, while General French, with a great force, was nearing Barberton. It appeared extremely likely that we should be surrounded very soon. We marched through the Godwan River and over the colossal mountain near Duivelskantoor, destroying the railway bridges behind us. The road we followed was swamped by the heavy rains and nearly impassable. Carts were continually being upset, breakdowns were frequent, and our guns often stuck in the swampy ground. To make matters worse, a burgher on horseback arrived about midnight to tell us that Buller's column had taken Nelspruit Station, and cut off our means of retreat. Yet we had to pass Nelspruit; there was no help for it. I gave instructions for the waggons and carts (numbering over a hundred), to push on as quickly as possible, and sent out a strong mounted advance guard to escort them.

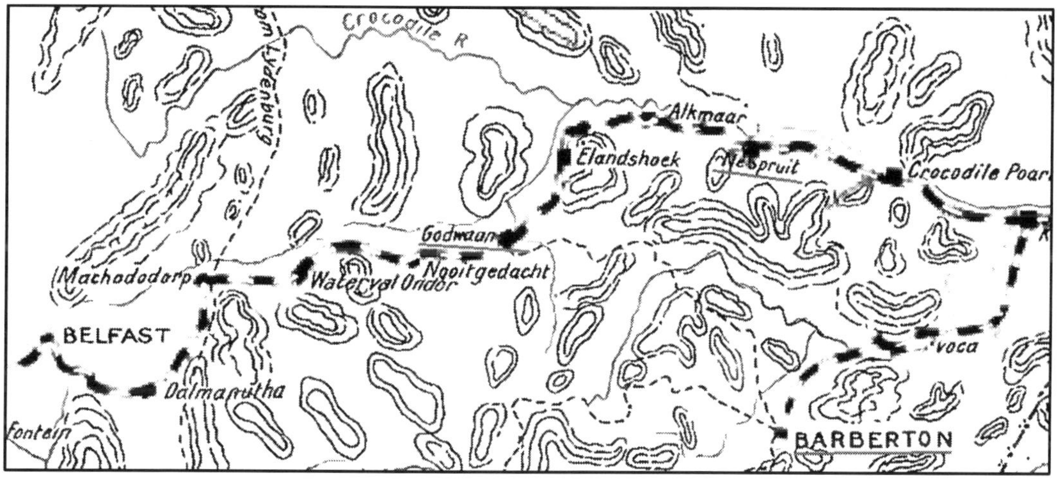

Godwan, Nelspruit and Baberton to the south.

I myself went out scouting with some burghers, for I wanted to find out before daybreak whether Nelspruit was really in the hands of the enemy or not. In that case our carts and guns would have to be destroyed or hidden, while the commando would have to escape along the footpaths. We crept up to the station, and just at dawn, when we were only a hundred paces away from it, a great fire burst out, accompanied by occasional loud reports. This somewhat reassured me. I soon found our own people to be in possession burning things, and the detonations were obviously not caused by the bursting of shells fired from field-pieces. On sending two of my adjutants—Rokzak and Koos Nel—to the station to obtain further details, they soon came back to report that there was nobody there except a nervous old Dutchman. The burgher, who had told me Nelspruit was in the hands of the enemy, must have dreamt it.

The conflagration I found was caused by a quantity of "kastions" and ammunition-waggons which had been set afire on the previous day, while the explosions emanated from the shells which had been left among their contents.

The enemy's advance guard had pushed on to Shamoham and Sapthorpe, about 12 miles from the railway, enabling the whole of my commando to pass. We arrived at Nelspruit by eight o'clock. That day we rested and discussed future operations, feeling that our prospects seemed to grow worse every day.

The station presented a sad spectacle. Many trucks loaded with victuals, engines, and burst gun-carriages—everything had been left behind at the mercy of the first-comer, while a large number of kaffirs were plundering and stealing. Only the day before the Government had had its seat there, and how desolate and distressing the sight was now! The traces of a fugitive Government were unmistakable. Whatever might have been our optimism before, however little inclination the burghers might have felt to surrender, however great the firmness of the officers, and their resolve to keep the beloved "Vierkleur" flying, scenes like those at Nooitgedacht, and again at Nelspruit, were enough to make even the strongest and most energetic lose all courage. Many men could not keep back their tears at the disastrous spectacle, as they thought of the future of our country and of those who had been true to her to the last.[112]

"General Louis Botha with his commandants …. Botha is seated in the centre."[208]

Kaffirs, as I said, had been making sad havoc among the provisions, clothes and ammunition, and I ordered them to be driven away. Amongst the many railway-waggons I found some loaded with clothes the fighting burghers had in vain and incessantly been asking for, also cannon and cases of rifle ammunition.

We also came across a great quantity of things belonging to our famous medical commission, sweets, beverages, etc. The suspicion which had existed for some considerable time against this commission was, therefore, justified. There was even a carriage which had been used by some of its members, beautifully decorated, with every possible comfort and luxury, one compartment being filled with bottles of champagne and valuable wines.

My officers, who were no saints, saw that our men were well provided for out of these. The remainder of the good things was shifted on to a siding, where about twenty engines were kept. By great good luck the Government commissariat stock, consisting of some thousands of sheep, and even some horses, had also been left behind. But we were not cheered.

[112] 8/28/1900 General Buller's troops occupied Machadodorp. Bergendal occupied.
8/29/1900 Kruger fled to Nelspruit. The Boers evacuated Helvetia, which was occupied by General Buller.
8/30/1900 British occupation of Waterval Boven. Release of about 2000 British prisoners at Nooitgedacht.
9/1/1900 Roberts annexed to the British Empire the South African Republic, which henceforth will be known as the Transvaal Colony.
9/4/1900 General Buller and Botha engaged at Lydenburg. Siege of Ladybrand raised.
9/6/1900 British occupied Lydenburg Botha retreated.
9/8/1900 Spitz Kop captured.
9/11/1900 Kruger, having fled from the Transvaal, arrived in Portuguese territory, and proceeded to Lorenzo Marques.
9/13/1900 Roberts issued a proclamation calling upon the Boers to surrender. General French occupied Barberton.
9/16/1900 British occupied Nelspruit. British occupation of Kaap Muiden.

Among the many questions asked regarding this sad state of affairs was one put by an old burger:

"Dat is nou die plan, want zooals zaken hier lyk, dan heeft die boel in wanhoop gevlug."

("Is that the plan, then? For from what I can see of it, they have all fled in despair.")

I answered, "Perhaps they were frightened away, Oom."

"Ja," he said, "but look, General, it seems to me as if our members of the Government do not intend to continue the war. You can see this by the way they have now left everything behind for the second time."

"No, old Oom," I replied, "we should not take any notice of this. Our people are wrestling among the waves of a stormy ocean; the gale is strong, and the little boat seems upon the point of capsizing, but, it has not gone down as yet. Now and then the boat is dashed against the rocks and the splinters fly, but the faithful sailors never lose heart. If they were to do that the dinghy would soon go under, and the crew would disappear for ever. It would be the last page of their history, and their children would be strangers in their own country. You understand, Oom?"

"Yes, General, but I shall not forget to settle up, for I myself and others with me have had enough of this, and the War has opened our eyes."

"All right, old man." I rejoined, "nobody can prevent you surrendering, but I have now plenty of work to do; so get along."

Burghers of different commandos who had strayed—some on purpose—passed us here in groups of two or ten or more. Some of them were going to their own districts, right through the English lines, others were looking for their cattle, which they had allowed to stray in order to evade the enemy. I could only tell them that the veldt between Nelspruit and Barberton up to Avoca, was, so far as I had been able to discover, full of cattle and waggons belonging to farmers who now had no chance of escaping. Everybody wanted some information from the General.

About half a score of burghers with bridle horses then came up. There was one old burgher among them with a long beard, a great veldt hat, and armed with a Mauser which seemed hardly to have been used. He carried two belts with a good stock of cartridges, a revolver, and a tamaai (long sjambok). This veteran strode up in grand martial style to where I was sitting having something to eat. As he approached he looked brave enough to rout the whole British army.

"Dag!" (Good morning.) "Are you the General?" asked the old man.

"Yes, I have the honour of being called so. Are you a field-marshal, a Texas Jack,[113] or what?"

"My name is Erasmus, from the Pretoria district," he replied, "and my nine comrades and myself, with my family and cattle, have gone into the bush. I saw them all running away, the Government and all. You are close to the Portuguese border, and my mates and I want to know what your plans are."

"Well," Mr. Erasmus, I returned, "what you say is almost true; but as you say you and your comrades have been hiding in the bush with your cattle and your wives, I should like to know if you have ever tried to oppose the enemy yet, and also what is your right to speak like this."

"Well, I had to flee with my cattle, for you have to live on that as well as I."

"Right," said I; "what do you want, for I do not feel inclined to talk any longer."

"I want to know," he replied, "if you intend to retire, and if there is any chance of making peace. If not, we will go straight away to Buller, and 'hands-up,' then we shall save all our property."

"Well, my friend," I remarked, "our Government and the Commandant-General are the people who have to conclude peace, and it is not for you or me, when our family and cattle are in danger, to surrender to the enemy, which means turning traitor to your own people."

"Well, yes; good-bye, General, we are moving on now."

[113] John Baker Omohundro (July 27, 1846–June 28, 1880), also known as "Texas Jack," was an American frontier scout, actor, and cowboy. Born in rural Virginia, he served the Confederate States of America during the American Civil War. He later served as a civilian scout for the US Army during the American Indian Wars. Before his untimely death, Texas Jack became a legendary figure in the American Old West as a Western showman performing dramas on the stage throughout the country, and was immortalized in dime novels published around the world. https://en.wikipedia.org/wiki/Texas_Jack_Omohundro#

I sent a message to our outposts to watch these fellows, and to see if they really were going over to the enemy. And, as it happened, that same night my Boers came to camp with the Mausers and horses Erasmus and his party had abandoned. They had gone over to Buller.

The above is but an instance illustrating what often came under my notice during the latter period of my command. This sort of burgher, it turned out, invariably belonged to a class that never meant to fight. In many cases we could do better without them, for it was always these people who wanted to know exactly what was "on the cards," and whenever things turned out unpleasantly, they only misled and discouraged others. Obviously, we were better off without them.

Image: My talk with Erasmus (Non-Combatant)

Chapter 22. An Ignominious Dispersal.

Commandant-General Botha, who was then invalided at Hector's Spruit Station, now sent word that we were to join him there without delay. He said I could send part of the commando by train, but the railway arrangements were now all disturbed, and everything was in a muddle. As nothing could be relied on in the way of transport, the greater number of the men and most of the draught beasts had to "trek."

At Crocodile poort Station the situation was no better than at Nelspruit, and the same might be said of Kaapmuiden. Many of the engine drivers, and many of the burghers even, who were helping in destroying the barrels of spirits at the stations, were so excited (as they put it) through the fumes of the drink, that the strangest things were happening. Heavily-laden trains were going at the rate of 40 miles an hour. A terrible collision had happened between two trains going in different directions, several burghers and animals being killed. Striplings were shooting from the trains at whatever game they saw, or fancied they saw, along the line, and many mishaps resulted. These things did not tend to improve matters.

Train accident 1900.[209]

Flour abandoned on Delagoa Bay Railway.[210]

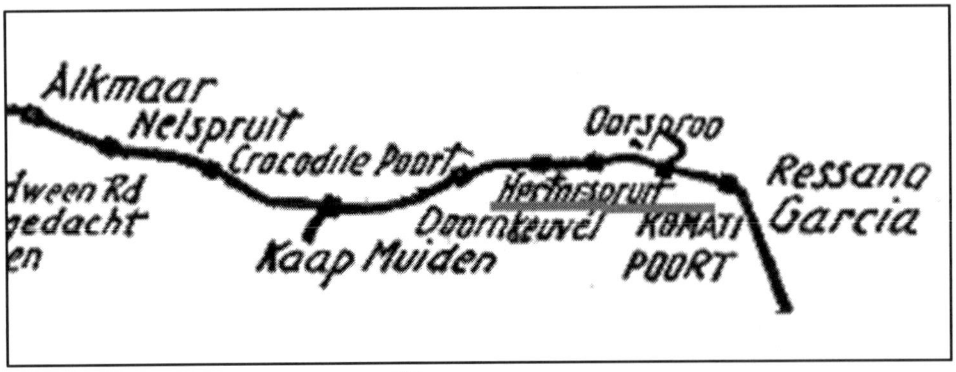

Pretoria to Komati Poort Railway.[211] Note Hectorspruit.

It was not so much that the officers had lost control over their men. It seemed as if the Evil Spirit had been let loose and was doing his very best to encourage the people to riotous enjoyment.

Hector's Spruit is the last station but one before you come to the Portuguese frontier, and about seventeen miles from Ressano Garcia. Here every commando stopped intending of course to push on to the north and then to cross the mountains near Lydenburg in a westerly direction.

The day when I arrived at Hector's Spruit, President Steyn, attended by an escort of 100 men, went away by the same route. Meanwhile General Buller was encamped at Glyn's mines near Spitskop and the Sabi River, which enabled him to command the mountain pass near Mac Mac and Belvedere without the slightest trouble, and to block the roads along which we meant to proceed. Although the late Commandant (afterwards fighting General) Gravett occupied one of the passes with a small commando, he was himself in constant danger of being cut off from Lydenburg by a flank movement.

On the 16th of September, 1900, an incident occurred which is difficult to describe adequately. Hector Spruit is one of the many unattractive stations along the Delegoa Bay railway situated between the great Crocodile river and dreary black "kopjes" or "randjes" with branches of the Cape mountains intervening and the "Low Veldts," better known as the "Boschveldt." This is a locality almost filled with black holly bushes, where you can only see the sky overhead and the spot of ground you are standing on. In September the "boschveldt" is usually dry and withered and the scorching heat makes the surroundings seem more lugubrious and inhospitable than ever.

The type of terrain described. This is the Modder River Battlefield area.[212]

The station was crowded with railway carriages loaded up with all sorts of goods, and innumerable passenger carriages, and the platform and adjoining places filled with agitated people. Some were packing up, others unpacking, and some, again, were looting. The majority were, however, wandering about aimlessly. They did not know what was happening; what ought to be done or would be done; and the only exceptions were the officers, who were busily engaged in providing themselves and their burghers with provisions and ammunition.

Merebank station after arrival of a refugee train.[213]

Loading Up Ammunition for The Transvaal Artillery.[214]

I now had to perform one of the most unpleasant duties I have ever known: that of calling the burghers together and telling them that those who had no horses were to go by train to Komati Poort, there to join General Jan

Coetser.[114] Those who had horses were to report themselves to me the next morning, and get away with me through the low fields.

Some burghers exclaimed: "We are now thrown over, left in the 'lurch,' because we have not got horses; that is not fair."

Others said they would be satisfied if I went with them, for they did not know General Coetser.

Commandant-General Botha did not see his way to let me go to Komati Poort, as he could not spare me and the other commandos. Those of the men who had to walk the distance complained very bitterly, and their complaints were well-founded. I did my best to persuade and pacify them all, and some of them were crying like babies when we parted.

Komati Poort was, of course, the last station, and if the enemy were to drive them any further they would have to cross the Portuguese border, and to surrender to the Portuguese; or they could try to escape through Swaziland (as several hundreds did afterwards) or along the Lebombo mountains, via Leydsdorp. But if they took the latter route then they might just as well have stayed with me in the first place. It was along this road that General Coetser afterwards fled with a small body of burghers, when the enemy, according to expectations, marched on Komati Poort, and met with no resistance, though there were over 1,800 there of our men with guns.

A certain Pienaar, who arrogated unto himself the rank of a general on Portuguese territory, fled with 800 men over the frontier. These, however, were disarmed and sent to Lisbon.

The end of the struggle was ignominious, as many a burgher had feared; and to this day I pity the men who, at Hector's Spruit, had to go to Komati Poort much against their will.

Komatie Poort Railway Bridge.[215]

Camp at Komatie Poort.[216]

Burned Boer Wagons at Komatie Poort.[217]

Fortunately they had the time and presence of mind to blow up the "Long Tom" and other guns before going; but a tremendous lot of provisions and ammunition must have fallen into the hands of the enemy.

At Hector's Spruit half a score of cannon of different calibre had been blown up, and many things buried which may be found some day by our progeny. Our carts were all ready loaded, and we were prepared to march next morning into the desert and take leave of our stores. How would we get on now? Where would we get our food, cut off as we were from the railway, and, consequently, from all imports and supplies? These questions and many others crossed our minds, but nobody could answer them.

[114] Not listed in the Military Officer Album. (MOA)

Destroyed Gun at Komati Poort.[115]

[Same] Boer Gun exhibited outside Glasgow Art Gallery.[218]

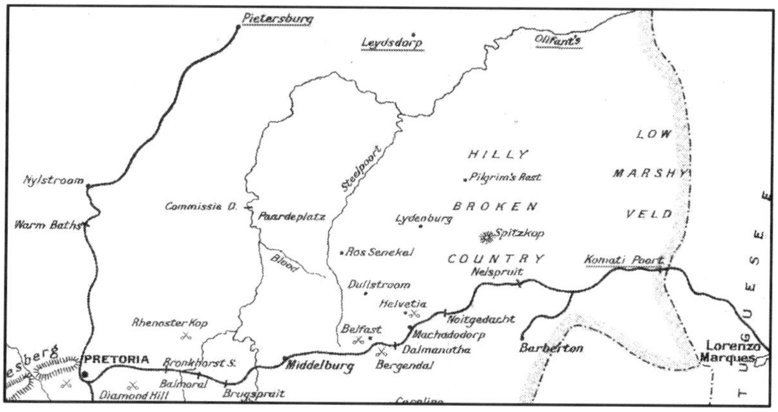

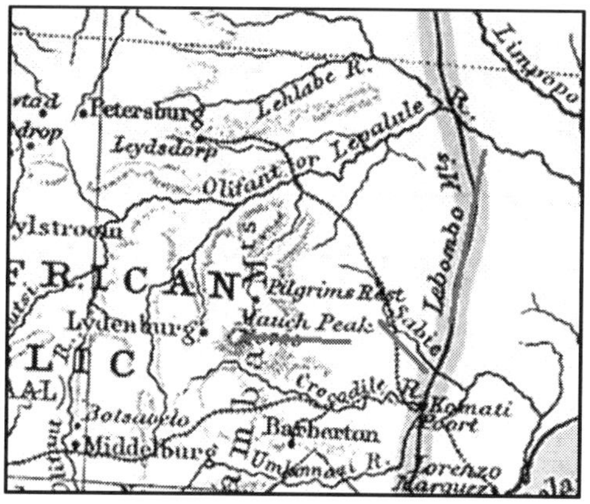

The journey is from the Komati Poort Area north through the low marshy veld to the Olifant's river then to Leydsdorp and finally Pietersburg. Geo. H. Walker Map of the general area. Note Mauch peak and Sabie River with Lebombo Mts. Running south to north on the right side.

Our convoys were ready waiting, and the following morning we trekked into the Hinterland Desert, saying farewell to commissariats and stores.

The prospect was melancholy enough. By leaving Hector's Spruit we were isolating ourselves from the outer world, which meant that Europe and civilisation generally could only be informed of our doings through English channels.

Once again our hopes were centred in our God and our Mausers.

Dr. Conan Doyle says about this stage of the war:—

"The most incredulous must have recognised as he looked at the heap of splintered and shattered gunmetal (at Hector's Spruit) that the long War was at last drawing to a close."

And here I am, writing these pages seventeen months later, and the War is not over yet. But Dr. Doyle is not a prophet, and cannot be reproached for a miscalculation of this character, for if I, and many with me, had been asked at the time what we thought of the future, we might have been as wide of the mark as Dr. Doyle himself.[116]

115 "Wonderboom Tom," photographed at Komati Poort (from the album of Lieutenant Alexander, Scots Guards). Destroyed by the Boers at Komati Poort on 22nd September 1900, Lord Roberts claimed the remains of "Wonderboom Tom" for his own private collection. After being loaned to the Glasgow Exhibition, the gun eventually found its way to Lord Roberts's new residence, Englemere, in 1906. Here it remained until the dark days of WWII, when Countess Roberts offered her father's entire collection of trophies to the country as part of the salvage scheme. All the guns, including the Creusot, were smelted to make armaments in June 1940. Courtesy of Neville Constantine: Anglo Boer War - Artillery and Ammunition - Page 19 - Boer War Forum And: Anglo Boer War - Artillery and Ammunition - Page 24 - Boer War Forum

116 Tuesday, September 4, 1900 General Buller and Botha engaged at Lydenburg.
Thursday, September 6, 1900 British occupied Lydenburg, Botha retreated.

Chapter 23. A Dreary Trek Through Feverland.

The 18th of September, 1900, found us trekking along an old disused road in a northerly direction. We made a curious procession, an endless retinue of carts, waggons, guns, mounted men, "voetgangers" nearly three miles long. The Boers walking comprised 150 burghers without horses, who refused to surrender to the Portuguese, and who had now joined the trek on foot. Of the 1,500 mounted Boers 500 possessed horses which were in such a parlous condition that they could not be ridden. The draught cattle were mostly poor and weak, and the waggons carrying provisions and ammunition, as also those conveying the guns, could only be urged along with great difficulty. In the last few months our cattle and horses had been worked hard nearly every day, and had to be kept close to our positions.

During the season the veldt in the Transvaal is in the very worst condition, and the animals are then poorer than at any other period. We had, moreover, the very worst of luck, kept as we were in the coldest parts of the country from June till September, [Winter in the Southern Hemisphere] and the rains had fallen later than usual. There was, therefore, scarcely any food for the poor creatures, and hardly any grass. The bushveldt through which we were now trekking was scorched by an intolerable heat, [in Spring and Summer] aggravated by drought, and the temperature in the daytime was so unbearable that we could only trek during the night.

Water was very scarce, and most of the wells which, according to old hunters with us, yielded splendid supplies, were found to be dried up. The veldt being burned out there was not a blade of grass to be seen, and we had great trouble in keeping our animals alive. From time to time we came across itinerant kaffir tribes from whom we obtained handfuls of salt or sugar, or a pail full of mealies, and by these means we managed to save our cattle and horses.

When we had got [North] through the Crocodile River the trek was arranged in a sort of military formation enabling us to defend ourselves, had we been attacked. The British were already in possession of the railway up to Kaapmuiden and we had to be prepared for pursuit; and really pursuit by the British seemed feasible and probable from along the Ohrigstad River towards Olifant's Nek and thence along the Olifant's River.

Our original plan was to cross the Sabi, along the Meritsjani River, over the mountains near Mac Mac, through Erasmus or Gowyn's Pass and across Pilgrim's Rest, where we might speedily have reached healthier veldt and better climatic conditions. President Steyn had passed there three days previously, but when our advance guard reached the foot of the high mountains, near Mac Mac, the late General Gravett sent word that General Buller with his force was marching from Spitskop along the mountain plateau and that it would be difficult for us to get ahead of him and into the mountains. The road, which was washed away, was very steep and difficult and contained abrupt deviations so that we could only proceed at a snail's pace.

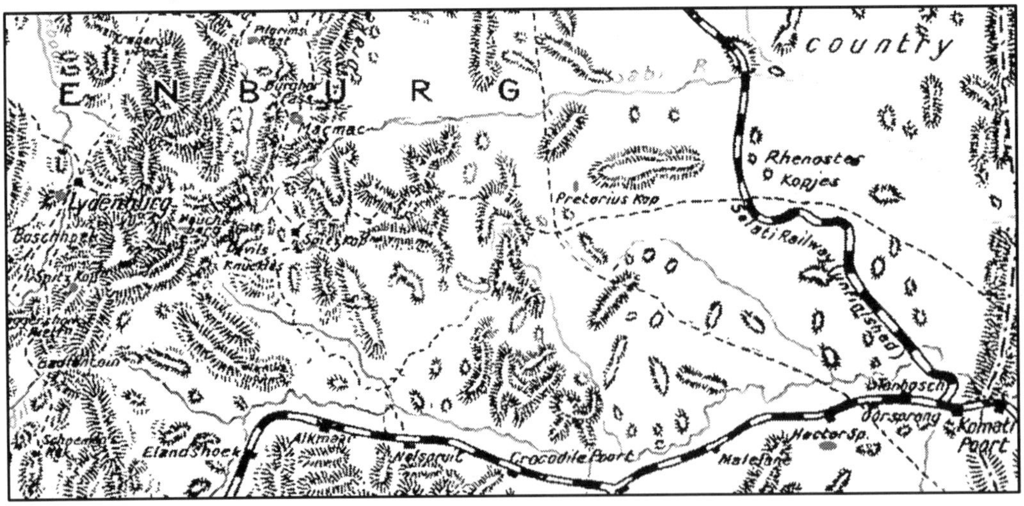

Note Right side Hector Spruit and Crocodile R. just north. Left side, two marked Spitz Kop(s), marked and Mac Mac and Pilgrims Rest.

Tuesday, September 11, 1900 Kruger, having fled from the Transvaal, arrived in Portuguese territory, and proceeded to Lorenzo Marques.
Thursday, September 13, 1900 General French occupied Barberton.
Sunday, September 16, 1900 British occupied Nelspruit. British occupation of Kaap Muiden.
Monday, September 24, 1900 Arrival of the British at the Portuguese frontier. Evacuation of all the Boer positions near the frontier.

Commandant-General Botha then sent instructions to me to take my commando along the foot of the mountains, via Leydsdorp, while he with his staff and the members of the Government would proceed across the mountains near Mac Mac. General Gravett was detailed to keep Buller's advance guard busy, and he succeeded admirably.

I think it was here that the British lost a fine chance of making a big haul. General Buller could have blocked us at any of the mountain roads near Mac Mac, and could also have swooped down upon us near Gowyn's Pass and Belvedere. At the time of which I write Buller was lying not 14 miles away at Spitskop. Two days after he actually occupied the passes, but just too late to turn the two Governments and the Commandant-General.

It might be said that they could in any case have, like myself, escaped along the foot of the mountains via Leydsdorp to Tabina and Pietersburg, but had the way out been blocked to them near Mac Mac, our Government and generalissimo would have been compelled to trek for at least three weeks in the low veldt before they could have reached Pietersburg, during which time all the other commandos would have been out of touch with the chief Boer military strategists and commanders, and would not have known what had become of their military leaders or of their Government. This would have been a very undesirable state of affairs, and would very likely have borne the most serious consequences to us. The British, moreover, could have occupied Pietersburg without much trouble by cutting off our progress in the low veldt, and barring our way across the Sabini and at Agatha.

This coup could indeed have been effected by a small British force. In the mountains they would, moreover, have found a healthy climate, while we should have been left in the sickly districts of the low veldt. And had we been compelled to stay there for two months we would have been forced to surrender, for about the middle of October the disease among our horses increased and so serious was the epidemic that none but salted horses survived. The enteric fever would also have wrought havoc amongst us.[117]

Another problem was whether all this would not have put an end to the war; we still had generals left, and strong commandos, and it was, of course, very likely that a great number of Boers driven to desperation would have broken through, although two-thirds of our horses were not fit for a bold dash. Perhaps fifteen hundred out of the two thousand Boers would have made good their escape, but in any case large numbers of wagons, guns, etc. would have fallen into the British hands and our leaders might have been captured as well. The moral effect would have caused many other burghers from the other commandos to have lost heart and this at a moment, too, when they already required much encouragement.

This was my view of the situation, and I think Lord Roberts, or whoever was responsible, lost a splendid opportunity.

As regards my commando at the foot of the Mauch Mountains we turned right about and I took temporary leave of Louis Botha. It was a very affecting parting; Botha pressed my hand, saying, "Farewell, brother; I hope we shall get through all right. God bless you. Let me hear from you soon and frequently."

That night we encamped at Boschbokrand, where we found a store unoccupied, and a house probably belonging to English refugees, for shop and dwelling had been burgled and looted. After our big laager had been arranged, Boer fashion, and the camp fire threw its lurid light against the weird dark outline of the woods, the Boers grouped themselves over the veldt. Some who had walked twenty miles that day fell down exhausted.

I made the round of the laager, and I am bound to say that in spite of the trying circumstances, my burghers were in fairly cheerful spirits.

I discussed the immediate prospects with the officers, and arranged for a different commando to be placed in the advance guard each day and a different field-cornet in the rear. Boers conversant with the locality were detailed to ride ahead and to scout and reconnoitre for water.

When I returned that night to my waggon the evening meal was ready, but for the first time in my life I could eat nothing. I felt too dejected. My cook, Jan Smith, and my messmates were curious to know the reason I did not "wade in," for they always admired my ferocious appetite.

It had been a tiring day, and I pretended I was not well; and soon afterwards I lay down to rest.

[117] Enteric fever is a medical term for two types of salmonellosis, which, specifically, are typhoid fever and paratyphoid fever.

I had been sitting up the previous evening till late in the night, and was therefore in hopes of dropping off to sleep. But whatever I tried—counting the stars, closing my eyes and doing my best to think of nothing—it was all in vain.

Insurmountable difficulties presented themselves to me. I had ventured into an unhealthy, deserted, and worst of all, unknown part of the country with only 2,000 men. I was told we should have to cover 300 miles of this enteric [Typhoid fever] stricken country.

The burghers without horses were suffering terribly from the killing heat, and many were attacked by typhoid and malarial fever through having to drink a lot of bad water; these enemies would soon decimate our commando and reduce its strength to a minimum. And for four or five weeks we should be isolated from the Commandant-General and from all white men.

Was I a coward, then, to lie there, dejected and even frightened? I asked myself. Surely, to think nothing of taking part in a fierce battle, to be able to see blood being shed like water, to play with life and death, one could not be without some courage? And yet I did not seem to have any pluck left in me here where there did not seem to be much danger.

These and many similar thoughts came into my head while I was trying to force myself to sleep, and I told myself not to waver, to keep a cool head and a stout heart, and to manfully go on to the end in order to reach the goal we had so long kept in view.

Ah, well, do not let anybody expect a general to be a hero, and nothing else, at all times; let us remember that "A man's a man for a' that," and even a fighting man may have his moments of weakness and fear.

The next morning, about four o'clock, our little force woke up again. The cool morning air made it bearable for man and beast to trek. This, however, only lasted till seven o'clock, when the sun was already scorching, without the slightest sign of a breeze. It became most oppressive, and we were scarcely able to breathe.

The road had not been used for twenty or thirty years, and big trees were growing in our path, and had to be cut down at times. The dry ground, now cut up by the horses' hoofs, was turned into dust by the many wheels, great clouds flying all round us, high up in the air, covering everything and everybody with a thick layer of ashy-grey powder.

About nine o'clock we reached Zand River, where we found some good water, and stayed till dusk. We exchanged some mealies against salt and other necessaries with some kaffirs who were living near by the water. Their diminutive, deformed stature was another proof of the miserable climate obtaining there.

There was much big game here; wild beasts, "hartebeest," "rooiboks" (sometimes in groups of from five to twenty at a time), and at night we heard the roaring of lions and the howling of wolves. Even by day lions were encountered.

Coke's Hartebeest.[219] [118]

Impala, Kruger National Park, South Africa [119] [120]

[118] The hartebeest (Alcelaphus buselaphus), also known as kongoni or kaama, is an African antelope A large antelope, the hartebeest stands just over 1 m (3 ft 3 in) at the shoulder, and has a typical head-and-body length of 200 to 250 cm (79 to 98 in). The weight ranges from 100 to 200 kg (220 to 440 lb).

Now, one of the weakest points, perhaps the weakest, of an Afrikander is his being unable to refrain from shooting when he sees game, whether such be prohibited or not. From every commando burghers had been sent out to do shooting for our commissariat, but a good many had slipped away, so that hundreds of them were soon hunting about in the thickly-grown woods. The consequence was that, whenever a group of them discovered game, it seemed as if a real battle were going on, several persons often being wounded, and many cattle killed. We made rules and regulations, and even inflicted punishments which did some good, but could not check the wild hunting instincts altogether, it being difficult to find out in the dark bush who had been the culprits.

Meanwhile the trek went on very slowly. On the seventh day we reached Blyde River,[121] where we had one of the loveliest views of the whole "boschveldt." The river, which has its source near Pilgrim's Rest and runs into the great Olifant's River near the Lomboba, owes its name to trekker pioneers, who, being out hunting in the good old times, had been looking for water for days, and when nearly perishing from thirst, had suddenly discovered this river, and called it Blyde (or "Glad") River. The stream at the spot we crossed is about 40 feet wide, and the water as pure as crystal. The even bed is covered with white gravel, and along both banks are splendid high trees. The whole laager could outspan under their shade, and it was a delightful, refreshing sensation to find oneself protected from the burning sun. We all drank of the delicious water, which we had seldom found in such abundance, and we also availed ourselves of it to bathe and wash our clothes.

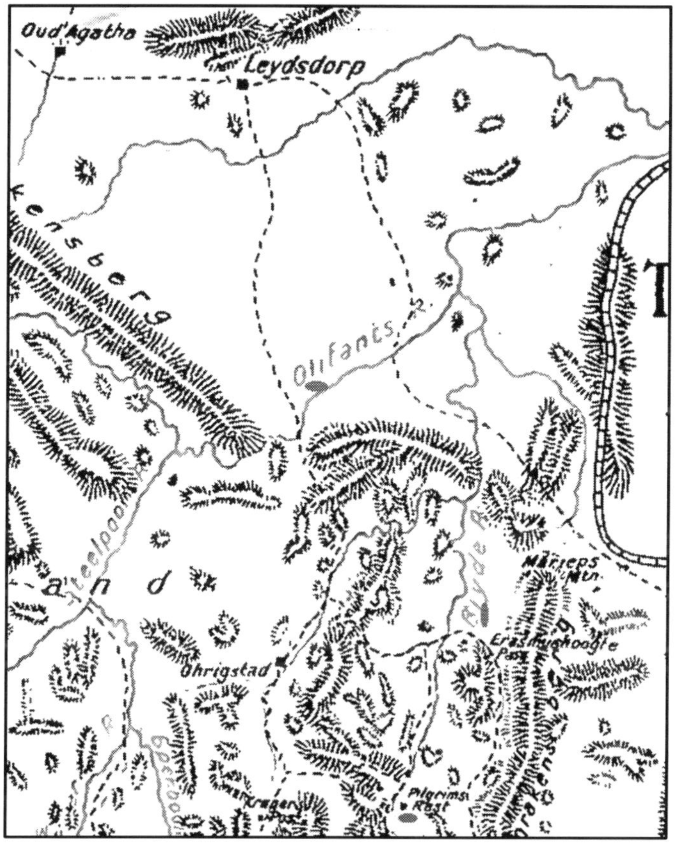

Pilgrims Rest, Blyde R. and Olifants R.

In the afternoon a burgher, whose name I had better not mention, came running up to us with his clothes torn to tatters, and his hat and gun gone. He presented a curious picture. I heard the burghers jeer and chaff him as he approached, and called out to him: "What on earth have you been up to? It looks as if you had seen old Nick with a mask on."

The affrighted Boer's dishevelled hair stood on end and he shook with fear.

[119] Impala (Aepyceros melampus) male, Kruger National Park, South Africa https://commons.wikimedia.org/wiki/File:Impala_(Aepyceros_melampus)_male_Kruger.jpg Author: Charles J. Sharp

[120] The impala or rooibok (Aepyceros melampus, lit. 'black-footed high-horn' in Ancient Greek) is a medium-sized antelope found in eastern and southern Africa. The impala reaches 70–92 cm (28–36 in) at the shoulder and weighs 40–65 kg (88–143 lb) https://en.wikipedia.org/wiki/Impala

[121] ~September 25th. 9/24/1900 Arrival of the British at the Portuguese frontier. Evacuation of all the Boer positions near the frontier.

He gasped: "Goodness gracious, General, I am nearly dead. I had gone for a stroll to do a bit of hunting like, and had shot a lion who ran away into some brushwood. I knew the animal had received a mortal wound, and ran after it. But I could only see a yard or so ahead through the thick undergrowth, and was following the bloodstained track. Seeing the animal I put down my gun and was stepping over the trunk of an old tree; but just as I put my foot down, lo! I saw a terrible monster standing with one paw on the beast's chest. Oh, my eye! I thought my last hour had come, for the lion looked so hard at me, and he roared so awfully. By Jove, General, if this had been an Englishman I should just have "hands-upped," you bet! But I veered round and went down bang on my nose. My rifle, my hat, my all, I abandoned in that battle, and for all the riches of England, I would not go back. General, you may punish me for losing my rifle, but I won't go back to that place for anything or anybody."

I asked him what the lion had done then, but he knew nothing more. Another burgher who stood by, remarked: "I think it was a dog this chap saw. He came running up to me so terrified that he would not have known his own mother. If I had asked him at that moment he would not have been able to remember his own name."

The poor fellow was roused to indignation, and offered to go with the whole commando and show them the lion's trail. But there was no time for that, and the hero had a bad time of it, for everybody was teasing and chaffing him, and henceforth he was called the "Terror of the Vaal."

We should have liked to have lingered a few days near that splendid and wholesome stream. We wanted a rest badly enough, but it was not advisable on account of the fever, which is almost invariably the penalty for sleeping near a river in the low veldt.

One of the regulations of our commando forbade the officers and men to spend the night by the side of any water or low spot. It would also have been fatal to the horses, for sickness amongst them and fever always coincide. But they did not always keep to the letter of these instructions. The burghers, especially those who had been walking, or arriving at a river, would always quickly undress and jump into the water, after which some of them would fall asleep on the banks or have a rest under the trees.

Both were unhealthy and dangerous luxuries. Many burghers who had been out hunting or had been sent out provisioning, stayed by the riverside till the morning, since they could dispense with their kit in this warm climate. They often were without food for twenty-four hours, unless we happened to trek along the spot where they were resting. To pass the night in these treacherous parts on an empty stomach was enough to give anybody the fever.

When we moved on from Blyde River many draught beasts were exhausted through want of food, and we were obliged to leave half a dozen carts behind. This caused a lot of trouble as we had to transfer all the things to other vehicles, and field-cornets did not like to take up the goods belonging to other field-cornets' burghers, the cattle being in such a weak condition that it made every man think of his own division. No doubt the burghers were very kind to their animals, but they sometimes carried it too far, and the superior officers had often to interfere.

The distance from Blyde River to the next stopping place could not be covered in one day, and we should have no water the next; not a very pleasant prospect. The great clouds of dust through which we were marching overnight and the scorching heat in the daytime made us all long for water to drink and to clean ourselves. So when the order came from the laager commandants: "Outspan! No water to-day, my boys, you will have to be careful with the water on the carts. We shall be near some stream to-morrow evening," they were bitterly disappointed.

When we got near the water the following day eight burghers were reported to be suffering badly from the typhoid fever, five of them belonging to the men who were walking. We had a very insufficient supply of ambulance waggons.

I had omitted to procure a great number of these indispensable vehicles on leaving Hector's Spruit, for there had been so many things to look after. We were lucky to have with us brave Dr. Manning, of the Russian Ambulance, who rendered us such excellent assistance, and we have every reason to be thankful to H.M. the Czarina of Russia for sending him out. Dr. Manning had the patients placed in waggons, which had been put at his disposal for this purpose, but notwithstanding his skilled and careful treatment, one of my men died the following day, while the number of those who were seriously ill rose to fifteen.

The symptoms of this fatal illness are: headache and a numb feeling in all the limbs, accompanied by an unusually high temperature very often rising to 104 and 106 degrees during the first 24 hours, with the blood running from the patient's nose and ears, which is an ominous sign. At other times the first symptom is what is commonly called "cold shivers."

Boer Ambulance Driver and attendants.[220]

Boer Doctors.[221]

We proceeded slowly until we came to the Nagout River,[122] where the monotony and dreariness of a trek through the "boschveldt" were somewhat relieved by the spectacle of a wide stream of good water, with a luxurious vegetation along the banks. It was a most pleasant and refreshing sight to behold. For some distance along the banks some grass was found, to which the half-starved animals were soon devoting their attention. It was the sort of sweet grass the hunters call "buffalo-grass," and which is considered splendid food for cattle. We pitched our camp on a hill about one mile from the river, and as our draught-beasts were in want of a thorough rest we remained there for a few days. We had been obliged to drive along some hundreds of oxen, mules, and horses, as they had been unfit to be harnessed for days, and had several times been obliged to leave those behind that were emaciated and exhausted.

From the Nagout River we had to go right up to the Olifant's River, a distance of about 20 miles, which took us three days. The track led all along through the immense bush-plain which extends from the high Mauch Mountains in the west to the Lebombo Mountains in the east; and yet one could only see a few paces ahead during all these days, and the only thing we could discern was the summit of some mountain on the westerly or easterly horizon, and even the tops of the Mauch and Lebombo Mountains one could only see by standing on the top of a loaded waggon, and with the aid of a field-glass. This thickly-wooded region included nearly one-third of the Transvaal, and is uninhabited, the white men fearing the unhealthy climate, while only some miserable little kaffir tribes were found about there, the bulk being the undisputed territory of the wild animals.

Mauch Peak, Lebombo Mts. to the East; and Olifant River.[123]

[122] Possibly: The Makhutswi River is a short river in Limpopo Province, South Africa. It flows northeastwards and is a tributary of the Olifants River, joining it in its central basin. https://en.wikipedia.org/wiki/Makhutswi_River

[123] See: https://en.wikipedia.org/wiki/List_of_rivers_of_South_Africa

The Olifant's River, which we had to cross, is over 100 feet wide. The old track leading down to it, was so thickly covered with trees and undergrowth that we had to cut a path through it. The banks of the river were not very high, thus enabling us to make a drift without much trouble. The bed was rocky, and the water pretty shallow, and towards the afternoon the whole commando had crossed. Here again we were obliged to rest our cattle for a few days, during which we had to fulfil the melancholy duty of burying two of our burghers who had died of fever. It was a very sad loss and we were very much affected, especially as one left a young wife and two little children, living at Barberton. The other one was a young colonial Afrikander who had left his parents in the Cradock district (Cape Colony) to fight for our cause. We could not help thinking how intensely sad it was to lose one's life on the banks of this river, far from one's home, from relatives and friends, without a last grasp of the hand of those who were nearest and dearest.

The Transvaaler's last words were:—

"Be sure to tell my wife I am dying cheerfully, with a clear conscience; that I have given my life for the welfare of my Fatherland."

We had now to leave some draught cattle and horses behind every day, and the number of those who were obliged to walk was continually increasing, till there were several hundred.

Near Sabini, the first river we came to after leaving Leydsdorp we secured twenty-four mules which were of very great use to us under the circumstances. But the difficulty was how to distribute them amongst the field-cornets. The men all said they wanted them very urgently, and at once found the cattle belonging to each cart to be too thin and too weak to move. Yet the twenty-four could only be put into two carts, and I had to solve the difficulty by asserting my authority.

It was no easy task to get over the Agatha Mountains and we had to rest for the day near the big Letaba, especially as we had to give the whole file of carts, guns, etc., a chance of forming up again. Here we succeeded in buying some loads of mealies, which were a real God-send to our half-starved horses. I also managed to hire some teams of oxen from Boers who had taken up a position with their cattle along the Letaba, which enabled us to get our carts out of the Hartbosch Mountains as far as practicable. The task would have been too fatiguing for our cattle. It took us two days before we were out of these mountains, when we camped out on the splendid "plateau" of the Koutboschbergen, where the climate was wholesome and pleasant.

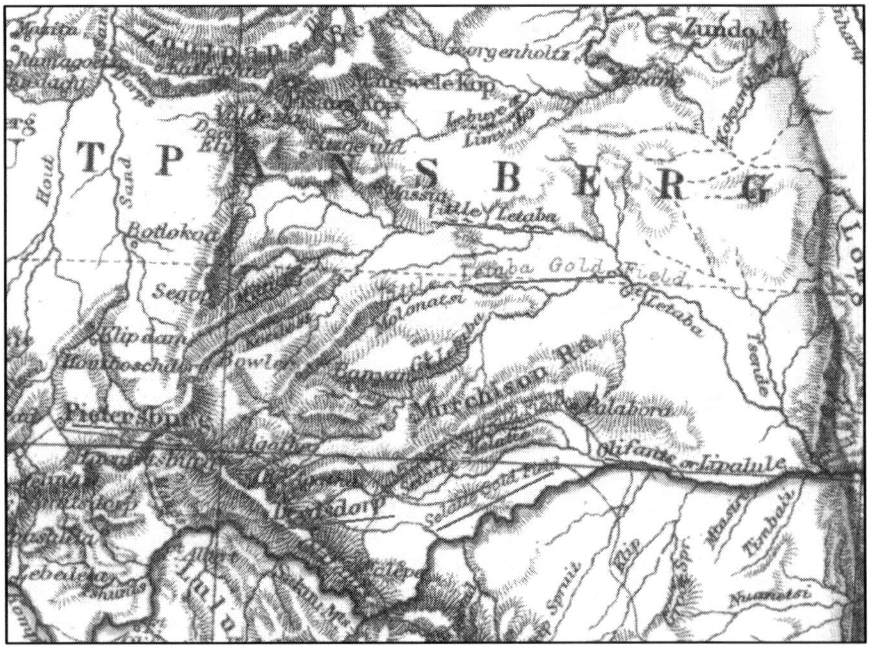

This map shows the topography of the area but does not identify some of the place names. Note: Leydsdorp, Selaite Gold Fields on the Olifants River; the Letaba Gold Field and Little Letaba River and Pietersburg.[222]

Here, after having passed a whole month in the wilderness of the low veldt, with its destructive climate, it was as though we began a new life, as if we had come back to civilisation. We again saw white men's dwellings, cultivated green fields, flocks of grazing sheep, and herds of sleek cows.

The inhabitants of the country were not a little surprised, not to say alarmed, to find, early one Sunday morning, a big laager occupying the plateau. A Boer laager always looks twice as large as it really is when seen from a little distance. Some Boer lads presently came up to ask us whether we were friends or enemies, for in these distant parts people were not kept informed of what happened elsewhere.

Boer Laager.[223]

Boers Preparing Supper.[224]

Scout Camp Life.[225]

A Boer picket in early morning.[226]

"A general," said a woman, who paid us a visit in a trap,[124] "is a thing we have all been longing to see. I have called to hear some news, and whether you would like to buy some oats; but I tell you straight I am not going to take "blue-backs" (Government notes), and if you people buy my oats you will have to pay in gold."

A burgher answered her: "There is the General, under that cart; 'tante' had better go to him."

Pony Trap, 1900.[227]

Of course I had heard the whole conversation, but thought the woman had been joking. The good lady came up to my cart, putting her cap a little on one side, probably to favour us with a peep at her beauty.

"Good morning. Where is that General Viljoen; they say he is here?"

[124] A trap, pony trap is a light, often sporty, two-wheeled or sometimes four-wheeled horse- or pony-drawn carriage, usually accommodating two to four persons in various seating arrangements. https://en.wikipedia.org/wiki/Trap_(carriage)

I thought to myself: "I wonder what this charming Delilah of fifty summers wants," and got up and shook hands with her, saying: "I am that General. What can I do for 'tante'?"

"No, but I never! Are you the General? You don't look a bit like one; I thought a General looked 'baing' (much) different from what you are like."

Much amused by all this I asked: "What's the matter with me, then, 'tante'?"[125]

"Nay, but cousin (meaning myself) looks like a youngster. I have heard so much of you, I expected to see an old man with a long beard."

I had had enough of this comedy, and not feeling inclined to waste any more civilities on this innocent daughter of Mother Eve, I asked her about the oats.

I sent an adjutant to have a look at her stock and to buy what we wanted, and the prim dame spared me the rest of her criticism.

We now heard that Pietersburg and Warmbad were still held by the Boers, and the road was therefore clear. We marched from here via Haenertsburg, a little village on the Houtboschbergrand, and the seat of some officials of the Boer Mining Department, for in this neighbourhood gold mines existed, which in time of peace give employment to hundreds of miners.

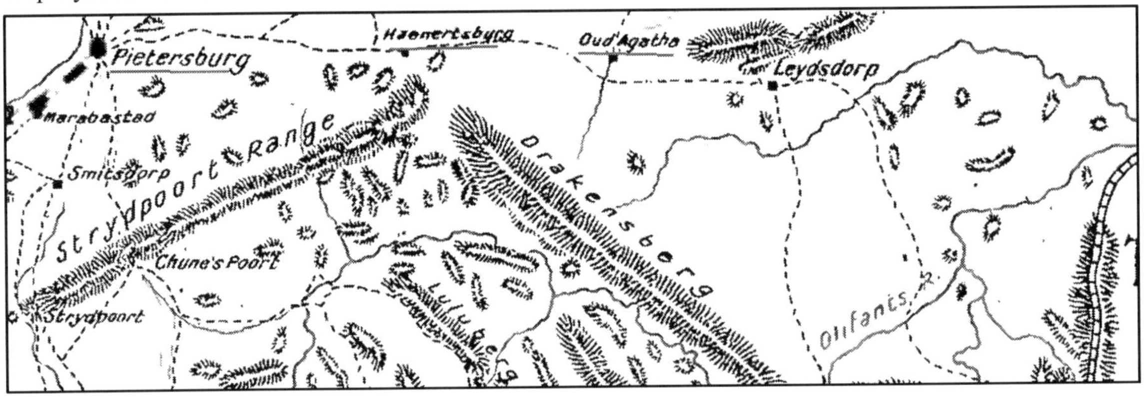

Note Oud Agatha and Haemertsburg at the top. Pietersburg top left.

Alluvial Gold Mine. This of Mr. Ignatius Ferreira.[126] Ferreira's Gold Mine in 1886.

Luckily, there was also a hospital at Haenertsburg, where we could leave half a dozen fever patients, under the careful treatment of an Irish doctor named Kavanagh, assisted by the tender care of a daughter of the local justice of the peace, whose name, I am sorry to say, I have forgotten.

About the 19th of October, 1900, we arrived at Pietersburg, our place of destination.[127]

[125] Tante is the French, Dutch, German, Danish, Yiddish and Latvian word for aunt.

[126] Ignatius Ferreira - Wikipedia

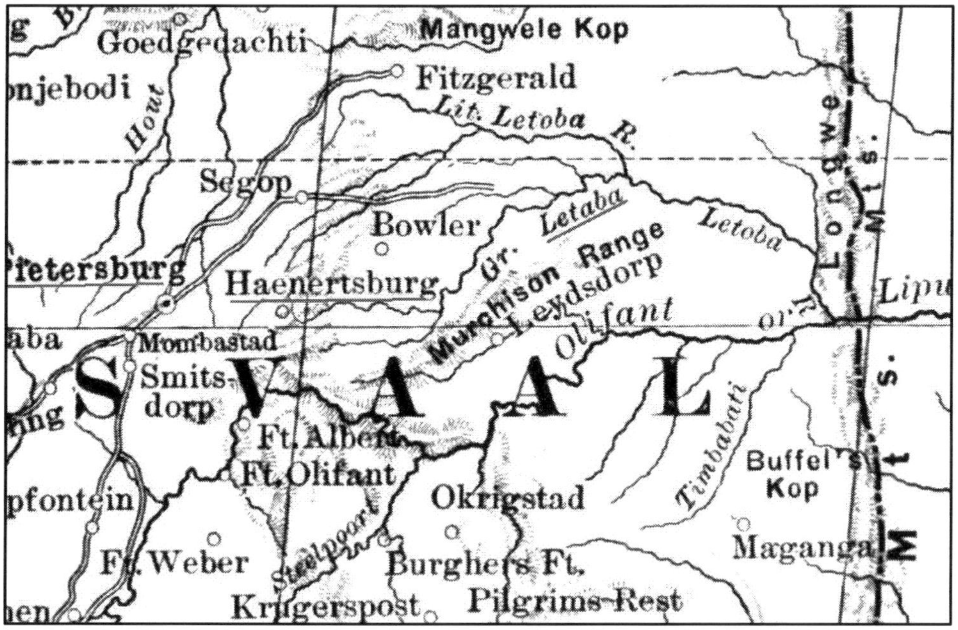

Another view of the same area showing Leydsdorp, Letaba River, Haenertsburg and Pietersburg.[228]

Chapter 24. Pains and Pleasures of Commandeering.

We found Pietersburg to be quite republican, all the officials, from high to low, in their proper places in the offices, and the "Vierkleur" flying from the Government buildings. The railway to Warmbad was also in Boer hands. At Warmbad were and his burghers and those of the Waterberg district. Although we had no coals left, this did not prevent us from running a train with a sufficient number of carriages from Pietersburg to Warmbad twice a week. We used wood instead, this being found in great quantities in this part of the country.

Of course, it took some time to get steam up, and we had to put in more wood all the time, while the boilers continually threatened to run dry. We only had two engines, one of which was mostly laid up for repairs. The other one served to keep the commandos at Warmbad provided with food, etc.

The Pietersburgers also had kept up telegraphic communication, and we were delighted to hear that clothes and boots could be got in the town, as we had to replace our own, which had got dreadfully torn and worn out on the "trek" through the "boschveldt."

Camp of The Field Telegraph Corps of The Transvaal Artillery Under First Lieutenant P. C. Paff. This corps was excellently equipped. It possessed the best and latest instruments, and did invaluable service to the Boer forces.[229]

"Boer Telegraphists at Tea."[230]

127 10/3/1900 Return of General Buller to Lydenburg after having marched through the whole of the hilly country to the north as far as Pilgrim's Rest, and having occupied the principal Boer positions.
10/9/1900 De Wet driven north, across the Vaal, at Venterstroom.
10/19/1900 Mr. Kruger left Lorenzo Marques for Europe.
10/24/1900 General Buller left Cape Town for England. Koffyfontein besieged.
10/25/1900 The Transvaal formally annexed.
11/3/1900 Koffyfontein relieved.

Each commandant did his best to get the necessary things together for his burghers, and my quarters were the centre of great activity from the early morning to late in the evening, persons who had had their goods commandeered applying to the General and lodging complaints.

After we had been at Pietersburg for eight days, a delay which seemed so many months to me, I had really had too much of it. The complaints were generally introduced by remarks about how much the complainants' ancestors had done for the country at Boomplaats, Majuba, etc., etc., and how unfairly they were now being treated by having their only horses, or mules, or their carriages, or saddles commandeered.

The worst of it was, that they all had to be coaxed, either with a long sermon, pointing out to them what an honour and distinction it was to be thus selected to do their duty to their country and their people, or by giving them money if no appeal to their generous feelings would avail; sometimes by using strong language to the timid ones, telling them it would have to be, whether they liked it or not.

Anyhow we got a hundred fine horses together at the cost of a good many imprecations. The complainants may be divided into the following categories:—

1st. Those who really believed they had some cause of complaint.

2nd. Those who did not feel inclined to part with anything without receiving the full value in cash—whose patriotism began and ended with money.

3rd. Those who had Anglophile tendencies and thought it an abomination to part with anything to a commando (these were the worst to deal with, for they wore a mask, and we often did not know whether we had got hold of the Evil One's tail or an angel's pinions), and

4th. Those who were complaining without reason. These were, as a rule, burghers who did not care to fight, and who remained at home under all sorts of pretexts.

The complaints from females consisted of three classes:—

1st. The patriotic ones who did all they could—sensible ladies as they were—to help us and to encourage our burghers, but who wanted the things we had commandeered for their own use.

2nd. The women without any national sympathy—a tiresome species, who forget their sex, and burst into vituperation if they could not get their way; and

3rd. The women with English sympathies, carefully hidden behind a mask of pro-Boer expressions.

The pity of it was that you could not see it written on their foreheads which category they belonged to, and although one could soon find out what their ideas were, one had to be careful in expressing a decided opinion about them, as there was a risk of being prosecuted for libel.

I myself always preferred an outspoken complaint. I could always cut up roughly refer him to martial law, and gruffly answer, "It will have to be like this, or you will have to do it!" And if that did not satisfy him I had him sent away. But the most difficult case was when the complaint was stammered under a copious flood of tears, although not supported by any arguments worth listening to.

There were a good many foreign subjects at Pietersburg but they were mostly British, and these persons, who also had some of their horses, etc., commandeered, were a great source of trouble, for many Boer officers and burghers treated them without any ceremony, simply taking away what they wanted for their commandos. I did not at all agree with this way of doing things, for so long as a foreign subject, though an Englishman, is allowed to remain within the fighting lines, he has a right to protection and fairness, and no difference ought to be made between him and the burghers who stay at home, when there is any fighting to be done.

From Pietersburg we went to Nylstroom, a village on the railway to which I had been summoned by telegram by the Commandant-General, who had arrived there on his way to the westerly districts, this being the first I had heard of him after we had parted at the foot of the Mauchberg, near Mac Mac.

I travelled by rail, accompanied by one of my commandants. The way they managed to keep up steam was delightfully primitive. We did not, indeed, fly along the rails, yet we very often went at the rate of nine miles an hour!

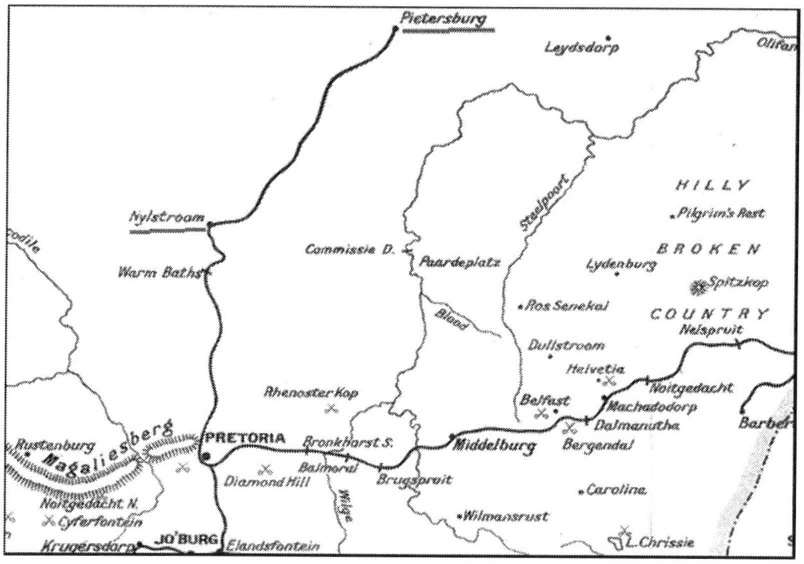

Pietersburg in far north and Nylstroom south on the Rail line.

When our supply of wood got exhausted, we would just stop the train, or the train would stop itself, and the passengers were politely requested to get out and take a hand at cutting down trees and carrying wood. This had a delicious flavour of the old time stage coach about it, when first, second, and third class passengers travelled in the same compartment, although the prices of the different classes varied considerably. When a coach came to the foot of a mountain the travellers would, however, soon find out where the difference between the classes lay, for the driver would order all first-class passengers to keep their seats, second-class passengers to get out and walk, and third-class passengers to get out and push.

We got to our destination, however, although the chances seemed to have been against it. I myself had laid any odds against ever arriving alive.

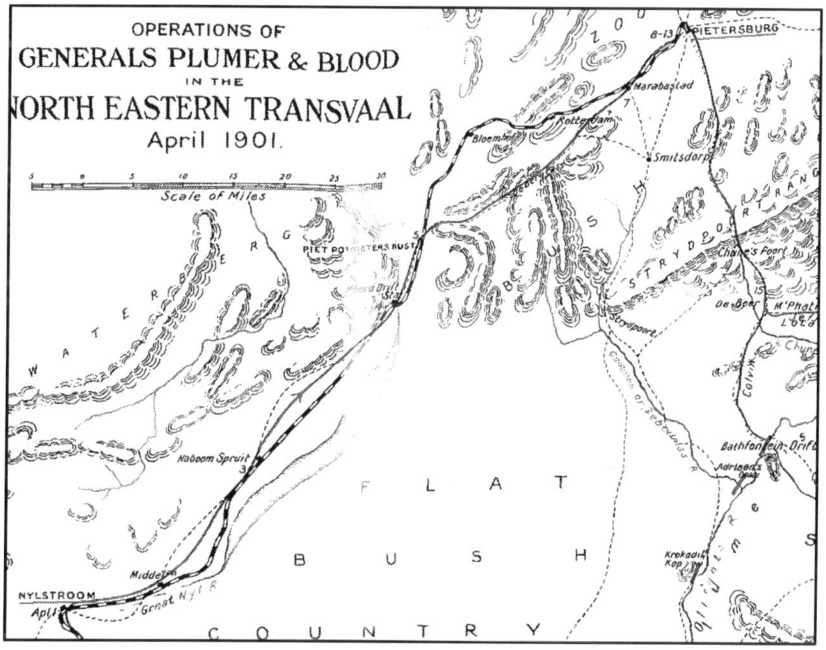

Same area with a little more detail.

At Nylstroom we found President Steyn and suite, who had just arrived, causing a great stir in this sleepy little village, which had now become a frontier village of the territory in which we still held sway.

A great popular meeting was held, which President Steyn opened with a manly speech, followed by a no less stirring one from our Commandant-General, both exhorting the burghers to do their duty towards their country and towards themselves by remaining faithful to the Cause, as the very existence of our nation depended on it.

In the afternoon the officers met in an empty hall of the hotel at Nylstroom to hold a Council of War, under the direction of the Commandant-General [Louis Botha].

Plans were discussed and arrangements made for the future. I was to march at once from Pietersburg to the north-westerly part of the Pretoria district, and on to Witnek, which would bring us back to our old battle-grounds. The state of the commandos, I was told, in those parts was very sad. The commandant of the Boksburg Commando had mysteriously fallen into the enemy's hands, and with his treacherous assistance nearly the whole commando had been captured as well. The Pretoria Commando had nearly shared this melancholy fate.

That same night we travelled to Pietersburg. After we had passed Yzerberg the train seemed to be going more and more slowly, till we came to a dead stop. The engine had broken down, and all we could do was to get out and walk the rest of the way. In a few hours' time, to our great joy, the second, and the only other train from Pietersburg there was, came up.

After having convinced the engine-driver that he had to obey the General's orders, he complied with our request to take us to Pietersburg, and at last, after a lot of trouble, we arrived the following day. Our cattle and horses were now sufficiently rested and in good condition. The commandos have been provided with the things they most urgently needed, and ordered to be ready within two days.

Chapter 25. Punishing the Pro-British.

During the first days of November, 1900, we went from Pietersburg to Witnek, about nineteen miles north of Bronkhorst Spruit, in the Pretoria district. We had enjoyed a fortnight's rest, which had especially benefited our horses, and our circumstances were much more favourable in every respect when we left Pietersburg than when we had entered it.

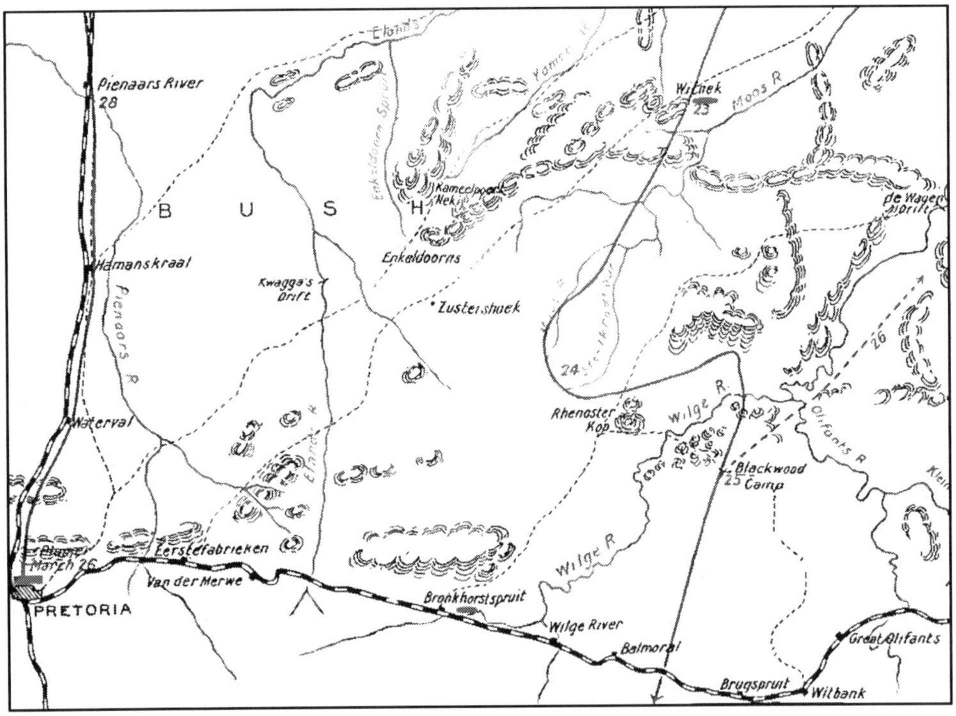

Witnek top and Bronkhorst Spruit on the rail line.

The Krugersdorp Commando had been sent to its own district, from Pietersburg via Warmbad and Rustenburg, under Commandant Jan Kemp, in order to be placed under General De la Rey's command. Most of the burghers preferred being always in their own districts, even though the villages scattered about were in the enemy's hands, the greater part of the homesteads burnt down and the farms destroyed, and nearly all the families had been placed in British Concentration Camps; and if the commanding officers would not allow the burghers to go to their own districts they would simply desert, one after the other, to join the commando nearest their districts.

I do not think there is another nation so fondly attached to their home and its neighbourhood, even though the houses be in ruins and the farms destroyed. Still the Boer feels attracted to it, and when he has at last succeeded in reaching it, you will often find him sit down disconsolately among the ruins or wandering about in the vicinity.

It was better, therefore, to keep our men somewhere near their districts, for even from a strategical point of view they were better there, knowing every nook and cranny, which enabled them to find exactly where to hide in case of danger. Even in the dark they were able to tell, after scouting, which way the enemy would be coming. This especially gave a commando the necessary self-reliance, which is of such great importance in battle. It has also been found during the latter part of the War to be easier for a burgher to get provisions in his own district than in others, notwithstanding the destruction caused by the enemy.

Commandant Muller,[128] of the Boksburg Commando, one of those who were lucky enough to escape the danger of being caught through the half-heartedness of the previous commandant (Dirksen), and had taken his place, arrived at Warmbad almost the same moment. He proceeded via Yzerberg and joined us at Klipplaatdrift near Zebedelestad.

I had allowed a field-cornet's company, consisting of Colonial Afrikanders, to accompany President Steyn to the Orange Free State, which meant a reduction of my force of 350 men, including the Krugersdorpers. But the junction with the Boksburg burghers, numbering about 200 men, somewhat made up for it.

We went along the Olifant's River, by Israelskop and Crocodile Hill, to the spot where the Eland's River runs into the Olifant's River, and thence direct to Witnek through Giftspruit.

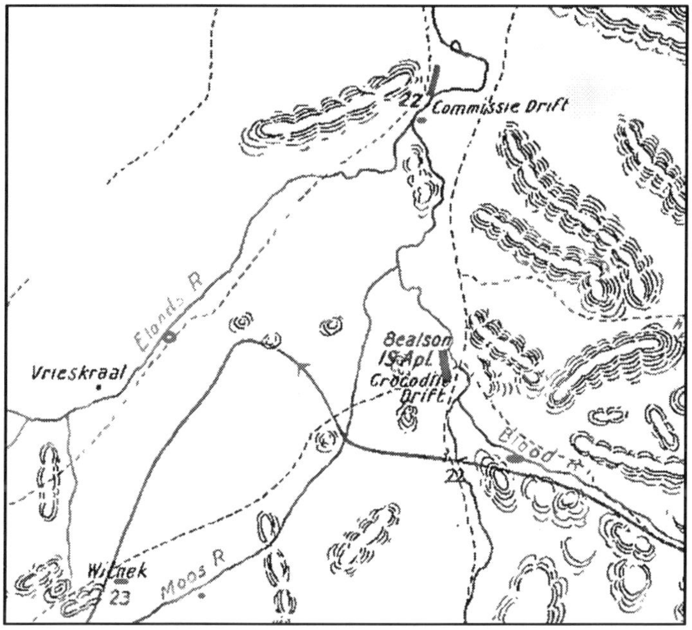

Witnek at the bottom. The Elands, Moos and Blood Rivers converge at Commissie Drift.

The grass, after the heavy rains, was in good condition and yielded plenty of food for our quadrupeds. Strange to say, nothing worth recording occurred during this "trek" of about 95 miles.

About the middle of November we camped near the "Albert" silver mines, south of Witnek.[129]

Commandant [Daniel] Erasmus was still in this part of the country with the remainder of the Pretoria Commando. Divided into three or four smaller groups, they watched in the neighbourhood of the railway, from

[128] General Chris Muller (1865–1945) was a Boer general during the Anglo-Boer War. He succeeded Gen. Ben Viljoen as commander of the Boer forces in the Eastern Transvaal after Viljoen's capture on 25 January 1902.
Christiaan Hendrik Muller was born on 4 January 1865 in the Cape Colony. He grew up in the Free State and later in the Transvaal where he fought in various indigenous campaigns. He fought in the Second War of Independence as a corporal in the Boxburg Commando and was later promoted to general because of his outstanding knowledge of weapons.
Muller was involved in the Battle of Helvetia with Ben Viljoen. The battle took place on 29 December 1900 at Helvetia, 10km north of Machadodorp. The Boers won a victory over the British Forces led by Major Cotton. Both the Boer force and the British force were 350 men strong, while the British also had a 120mm cannon at their disposal. Muller would later become best known for exploiting this cannon, dubbed the 'Lady Roberts.'
He took part in the peace negotiations at Vereeniging. In later years he was the director of the Land Bank. In 1914, he was involved in the Maritz Rebellion. He was also a Member of Parliament for a number of years. https://en.wikipedia.org/wiki/Chris_Muller

[129] 11/7/1900 The Battle of Leliefontein (also known as the Battle of Witkloof) was an engagement between British-Canadian and Boer forces during the Second Boer War on 7 November 1900, at the Komati River 30 kilometres (19mi) south of Belfast at the present day Nooitgedacht Dam. During the engagement, the Canadian rearguard successfully repelled Boer assaults against the column, facilitating the successful withdrawal of the British-Canadian force from the area. Battle of Leliefontein - Wikipedia See also: Joachim Christoffel Fourie https://en.wikipedia.org/wiki/Joachim_Christoffel_Fourie (1 February 1845–7 November 1900) And: Hendrik Frederik Prinsloo (10 May 1861–7 November 1900) https://en.wikipedia.org/wiki/Hendrik_Frederik_Prinsloo_(1861%E2%80%931900)

Donkerhoek till close to Wilgeriver Station, and whenever the enemy moved out, the men on watch gave warning and all fled with their families and cattle into the "boschveldt" along Witnek.

It was these tactics which enabled the British Press to state that the Generals Plumer[130] and Paget[131] had a brilliant victory over Erasmus the previous month; for, with the exception of a few abandoned carts at Zusterhoek, they could certainly not have seen anything of Erasmus and his commando except a cloud of dust on the road from Witnek to the "boschveldt."

An open veld landscape.[132] Colonel Plumer, Column Commander, South African Field Force, 1901.[133]

I had instructions to reorganise the commandos in these regions and to see that law and order were maintained. The reorganisation was a difficult work, for the burghers were divided amongst themselves.

Some wanted a different commando, while others wanted to keep to Erasmus, who was formerly general and who had been my superior, round Ladysmith. He, one of the wealthiest and most influential burghers in the Pretoria district, did not seem inclined to carry out my instructions, and altogether he could not get accustomed to the altered conditions. I did all I could in the matter, but, so far as the Pretoria Commando was concerned, the result of my efforts was not very satisfactory. Nor did the generals who tried the same thing after me get on with the reorganisation while Erasmus remained in control as an officer.

A dangerous element, which he and his clique tolerated, was formed by some families (Schalkwyk and others) who, after having surrendered to the enemy, were allowed to remain on their holdings, with their cattle, and to go on farming as if nothing had happened. They generally lived near the railway between our sentry stations and those of the enemy. These "voluntarily disarmed ones," as we called them, had got passes from the enemy, allowing them free access to the British camps, and in accordance with one of Lord Roberts' proclamations, their duty, on seeing any Boers or commandos, was, to notify this at once to the nearest English picket, and also to communicate all information received about the Boers. All this was on penalty of having their houses burnt down and their cattle and property confiscated.

[130] Herbert Charles Onslow Plumer, 1st Viscount Plumer (13 March 1857–16 July 1932) He was promoted to colonel on 29 November 1900 and was then given command of a mixed force which captured General Christiaan de Wet's wagon train at Hamelfontein in February 1901.
Later he was a senior British Army officer who fought in the First World War, being perhaps most notable for commanding the Second Army of the British Expeditionary Force on the Western Front from 1915 to 1918.

[131] Sir Arthur Henry Fitzroy Paget, (1 March 1851–8 December 1928) was a British army general who was, in the later stages of the Boer War, put in command of an independent column hunting Boer commandos. Many of Paget's Australian and New Zealand volunteers were disenchanted and wanted to go home before the expiry of their one-year term of service. Paget gave what Ian Beckett describes as a "blustering speech" to his whole assembled force in November 1900, refusing their request. Herbert Plumer had to pacify the men. Subsequently, they blamed Paget for ordering a costly frontal assault on Rhenoster Kop on 29 November 1900.
In June 1901, Paget resigned his command, intending to give up his career, after quarrelling with his superior, the Hon. Neville Lyttelton. Paget was persuaded by Edward VII, whom he knew well, to continue, Lord Roberts suggesting Paget was 'difficult with those above him'. Paget wrote to Sir John French praising his leadership in South Africa, and claiming that respect for him had been his reason for remaining in the Army. https://en.wikipedia.org/wiki/Arthur_Paget_(British_Army_officer)
Later, in 1914 he served as Commander-in-Chief, Ireland, where he was partly responsible for the Curragh Incident. [The Curragh incident of 20 March 1914, sometimes known as the Curragh mutiny, in Ireland in 1914. https://en.wikipedia.org/wiki/Curragh_incident

[132] A veld landscape. https://commons.wikimedia.org/wiki/File:OK_Vrystaat4.JPG Author: Katangais.

[133] Colonel Plumer, South African Field Force, 1901.
https://commons.wikimedia.org/wiki/File:StateLibQld_2_187251_Colonel_Plumer,_South_African_Field_Force,_1901.jpg

Sometimes a brother or other relative of these "hands-uppers" would call on them. The son of one of them was adjutant to Commandant Erasmus, and shared his tent with him, while the adjutant often visited his parents during the night and sometimes by day; the consequence being that the English always knew exactly what was going on in our district. This situation could not be allowed to go on, and I instructed one of my officers to have all these suspected families placed behind our commandos. Any male persons who had surrendered to the enemy out of cowardice were arrested.

Most of them were court-martialed for high treason and desertion, and giving up their arms, and fifteen were imprisoned in a school building at Rhenosterkop, which had been turned into a gaol for the purpose. The court consisted of a presiding officer selected from the commandants by the General, and of four members, two of whom had been chosen by the General and the President, and two by the burghers.

In the absence of our "Staats-procureur," a lawyer was appointed public prosecutor.

Before the trial commenced the President was sworn by the General and the other four members by the President. The usual criminal procedure was followed, and each sentence was submitted for the General's ratification.

The court could decree capital punishment, in which case there could be an appeal to the Government.

There were other courts, constituted by the latter, but as they were moving about almost every day, they were not always available, and recourse had then to be taken to the court-martial.

The fifteen prisoners were tried in Rhenosterkop churchyard. The trial lasted several days, and I do not remember all the particulars of the various sentences, which differed from two and a half to five years' imprisonment, I believe with the option of a fine. The only prison we could send them to was at Pietersburg, and there they went.

The arresting and punishing of these people caused a great sensation in the different commandos.

It seems incredible, but it is a fact that many members of these traitors' families were very indignant about my action in the matter, even sending me anonymous letters in which they threatened to shoot me.

Although there was less treason after the conviction of these fifteen worthies had taken place, there always remained an easy channel in the shape of correspondence between burghers from the commandos and their relatives within the English fighting lines, carried by kaffir runners. This could not be stopped so easily.

On the 19th of November, 1900, I attacked the enemy on the railway simultaneously at Balmoral and Wilgeriver, and soon found that the British had heard of our plan beforehand.

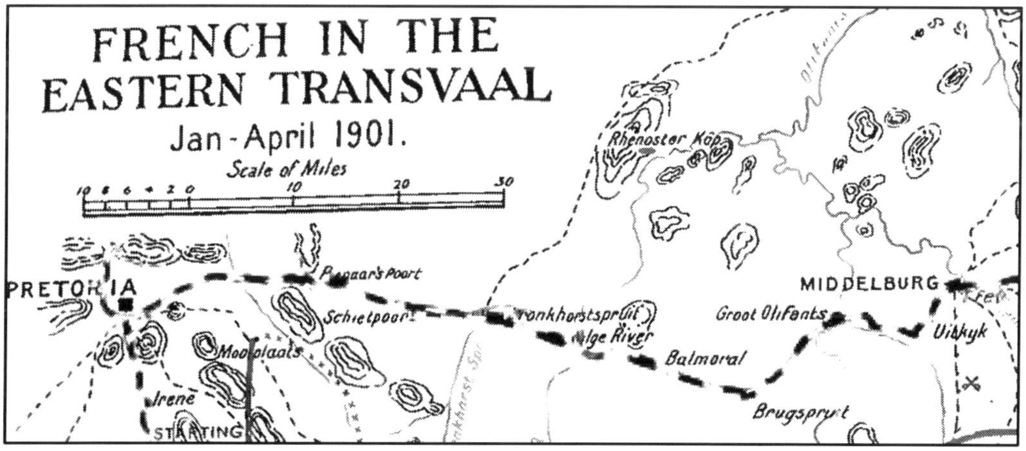

PF: This action is not recorded in the Chronology. Note Rhenoster Kop.

Commandant Muller, who was cautiously creeping up to the enemy at Wilgeriver with some of his burghers, and a Krupp gun, met with a determined resistance early in the morning. He succeeded, indeed, in taking a few small forts, but the station was too strongly fortified, and the enemy used two 15-pounders in one of the forts with such precision as to soon hit our Krupp gun, which had to be cleared out of the fighting line.

The burghers, who had taken the small forts in the early morning, were obliged to stop there till they could get away under protection of the darkness, with three men wounded. We did not find out the enemy's losses.

We were equally unfortunate near Balmoral Station, where I personally led the attack.

At daybreak I ordered a fortress to be stormed, expecting to capture a gun, which would enable us to fire on the station from there, and then storm it. In fact we occupied the fort with little trouble, taking a captain and 32 men prisoners, besides inflicting a loss of several killed and wounded, while a score more escaped. These all belonged to the "Buffs," the same regiment which now takes part in watching us at St. Helena. But, on the whole, we were disappointed, not finding a gun in the fort, which was situated to the west of the station. Two divisions of burghers with a 15-pounder and a pom-pom were approaching the station from north and east, while a commando, under Field-Cornet Duvenhage, which had been called upon to strengthen the attack, was to occupy an important position in the south before the enemy could take it up, for during the night it was still unoccupied.

Our 15-pounder, one of the guns we had captured from the English, fired six shells on the enemy at the station, when it burst, while the pom-pom after having sent some bombs through the station buildings, also jammed. We tried to storm over the bare ground between our position and the strongly barricaded and fortified station, and the enemy would no doubt have been forced to surrender if they had not realised that something had gone wrong with us, our guns being silent, and Field-Cornet Duvenhage and his burghers not turning up from the south. The British, who had taken an important position from which they could cover us with their fire, sent us some lyddite shells from a howitzer in the station fort. Although there was a good shower of them, yet the lyddite-squirt sent the shells at such a slow pace, that we could quietly watch them coming and get under cover in time and therefore they did very little harm.

At eight o'clock we were forced to fall back, for although we had destroyed the railway and telegraphic communications in several places overnight, the latter were repaired in the afternoon, and the enemy's reinforcements poured in from Pretoria as well as from Middelburg. I observed all this through my glass from the position I had taken up on a high point near the Douglas coal mines.

Douglas Mine (miners in coal trollies).[231]

Douglas Mine (Colliery) and Jansen's House.[232] [134]

The area around the colliery & Witbank.[233]

[134] From Haldane's account: Mr. Moore, the manager of the Douglas Colliery store, now took complete charge of us. He led us to his house, which stood in a garden quite apart from any others, and soon we were seated at his hospitable table eating as if our lives depended on it. Our first civilised meal for over three weeks consisted of tinned salmon, cocoa, and the usual adjuncts. I think we horrified our host by our indecent rapacity, but it gave him a good idea of what we should require on the morrow.
I regret to say that, many months later, when Botha and his men were retiring eastward from Pretoria after Lord Roberts' entry into the capital, some of his force happened to camp at the Douglas mine.
As I have mentioned before, the secret of our having been hidden there was known to many, and came to the knowledge of the commando which was passing through. Whereupon some of them without hesitation maliciously destroyed Moore's house and property, and he probably only escaped with his life by happening to be in Johannesburg at the time.

Amongst the prisoners we had made in the morning was a captain of the "Buffs," whose collar stars had been stripped off for some reason, the marks showing they had only recently been removed. At that time there were no orders to keep officers as prisoners-of-war, and this captain was therefore sent back to Balmoral with the other "Tommies," after we had relieved them of their weapons and other things which we were in want of. I read afterwards, in an English newspaper, that this captain had taken the stars off in order to save himself from the "cruelties of the Boers."

This, I considered, an unjust and undeserved libel.

Chapter 26. Battle of Rhenosterkop.

On the 27th of November, 1900, our scouts reported that a force of the enemy was marching from the direction of Pretoria, and proceeding along Zustershoek. I sent out Commandant [Chris] Muller with a strong patrol, while I placed the laager in a safe position, in the ridge of kopjes running from Rhenosterkop some miles to the north. This is the place, about 15 miles to the north-east of Bronkhorst Spruit, where Colonel Anstruther with the 94th regiment was attacked in 1881 by the Boers and thoroughly defeated.[135] Rhenosterkop is a splendid position, rising several hundred feet above the neighbouring heights, and can be seen from a great distance. Towards the south and south-east this kopje is cut off from the Kliprandts (known by the name of Suikerboschplaats) by a deep circular cleft called Rhenosterpoort.

On the opposite side of this cleft the so-called "banks" form a "plateau" about the same height as the Rhenosterkop, with some smaller plateaux, at a lesser altitude, towards the Wilge River. These plateaux form a crescent running from south-east to north of the Rhenosterkop. Only one road leading out of the "bank" near Blackwood Camp and crossing them near Goun, gives access to this crescent. On the west side is a great gap up to Zustershoek, only interrupted by some "randjes," or ridges, near the Albert silver mines and the row of kopjes on which I had now taken up a position.

The enemy's force had been estimated at 5,000 men, mostly mounted, who, quite against their usual tactics, charged us so soon as they noticed us. Muller had to fall back again and again. The enemy under General Paget, pursued us as if we were a lot of game, and it soon became apparent that they had made up their mind to catch us this time. I sent our carts into the forest along Poortjesnek to Roodelaager, and made a stand in the kopjes near Rhenosterkop.

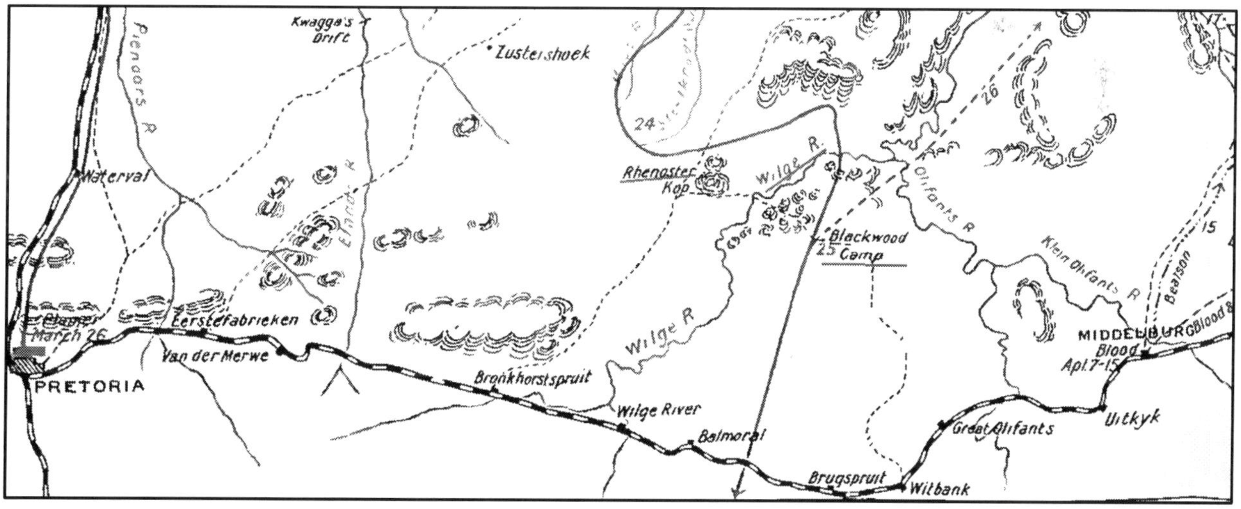

Blackwood Camp and Wilge River.

[135] On 20 December 1880, Lieutenant-Colonel Philip Robert Anstruther and men of the 94th Foot (Connaught Rangers) were on their way to quell the Boer uprising in Pretoria, when they were ambushed by Boer Forces near the Bronkhorstspruit Stream. This mainly Irish regiment was marching westward toward Pretoria, led by Lieutenant-Colonel Anstruther, when halted by a Boer commando group as they approached the Bronkhorstspruit, 38 miles from Pretoria. The Boer leader, Commandant Frans Joubert (brother of General Piet Joubert), demanded that Anstruther and the column turn back, stating that the territory was now again a Boer Republic and therefore any further advance by the British would be deemed an act of war. Anstruther refused and ordered that ammunition be distributed, and the Boers opened fire: about 120 soldiers were killed or wounded within minutes of the first shots, with total losses 56 men dead and 92 wounded. Boer casualties totaled two killed and five wounded. With the majority of his troops dead or wounded, the dying Anstruther ordered surrender. https://en.wikipedia.org/wiki/First_Boer_War And https://en.wikipedia.org/wiki/Battle_of_Bronkhorstspruit

On the 28th—the next day—General Paget pitched his camp near our positions, shelling us with some batteries of field guns till dusk. The same evening I received information that a force under General Lyttelton had marched from Middelburg and arrived near Blackwood Camp. This meant that our way near Gourjsberg had been cut off. All we could do was to keep the road along Poortjesnek well defended, for if the enemy were to succeed in blocking that as well, we would be in a trap and be entirely cut up.

There was General Paget against us to the west, to the south there was Rhenosterkop with no way out, and General Lyttelton to the east, while to the north there was only one road, running between high chains and deep clefts. If General Paget were to make a flanking movement threatening the road to the north, I should have been obliged to retire in hot haste, but we were in hopes the General would not think of this.

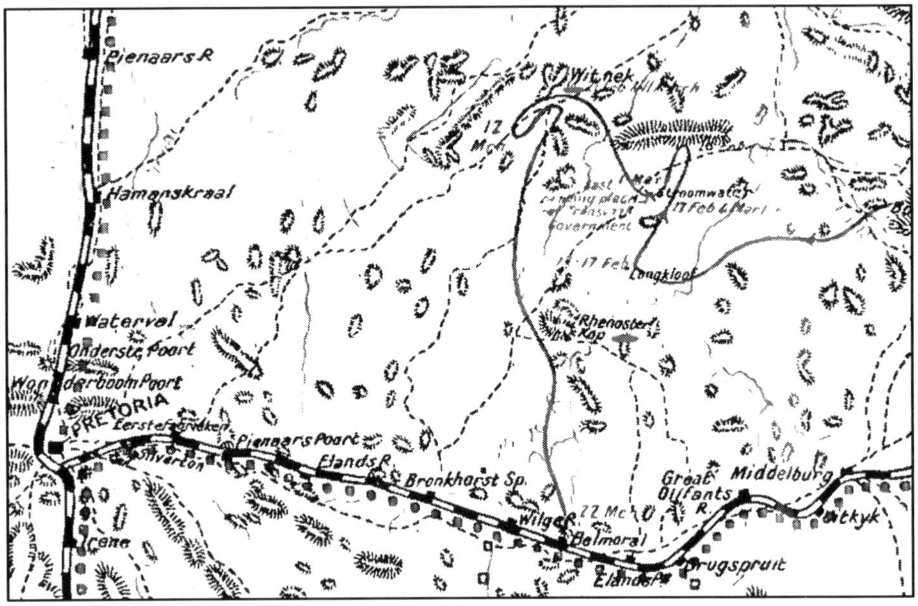

Rhenoster Kop and Witnek. No other landmarks are available. This map from later in the war shows the completed blockhouse system along the line.

General Lyttelton only needed to advance another mile, right up to the first "randts" of the mountain near Blackwood Camp, for his guns to command our whole position, and to make it impossible for us to hold it. I had, however, a field-cornet's company between him and my burghers, with instructions to resist as long as possible, and to prevent our being attacked from behind, which plan succeeded, as luck would have it. My Krupp and pom-pom guns had been repaired, or rather, patched up, though the former had only been fired fourteen times when it was done up.

I placed the Johannesburgers on the left, the Police in the centre, and the Boksburgers on the right. As I have already pointed out, these positions were situated in a row of small kopjes strewn with big "klips," while the assailant would have to charge over a bare "bult," and we should not be able to see each other before they were at 60 to 150 paces distant.

Next morning, when the day dawned, the watchmen gave the alarm, the warning we knew so well, "The Khakis are coming!" The horses were all put out of range of the bullets behind the "randts." I rode about with my officers in front of our positions, thus being able to overlook the whole ground, just at daybreak.

It gave me a turn when I suddenly saw the gigantic army of "Khakis" right in front of us, slowly approaching, in grand formation, regiment upon regiment, deploying systematically, in proper fighting order, and my anxiety was mingled with admiration at the splendid discipline of the adversary. This, then, was the first act in the bloody drama which would be played for the next fifteen hours. The enemy came straight up to us, and had obviously been carefully reconnoitring our positions.

General Paget seemed to have been spoiling for a fight, for it did not look as if he simply meant to threaten our only outlet. His heavy ordnance was in position near his camp, behind the soldiers, and was firing at us over their heads, while some 15-pounders were divided amongst the different regiments. The thought of being involved in such an unequal struggle weighed heavily on my mind. Facing me were from four to five thousand soldiers, well equipped, well disciplined, backed up by a strong artillery; just behind me my men, 500 at the outside, with some patched-up guns, almost too shaky for firing purposes.

English 15-Pounder Gun, captured by Boers and retaken by British troops at Bethlehem.[234]

The Royal Artillery crew in action.[235] Navy Bluejacket servicing the gun, probably holding the next shell.

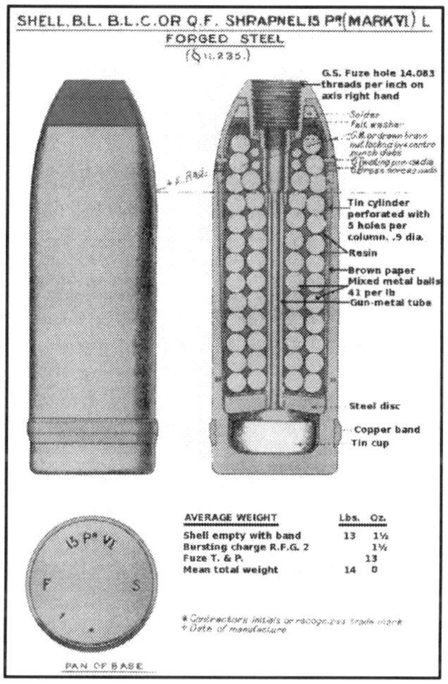

15 Lb. Shrapnel Shell with timed fuse on top.[236] Diagram of Construction.[237]

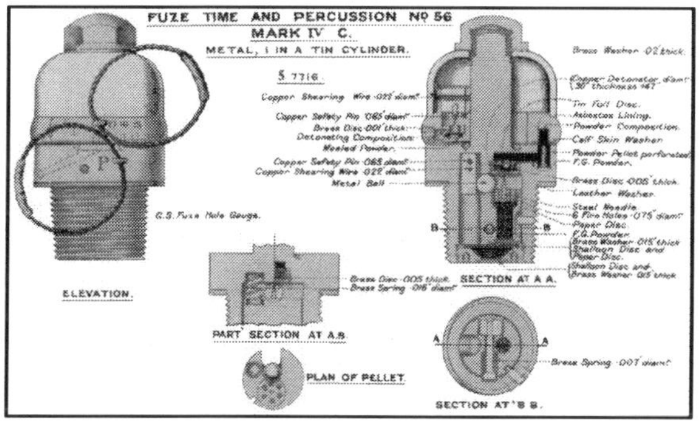

Fuse design.[238]

Setting the fuse times.[136] [239]

But I could rely on at least 90 per cent. of my burghers being splendid shots, each man knowing how to economise his store of ammunition, while their hearts beat warmly for the Cause they were fighting.

The battle was opened by our Krupp gun, from which they had orders to fire the fourteen shells we had at our disposal, and then "run." The enemy's heavy guns soon answered from the second ridge. When it was broad

[136] Photograph of naval ratings fitting primers / primer adaptors during the Tugela operations. For land service adaptors had to be inserted into the bases of the cartridges, enabling the electric primers to be replaced with percussion primers.

daylight the enemy tried his first charge on the Johannesburg position, over which my brother [Phillip] had the command, and approached in skirmishing order. They charged right up to seventy paces, when our men fired for the first time, so that we could not very well have missed our aim at so short a distance, in addition to which the assailants' outline was just showing against the sky-line as he was going over the last ridge. Only two volleys and all the Khakis were flat on the ground, some dead, others wounded, while those who had not been hit were obliged to lie down as flat as a pancake.

The enemy's field-pieces were out of our sight behind the ridge which the enemy had to pass in charging, and they went on firing without any intermission. Half an hour later the position of the Johannesburg Police, under the late Lieutenant D. Smith, was stormed again, this time the British being assisted by two field-pieces which they had brought up with them in the ranks and which were to be used as soon as the soldiers were under fire. They came to within a hundred paces. One of these guns, I think, I saw put up, but before they could get the range it had to be removed into safety, for the attacking soldiers fared equally badly here as on our left flank.

Then, after a little hesitation, they tried the attack on our right flank again, when Commandant Muller and the Boksburgers and some Pretoria burghers, under Field-Cornet Opperman held the position, but with the same fatal result to the attackers. Our fifteen-pounder, after having been fired a few times, had given out, while our pom-pom could only be used from time to time after the artilleryman had righted it.

I had a heliograph post near the left-hand position, one near the centre and the one belonging to my staff on our extreme right. I remained near this, expecting a flank movement by General Paget after his front attacks had failed. From this coign of vantage I was able to overlook the whole of the fighting ground, besides which I was in constant touch with my officers, and could tell them all the enemy's movements.

Corps of the State Artillery, Orange Free State. Telegraphists, heliographers and artillerymen commanded by Lieut. Baay.[240]

Using a Heliograph.[241] [242]

About 10 o'clock they charged again, and so far as I could see with a fresh regiment. We allowed them to come up very closely again and once more our deadly Mauser fire mowed them down, compelling those who went scot-free to go down flat on the ground, while during this charge some who had been obliged to drop down, now jumped up and ran away. If I remember rightly, it was during this charge that a brave officer, who had one of his legs smashed, leant on a gun or his sword, and kept on giving his orders, cheering the soldiers and telling them to charge on. While in this position, a second bullet struck him, and he fell mortally wounded. We afterwards heard it was a certain Colonel Lloyd of the West Riding Regiment. A few months after, on passing over this same battlefield, we laid a wreath of flowers on his grave, with a card, bearing the inscription: "In honour of a brave enemy."

General Paget seemed resolved to take our positions, whatever the sacrifice of human lives might be. If he succeeded at last, at this rate, he might find half a score of wounded burghers and, if his cavalry hurried up, perhaps a number of burghers with horses in bad condition, but nothing more.

Whereas, if he had made a flanking movement, he might have attained his end, perhaps without losing a single man.

Pride or stupidity must have induced him not to change his tactics. Nothing daunted by the repeated failures in the morning, our assailant charged again, now one position and then another, trying to get their field-pieces in position, but each time without success. At their wits' end, the enemy tried another dodge, bringing his guns right up to our position under cover of some Red Cross waggons. The officer who perceived this, reported to me by heliograph, asking for instructions. I answered: 'If a Red Cross waggon enters the fighting lines during the battle, it is there on its own responsibility.' Besides, General Paget, under protection of the white flag, might have asked any moment or an hour, or longer, to carry away his many unfortunate wounded, who were lying between two fires in the burning sun.

When the Red Cross waggon was found to be in the line of fire, it was put right-about face, while some guns remained behind to fire shrapnel at us from a short distance. They could only fire one or two shots, for our burghers soon put out of action the artillerists who were serving them. Towards the afternoon some of my burghers began to run short of ammunition, I had a field-cornet's force in reserve, from which five to ten men were sent to the position from time to time, and this cheered the burghers up again.

The same attacking tactics were persisted in by General Paget all day long, although they were a complete failure. When the sun disappeared behind the Magaliesbergs, the enemy made a final, in fact, a desperate effort to take our positions, the guns booming along while we were enveloped by clouds of dust thrown up by the shells.

The soldiers charged, brave as lions, and crept closer to our positions than they had done during the day.

But it seemed as if Fate were favouring us, for our 15-pounder had just got ready, sending his shells into the enemy's lines in rapid succession, and finding the range most beautifully. The pom-pom too—which we could only get to fire one or two shells all day long, owing to the gunner having to potter about for two or three hours after each shot to try and repair it—to our great surprise suddenly commenced booming away, and the two pieces—I was going to say the "mysterious" pieces—poured a stream of murderous steel into the assailants, which made them waver and then retire, leaving many comrades behind.

ZAR Staats Artillery using Maxim Pom Pom.[243]

ZAR Norden Maxim Mafeking.[244]

On our side only two burghers were killed, while 22 were wounded. The exact loss of the enemy was difficult to estimate. It must, however, have amounted to some hundreds.

Again night spread a dark veil over one of the most bloody dramas of this war. After the cessation of hostilities, I called my officers together and considered our position. We had not lost an inch of ground that day, while the enemy had gained nothing. On the contrary, they had suffered a serious repulse at our hands. But our ammunition was getting scarce, our waggons, with provisions, were 18 miles away. All we had in our positions was mealies and raw meat, and the burghers had no chance of cooking them. We therefore decided, as we had no particular interest in keeping these positions, to fall back that night on **Poortjesnek**, which was a "half-way house" between the place we were leaving and our carts, from which we should be able to draw our provisions and reserve ammunition.

We therefore allowed General Paget to occupy these positions without more ado.

I have tried to describe this battle as minutely as possible in order to show that incompetence of generals was not always on our side only.[137] [138]

I have seen from the report of the British Commander-in-Chief, published in the newspapers, that this battle had been a most successful and brilliant victory, gained by General Paget. People will say, perhaps, that it was silly on my part to evacuate the positions, and that I should have gone on defending them the next day. Well, in the old days this would have been done by European generals, but no doubt they were fighting under different circumstances. They were not faced by a force ten times their own strength; not restricted to a limited quantity of ammunition; nor were they in want of proper food or reinforcements.

The nearest Boer commando was at Warmbad, about 60 miles distant. Besides, there was no necessity, either for military or strategical reasons, for us to cling to these positions. It had already become our policy to fight whenever we could, and to retire when we could not hold on any longer. The Government had decided that the War should be continued and it was the duty of every general to manoeuvre so as to prolong it. We had no reserve troops, so my motto was: "Kill as many of the enemy as you possibly can, but see you do not expose your own men, for we cannot spare a single one."

On the 30th of November, the day after the fight, I was with a patrol on the first "randts," north-east of Rhenosterkop, just as the sun rose, and had a splendid view of the whole battlefield of the previous day. I saw the enemy's scouts, cautiously approaching the evacuated positions, and concluded from the precautions they were taking that they did not know we had left overnight. Indeed, very shortly after I saw the Khakis storming and occupying the kopjes. How great must have been their astonishment and disappointment on finding those positions deserted, for the possession of which they had shed so much blood. A number of ambulance waggons were brought up and were moving backwards and forwards on the battlefield, taking the wounded to the hospital camp, which must have assumed colossal proportions. Ditches were seen to be dug, in which the killed soldiers were buried. A troop of kaffirs carried the bodies, as far as I could distinguish, and I could distinctly see some heaps of khaki-coloured forms near the graves.

Battle of Rhenosterkop—How Colonel Lloyd Died. Waiting for the Wounded. "An Ambulance Waggon Receiving its Freight."[245]

[137] The Battle of Renosterkop [Rhinoceros Head] took place on 29 November 1900. The British were determined to wipe out a Boer commando, commanded by Viljoen, that was operating northeast of Pretoria. Paget launched a frontal attack. It began with an infantry attack on Viljoen's left flank, but it was repulsed by the Boers with accurate rifle and fire from the Maxim Nordenfeldt. The second attack was against the center of the front and was also repulsed by the Boer forces. The third attack was also towards the middle of the front, this time more on the right side where Viljoen himself was waiting. Once again, the attack was repulsed with heavy losses on the British side. The British forces fell back to continue the next day. However, the Boer forces withdrew during the night because they had run out of ammunition and other supplies.
The British suffered 106 casualties, of which 15 were casualties, while the Boers had 24 casualties, of which only 2 were casualties. https://af.wikipedia.org/wiki/Slag_van_Renosterkop

[138] 11/29/1900 A Boer force commanded by Gen. Ben Viljoen defeats a British force commanded by General Arthur Paget at Renosterkop (also Rhenosterkop) about 30 km north of Bronkhorstspruit.

As the battlefield looked now, it was a sad spectacle. Death and mutilation, sorrow and misery, were the traces yesterday's fight had left behind. How sad, I thought, that civilised nations should thus try to annihilate one another. The repeated brave charges made by General Paget's soldiers, notwithstanding our deadly fire, had won our greatest admiration for the enemy, and many a burgher sighed even during the battle. What a pity such plucky fellows should have to be led on to destruction like so many sheep to the butcher's block!

Meanwhile, General Lyttelton's columns had not got any nearer, and it appeared to us that he had only made a display to confuse us, and with the object of inducing us to flee in face of their overwhelming strength.

On the 1st of December General Paget sent a strong mounted force to meet us, and we had a short, sharp fight, without very great loss on either side.

This column camped at Langkloof, near our positions, compelling us to graze and water our horses at the bottom of the "neck" in the woods, where horse-sickness was prevalent. We were, therefore, very soon obliged to move.

About this time I received a report to the effect that a number of women and children were wandering about near Rhenosterkop along the Wilge River. Their houses had been burnt by order of General Paget, and we were asked to protect these unfortunate people.

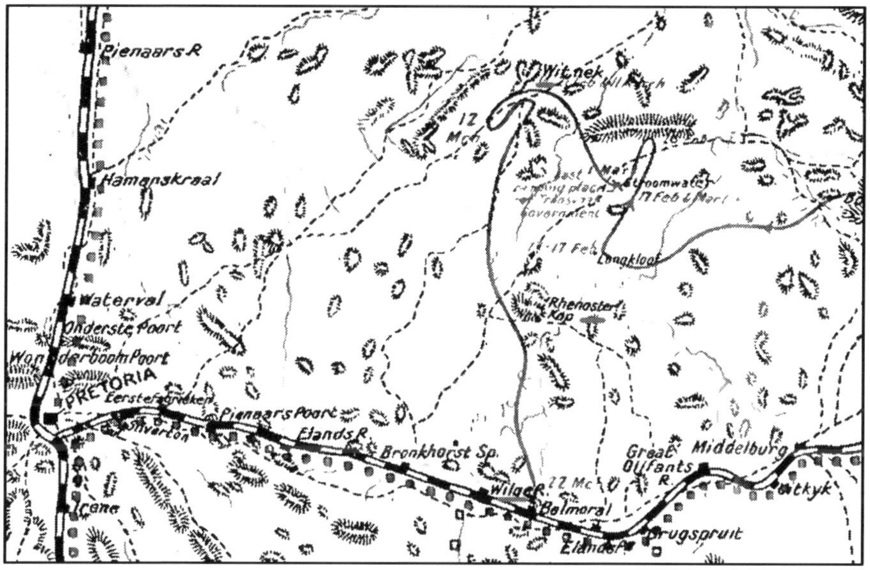

Wilge R., Rhenoster Kop and Witnek.

Some burghers offered to ride out at night time to try and find them, and the next morning they brought several families into our camp. The husbands of these poor sufferers were on duty in the neighbourhood, so that they were now enabled to do the needful for their wives and children. I put some questions to some of the women, from which it appeared that although they had besought the English not to burn their clothes and food, yet this had been done. Some Australians and Canadians, who had been present, had done their best to save some of the food and clothes, and these Colonials had shown them much consideration in every respect, but, the women added, a gang of kaffirs, who were ordered to cause this destruction, were behaving in the most barbarous and cruel manner, and were under no control by the British soldiers.

I felt bound to protest against these scandalous acts of vandalism, and sent two of my adjutants to the English camp next day with a note of about the following tenour:—

Letter to Paget.

"To General [Arthur] Paget, commanding H.M's. forces at Rhenosterkop.

"It is my painful duty to bring under your Honour's notice the cruel way in which the troops under your command are acting in ill-treating defenceless women and children. Not only their homes, but also their food and clothes, are being burnt. These poor creatures were left in the open veldt, at the mercy of the kaffirs, and would have died of starvation and exhaustion but for our assistance.

This way of treating these unfortunate people is undoubtedly against the rules of civilised warfare, and I beg to emphasise that the responsibility for this cruelty will be entirely yours. You may rest assured that a similar treatment of our families will not shorten the duration of the War, but that, on the contrary, such barbarities will force the burghers to prolong the struggle and to fight on with more bitterness and determination than ever."

The two despatch carriers whom I sent to the British General under a white flag were taken for spies, and however much they tried to establish their identity, General Paget was not to be convinced, and had them arrested, detaining them for three days. Their horses were used every day by the English officers, which I consider far from gentlemanly. On the third day my two adjutants were again taken before the general, and cross-examined, but no evidence could be found against their being bona-fide messengers. Paget told them that my despatch was all nonsense, and did not give them the right to enter his lines under the white flag, adding, while he handed them a letter addressed to me:

"You can go now; tell your General that if he likes to fight I shall be pleased to meet him at any time in the open. You have killed some of my Red Cross people, but I know it was done by those 'damned' unscrupulous Johannesburgers. Tell them I shall pay them for this!"

Before my adjutants left, a certain Captain —— said to one of them:

"I say, what do your people think of the fight?"

"Which fight do you mean?" asked the adjutant.

"The fight here," returned the captain.

"Oh," remarked the adjutant, "we think it was rather a mismanagement." To which the captain replied: "By Jove! you are not the only people who think so."

The contents of General Paget's letter were short and rough; "The responsibility for the suffering of women and children rests on the shoulders of those who blindly continue the helpless struggle," etc., etc.

I may say here that this was the first time in this War the English officers treated my despatch riders under the white flag in such a manner, giving me at the same time such a discourteous answer.

No doubt we have had generals acting like this on our side, and I admit that we did not always stand on etiquette.

As already stated, part of the enemy's forces were camping out near Poortjesnek, so close by that we had to shift our laager and commando to a more healthy part on account of the horse-sickness. The enemy installed a permanent occupation at Rhenosterkop, and we moved [East] into the Lydenberg district, where we knew we should find some wholesome "veldt" on the Steenkamps Mountains. We went through the forest near Maleemskop via Roodekraal, to the foot of Bothasberg, where we had a few weeks' rest.[139]

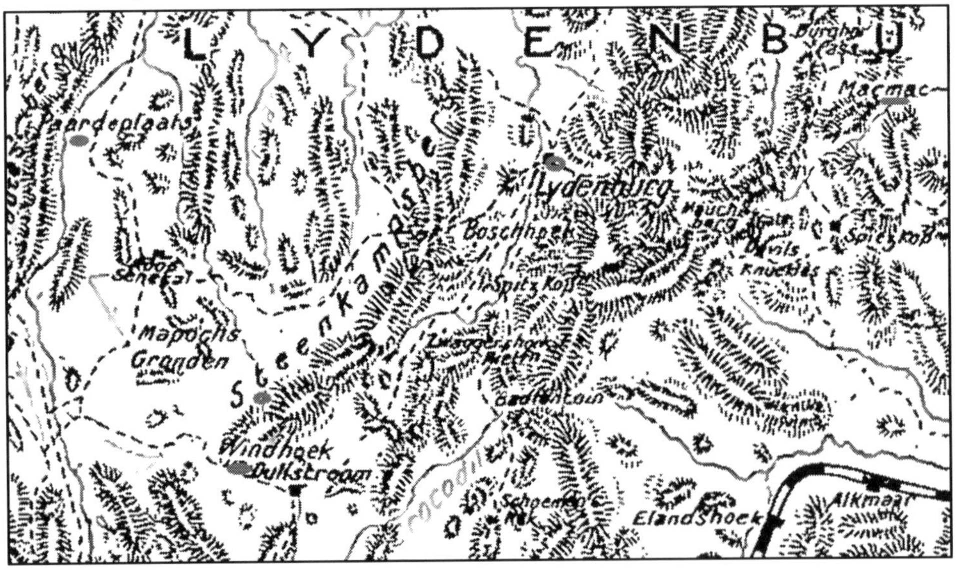

Steenkampsberg range, Windhoek/Dusllstroom at bottom and Paardeplaats top left. See: Steenkampsberg Pass.[140 246]

[139] 12/13/1900 Reverse to General Clements De la Rey, J.C. Smuts and C.F. Beyers overran a British camp under the control of General Clements near the Magaliesberg. Brabant's Horse mishap near Zastron. https://en.wikipedia.org/wiki/Battle_of_Nooitgedacht

[140] Steenkampsberg: The mountain ranges between 1,700 and 2,274 metres in altitude, and is covered largely by high-altitude grassland, broken in places by rocky outcrops. The northern slope of the mountain is drained by the Groot-Dwarsrivier, which enters a deep linear valley, to eventually join the Steelpoort River. https://en.wikipedia.org/wiki/Steenkampsberg,_Mpumalanga

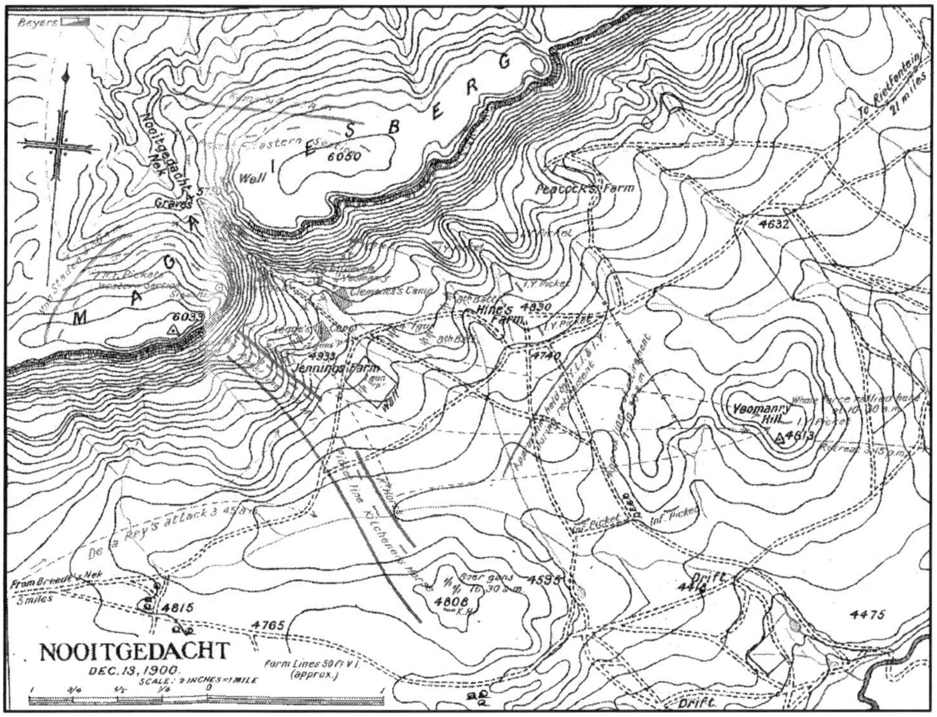

DIRECTIONS

N. F. Northumberland Fusiliers

Y. L. I. King's Own Yorkshire Light Infantry

I. Y. Imperial Yeomanry

K. H. Kitchener's Horse

2nd M. I. 2nd Battn Mounted Infantry

Vryheid Attack Dec 11-12-1900

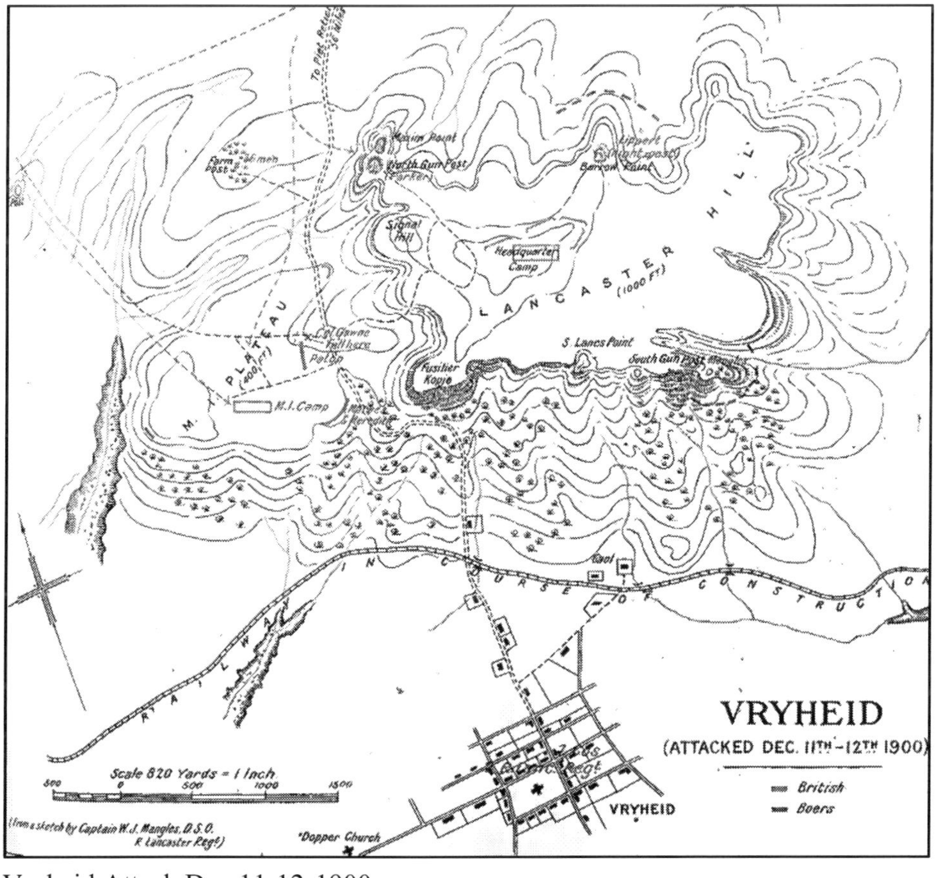

Vryheid Attack Dec 11-12-1900

Chapter 27. The Second Christmas at War.

The veldt was in splendid condition at the foot of Bothasberg, where we had pitched our camp. We found mealies and cattle left everywhere. The enemy did not know where we really were, and could not, therefore, bother us for the time being. Our Government was at Tautesberg,[141] about 12 miles north of Bothasberg, and we received a visit from Acting-President Burger, who brought with him the latest news from Europe, and the reports from the other commandos. Mr. Burger said he was sorry we had to leave the Pretoria district, but he could understand our horses would have all been killed by the sickness if we had stopped at Poortjesnek. As regards the Battle of Rhenosterkop, he expressed the Government's satisfaction with the result.

On the 16th of December we celebrated Dingaan's Day[142] in a solemn manner. Pastor J. Louw, who had faithfully accompanied us during these fatiguing months of retreats and adversity, delivered a most impressive address, describing our position. Several officers also spoke, and I myself had a go at it, although I kept to politics. In the afternoon the burghers had sports, consisting of races on foot and on horseback. The prizes were got together by means of small contributions from the officers. All went well, without any mishaps, and it was unanimously voted to have been very entertaining.

It was a peculiar sight—taking into consideration the circumstances—to see these people on the "veldt" feasting and of good cheer, each trying to amuse the other, under the fluttering "Vierkleur"—the only one we possessed—but the look of which gladdened the hearts of many assisting at this celebration in the wilderness. How could we have been in a truly festive mood without the sight of that beloved banner, which it had cost so many sacrifices to protect, and to save which so much Afrikander blood had been shed.

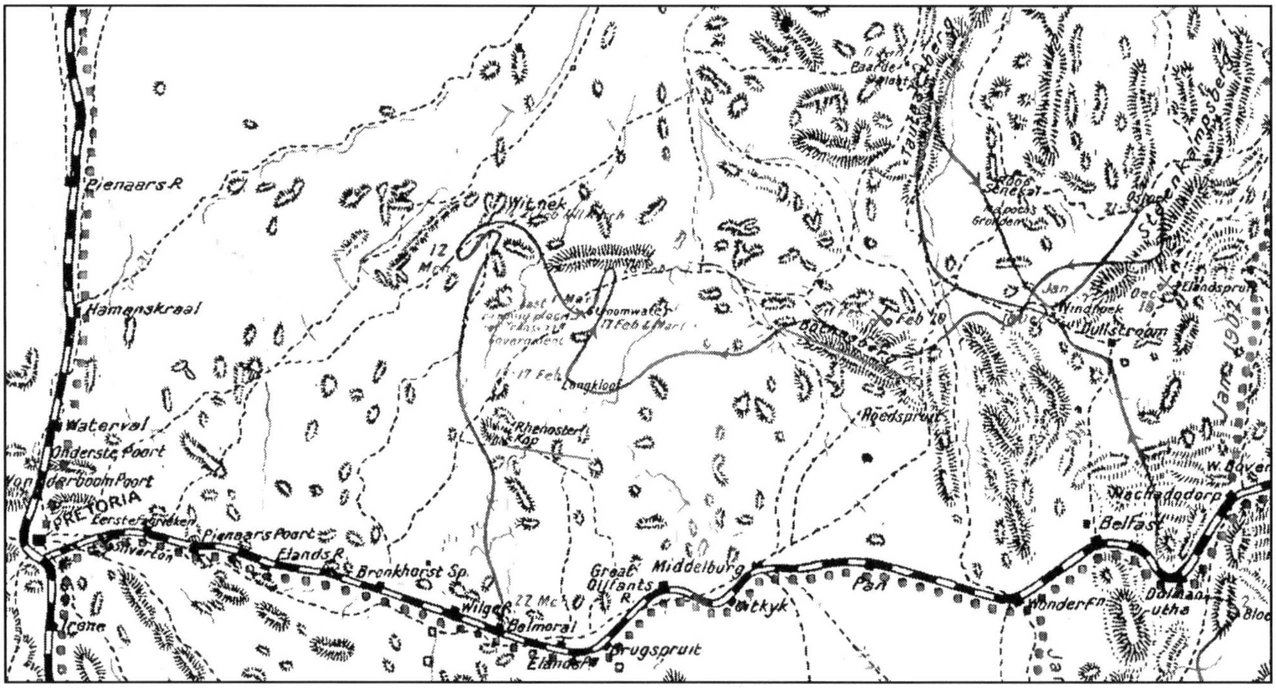

Rhenoster Kop and the Bothasberg 'mountain' and the same for Tautesberg.

And in many of us the thought suggested itself: "O, Vierkleur of our Transvaal, how much longer shall we be allowed to see you unfurled? How long, O Lord, will a stream of tears and blood have to flow before we are again the undisputed masters of our little Republic, scarcely visible on the world's map? For how long will our adored Vierkleur be allowed to remain floating over the heads of our persecuted nation, whose blood has stained and soaked your colours for some generations? We hope and trust that so sure as the sun shall rise in the east and set in the west, so surely may this our flag, now wrapped in sorry mourning, soon flutter aloft again in all its glory, over the country on which Nature lavishes her most wondrous treasures."

[141] The Times' Chronology indicates the Govt. was close by at Paardeplaats, from December 1900 through April of 1901.

[142] The Day of the Vow (Afrikaans: Geloftedag) is a religious public holiday in South Africa. It is an important day for Afrikaners, originating from the Battle of Blood River on 16 December 1838, before which 464 Voortrekkers made a promise to God that if he rescued them out of the hands of the approximately 16,000 Zulu warriors they were facing, they would honour that day as a sabbath day in remembrance of what God did for them. https://en.wikipedia.org/wiki/Day_of_the_Vow

The Afrikander character may be called peculiar in many respects. In moments of reverse, when the future seems dark, one can easily trace its pessimistic tendencies. But once his comrades buried, the wounded attended to, and a moment's rest left him by the enemy, the cheerful part of the Boer nature prevails, and he is full of fun and sport. If anybody, in a sermon or in a speech, try to impress on him the seriousness of the situation, pointing out how our ancestors have suffered and how we have to follow in their steps, our hero of yesterday, the jolly lad who was laughing boisterously and joking a minute ago, is seen to melt, and the tears start in his eyes. I am now referring to the true Afrikander. Of course, there are many calling themselves Afrikanders who during this War have proved themselves to be the scum of the nation. I wish to keep them distinguished from the true, from the noble men belonging to this nationality of whom I shall be proud as long as I live, no matter what the result of the War may be.

Our laagers were not in a very satisfactory position, more as regards our safety than the question of health, sickness being expected to make itself felt only later in the year.

We therefore decided to "trek" another 10 miles, to the east of Witpoort, through Korfsnek, to the Steenkampsbergen, in order to pitch or camp at Windhoek. Windhoek (wind-corner) was an appropriate name, the breezes blowing there at times with unrelenting fury.

Here we celebrated Christmas of 1900, but we sorely missed the many presents our friends and lady acquaintances sent us from Johannesburg on the previous festival, and which had made last year's Christmas on the Tugela such a success.

No flour, sugar or coffee, no spirits or cigars to brighten up our festive board. This sort of thing belonged to the luxuries which had long ceased to come our way, and we had to look pleasant on mealie-porridge and meat, varied by meat and mealie-porridge.

Yet many groups of burghers were seen to be amusing themselves at all sorts of games; or you found a pastor leading divine service and exhorting the burghers. Thus we kept our second Christmas in the field.

About this time the commandos from the Lydenburg district (where we now were) as well as those from the northern part of Middelburg, were placed under my command, and I was occupied for several days in reorganising the new arrivals. The fact of the railway being almost incessantly in the hands of the enemy, and the road from Machadodorp to Lydenburg also blocked by them (the latter being occupied in several places by large or small garrisons) compelled us to place a great number of outposts to guard against continual attacks and to report whenever some of the columns, which were always moving about, were approaching.

The spot where our laagers were now situated was only 13 miles from Belfast and Bergendal, between which two places General Smith-Dorrien's strong force was posted; while a little distance behind Lydenburg was General Walter Kitchener with an equally strong garrison. We were, therefore, obliged to be continually on the alert, not relaxing our watchfulness for one single moment. One or two burghers were still deserting from time to time, aggravating their shameful behaviour by informing the enemy of our movements, which often caused a well-arranged plan to fail. We knew this was simply owing to these very dangerous traitors.

The State Artillerymen, who had now been deprived of their guns, were transformed into a mounted corps of 85 men, under Majors Wolmarans[143] and Pretorius, and placed under my command for the time being.

It was now time we should assume the offensive, before the enemy attacked us. I therefore went out scouting for some days, with several of my officers, in order to ascertain the enemy's positions and to find out their weakest spot. My task was getting too arduous, and I decided to promote Commandant Muller to the rank of a fighting-general. He turned out to be an active and reliable assistant.

[143] See MOA for Major Jan Francois Wolmarans and also Commandant Willem Francois Pretorius who is probably the individual referred to but also see, Major Johannes Lodewijk Pretorius who is captured in December 1901 and sent to St. Helena.

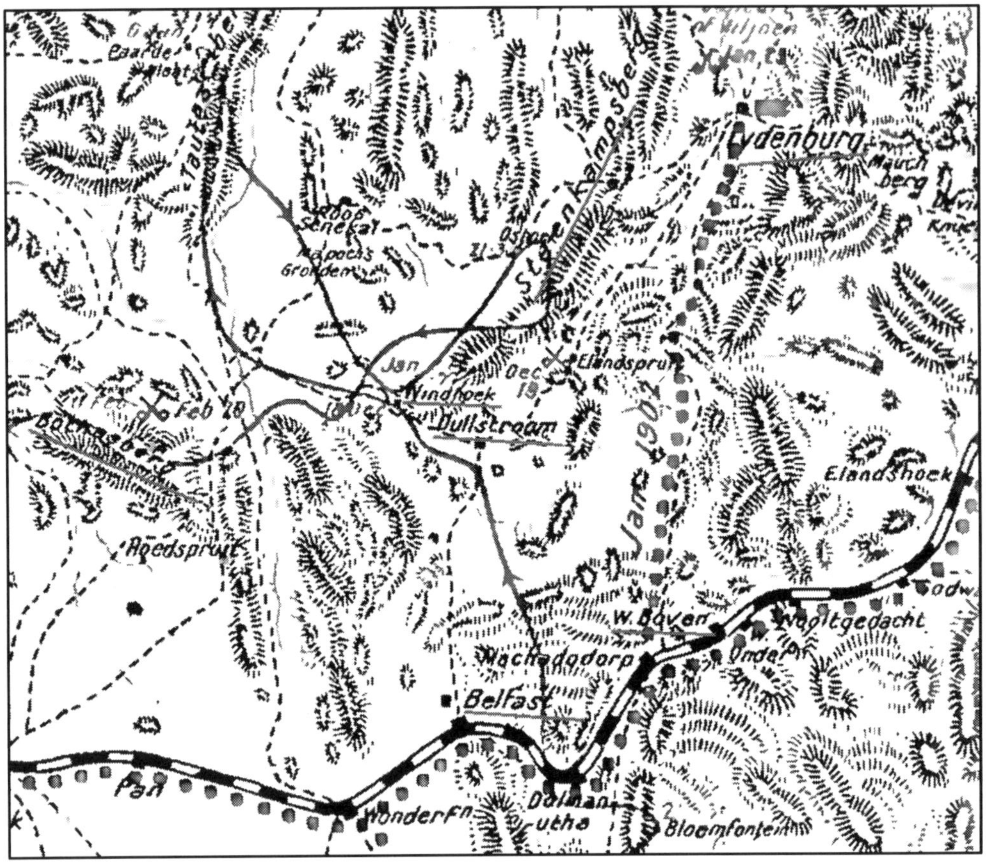

Belfast, Windhoek and Dullstroom as well as both the Bothasberg and Steenkampsberg.

Chapter 28. Capture of "Lady Roberts."

After I had carefully reconnoitred the enemy's positions, I resolved, after consulting my fighting-general, Muller, to attack the Helvetia garrison, one of the enemy's fortifications or camps between Lydenburg and Machadodorp. Those fortifications served to protect the railway road from Machadodorp Station to Lydenburg, along which their convoys went twice a week to provision Lydenburg village. Helvetia is situated three miles east of Machadodorp, four miles west of Watervalboven Station, where a garrison was stationed, and about three miles south of a camp near Zwartkoppies. It was only protected on the north side. Although it was difficult to approach this side on account of a mountainous rand through which the Crocodile River runs, yet this was the only road to take. It led across Witrand or Bakenkop; the commandos were therefore obliged to follow it, and had to do this at night time, for if they had passed the Bakenkop by day they would have exposed themselves to the enemy's artillery fire from the Machadodorp and Zwartkoppies garrisons.

During the night of the 28th of December 1900, we marched from Windhoek, past Dullstroom, up to the neighbourhood of Bakenkop, where we halted and divided the commandos for the attack, which was to be made in about the following order:—

Fighting-General Muller was to trek with 150 men along the convoy-road between Helvetia and Zwartkoppies up to Watervalboven, keeping his movements concealed from the adversary. Commandant W. Viljoen (my brother), would approach the northerly and southerly parts of Helvetia within a few hundred paces, with part of the Johannesburgers and Johannesburg Police. This commando numbered 200 men.

General “Fighting Bill” Muller and his assistant commandants and veld-cornets near Lydenburg, just before the general surrender.[247]

Christiaan H. Muller.[144]

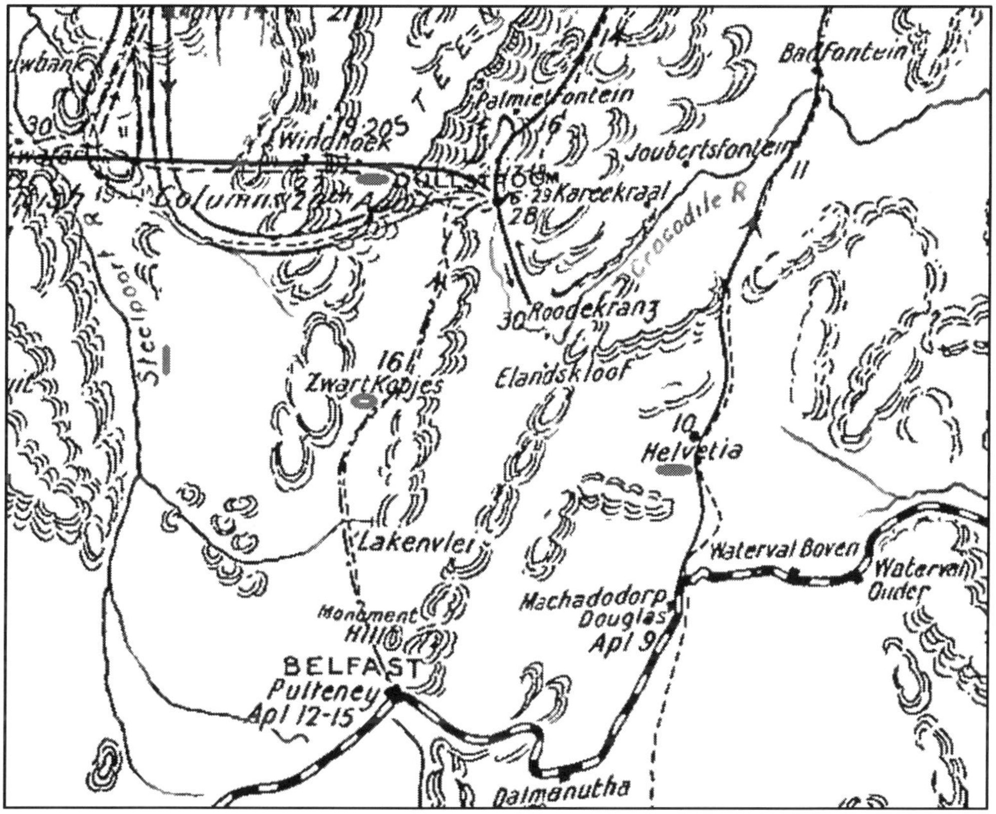

Helvetia and Zwart Kopjes as well as the Steelpoort River on left.

In order to be able to storm the different forts almost simultaneously we were all to move at 3.30 a.m., and I gave the men a password, in order to prevent confusion and the possibility of our hitting one another in the general charge. There being several forts and trenches to take the burghers were to shout “Hurrah!” as loudly as they could in taking each fort, which would show us it was captured, and at the same time encourage the others. Two of our most valiant field-cornets, P. Myburgh and J. Cevonia, an Italian Afrikander, were sent to the left, past Helvetia, with 120 men, to attack Zwartkoppies the moment we were to storm Helvetia, while I kept in reserve the State Artillerists and a field-cornet’s posse of Lydenburgers to the right of the latter place, near Machadodorp, which would enable me to stop any reinforcements sent to the other side from that place or from Belfast. For if the British were to send any cavalry from there they would be able to turn our rear, and by marching up as soon as they heard the first report of firing at Helvetia, they would be in a position to cut me up with the whole of my commando. I only suggest the possibility of it, and cannot make out why it was not attempted. I can only be thankful to the British officers for omitting to do this.

[144] See Military Officer Album.

I had taken up a position, with some of my adjutants, between the commandos as arranged, and stood waiting, watch in hand, for the moment the first shot should be fired. My men all knew their places and their duties, but unfortunately a heavy fog rose at about 2 o'clock, which made the two field-cornets who were to attack the Zwartkoppies lose their way and the chance of reaching their destination before daybreak.

I received the news of this failure at 3.20, i.e., ten minutes before the appointed time of action. A bad beginning, I thought, and these last ten minutes seemed many hours to me.

I struck a match every moment, under cover of my Macintosh,[145] to see if it were yet half past three. Another minute and it would soon be decided whether I should be the vanquished or the victor. How many burghers, who were now marching so eagerly to charge the enemy in his trenches, would be missed from our ranks to-morrow? It is these moments of tension which make an officer's hair turn grey. The relation between our burgher and his officers is so entirely different from that which exists between the British officer and his men or between these ranks perhaps in any other standing army. We are all friends. The life of each individual burgher in our army is highly valued by his officer and is only sacrificed at the very highest price. We regret the loss of a simple burgher as much as that of the highest in rank. And it was the distress and worry of seeing these lives lost, which made me ponder before the battle.

Suddenly one of my adjutants called out: "I hear some shouting. What may this be?"

I threw my waterproof over my head and struck a match, then cried: "It is time, my lads!" And in a few seconds a chain of fire flamed up round the forts, immediately followed by the rattling and crackling of the burghers' Mausers. The enemy was not slow in returning our fire.

It is not easy to adequately render the impression a battle in the dark makes. Each time a shot is fired you see a flash of fire several yards long, and where about 500 or 600 rifles are being fired at a short distance from you, it makes one think of a gigantic display of fireworks.

Although it was still dusk, I could easily follow the course of the fight. The defenders' firing slackened in several places, to subside entirely in others, while from the direction of the other reports and flashes, our men were obviously closing up, drawing tighter the ring round the enemy.

So far, according to my scouts, no stir had been made from Belfast, which encouraged me to inform the officers that we were not being cut off. At daybreak only a few shots were falling, and when the fog cleared up I found Helvetia to be in our hands.

General Muller reported that his part of the attack had been successfully accomplished, and that a 4·7 [inch] naval gun had been found in the great fortress. I gave orders to fetch this gun out of the fort without delay, to take away the prisoners we had made and as much of the commissariat as we could manage to carry, and to burn the remainder.

Towards the evening we were fired at by two guns at Zwartkoppies, making it very difficult for us to get the provisions away.

A great quantity of rum and other spirits was found among the enemy's commissariat, and as soon as the British soldiers made prisoners were disarmed, they ran up to it, filled their flasks, and drank so freely that about thirty of them were soon unable to walk. Their bad example was followed by several burghers, and many a man who had not been given to drinking used this opportunity to imbibe a good quantity, making it very difficult for us to keep things in order.

About 60 men of the garrison had been killed or wounded, and their commanding officer had received some injuries, but fortunately there was a doctor there who at once attended to these cases. On our side we had five men killed and seven wounded—the brave Lieutenant Nortje and Corporal J. Coetzee being amongst them.

A small fort, situated between the others, had been overlooked, through a misunderstanding, and a score of soldiers who were garrisoning it had been forgotten and omitted to be disarmed.

An undisciplined commando is not easily managed at times. It takes all the officers' tact and shrewdness to get all the captured goods—like arms, ammunition, provisions, &c.—transported, especially when drink is found in a captured camp.

[145] The Mackintosh raincoat (abbreviated as mac) is a form of waterproof raincoat, first sold in 1824, made of rubberized fabric.

When we discussed the victory afterwards, it became quite clear that our tactics in storming the enemy's positions on the east and south sides had been pregnant of excellent results, for the English were not at all prepared at these points, though they had been on their guard to the north. In fact it had been very trying work to force them to surrender there. The officer in command, who was subsequently discharged from the British Army, had done his best, but he was wounded in the head at the beginning of the fight, and so far as I could ascertain there had been nobody to take his place. Three lieutenants were surprised in their beds and made prisoners-of-war.

In the big fort where we found the naval gun, a captain of the garrison's artillery was in command. This fortress had been stormed, as already stated, from the side on which the attack had not been expected and the captain had not had an opportunity of firing many shots from his revolver, when he was wounded in the arm and compelled to surrender to the burghers who rushed up. Two hundred and fifty prisoners, including four officers, were made, the majority belonging to the Liverpool regiment and the 18th regiment of Hussars. They were all taken to our laager.[146] [147]

We succeeded in bringing away the captured gun in perfect order, also some waggons. Unfortunately the cart with the projectiles or shell, stuck in the morass and had to be left behind.

I gave orders to have a gun which we had left with the reserve burghers at Bakenkop, brought up, to open fire on the two pieces which were firing at us from Zwartkoppies, and to cover our movements while we were taking away the prisoners-of-war and the captured stores. I was in hopes of getting an opportunity of releasing the carts which stuck. But Fate was against us. A heavy hailstorm accompanied by thunder and lightning, fiercer than I have ever witnessed in South Africa before, broke over our heads. Several times the lightning struck the ground around us, and the weather became so alarming that the drunken "Tommies" began to talk about their souls, and further efforts to save the carts had to be abandoned.

Whoever may have been the officer in command at Zwartkoppies he really deserved a D.S.O., which he obtained, too.

What that order really means I wot[148] not, but I know that an English soldier is quite prepared to risk his life to deserve one, and as the decoration itself cannot be very expensive, it pays the British Government to be very liberal with it. A Boer would be satisfied with nothing less than promotion as a reward for heroism.

When the storm subsided we went on. It was a remarkable sight—a long procession of "Tommies," burghers, carts, and the naval gun, 18 feet long, an elephantine one when compared with our small guns.

It struck me again on this occasion what little bad feeling there was really between Boer and Briton, and how they both fight simply to do their duty as soldiers. As I rode along the stream of men I noticed several groups of burghers and soldiers sitting together along the road, eating from one tin of jam and dividing their loaf between them, and drinking out of the same field flask.

I remember some snatches of conversation I overheard:—

Tommy: By Jove, but you fellows gave us jip. If you had come a little later you wouldn't have got us so easy, you know.

Burgher: Never mind, Tommy, we got you. I suppose next time you will get us. Fortunes of war, you know. Have some more, old boy. Oh, I say, here is the general coming.

Tommy: Who's he? Du Wyte or Viljohn?

And then as I passed them the whole group would salute very civilly.

We stopped at Dullstroom that night, where we found some lodgings for the captured British officers. We were sorry one of the Englishmen had not been given time to dress himself properly, for we had a very scanty stock of clothes, and it was difficult to find him some.

[146] The Battle of Helvetia took place on 29 December 1900 during the Second War of Independence at Helvetia, 10km north of Machadodorp. Here the Boer forces led by Generals Ben Viljoen and Chris Muller won a victory over the British Forces, led by Major Cotton. The Boer force was 350 men and the British force 350 men strong and the British had one 120mm (4.7 in) gun at their disposal. Casualties: British 11 killed, 29 wounded, 253 captured. Boers: 3 killed, 5 wounded https://af.wikipedia.org/wiki/Slag_van_Helvetia

[147] 12/29/1900 Boer Forces led by Generals Ben Viljoen and Chris Muller won a victory over the British Forces, led by **Major Cotton**. British garrison at Helvetia was captured.

[148] (archaic) To know (in the sense of knowing a fact).

The next morning I found half a dozen prisoners-of-war had sustained slight flesh wounds during the fight, and I sent them on a trolley to Belfast with a dispatch to General Smith-Dorrien,[149] informing him that four of his officers and 250 men were in our hands, that they would be well looked after, and that I now sent back the slightly wounded who had been taken away by mistake.

I will try to give the concluding sentence of my communication as far as I remember it, and also the reply to it.

I may add that the words "The Lady Roberts" had been chiselled on the naval gun, and that many persons had just been expelled from Pretoria and other places as being considered "undesirables."

My letter wound up as follows:—

"I have been obliged to expel "The Lady Roberts" from Helvetia, this lady being an "undesirable" inhabitant of that place. I am glad to inform you that she seems quite at home in her new surroundings, and pleased with the change of company."

To which General Smith-Dorrien replied:

"As the lady you refer to is not accustomed to sleep in the open air, I would recommend you to try flannel next to the skin."

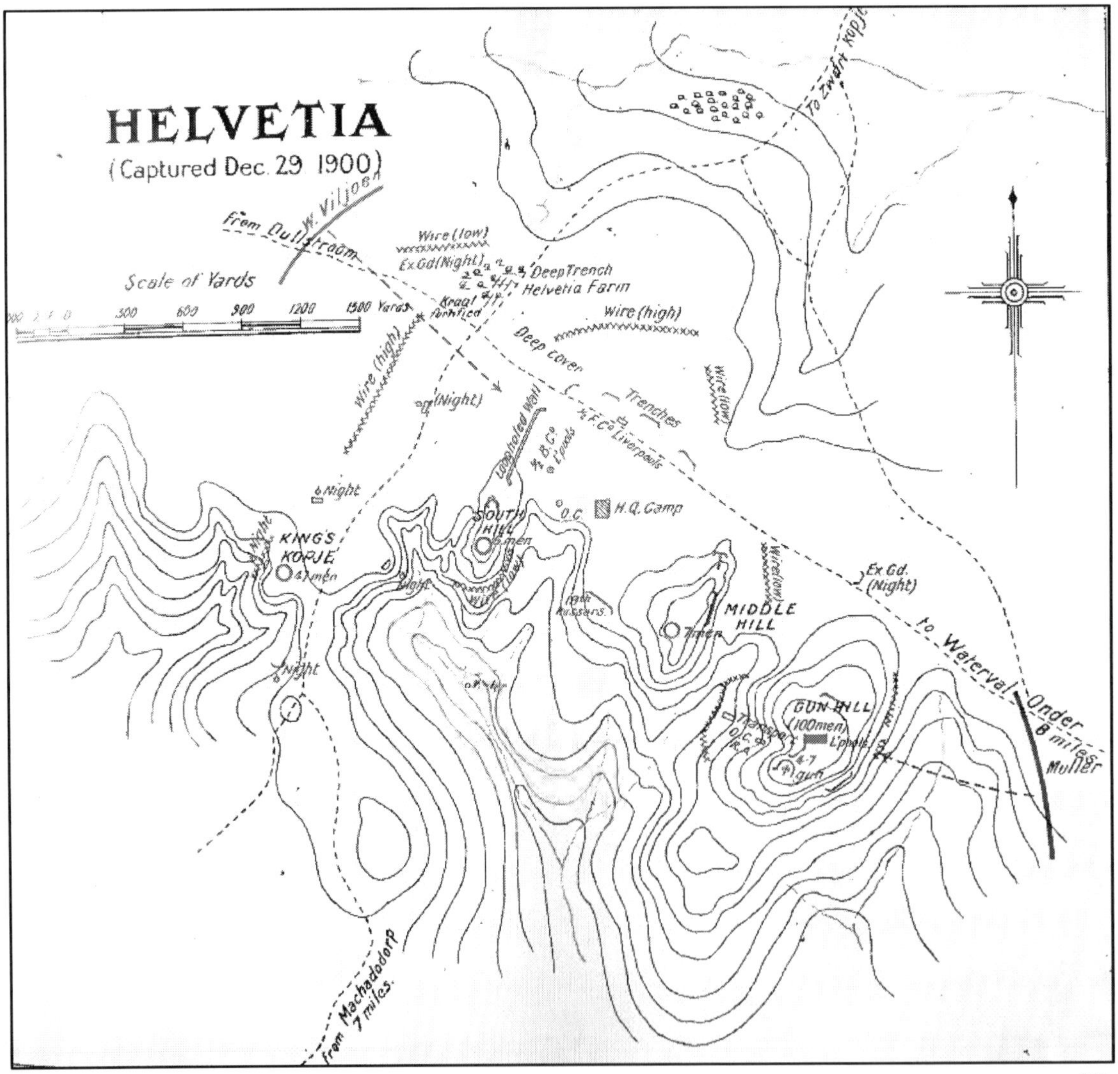

Helvetia 12/29/1900. Top Image: Col H. L. Smith-Dorian.[248]

[149] General Sir Horace Lockwood Smith-Dorrien, GCB, GCMG, DSO, ADC (26 May 1858–12 August 1930) was a British Army General. One of the few British survivors of the Battle of Isandlwana as a young officer, he also distinguished himself in the Second Boer War. https://en.wikipedia.org/wiki/Horace_Smith-Dorrien

PF: I am unable to find an image of the 'Lady Roberts' herself but Lady Randolph Churchill had a similar large calibre gun named after her. In part due to her service to the troops. Much like ships may be named or christened after their sponsors. "My Lady's Carriage Stops the Way." 4.7" gun named Lady Randolph Churchill in action towards, Hlawange Hill.[249] [250]

I had been instructed to keep the officers we had taken prisoners until further orders, and these four were therefore lodged in an empty building [North] near Roos Senekal under a guard. The Boers had christened this place "Ceylon," but the officers dubbed it "the house beautiful" on account of its utter want of attractiveness.

They were allowed to write to their relatives and friends, to receive letters, and food and clothes, which were usually sent through our lines under the white flag. The company was soon augmented by the arrivals of many other British officers who were taken prisoners from time to time.

The 250 captured rank and file were given up to the British authorities at Middelburg some days after, for military reasons.

"The Lady Roberts" was the first and so far the last big gun taken from the English, and we are proud to say that never during this War, notwithstanding all our vicissitudes and reverses, have the British succeeded in taking one of our big guns.

One might call this bragging, but that is not my intention and I do not think I am given to boasting. We only relate it as one of the most remarkable incidents of the War, and as a fact which we may recall with satisfaction.

As already related, the cart with the shells for "The Lady Roberts" had to be left behind after the battle. Nothing would have given us greater pleasure than to send some shells from "Her Ladyship" into the Belfast camp on the last day of 1900, with the "Compliments of the Season." Not of course, in order to cause any destruction, but simply as a New Year's greeting. We would have sent them close by like the Americans in Mark Twain's book: "Not right in it, you know, but close by or near it." Only the shells were wanting, for with the gun were 50 charged "hulzen" and a case of cordite "schokbuizen."

We tried to make a shell from an empty "Long Tom" one, by cutting the latter down, for the "Long Toms" shells were of greater calibre, and after having it filled with four pom-pom bullets, some cordite etc., we made it tight with copper wire, and soldered the whole together.

But when the shell was fired it burst a few steps away from the mouth of the cannon, and we had to abandon all hope of ever hearing a shout from the distinguished "Lady's" throat.[150]

It was stowed away safely in the neighbourhood of Tautesberg [mountain] and guarded by a group of cattle-farmers, or rather "bush-lancers,"[151] as they were afterwards called, in case we should get hold of the proper shells some day or other.

[150] During the siege of Ladysmith, the Boer forces had sent a hollowed out Long Tom shell into the town with note, a plum pudding and other compliments of the season on Christmas Day.

[151] PF: This may be something of a derogatory term. These 'farmers' are not taking an active part in the conflict. They are hiding out in the bush with their property to avoid the British. They support the Republican cause but protecting their property comes first. Lancers are generally known as very brave and aggressive soldiers who close to close quarters with the enemy.

In connection with the attack on Helvetia I should like to quote the following lines, written by one of our poetasters, State-Secretary Mr. F. W. Reitz, in the field, although the translation will hardly give an adequate idea of the peculiar treatment of the subject:—

Hurrah for General Muller, hurrah for Ben Viljoen, They went for 'Lady Roberts' and caught her very soon. They caught her at Helvetia, great was Helvetia's fall! Come up and see 'The Lady,' you Ooms and Tantes all.

It was a Christmas present (they made a splendid haul), And sent 'The Lady Roberts,' a present to Oom Paul. It cheered the poor Bush-lancers, it cheered the 'trek boers' all, It made them gladly answer to freedom's battle call.

Lord Roberts gave up fighting, he did not care a rap, But left his dear old 'Lady,' who's fond of mealie-pap. Of our dear wives and children he burned the happy homes, He likes to worry Tantes but fears the sturdy Ooms.

But his old 'Lady Roberts' (the lyddite-spitting gun), He sent her to Helvetia to cheer the garrison; He thought she would be safe there, in old Smith-Dorrien's care; To leave the kopjes' shelter the Boers would never dare.

Well done, Johannesburgers, Boksburgers, and police, Don't give them any quarter, don't give them any peace; Before the sleepy "Tommies" could get their stockings on, The forts were stormed and taken, and all the burghers gone.

We took 300 soldiers, provisions, and their guns, And of their ammunition we captured many tons. 'This is guerilla warfare,' says Mr. Chamberlain, But those we have bowled over will never fight again.

Let Roberts of Kandahar, and Kitchener of Khartoum, Let Buller of Colenso make all their cannon boom. They may mow down the kaffirs, with shield and assegai, But on his trusty Mauser the burgher can rely.

For now the white man's fighting, these heroes dare not stay, Lord Kitchener's in Pretoria, the others ran away. Lord Roberts can't beat burghers, although he Candahar, The Lords are at a distance, the Generals few and far!

They may annex and conquer, have conquered and annexed, Yet when the Mauser rattles the British are perplexed. Stand firm then, Afrikanders, prolong the glorious fight, Unfurl the good old 'Vierkleur.' Stand firm, for right is might!

What though the sky be clouded, what though the light be gone; The day will dawn to-morrow, the sun will shine anon; And though in evil moments a hero's hand may fail, The strong will be confounded and right will yet prevail!"

Giving some idea of what it takes to move such a piece of artillery.[251]

THE BOER WAR.

THE REVERSE AT HELVETIA.

FURTHER DETAILS.

A NAVAL GUN CAPTURED.

LONDON, Dec. 31.

Further details have been received of the reverse which on Saturday last befell the British garrison at Helvetia, a strong position on the branch railway to Lydenburg from Machadodorp, on the Delagoa Bay line, when a detachment of the Liverpool Regiment was attacked by a force of Boers, who succeeded in completely overwhelming them and capturing the town, the Liverpools sustaining a loss of 50 killed and wounded, 200 of them being taken prisoners.

Early on Saturday morning the Boers surprised the town, and rushing upon a 4.7 naval gun captured it.

Subsequently a British force, under Colonel Reeves, of the Royal Irish Fusileers, opened a heavy fire upon the Boers from Swartz Koppes, driving the enemy out of Helvetia.

The Boers surrounded the gun with prisoners, thus preventing the British from firing upon it. By this means they were enabled to withdraw it, though they failed to take away with them any ammunition.

The Reverse at Helvetia.[252]

Firing into Colenso..[253] Note the size of the compliment necessary to service the gun.

Chapter 29. A Dismal "Happy New Year."

This is the 31st of December, 1900, two days after the victory gained by our burghers over the English troops at Helvetia, at the same time the last day of the year, or, as they call it, "New Year's Eve"; which is celebrated in our country with great enjoyment. The members of each family used to meet on that day, sometimes coming from all parts of the country. If this could not be done they would invite their most intimate friends to come and see the Old Year out—to "ring out the old, and ring in the new," for "Auld Lang Syne." This was one of the most festive days for everybody in South Africa. On the 31st of December, 1899, we had had to give up our time-honoured custom, there being no chance of joining in the friendly gathering at home, most of us having been at the front since the beginning of October, 1899, while our commandos were still in the very centre of Natal or in the northern part of Cape Colony; Ladysmith, Kimberley, and Mafeking were still besieged, and on the 15th of December the great victory of Colenso over the English Army had been won.

It is true that even then we were far from our beloved friends, but those who had not been made prisoners were still in direct communication with those who were near and dear to them. And although we were unable to pass the great day in the family circle, yet we could send our best wishes by letter or by wire. We had then hoped it would be the last time we should have to spend the last day of the year under such distressing circumstances, trusting the war would soon be over.

Now 365 days had gone by—long, dreary, weary days of incessant struggle; and again our expectations had not been realised, and our hopes were deferred. We were not to have the privilege of celebrating "the Old and the New" with our people as we had so fervently wished the previous year on the Tugela.

The day would pass under far more depressing circumstances. In many homes the members of the family we left behind would be prevented from being in a festive mood, thinking as they were of the country's position, while mourning the dead, and pre-occupied with the fate of the wounded, of those who were missing, or known to be prisoners-of-war.

It was night-time, and everybody was under the depression of the present serious situation. Is it necessary to say that we were all absorbed in our thoughts, reviewing the incidents of the past year? Need we say that everyone of us was thinking with sadness of our many defeats, of the misery suffered on the battlefields, of our dead and wounded and imprisoned comrades; how we had been compelled to give up Ladysmith, Kimberley, and Mafeking, and how the principal towns of our Republics, Bloemfontein and Pretoria, where our beloved flag had been flying for so many long years, over an independent people, were now in the hands of the enemy? Need we say we were thinking that night more than ever of our many relatives who had sacrificed their blood and treasure in this melancholy War for the good Cause; of our wives and children, who did not know what had become of us, and whom most of us had not seen for the last eight months. Were they still alive? Should we ever see them alive? Such were the terrible thoughts passing through our minds as we silently sat round the fires that evening.

Nor did anything tend to relieve the sombre monotony. This time we should not have a chance of receiving some little things to cheer us up and remind us that our dearest friends had thought of us. Our fare would that day be the eternal meat and mealies—mealies and meat.

But why call to mind all these sombre memories of the past? Sufficient unto the day it seems was the evil thereof. Why sum up the misery of a whole year's struggles? And thus we "celebrated" New Year's Eve of 1900, till we found our consolation in that greatest of blessings to a tired-out man—a refreshing sleep.

But no sooner had we risen next morning than the cheerful compliments: "A Happy New Year!" or "My best wishes for the New Year" rang in our ears. We were all obviously trying to lay stress on the possible blessings of the future, so as to make each other forget the past, but I am afraid we did not expect the fulfilment of half of what we wished.

For well we knew how bad things were all round, how many dark clouds were hanging over our heads, and how very few bright spots were visible on the political horizon.

Chapter 30. General Attack on British Forts.

My presence was requested on the 3rd of January, 1901, by the Commandant-General at a Council of War, which was to be held two days after at Hoetspruit, some miles east of Middelburg. General Botha would be there with his staff, and a small escort would take him from Ermelo over the railway through the enemy's lines. My commandos were to hold themselves in readiness. There was no doubt in my mind as to there being some great schemes on the cards, and that the next day we should have plenty to do, for the Commandant-General would not come all that way unless something important was on. And why should my commandos have to keep themselves in readiness?

On the morning of the 5th I went to the place of destination, which we reached at 11 o'clock, to find the Commandant-General and suite had already arrived. General Botha had been riding all night long in order to get through the enemy's lines, and had been resting in the shadow of a tree at Hoetspruit. The meeting of his adjutants and mine was rather boisterous, and woke him up, whereupon he rose immediately and came up to me with his usual genial smile.

We had often been together for many months in the War, and the relations between us had been very cordial. I therefore do not hesitate to call him a bosom-friend, with due respect to his Honour as my chief.

"Hullo, old brother, how are you?" was Botha's welcome.

"Good morning, General, thank you, how are you?" I replied.

My high appreciation of, and respect for his position, made me refrain from calling him Louis, although we did not differ much in age, and were on intimate terms.

"I must congratulate you upon your successful attack on Helvetia. You made a nice job of it," he said. "I hope you had a pleasant New Year's Eve. But," he went on, "I am sorry in one way, for the enemy will be on his guard now, and we may not succeed in the execution of the plans we are going to discuss to-day, and which concern those very districts."

"I am sorry, General," I replied, "but of course I know nothing of those plans."

"Well," rejoined the Commandant-General, "we will try anyhow, and hope for the best."

Gen. Viljoen Meeting Gen. Botha at Hoedspruit, Near Middleburg. Commandant General Louis Botha who organized the successful defense of Spion Kop.[254]

An hour later we met in council. Louis Botha briefly explained how he had gone with General C. Botha[152] and Tobias Smuts, with 1,200 men, to Komatiboven, between Carolina and Belfast, where they had left the commandos to cross the line in order to meet the officers who were to the north of it with the object of going into the details of a combined attack on the enemy's camps.

All were agreed and so it was decided that the attack would be made during the night of the 7th of January, at midnight, the enemy's positions being stormed simultaneously.

The attack was to be made in the following way: The Commandant-General and General C. [Christiaan] Botha along with F.[sic?] Smuts, would attack on the southern side of the garrisons, in the following places: Pan Station, Wonderfontein Station, Belfast Camp and Station, Dalmanutha and Machadodorp, while I was to attack these places from the north. The commandos would be divided so as to have a field-cornet's force charge at each place.

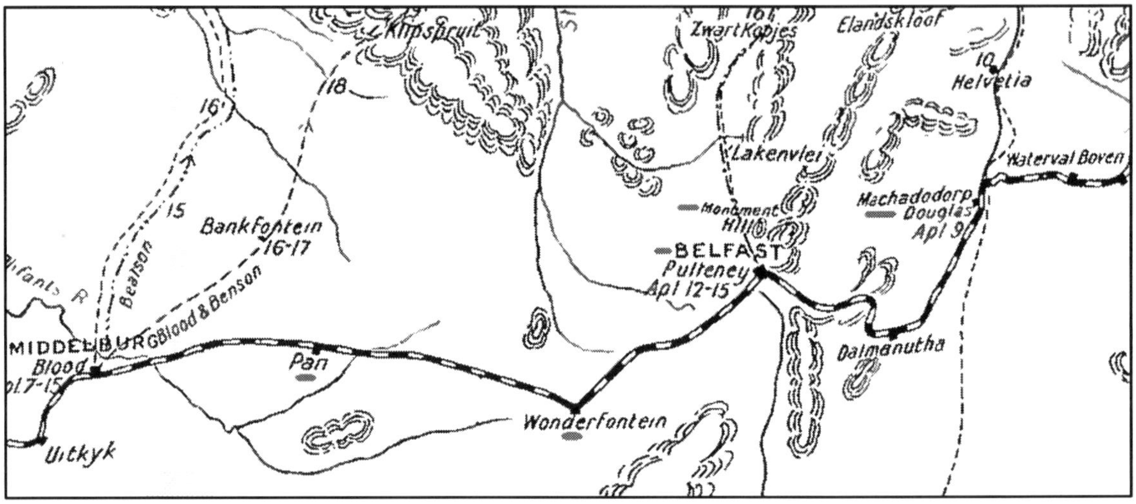

Monument hill above Belfast.

General Tobias Smuts.[255] [153]

General Christian Botha.[256]

[152] Christiaan Botha (also Chris Botha and in English language literature mistakenly Christian Botha, 6 October 1864–29 October 1902) was a younger brother of Louis Botha (1862–1919) and Philip Botha (1851–1901), but an older brother of Theunis Jacobus Botha (1867-1930), and likewise a Boer general in the Second Boer War (1899–1902) who then both fought the British to the end. https://en.wikipedia.org/wiki/Chris_Botha And see MOA.

[153] Tobias Smuts was made an assistant commander of the Ermelo Commando, and after the Battle of Colenso was promoted to fighting general (Afrikaans: veggeneraal) on the Upper Tugela. When the Transvaal Commando's transferred from Natal to the Orange Free State, Smuts led them as Assistant Commander General and joined in the fighting near Brandfort. Afterwards he returned to Pretoria to take part in the last session of the Volksraad. In the army officer's meeting of 2 June 1900 in Pretoria Smuts agreed with the proposal of Danie Theron to fight on. Presidents Kruger and Steyn, who first considered peace negotiations with the British, concurred.
Tobias Smuts fought in the Battle of Diamond Hill, and afterwards was sent to the Wakkerstroom area. Under the command of general Chris Botha. Smuts led the remaining Ermelo, Carolina, Standerton, and Swaziland Commandos. Breytenbach II (1971), III (1973), V (1983), and VI (1996).

I must say that I had considerable difficulty in trying to make a little go a long way in dividing my small force along such a long line of camps, but the majority were in favour of this "frittering-away" policy, and so it had to be done.

The enemy's strength in different places was not easy to ascertain. I knew the strongest garrison at Belfast numbered over 2,500 men, and this place was to be made the chief point of attack, although the Machadodorp garrison was pretty strong too. The distance along which the simultaneous attack was to be made was about 22 miles and there were at least seven points to be stormed, viz., Pan Station, Wonderfontein, Belfast Village, Monument Hill (near Belfast), the coal mines (near Belfast), Dalmanutha Station and Machadodorp. A big programme, no doubt.

I can only, of course, give a description of the incidents on my side of the railway line, for the blockhouses and the forts provided with guns, which had been built along the railway, separated us entirely from the commandos to the south. The communication between both sides of the railway could be only kept up at night time and with a great amount of trouble, by means of despatch-carriers. We, therefore, did not even know how the attacking-parties on the southern side had been distributed. All we knew was, that any place which was to be attacked from the north would also be stormed from the south at the same time, except the coal mine west of Belfast, occupied by Lieutenant Marshall with half a section of the Gloucester Regiment, which we were to attack separately, as it was situated some distance north of the railway line.

I arranged my plans as follows: Commandant Trichardt,[154] with two field-cornets posses of Middelburgers and one of Germiston burghers, were to attack Pan and Wonderfontein; the State Artillery would go for the coal mine; the Lydenburgers look after Dalmanutha and Machadodorp; while General Muller with the Johannesburgers and Boksburgers would devote their attention to Monument Hill.

I should personally attack Belfast Village, with a detachment of [Johannesburg] police, passing between the coal mine and Monument Hill. My attack could only, of course, be commenced after that on the latter two places had turned out successfully, as otherwise I should most likely have my retreat cut off.

In the evening of the 7th of January all the commandos marched, for the enemy would have been able to see us from a distance on this flat ground if we had started in the daytime, and would have fired at us with their 4·7 guns, one of which we knew to be at Belfast. We had to cover a distance of 15 miles between dusk and midnight. There was therefore no time to be lost, for a commando moves very slowly at night time if there is any danger in front. If the danger comes from the rear, things very often move quicker than is good for the horses. Then the men have to be kept together, and the guides are followed up closely, for if any burghers were to lag behind and the chain be broken, 20 or 30 of them might stray which would deprive us of their services.

It was one of those nights, known in the Steenkamp Mountains as "dirty nights," very dark, with a piercing easterly wind, which blew an incessant, fine, misty rain into our faces. About nine o'clock the mist changed into heavy rains, and we were soon drenched to the skin, for very few of us wore rainproof cloaks.

Picket on the guard at night.[257]

[154] Stephanus P. E. Trichardt. Trichardt received an appointment as Commandant of that part of the Middelburg commando that is north of the railroad operated, and as such he served until the close of the war. See MOA.

At ten the rain left off, but a thick fog prevented us from seeing anything in front of us, while the cold easterly wind had numbed our limbs, almost making them stiff. Some of the burghers had therefore to be taken up by the ambulance in order to have their circulation restored by means of some medicine or artificial treatment. The impenetrable darkness made it very difficult to get on, as we were obliged to keep contact by means of despatch-riders; for, as already stated, I had to wait with the [Johannesburg] police [Commando] for the result of the attack on the two positions to the right and left of me.

Exactly at midnight all had arrived at the place of destination. Unfortunately the wind was roaring so loudly as to prevent any firing being heard even at a hundred paces distant.

The positions near Monument Hill and the coal mine were attacked simultaneously, but unfortunately our artillerymen could not distinctly see the trenches on account of the darkness, and they charged right past them, and had to turn back when they became aware of the fact, by which time the enemy had found out what was up, and allowed their assailants to come close up to them (it was a round fort about five feet high with a trench round it), and received them with a tremendous volley. The artillerymen, however, charged away pluckily, and before they had reached the wall four were killed and nine wounded. The enemy shot fiercely and aimed well.

British Army piquet post.[258]

British Army piquet post.[259]

Our brave boys stormed away, and soon some of them jumped over the wall and a hand-to-hand combat ensued. The commanding officer of the fortress, Lieutenant Marshall, was severely wounded in the leg, which fact must have had a great influence on the course of the fight, for he surrendered soon after. Some soldiers managed to escape, some were killed, about 10 wounded, and 25 were taken prisoners. No less than five artillerymen were killed and 13 wounded, amongst the latter being the valiant Lieutenant Coetsee who afterwards was cruelly murdered by kaffirs near Roos Senekal. The defenders as well as the assailants had behaved excellently.

Near Monument Hill, at some distance from the position, the burghers' horses were left behind, and the men marched up in scattered order, in the shape of a crescent. When we arrived at the enemy's outposts they had formed up at 100 paces from the forts, but in the dark the soldiers did not see us till we almost ran into them. There was no time to waste words. Fortunately, they surrendered without making any defence, which made our task much lighter, for if one shot had been fired, the garrison of the forts would have been informed of our approach. Only at 20 paces distance from the forts near the Monument (there were four of them), we were greeted with the usual "Halt, who goes there." After this had been repeated three times without our taking any notice, and as we kept coming closer, the soldiers fired from all the forts. Only now could we see how they were situated. We found them to be surrounded by a barbed wire fence which was so strong and thick that some burghers were soon entangled in it, but most of them got over it.

The first fort was taken after a short but sharp defence, the usual "hurrah" of the burghers jumping into the fort was, like a whisper of hope in the dark, an encouragement to the remainder of the storming burghers, who now soon took the other forts, not without having met with a stout resistance. Many burghers were killed, amongst whom the brave Field-Cornet John Ceronie, and many were wounded.

It had looked at first as if the enemy did not mean to give in, but we could not go back, and "onward" was the watchword. In several instances there was a struggle at a few paces' distance, only the wall of the fort intervening between the burghers and the soldiers. The burghers cried: "Hands up, you devils," but the soldiers replied: "Hy kona," a kaffir expression which means "shan't."

"Jump over the walls, my men!" shouted my officers, and at last they were in the forts: not, of course, without the loss of many valuable lives. A "melée" now followed; the English struck about with their guns and with their fists, and several burghers lay on the ground wrestling with the soldiers. One "Tommy" wanted to thrust a bayonet through a Boer, but was caught from behind by one of the latter's comrades, and knocked down and a general hand-to-hand fight ensued, a rolling over and over, till one of the parties was exhausted, disarmed, wounded, or killed. One of the English captains (Vosburry) and 40 soldiers were found dead or wounded, several having been pierced by their own bayonets.

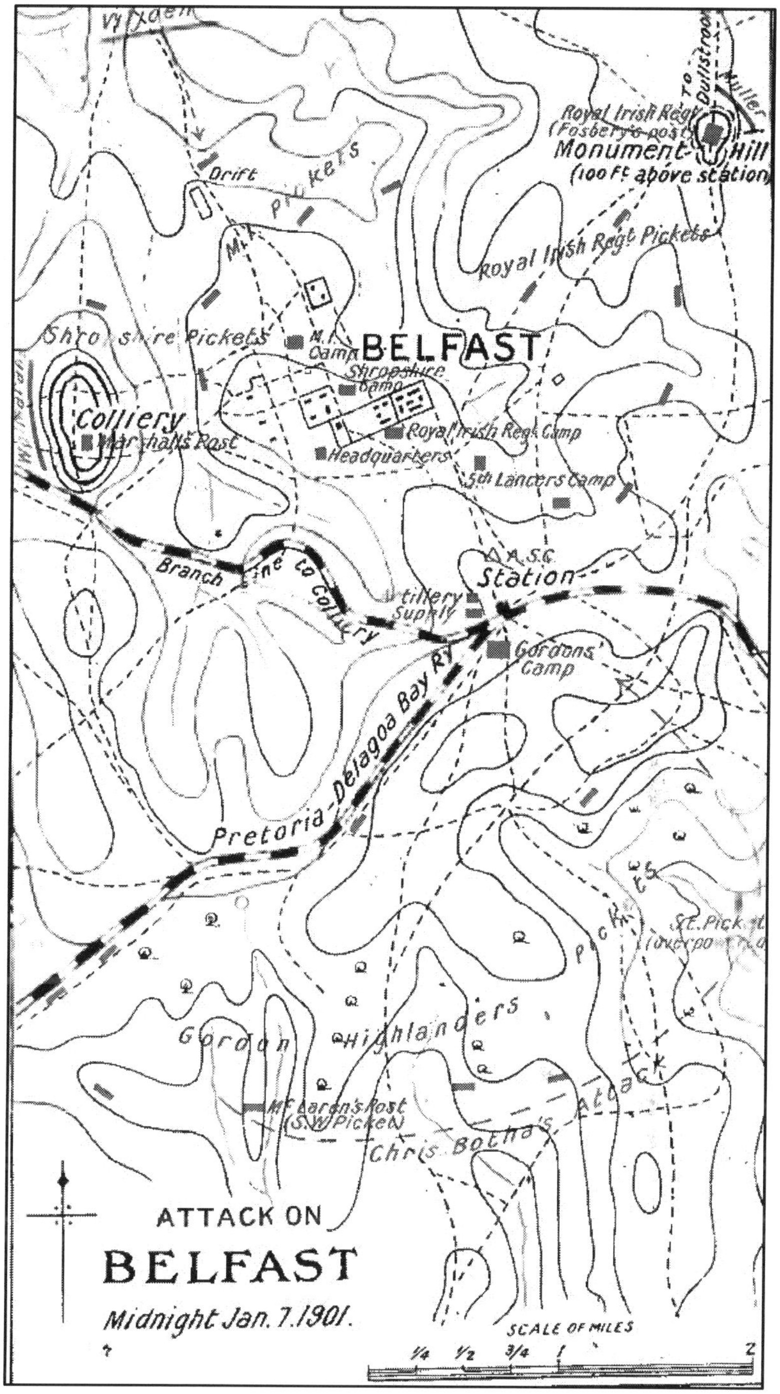

ATTACK ON BELFAST
Midnight Jan. 7.1901.

Some burghers had been knocked senseless with the butt-end of a rifle in the struggle with the enemy.

This carnage had lasted for twenty minutes, during which the result had been decided in our favour, and a "hurrah," full of glory and thankfulness, came from the throats of some hundreds of burghers. We had won the day, and 81 prisoners-of-war had been made, including two officers—Captain Milner and Lieutenant Dease—both brave defenders of England's flag.

They belonged to the Royal Irish Regiment, of which all Britons should be proud.

In the captured forts we found a Maxim, in perfect order, 20 boxes of ammunition, and other things, besides provisions, also a quantity of spirits, which was, however, at once destroyed, to the disappointment of many burghers.

We now pushed on to Belfast village, but found every cliff and ditch occupied. All efforts to get in touch with the commandos which meant to attack the village from the south were without avail. Besides, we did not hear a single shot fired, and did not know what had become of the attack from the south. In intense darkness we were firing at each other from time to time, so that it was not advisable to continue our operations under the circumstances, and at daybreak I told all my commandos to desist.

The attacks on Wonderfontein, Pan Station, Dalmanutha, and Machadodorp had failed.

I afterwards received a report from the commandos on the other side of the line, that, owing to the dark night, their attacks, although they were made with deliberation and great bravery, had all been unsuccessful. They had repeatedly missed the forts and had shot at one another.

General Christian Botha had succeeded in capturing some of the enemy's outposts, and in pushing on had come across a detachment of Gordon Highlanders and been obliged to retire with a loss of 40 killed and wounded.

We found, therefore, these forts in the hands of the soldiers, who, in my opinion, belonged to the best regiments of the English army.[155]

The guests of our Government, at "the house beautiful" near Roos Senekal were thus added to by two gentlemen, Captain Milner and Lieutenant Dease, and they were my prisoners-of-war for four months, during which time I found Captain Milner one of the most worthy British officers whom it had been my privilege to meet in this War. Not only in his manly appearance, but especially by his noble character he stood head and shoulders above his fellow-officers.

Lieutenant Dease bore a very good character but was young and inexperienced. For several reasons I am pleased to be able to make publicly these statements.

The soldiers we had made prisoners during this fight, as well as those we took at Helvetia, were given up to the British officers a few days afterwards, as we were not in a position to feed them properly, and it would not be humane or fair to keep the soldiers who had the misfortune of falling into our hands without proper food. This, of course, was a very unsatisfactory state of affairs, for we had to fight fiercely, valuable lives had to be sacrificed, every nerve had to be strained to force the enemy to surrender, and to take his positions; and then, when we had captured them, the soldiers were merely disarmed and sent back to the English lines after a little while, only to find them fighting against us once more in a few days.

The Boers asked, "Why are not these "Tommies" required to take the oath before being liberated not to fight against us again?" I believe this would have been against the rules of civilised warfare, and we did not think it chivalrous to ask a man who was a prisoner to take an oath in return for his release.

A prisoner-of-war has no freedom of action, and might have promised under the circumstances what he would not have done if he had been a free man.

Chapter 31. A "Bluff" and a Battle.

The last days of February, 1901, were very trying for our commandos on the "Hoogeveld," [High-field] south of the railway. General French, assisted by half a dozen other generals, with a force of 60,000 men, crossed the "Hoogeveld," between the Natal border and the Delagoa Railway, driving all the burghers and cattle before him, continually closer to the Swazi frontier, in order to strike a "final blow" there.

These operations the English called "The Great Sweep of February, 1901."[156]

Commandant-General Botha sent word that he was in a bad plight on the "Hoogeveld," the enemy having concentrated all his available troops upon him. I was asked to divert their attention as much as possible by repeated attacks on the railway line, and to worry them everywhere.

[155] 1/7/1901 Boers attacked Belfast, Wonderfontein, Nooitgedacht, Widfontein, and Pan, and after sharp fighting were dispersed.
1/10/1901 Machadodorp attacked by night. Post gallantly defended.

[156] 1/12/1901 Boers driven eastward from Witwatersberg by General French. Activities in Cape Colony to frustrate Hertzog's advance.
1/22/1901 Death of Queen Victoria.
1/23/1901 Colonels De Lisle, Scobell, and Collenbrander drove the enemy out of Calvinia and Van Rhynsdorp, pursued north to Carnarvon.
1/28/1901 General French marched eastward, clearing the valley of the Wilge River.
1/30/1901 De Wet breaks through the Bloemfontein-Ladybrand line going south.
2/1/1901 General French continued to operate against Botha in the Eastern Transvaal.
2/6/1901 General French, after encountering little resistance, entered Ermelo. General Smith-Dorrien repulsed 2000 of the enemy. His losses were 23 killed and 52 wounded. The War Office decided to reinforce Kitchener by 30,000 mounted troops beyond those already landed in Cape Colony.
2/9/1901 Eastern movement continued in deluges of rain, but invasion of Natal by Botha eventually frustrated.

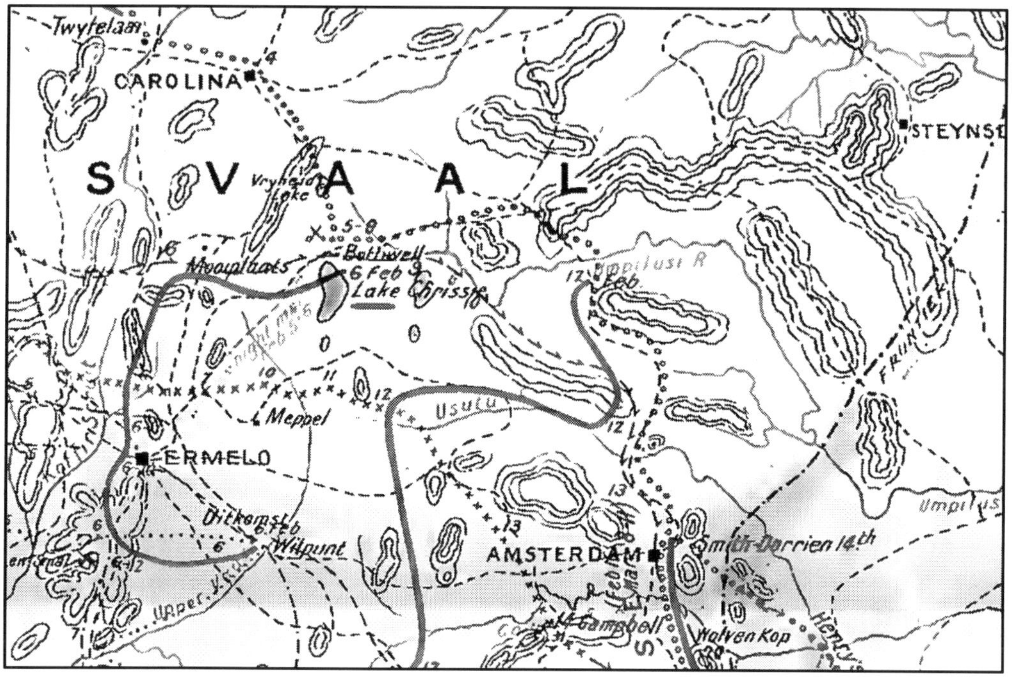

Belfast Area and Lake Chrssie identified. See footnote for the battle overview.

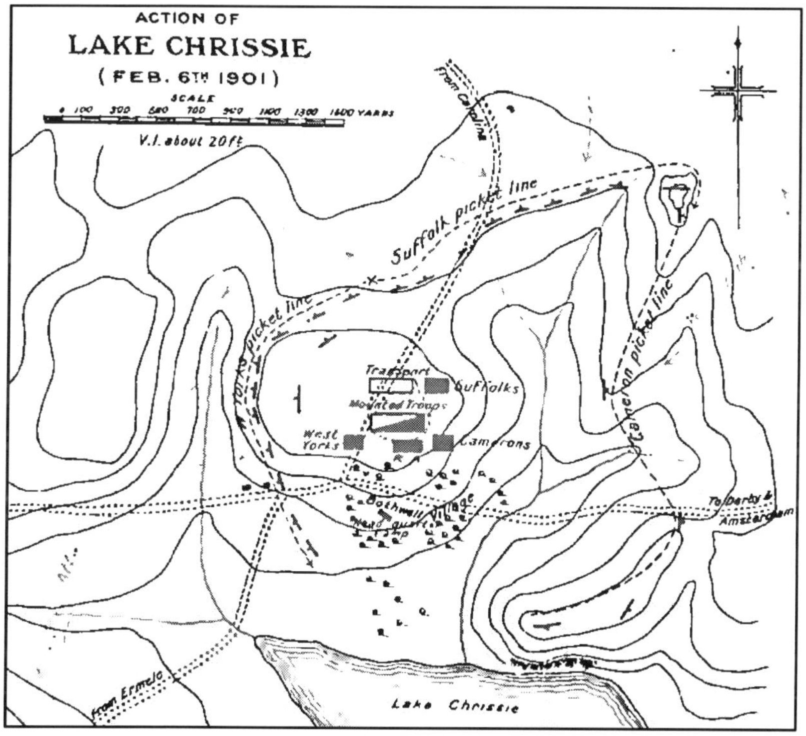

Lake Chrissie. From Times History, French in the Eastern Transvaal Jan-April 1901. [157]

[157] February 6, 190. The Battle of Chrissiesmeer (Lake Chrissie) during Anglo-Boer War 2 started with an attack by General Louis Botha on the British forces under General Smith-Dorrien. Botha's main aim was to cripple the advance of Smith-Dorrien into the eastern Transvaal. According to oral tradition Smith-Dorrien's march had been kept under observation by San (Bushmen) who also acted as messengers. The Boers capitalised on the San's knowledge of the terrain and attacked at 02:50 in pitch darkness. The burgers achieved initial success, but could not gain a foothold on the slopes and were not able to open direct fire on the main camp situated on a plateau. At 04:30, covered by thick morning mist, Botha ordered them to retreat. The Boer commando suffered about eighty casualties and the British seventy-five. About 300 horses of the British force were killed or stampeded. Despite heavy losses on the Boer side the British advance into the eastern Transvaal was delayed by the loss of their horses, which gave the Boer commandos time to re-group.
South African History Online. (2011). San involvement in the Battle of Chrissiesmeer. South African History Online. https://sahistory.org.za/dated-event/san-involvement-battle-chrissiesmeer And: Lake Chrissie: 26° 20′ 0″ S, 30° 13′ 0″ E.

To attack the fortified entrenchments in these parts, where we had only just been taking the offensive, causing the enemy to be on his guard, would not have been advisable. I therefore decided to make a feint attack on Belfast.

One night we moved with all the burghers who had horses, about 15 carts, waggons, and other vehicles, guns and pom-pom, to a high "bult," near the "Pannetjes." When the sun rose the next morning we were in full sight of the enemy at Belfast, from which we were about ten miles away.

Here our commando was split into two parts, and the mounted men spread about in groups of fifty men each, with carts scattered everywhere among the ranks. We slowly approached Belfast in this order. Our commando numbered about 800 men, and considering the way we were distributed, this would look three times as many. We halted several times, and the heliographers, who were posted everywhere in sight of the enemy, made as much fuss as possible. Scouts were riding about everywhere, making a great display by dashing about all over the place, from one group of burghers to another.

After we had waited again for some little time we moved on, and thus the comedy lasted till sunset; in fact, we had got within range of the enemy's guns. We had received information from Belfast to the effect that General French had taken all the guns with him to Belfast, leaving only a few of small calibre, which could not reach us until we were at about 4,000 yards from the fort. Our pom-pom and our 15-pounder were divided between the two divisions, and the officers had orders to fire a few shots on Belfast at sunset. We could see all day long how the English near Monument Hill were making ditches round the village and putting up barbed wire fences.

Trains were running backwards and forwards between Belfast and the nearest stations, probably to bring up reinforcements.

At twilight we were still marching, and by the light of the last rays of the sun we fired our two valuable field-pieces simultaneously, as arranged. I could not see where the shells were falling, but we heard them bursting, and consoled ourselves with the idea that they must have struck in near the enemy. Each piece sent half a dozen shells, and some volleys were fired from a few rifles at intervals. We thought the enemy would be sure to take this last movement for a general attack.

Pattern 2 train.[260] No. 3 Artillery Train at the railway station at Bronkhorstspruit, Gauteng Province.[261]

What he really did think, there is no saying. As the burghers put it, "We are trying to make them frightened, but the thing to know is, did they get frightened?" For this concluded our programme for the day, and we retired for the night, leaving the enemy in doubt as to whether we meant to give him any further trouble, yet without any apology for having disturbed his rest.

The result of this bloodless fight was nil in wounded and killed on both sides.

On the 12th of February, 1901, the first death-sentence on a traitor on our side was about to be carried out, when suddenly our outposts round Belfast were attacked by a strong British column under General Walter Kitchener. When the report was brought to our laager, all the burghers went to the rescue, in order to keep the enemy as far from the laager as possible, and beat them back. Meanwhile the outposts retired fighting all the while. We took up the most favourable positions we could and waited. The enemy did not come up close to us that evening, but camped out on a round hill between Dullstroom and Belfast and we could distinctly see how the soldiers were all busy digging ditches and trenches round the camp and putting up barbed wire enclosures. They were very likely afraid of a night attack and did not forget the old saying about being "wise in time."

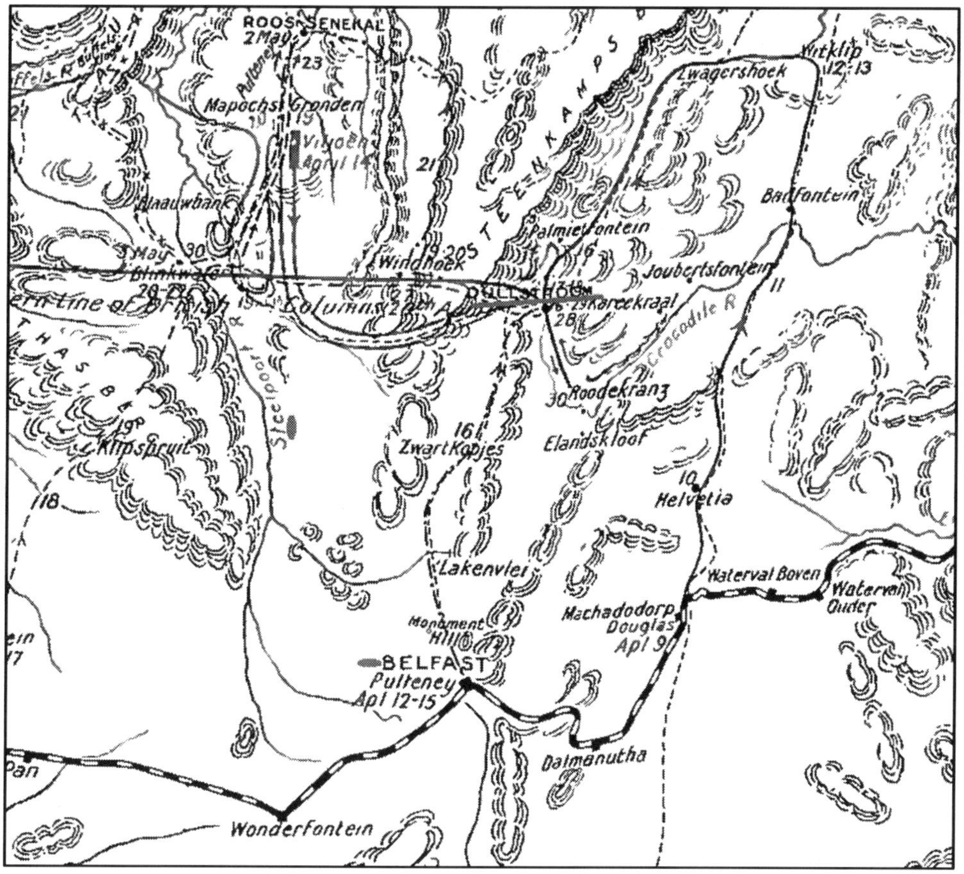

Dullstruoom is underlined and Belfast is near the bottom.

Near the spot where their camp was situated were several roads leading in different directions which left us in doubt as to which way they intended to go, and whether they wanted to attack us, or were on their way to Witpoort-Lydenburg.

The next morning, at sunset, (sic) the enemy broke up his camp and made a stir. First came a dense mass of mounted men, who after having gone about a few hundred paces, split up into two divisions. One portion moved in a westerly direction, the other to the north, slowly followed by a long file, or as they say in Afrikander "gedermte" (gut) of waggons and carts which, of course, formed the convoy. Companies of infantry, with guns, marched between the vehicles.

Royal Artillery Column.[262]

I came to the conclusion that they intended to attack from two sides, and therefore ordered the ranks to scatter. General Muller, with part of the burghers, went in advance of the enemy's left flank and, as the English spread out their ranks, we did the same.

At about 9 a.m. our outposts near the right flank of the English were already in touch with the enemy, and rifle-fire was heard at intervals.

I still had the old 15-pounder, but the stock of ammunition had gone down considerably and the same may be said of the pom-pom of Rhenosterkop fame. We fired some shots from the 15-pounder at a division of cavalry at the foot of a kopje. Our worthy artillery sergeant swore he had hit them right in the centre, but even with my strong spy-glass I could not see the shells burst, although I admit the enemy showed a little respect for them, which may be concluded from the fact that they at once mounted their horses and looked for cover.

A British soldier is much more in awe of a shell than a Boer is, and the enemy's movements are therefore not always a criterion of our getting the range. We had, moreover, only some ordinary grenades left, some of which would not burst, as the "schokbuizen"[158] were defective, and we could not be sure of their doing any harm.

The other side had some howitzers, which began to spit about lyddite indiscriminately. They also had some quick-firing guns of a small calibre, which, however, did not carry particularly far. But they were a great nuisance, as they would go for isolated burghers without being at all economical with their ammunition.

Meanwhile, the enemy's left reached right up to Schoonpoort, where some burghers, who held good positions, were able to fight them. This caused continual collisions with our outposts. Here, also, the assailants had two 15-pounder Armstrong's, which fired at any moving target, and hardly ever desisted, now on one or two burghers who showed themselves, then on a tree, or an anthill, or a protruding rock. They thus succeeded in keeping up a deafening cannonade, which would have made one think there was a terrific fight going on, instead of which it was a very harmless bombardment.

It did no more harm than at the English maneuvres, although it was no doubt a brilliant demonstration, a sort of performance to show the British Lion's prowess. I could not see the practical use of it, though.

It was only on the enemy's right wing that we got near enough to feel some of the effect of the artillery's gigantic efforts, which here forced us to some sharp but innocent little fights between the outposts. At about 4 o'clock in the afternoon, the British cavalry stormed our left, which was in command of General Muller. We soon repulsed them, however. Half an hour after we saw the enemy's carts go back.

I sent a heliographic message to General Muller, with whom I had kept in close contact, to the effect that they were moving away their carts and that we ought to try and charge them on all points as well as we could.

"All right," he answered; "shall we start at once?" I flashed back "Yes," and ordered a general charge.

The burghers now appeared all along the extended fighting line.

The enemy's guns, which were just ready to be moved, were again placed in position and opened fire, but our men charged everywhere, a sort of action which General Kitchener did not seem to like, for his soldiers began to flee with their guns, and a general confusion ensued. Some of these guns were still being fired at the Boers but the latter stormed away determinedly. The British lost many killed and wounded.

The cavalry fled in such a hurry as to leave the infantry as the only protection of the guns, and although these men also beat a retreat they, at least, did it while fighting.

I do not think I overstate the case by declaring that General Walter Kitchener owed it to the stubborn defence of his infantry that his carts were not captured by us that day.

Their ambulance, in charge of Dr. Mathews and four assistants, and some wounded fell into our hands, and were afterwards sent back.

We pursued the enemy as well as we could, but about nine miles from Belfast, towards which the retreating enemy was marching, the forts opened fire on us from a 4·7 naval gun and they got the range so well that lyddite shells were soon bursting about our ears.

We were now in the open, quite exposed and in sight of the Belfast forts. Two of our burghers were wounded here.

Field-Cornet Jaapie Kriege, who was afterwards killed, with about 35 burghers, was trying to cut off the enemy from a "spruit"-drift; the attack was a very brave one, but our men ventured too far, and would all have been captured had not the other side been so much in a hurry to get away from us. Luckily, too, another field-cornet realised the situation, and kept the enemy well under fire, thus attracting Kriege's attention, who now got out of this scrape.

[158] Shock Tubes. Igniters.

Walter Kitchener.[263] [159]

Ambulance Parties at Work.[264]

When night fell we left the enemy alone, and went back to our laager. The next morning the outposts reported that the would-be assailants were all gone.

How much this farce had cost General Kitchener we could not tell with certainty. An English officer told me afterwards he had been in the fight, and that their loss there had been 52 dead and wounded, including some officers. He also informed me that their object that day had been to dislodge us. If that is so, I pity the soldiers who were told to do this work.

Our losses were two burghers wounded, as already stated.[160]

[159] Lieutenant-General Sir Frederick Walter Kitchener KCB (26 May 1858–6 March 1912), known as Walter Kitchener, was a British soldier and colonial administrator. He was the younger brother of Herbert Kitchener. https://en.wikipedia.org/wiki/Walter_Kitchener

[160] The Chronology does not mention this specific engagement.

2/10/1901 "Call to arms" at Cape Town. De Wet, after many contests with the British forces in Orange River Colony, succeeded in crossing the river at Sand Drift.

2/14/1901 Animated chases after De Wet.

2/17/1901 P J de Villiers defended Klerksdorp from Methuen's forces for 16 hours, before retreating. Battle of Hartbeesfontein - Wikipedia

2/22/1901 Extraordinary proclamation signed by Steyn and De Wet published. The news is the breakup of the Commandos. The daily morning Alaskan. (Skagway, Alaska) 1899-1904, March 03, 1901, Image 3 « Chronicling America « Library of Congress

2/23/1901 De Wet succeeded in recrossing the river after losing 200 prisoners, all his guns, ammunition, and waggons.

Chapter 32. Execution of a Traitor.

As briefly referred to in the last chapter, there occurred in the early part of February, 1901, what I always regard as one of the most unpleasant incidents of the whole Campaign, and which even now I cannot record without awakening the most painful recollections. I refer to the summary execution of a traitor in our ranks, and inasmuch as a great deal has been written of this tragic episode, I venture to state the particulars of it in full. The facts of the case are as follows:—

At this period of the War, as well as subsequently, much harm was done to our cause by various burghers who surrendered to the enemy, and who, actuated by the most sordid motives, assisted the British in every possible way against us. Some of these treacherous Boers occasionally fell into our hands, and were tried by court martial for high treason; but however damning the evidence brought against them they usually managed to escape with some light punishment. On some occasions sentence of death was passed on them, but it was invariably commuted to imprisonment for life, and as we had great difficulty in keeping such prisoners, they generally succeeded, sooner or later, in making their escape. This mistaken leniency was the cause of much dissatisfaction in our ranks, which deeply resented that these betrayers of their country should escape scot-free.

About this time a society was formed at Pretoria, chiefly composed of surrendered burghers, called the "Peace Committee," but better known to us as the "Hands-uppers." Its members surreptitiously circulated pamphlets and circulars amongst our troops, advising them to surrender and join the enemy. The impartial reader will doubtless agree that such a state of things was not to be tolerated. Imagine, for example, that English officers and soldiers circulated similar communications amongst the Imperial troops! Would such proceedings have been tolerated?

The chairman of this society was a man by the name of Meyer De Kock, who had belonged to a Steenkampsberg field-cornet's force and had deserted to the enemy. He was the man who first suggested to the British authorities the scheme of placing the Boer women and children in Concentration Camps—a system which resulted in so much misery and suffering—and he maintained that this would be the most effective way of forcing the Boers to surrender, arguing that no burgher would continue to fight when once his family was in British hands.

One day a kaffir, bearing a white flag, brought a letter from this person's wife addressed to one of my field-cornets, informing him that her husband, Mr. De Kock, wished to meet him and discuss with him the advisability of surrendering with his men to the enemy. My field-cornet, however, was sufficiently sensible and loyal to send no reply.

And so it occurred that one morning Mr. De Kock, doubtlessly thinking that he would escape punishment as easily as others had before him, had the audacity to ride coolly into our outposts. He was promptly arrested and incarcerated in Roos Senekal Gaol, this village being at the time in our possession. Soon afterwards he was tried by court-martial, and on the face of the most damning evidence, and on perusal of a host of incriminating documents found in his possession, was condemned to death.

About a fortnight later a waggon drove up to our laager at Windhoek, carrying Lieutenant De Hart, accompanied by a member of President [Schalk] Burger's bodyguard, some armed burghers, and the condemned man De Kock. They halted at my tent, and the officer handed me an order from our Government, bearing the President's ratification of the sentence of death, and instructing me to carry it out within 24 hours. Needless to say I was much grieved to receive this order, but as it had to be obeyed I thought the sooner it was done the better for all concerned. So then and there on the veldt I approached the condemned man, and said:—

"Mr. De Kock, the Government has confirmed the sentence of death passed on you, and it is my painful duty to inform you that this sentence will be carried out to-morrow evening. If you have any request to make or if you wish to write to your family you will now have an opportunity of doing so."

At this he turned deadly pale, and some minutes passed before he had recovered from his emotion. He then expressed a wish to write to his family, and was conducted, under escort, to a tent, where writing materials were placed before him. He wrote a long communication to his wife, which we sent to the nearest British officers to forward to its destination. He also wrote me a letter thanking me for my "kind treatment," and requested me to forward the letter to his wife. Later on spiritual consolation was offered and administered to him by our pastor.

Next day, as related in the previous chapter, we were attacked by a detachment of General Kitchener's force from Belfast. This kept me busy all day, and I delegated two of my subaltern officers to carry out the execution.

At dusk the condemned man was blindfolded and conducted to the side of an open grave, where twelve burghers fired a volley, and death was instantaneous. I am told that De Kock met his fate with considerable fortitude.

So far as I am aware, this was the first Boer "execution" in our history. I afterwards read accounts of it in the English press, in which it was described as murder, but I emphatically repudiate this description of a wholly justifiable act. The crime was a serious one, and the punishment was well deserved, and I have no doubt that the same fate would have awaited any English soldier guilty of a similar offence. It seems a great pity, however, that no war can take place without these melancholy incidents.

Execution of a Traitor. [161]

Piet de Wet

'Which is better, for the Republics to continue the struggle and run the risk of total ruin as a nation, or to submit? Could we for a moment think of taking back the country, if it were offered to us, with thousands of people to be supported by a Government which has not a farthing. Put passionate feeling aside for a moment and use common-sense, and you will then agree with me that the best thing for the people and the country is to give in, to be loyal to the new Government, and to get responsible government.'

Such were the sentiments of many of the best of the burghers, and they endeavoured to persuade their fellows. Both in the Transvaal and in the Free State, Peace Committees were formed among the burghers, who sent deputies to lay the facts of the situation before their brethren on commando.

The results were tragic. Two of the envoys, Morgendaal and de Koch, were shot in cold blood, the former having been first beaten. Several of the others were beaten, and all were ill-used. Doyle, Arthur Conan, 1859-1930 The war in South Africa, its cause & conduct; 1902

[161] Meyer de Kock (5 October 1849–12 February 1901) was a citizen of the Zuid-Afrikaansche Republiek ("ZAR") during the Second Anglo-Boer War. After the British had captured the capital Pretoria and annexed the Transvaal, he surrendered on 10 December 1900. In January 1901, he went to the eastern Transvaal to persuade burghers still in the field to surrender. He was arrested, court-martialed, and later, on 12 February 1901, executed by firing squad.
When De Kock arrived at the Boers' pickets, he was arrested immediately and taken to General Muller. Muller sent him to the government laager on Tautesberg. On 29 January 1901, De Kock appeared before a special military court consisting of a magistrate (landdrost) Gideon F. Joubert, as chairman and two additional members. H.L.J. Jacobs, an Assistant State Attorney, prosecuted while a certain Brugman acted as Registrar. De Kock did not have legal representation. The charge against De Kock was one of high treason and the prosecution relied on four instances. In the first instance, it was alleged that he had left his commando and handed his weapons to the enemy. To this charge, he pleaded guilty but added that if he had not done so, his family would have been ejected from their home or suffered in some other way. He pleaded not guilty on the second count that he had unlawfully cooperated with the enemy to persuade burghers to lay their weapons down. Thirdly it was alleged that he had left the enemy's lines with Peace Committee documents, in which burghers were encouraged to lay down their arms. He pleaded guilty but explained that it was done with good intentions to promote peace. He also pleaded guilty to the fourth charge that he tried to persuade two named burghers to lay down their arms. A number of witnesses were called in support of the charges. De Kock handed in an affidavit and he was cross-examined. At the end of the trial, the court found De Kock guilty of high treason and sentenced him to death. The government confirmed the conviction and sentence.
De Kock was sent back to the commandos to be executed by firing squad. On 12 February 1901, he was shot next to his open grave. The location of the gravestone is 25°22.89'S, 30°00.536'E (WGS84). PF: The Boer Governments and its leaders were determined to be the arbiter of any peace negotiations and not to tolerate informal approaches to Burghers however well meaning.
https://en.wikipedia.org/wiki/Meyer_de_Kock

Middleburg Conference

Viljoen only mentions this in passing.[162] See Volume 4 'Commando' for additional information.

In early February of 1901, General Kitchener, with the consent of the Government decided to make peace overtures to General Botha well before the February 14th debate in the House of Commons. The actual conference took place from February 28th, until its unsuccessful conclusion on March 16th 1901.

On February 13, [163] a week after the failure of his night attack at Lake Chrissie,[164] Louis Botha received a message through his wife, a verbal message, from Kitchener.

If General Botha desired Kitchener, it, so the message ran, a meeting could be held with the British Commander to arrange terms of peace; the one condition being that the question of the independence of the two republics was not to be discussed in any way.

The invitation arrived at an opportune moment; for Botha, though he had effectively countered the converging movement on Ermelo, had nevertheless seen French's columns sweeping unchecked across the high veld, and was fully aware of the searchings of heart which this great display of force had caused among his burghers.

The reply, despatched the same day, but not received by Kitchener until the 22nd, was an acceptance. In the same letter Botha suggested Middelburg as a meeting-place, and requested a safe conduct to and from that place. Kitchener agreed, and February 28 was fixed for the conference.

Middleburg stands on a tributary of Olifant's River and is 94 miles from Pretoria.[265]

Kitchener had no illusions as to the character of the struggle in which he was engaged. Even without the eloquent policy examples of history, events in South Africa since his accession to the command had shown the enormous capacity for resistance possessed by a pastoral race skilled in arms. He was conscious, too, that his mounted army was passing through a period of transition and for a time, at any rate, must inevitably decline in usefulness. The efforts of the Burgher Peace Committee[165] [166] having totally failed, there were two courses open:

[162] 2/27/1901 Lengthy negotiations for the promotion of peace took place between Horatio Herbert Kitchener and Commandant Botha, which negotiations eventually fell to the ground.

[163] The Times' History of the War in South Africa 1899-1902 by Amery, L. S. (Leopold Stennett), Vol 5 Ch. VII pages 183–193.

[164] In an attempt to invade Natal in force and later his unsuccessful attempt to overwhelm a British Column. He eventually did break out back to his base area which the British operation had been designed to deprive him of access to.

[165] Peace Committee Times History Vol 5. P. 91-93 In December of 1900 Kitchener placed himself in communication with influential men among the surrendered Boers, and it was partly on their advice that the proclamation of December 20 was framed. Nor did he stop here. At a meeting held early in December, under the presidency of Mr. Meyer de Kock, of Belfast, a Peace Committee was formed, with de Kock as secretary, for the purpose of urging upon the commandos the hopelessness of resistance and the need for a general surrender. A meeting of surrendered Boers convened by this Committee was held in Pretoria on December 21, and was addressed sympathetically by Kitchener. He promised the Committee his advice and support, spoke of the military conditions which exacted the formation of concentration camps, of his determination to give the enemy every possible chance of submitting voluntarily and of the just and progressive government which at the end of the war it was proposed to establish. The speech was printed in Dutch and English, and steps were promptly taken to give it and the proclamation of December 20 wide circulation. To this end emissaries were appointed to visit the commandos and preach the gospel of peace.

Like the proclamation itself the peace propaganda produced the contrary effect to that hoped for. It exasperated the fighting burghers, who viewed its promoters, sincerely patriotic as they were, as cowards and traitors, and, when they obtained the opportunity, treated them as such. Though most of the emissaries escaped in the course of time to the British lines, many were tried by court-martial and sentenced to fines and imprisonment. Meyer de Kock himself, having ridden into Viljoen's lines at the end of January, was tried, sentenced to death for high treason, and executed. In one notorious case there is, at least, a grave suspicion that an irregular and violent course was taken.

to maintain an attitude of unbending reserve, and accept with philosophy the prospect of an incalculably long and costly struggle; or, should a favourable opportunity arise, to persuade the enemy, on the promise of certain terms from which independence was rigidly excluded, to abandon a ruinous and futile resistance.

In the middle of February such an opportunity appeared to have arrived. De Wet and Steyn, the two most stubborn foes of British supremacy, were riding for their lives in Cape Colony. In the Transvaal French's great movement was throwing into panic the whole of a very important district. Kitchener, ...resolved to seize the opportunity; and, while de Wet and Steyn were incapable of interference, to effect a settlement with Botha. ....On February 28 Kitchener and Botha met at Middelburg. By a stroke of dramatic irony this was the very day on which de Wet and Steyn, after a period of desperate peril, crossed the Orange and won their freedom. It might have been read as a bad omen for peace.

The Middleburg[167] Conference. Group photograph of the participants in the Middelburg peace conference. (February-March 1901)[266 267] Standing (left to right); Colonel G. F. R. Henderson, Dirk Overgauwe ("Dick") van Velden,[168] Major Watson, H. Fraser, Captain Maxwell, H. de Jager. Sitting (left to right); Christiaan de Wet, (sic. Looks like Wynand Viljoen.) General Louis Botha, General Lord Kitchener, Colonel Ian S. M. Hamilton. Note the black armbands worn by all five British officers to commemorate Queen Victoria's death a month earlier. (January 22nd, 1901).

Annie Botha, the wife of general Louis Botha.[268 169] "Mrs. Botha is descended from Robert Emmet,[170] the famous "United Irishman" who was executed for high treason, September 20, 1803."

On the whole, however, the treatment accorded to the peace emissaries was regular and not, under the circumstances, unduly harsh. In taking the risk of attempting to seduce the men on commando, they acted with their eyes open according to their own sense of patriotism. The fighting burghers, holding a different standard of patriotism, were within their rights in applying even extreme penalties.

[166] Meyer de Kock (5 October 1849–12 February 1901) After the British had captured the capital Pretoria and annexed the Transvaal, he surrendered on 10 December 1900. In January 1901, he went to the eastern Transvaal to persuade burghers still in the field to surrender. He was arrested, court-martialed, and later, on 12 February 1901, executed by firing squad. Meyer de Kock - Wikipedia

[167] It was initially established as a halfway station between Lydenburg and Pretoria. The British built a large concentration camp in Middelburg during the Second Boer War. A Memorial Museum was built at the site next to the 1,381 graves of women and children who died in the concentration camp. Middelburg, Mpumalanga - Wikipedia

[168] Van Velden Dirk Overgauwe Kaptein Commando: Staf General Botha. Awarded the DTD.

[169] Annie Botha (née Emmett; 3 July 1864–20 May 1937) was a South African civic leader and political hostess. She was the wife of Louis Botha, who served as the first Prime Minister of South Africa. She established an orphanage in South Africa and, with Georgiana Solomon, co-founded and chaired the South African Women's Federation. Annie Botha - Wikipedia

[170] Robert Emmet (4 March 1778–20 September 1803) was an Irish Republican, orator and rebel leader. Emmet's Proclamation of the Provisional Government to the People of Ireland, his Speech from the Dock, and his "sacrificial" end on the gallows inspired later generations of Irish republicans. Robert Emmet - Wikipedia

Newspaper Coverage.

There were several fragmented reports that a meeting was taking place published at the time.

Several of the reports were of this brief nature.

Highland News - Saturday 02 March 1901—The Fighting in South Africa—Reported Surrender of Botha.

> Various reports were current on Thursday [February 28th] that General Louis Botha had formally surrendered to Lord Kitchener at Middleburg, but they remain totally unconfirmed. The War Secretary, replying to a question on the subject in the House of Commons, said no official report of the surrender had come to his Department.

The most accurate summary appeared in the Weekly Irish Times on Saturday 09 March 1901.[269]

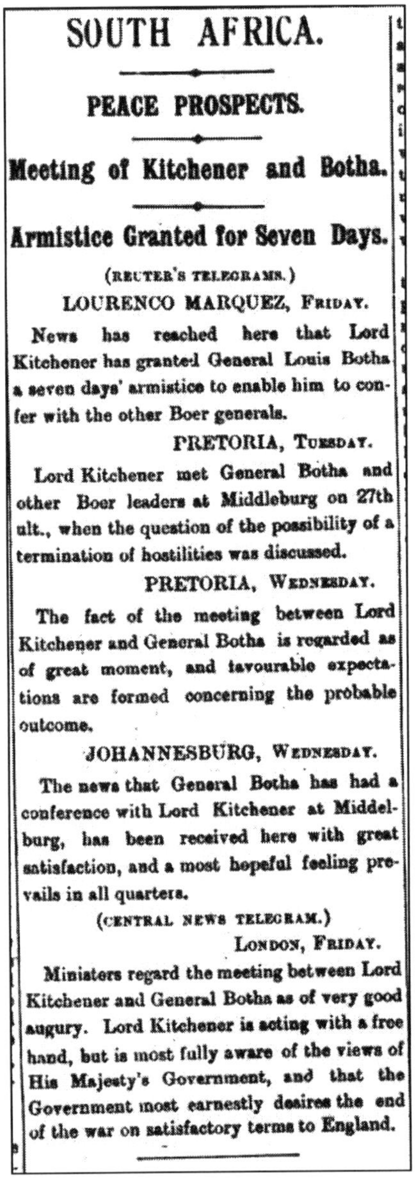

SOUTH AFRICA.

PEACE PROSPECTS.

Meeting of Kitchener and Botha.

Armistice Granted for Seven Days.

(REUTER'S TELEGRAMS.)

LOURENCO MARQUEZ, FRIDAY.

News has reached here that Lord Kitchener has granted General Louis Botha a seven days' armistice to enable him to confer with the other Boer generals.

PRETORIA, TUESDAY.

Lord Kitchener met General Botha and other Boer leaders at Middleburg on 27th ult., when the question of the possibility of a termination of hostilities was discussed.

PRETORIA, WEDNESDAY.

The fact of the meeting between Lord Kitchener and General Botha is regarded as of great moment, and favourable expectations are formed concerning the probable outcome.

JOHANNESBURG, WEDNESDAY.

The news that General Botha has had a conference with Lord Kitchener at Middelburg, has been received here with great satisfaction, and a most hopeful feeling prevails in all quarters.

(CENTRAL NEWS TELEGRAM.)

LONDON, FRIDAY.

Ministers regard the meeting between Lord Kitchener and General Botha as of very good augury. Lord Kitchener is acting with a free hand, but is most fully aware of the views of His Majesty's Government, and that the Government most earnestly desires the end of the war on satisfactory terms to England.

Peace Prospects. Meeting Kitchener and Botha. Armistice Granted for Seven Days. (Reuter's telegrams.) Lourenco Marquez, Friday.

New has reached here that Lord Kitchener has granted General Louis Botha seven days' armistice to enable him to confer with the other Boer generals.

PRETORIA, Tuesday.

Lord Kitchener met General Botha and other Boer leaders Middleburg on 27th inst., when the question the possibility of termination of hostilities was discussed.

Pretoria, Wednesday.

The fact of the meeting between Lord Kitchener and General Botha is regarded as of great moment, and favourable expectations are formed concerning the probable outcome.

Johannesburg, Wednesday

The news that General Botha has had a conference with Lord Kitchener Middelburg, has been received here with great satisfaction, and a most hopeful feeling prevails in all quarters.

(Central News Telegram.) London, Friday. Ministers regard the meeting between Lord Kitchener and General Botha as of very good augury. Lord Kitchener is acting with free hand, but is most fully aware the views of His Majesty's Government, and that the Government most earnestly desires the end of the war on satisfactory terms to England.

In spite of the express condition laid down by Kitchener draft terms. and implicitly accepted in the letter of the Boer general, Botha began by asking that the Republics should retain some sort of independence. Kitchener having refused to discuss the point, a conference began upon the details of a possible settlement. At the end of the discussion, which was conducted in a very friendly and reasonable spirit, Botha took his departure, and the terms of a proposed settlement were thrown into the shape of a draft letter from the British to the Boer general. This letter was submitted to Milner, who met Kitchener at Bloemfontein, and was sent by him for approval to the Home Government.

The terms subsequently published in England for general readership.

THE BOER WAR.

THE RECENT PEACE NEGOTIATIONS.

THE TERMS REJECTED.

The terms were published in England on March 23rd.[270]

Previously the Government made changes to Kitchener's proposed language and this was sent to Botha on the 7th. (March) In its final form it ran as follows:—

Pretoria, March 7, 1901.

To his Honour Commandant-General Louis Botha.

Your Honour,

With reference to our conversation at Middelburg on the 28th February, I have the honour to inform you that, in the event of a general and complete cessation of hostilities, and the surrender of all rifles, ammunition, cannon, and other munitions of war in the hands of the burghers, or in Government depots, or elsewhere, His Majesty's Government is prepared to adopt the following measures. His Majesty's Government will at once grant an amnesty in the Transvaal and Orange River Colony for all bona fide acts of war committed during the recent hostilities. British subjects belonging to Natal and Cape Colony, while they will not be compelled to return to those Colonies, will, if they do so, be liable to be dealt with by the laws of those Colonies specially passed to meet the circumstances arising out of the present war. As you are doubtless aware the special law in the Cape Colony has greatly mitigated the ordinary penalties for High Treason in the present case. All prisoners of war, now in St. Helena, Ceylon, or elsewhere, being burghers or Colonists, will, on the completion of the surrender, be brought back to their country as quickly as arrangements can be made for their transport.

At the earliest practicable date military administration will cease, and will be replaced by civil administration in the form of Crown Colony Government.

There will, therefore, be, in the first instance, in each of the new Colonies, a Governor and an Executive Council, composed of the principal officials, with a Legislative Council consisting of a certain number of official members to whom nominated unofficial element will be added.

But it is the desire of His Majesty's Government, as soon as circumstances permit, to introduce a representative element, and ultimately to concede to the new Colonies the privilege of self-government. Moreover, on the cessation of hostilities a High Court will be established in each of the new Colonies to administer the laws of the land, and this Court will be independent of the Executive.

Church property, public trusts, and orphan funds will be respected. Both the English and Dutch languages will be used and taught in public schools when the parents of the children desire it, and allowed in Courts of Law.

As regards the debts of the late Republican Governments, His Majesty's Government cannot undertake any liability. It is, however, prepared, as an act of grace, to set aside a sum not exceeding one million pounds sterling to repay inhabitants of the Transvaal and Orange River Colony for goods requisitioned from them by the late Republican Governments, or subsequent to annexation, by Commandants in the field being in a position to enforce such requisitions.

But such claims Will have to be established to the satisfaction of a Judge or Judicial Commission, appointed by the Government, to investigate and assess them, and, if exceeding in the aggregate one million pounds, they will be liable to reduction pro rata.

I also beg to inform your Honour that the new Government will take into immediate consideration the possibility of assisting by loan the occupants of farms, who will take the oath of allegiance, to repair any injuries sustained by destruction of buildings or loss of stock during the war, and that no special war tax will be imposed upon farms to defray the expense of the war.

When burghers require the protection of fire-arms, such will be allowed to them by licence, and on due registration, provided they take the oath of allegiance. Licences will also be issued for sporting rifles, guns, &c., but military fire-arms will only be allowed for purposes of protection.

As regards the extension of the franchise to Kaffirs in the Transvaal and Orange River Colony, it is not the intention of His Majesty's Government to give such franchise before representative Government is granted to those Colonies, and if then given it will be so limited as to secure the just predominance of the white race. The legal position of coloured persons will, however, be similar to that which they hold in the Cape Colony.

In conclusion, I must inform your Honour that, if the terms now offered are not accepted after a reasonable delay for consideration they must be regarded as cancelled.

I have, &c.,

Kitchener, General,
Commander-in-Chief, British Forces, South Africa.

But in the meantime the prospects of peace had been declining. French, hampered by rain and scarcity, had for some time been at a complete standstill on the Swazi frontier, and the burghers behind him had had time to recover their spirits. Botha had consulted his Government and found it stubborn; he had corresponded with De la Rey and had met with no encouragement. Before the time came to reply, it was known that de Wet and Steyn had escaped capture, and their opinion of the negotiations scarcely needed asking.[171] Nor would they have held themselves bound by any decision of the Transvaal.

In a letter dated March 16th Botha, without Botha giving any reasons, declined to go any further with the negotiations.

Botha's Reply. These terms were refused by Botha and reported on March 23rd, in the following message:[271]

> I have the honour to acknowledge the receipt of your Excelleney's letter stating what steps your Excellency's Government is prepared to take in the event of a general and total cessation of hostilities. I have advised my Government of your Excellency's said letter, and, after. mutual exchanges of views at Middleburg on the 28th February, it will certainly not surprise your Excellence to know that I do not feel disposed to recommend that the terms of the said letter shall have the earnest consideration of my Government. I may add also that my Government and my chief officers here entirely agree to my views.

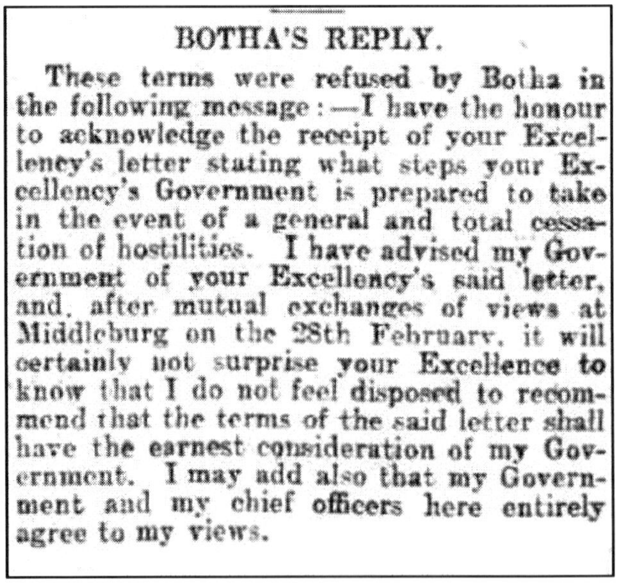
BOTHA'S REPLY.

These terms were refused by Botha in the following message:—I have the honour to acknowledge the receipt of your Excelleney's letter stating what steps your Excellency's Government is prepared to take in the event of a general and total cessation of hostilities. I have advised my Government of your Excellency's said letter, and, after mutual exchanges of views at Middleburg on the 28th February, it will certainly not surprise your Excellence to know that I do not feel disposed to recommend that the terms of the said letter shall have the earnest consideration of my Government. I may add also that my Government and my chief officers here entirely agree to my views.

[171] Original footnote: * Botha. wrote to de Wet and Steyn, asking them to come north and confer with him, but the letter arrived too late and the meeting did not take place till April. De Wet's secretary, however, went by train to Pretoria, only to find that negotiations had been broken off. He could have been of no use, for his master's temper, as shown by a circular to his burghers, was as unbending as ever. In this circular, which is dated April 1, he inveighs against various points in the proposed settlement, and asks finally, what is the use of examining it in detail, since nothing short of independence will induce the Boers to make peace.

Kitchener was always inclined to feel that with a little a little stronger support, especially on the question of amnesty, he might have obtained peace. However long the war lasted, it must, he held, eventually end in some such settlement as this. There was much truth in that; but the time had not yet come. His overtures were a bold but premature experiment. The Boer spirit-that new spirit born amidst the disasters of the previous year and matured in an offensive revival still quite recent, was far from being broken.

Although exhaustion and despondency existed in some quarters, when it was put to the Boers in black and white that their cause was lost irretrievably and that it only remained to settle subordinate issues, the national will revolted. The conditions which favoured this revolt are conditions which it can never be easy for a highly civilised industrial nation fully to understand.

The causes which bring such a nation to its knees were absent here. There were no vital industries to dislocate, no teeming population of peaceful workers to plunge into starvation. The people lived on the land, and though its fruits could be destroyed and the farms upon it gutted, the land could not be damaged. A year's crops were a relatively small matter, farm-buildings were of little value, and when a farm was once gutted it signified little whether it was rebuilt a year sooner or later.

Meanwhile, the male population, in other words the army, was inured by habit and inherited instinct to a life of peril and adventure under the stars. These men, moreover, were scattered over a vast expanse of territory under circumstances so varied as to render a common judgment unattainable. A few were suffering acutely; some were fighting with fair success; some were living in perfect safety and comfort. In short, there had never come to the Boers one of those moments of national agony which wring from a people a cry for peace.

As events fell out, Botha had effected a very useful reconnaissance in two directions. He had tested the sentiment of his burghers, made them familiar-and that was something-with the idea of negotiation; and he had found out on what lines an ultimate settlement was to be framed. ....

Commander-in-Chief by right of undoubted military merit, he nevertheless had not risen without effort. A keen ambitious purpose had helped him to gain the high position which he occupied. To crown its tenure by signing away independence as long as there remained the faintest hope of carrying his career to a triumphant military climax, was a hard step to take.

He knew very well that it was impossible to recover the lost territories with the forces at his disposal, but there always remained the chance of a general rising in Cape Colony. The chimera of foreign intervention still survived, as did the dim hope that pro-Boer sentiment in Great Britain would grow and triumph.

Meanwhile he could salve his conscience with the knowledge that the British policy of concentration camps freed him from responsibility for the women and children, whose misery was the most cogent argument for submission. There was little to lose; there was much to gain. For Botha had the shrewd perception that, whatever the ultimate issue, the Boers could fare no worse than they would fare under the proposed settlement.

His mental attitude is revealed clearly in a letter addressed to his burghers on March 15.

"Virtually, "he writes, "Lord Kitchener's letter contains nothing more, but rather less than what the British Government will be obliged to do should our cause go wrong." ....

"The cause is not yet lost," he writes, "and since nothing worse than this can befall us, it is well worth while to fight on." ....

Such were the abortive negotiations of 1901. Their failure caused much disappointment in England.

On public platforms and in Parliament the Government was bitterly attacked by a section of the Opposition for having thrown away the chances of peace by attempting to exact too much, and the differences of opinion, frankly set forth in the despatches, between Mr. Chamberlain, Lord Milner and Lord Kitchener, were used to further the argument. ....

The nation at large instinctively grasped the truth, and resigned itself with unflinching determination to a further struggle of indefinite length and severity.

Philip Botha

General Botha's brother, Philip,[172] was killed in action and his two nephews wounded on March 6th, 1901,[272] while the negotiations at Middleburg were proceeding. It was reported: "From Lord Kitchener to Secretary of State for War. Pretoria, Wednesday, 4.30 p.m.—"General P. Botha, brother of the Commandant-General, was killed on the Doornberg. His two sons were wounded."

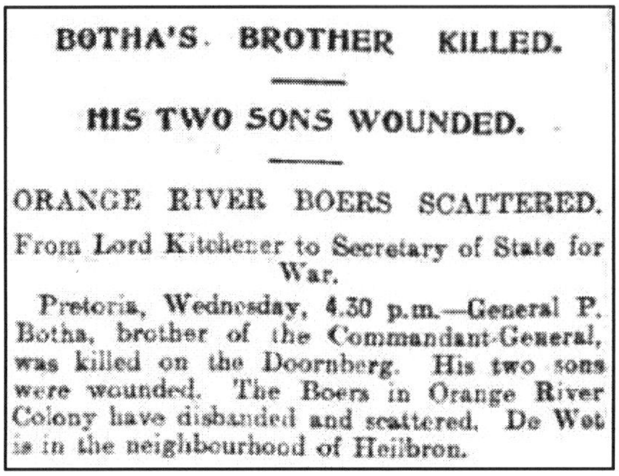

BOTHA'S BROTHER KILLED.

HIS TWO SONS WOUNDED.

ORANGE RIVER BOERS SCATTERED.

From Lord Kitchener to Secretary of State for War.

Pretoria, Wednesday, 4.30 p.m.—General P. Botha, brother of the Commandant-General, was killed on the Doornberg. His two sons were wounded. The Boers in Orange River Colony have disbanded and scattered. De Wet is in the neighbourhood of Heilbron.

Philip Rudolph Botha.[273]

Treaty of Vereeniging Terms

On May 31st, 1902 the decision of the Boer Assembly was transmitted to Kitchener regarding their decision to sign the treaty as proposed. At half-past ten of the same night the delegates arrived at Pretoria and authorize representatives of both countries signed the treaty returning peace to South Africa.

The following were in brief the terms of surrender:—

1. That the burghers lay down their arms and acknowledge themselves subjects of Edward VII.
2. That all prisoners taking the oath of allegiance be returned.
3. That their liberty and property be inviolate.
4. That an amnesty be granted—save in special cases.
5. That the Dutch language be allowed in schools and law-courts.
6. That rifles be allowed if registered.
7. That self-government be granted as soon as possible.
8. That no franchise be granted for natives until after self-government.
9. That no special land tax be levied.
10. That the people be helped to reoccupy the farms.
11. That 3,000,000 pounds be given to help the farmers.
12. That the rebels be disfranchised and their leaders tried, on condition that no death penalty be inflicted.

These terms were practically the same as those which had been refused by Botha in March 1901. [At the Middleburg Conference] Thirteen months of useless warfare had left the situation as it was."
Quote from The Great Boer War by Arthur Conan Doyle.[173 174]

[172] Botha Philip Rudolf, was born in 1851. While in command of the Harrismith burghers, he made a breakthrough at Sprinkaan's Nek. He was killed in action while attacking the British position near Ventersburg on March 6th, 1901. See Also: Philip Botha - Wikipedia
[173] Doyle, Arthur Conan. The Great Boer War 17th Edition, 1902 with the two volumes as one document: The Great Boer War by Arthur Conan Doyle - Free Ebook (gutenberg.org) Chapter 39. Pages 232–234 of the Word document.
[174] General Article: Treaty of Vereeniging - Wikipedia

Chapter 33. In a Tight Corner.

It was now March, 1901. For some time our burghers had been complaining of inactivity, and the weary and monotonous existence was gradually beginning to pall on them. But it became evident that April would be an eventful month, as the enemy had determined not to suffer our presence in these parts any longer. A huge movement, therefore, was being set on foot to surround us and capture the whole commando en bloc.

It began with a night attack on a field-cornet's force posted at Kruger's Post, north of Lydenburg, and here the enemy succeeded in capturing 35 men and a quantity of "impedimenta;" the field-cornet in question, although warned in time, having taken no proper precautions. By the middle of April the enemy's forward movement was in full swing. General Plumer came from Pietersburg, General Walter Kitchener from Lydenburg, and General Barber from Middelburg. They approached us in six different directions, altogether a force of 25,000 men, and the whole under the supreme command of General Sir Bindon Blood.

No escape was available for us through Secoekuniland on the north, as the natives here, since the British had occupied their territory, were avowedly hostile to us.[175] [176]

To escape, therefore, we would have to break through the enemy's lines [to the south] and also to cross the railway, which was closely guarded.

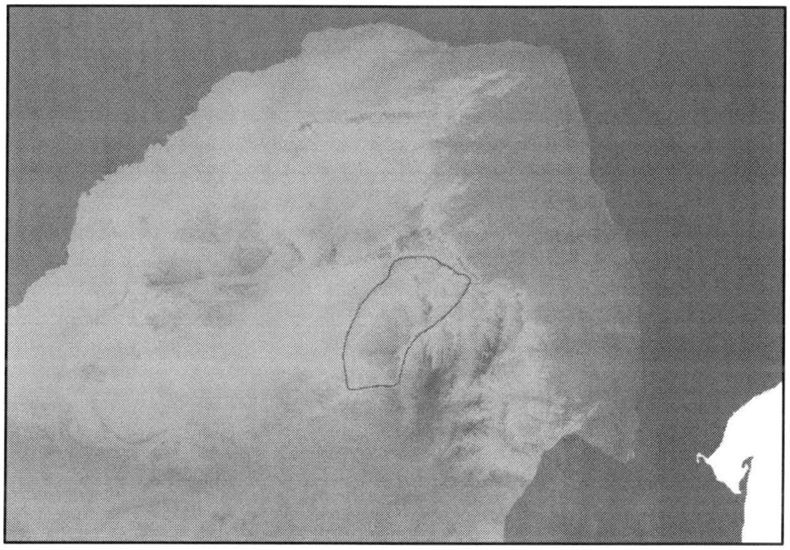

Map of Sekukuniland.[274] [177]

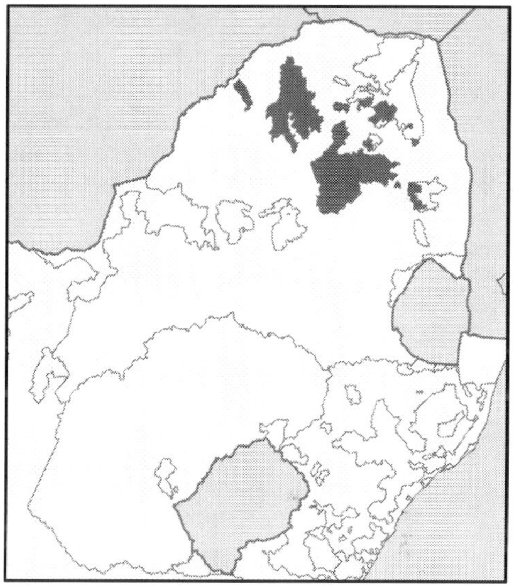

Map showing the location of the former bantustan of Lebowa.[275]

[175] Sekhukhune I (Matsebe; circa 1814–13 August 1882) was the paramount King of the Marota, more commonly known as the Bapedi (Pedi people), from 21 September 1861 until his assassination on 13 August 1882 by his rival and half-brother, Mampuru II. As the Pedi paramount leader he was faced with political challenges from Voortrekkers (Boer settlers), the independent South African Republic the British Empire, and considerable social change caused by Christian missionaries. In 1882, Sekhukhune was assassinated by conspirators, leading to the dismantling of the monarchy and statehood. In 1885, the Transvaal government only allocated a small territory to the Pedi, with the majority of the people living outside of it. https://en.wikipedia.org/wiki/Sekhukhune

[176] The Pedi or Bapedi - also known as the Northern Sotho, Basotho ba Lebowa, bakgatla ba dithebe, Transvaal Sotho, Marota, or Dikgoshi- are a Sotho-Tswana ethnic group native to South Africa, Botswana, and Lesotho. Throughout history, the Pedi actively participated in the struggle against colonization and apartheid in South Africa, joining the broader movement of African peoples fighting for their rights and freedom. In 1885, an area of 1,000 square kilometres (390 sq mi) was set aside for the Pedi, known as Geluk Location created by the Transvaal Republic's Native Location Commission. Later, according to apartheid segregation policy, the Pedi would be assigned the homeland of Lebowa. https://en.wikipedia.org/wiki/Pedi_people See also: https://en.wikipedia.org/wiki/Lebowa And: https://en.wikipedia.org/wiki/Sekhukhuneland

[177] Sekhukhuneland or Sekukuniland (Afrikaans: Sekoekoeniland) is a natural region in north-east South Africa, located in the historical Transvaal zone, former Transvaal Province, also known as Bopedi (meaning "land of Bapedi"). The region is named after the 19th-century King, Sekhukhune I. https://en.wikipedia.org/wiki/Sekhukhuneland

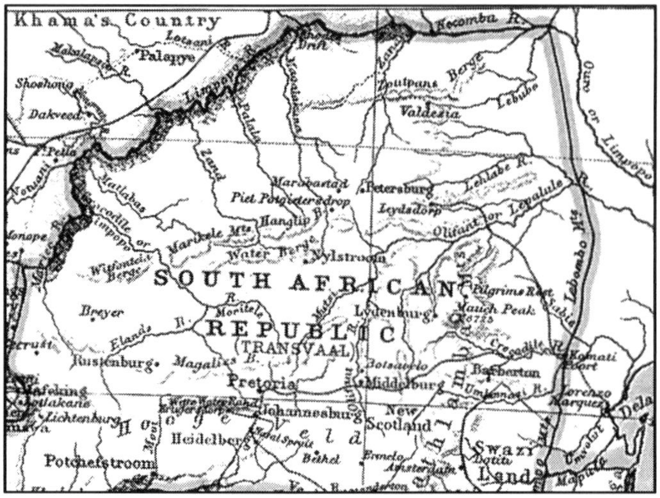

The shaded area is Bechuanaland Protectorate.[276] [178] Note Matabele Land to the north. The latitude line is 22° South. Walker Map, Note Khama's Country to the north.

"The Chief Khama." Son of Sekgoma I.[277] [179] Lekuku, Khama's fighting General, who has offered to help British against the Boers.[278]

Portrait of Sebele I in his twenties taken in 1865.[279] [180]

Bathoen I in London, 1895.[280] [181]

178 The Bechuanaland Protectorate was a protectorate established in 1885. It became the Republic of Botswana on 30 September 1966. Cecil Rhodes had wanted Bechuanaland incorporated in the BSAC charter. But three Tswana kings, including Khama III, travelled to Britain and won over British public opinion for it to remain governed by the British Colonial Office in London. https://en.wikipedia.org/wiki/Bechuanaland_Protectorate

179 Khama III (c. 1837–21 February, 1923), referred to by missionaries as Khama the Good also called Khama the Great, was the Kgosi (meaning king) of the Bangwato people. https://en.wikipedia.org/wiki/Khama_III

Natal Native Volunteer Corps.[281]

African soldiers prepared to repel raiders into Basutoland/Lesotho.[282] Black citizens serving as scouts for the British Forces.[283]

The enemy were advancing slowly from various directions. All our roads were carefully guarded, and the cordon was gradually tightening around us. We were repeatedly attacked, now on this side, now on that, the British being clearly anxious to discover our position and our strength. In a sharp skirmish with a column from Lydenburg my faithful Fighting-General Muller was severely wounded in his shoulder, and a commando of Lydenburgers had been isolated from me and driven by the enemy along Waterfal River up to Steelpoort, where they encountered hostile tribes of kaffirs. The commandant of the corps after a short defence was obliged to destroy his guns, forsake his baggage, and escape with his burghers in small groups into the mountains.

Our position was growing more critical, but I resolved to make a stand before abandoning our carts and waggons, although there seemed little hope of being able to save anything. In fact the situation was extremely perilous. As far as I could see we were entirely hemmed in, all the roads were blocked, my best officer wounded, I had barely 900 men with me, and our stock of ammunition was very limited.

[180] Sebele I was a chief (kgosi) of the Kwena —a major Tswana tribe (morafe) in modern-day Botswana— who ruled from 1892 until his death in 1911. During his lifetime, he resisted the 1885 Bechuanaland Protectorate as well as the control of his domains by Cecil Rhodes' British South African Company, which was administering, by a royal charter signed in October 1889, his homeland in the Bechuanaland Protectorate and other regions of Central Africa. With support from Christian missionaries, Sebele traveled to Britain in 1895 along with Bathoen I and Khama III to argue against the latest attempts to incorporate the protectorate into the Cape Colony. They managed to secure support from Queen Victoria in exchange for an eastern strip of territory. Between 1908 and 1909 he also resisted the incorporation of Bechuanaland into the Union of South Africa. https://en.wikipedia.org/wiki/Sebele_I

[181] Bathoen I (c. 1845–1 July 1910) was a kgosi (paramount chief) of the Ngwaketse people (1889–1910). Together with Khama III and Sebele I he is credited with saving the young British Bechuanaland Protectorate, a predecessor of Botswana, from being absorbed by expansionist forces in the 1890s.

In 1889, the British South Africa Company founded by Cecil Rhodes started to expand north, and the Tswana people became afraid that they would be eventually deposed from their lands. In response, Khama III, Sebele I, and Bathoen I, being mandated by their people, travelled to London, made public speeches in support of their causes, and finally convinced Queen Victoria to ring-fence the Bechuanaland Protectorate, which would preserve the self-government institutes of Tswana and the British would only have limited authority, such for example, control over the railway to be built. In 1908, he led the protests by the Tswana against the planned incorporation of Bechuanaland into South Africa. https://en.wikipedia.org/wiki/Bathoen_I

I have omitted to mention that early in April, when we first got an inkling of this move I had liberated all the British officers whom I had kept as prisoners at Middelburg, and thus saved the British authorities many a D.S.O. which would otherwise have been claimed by their rescuers.

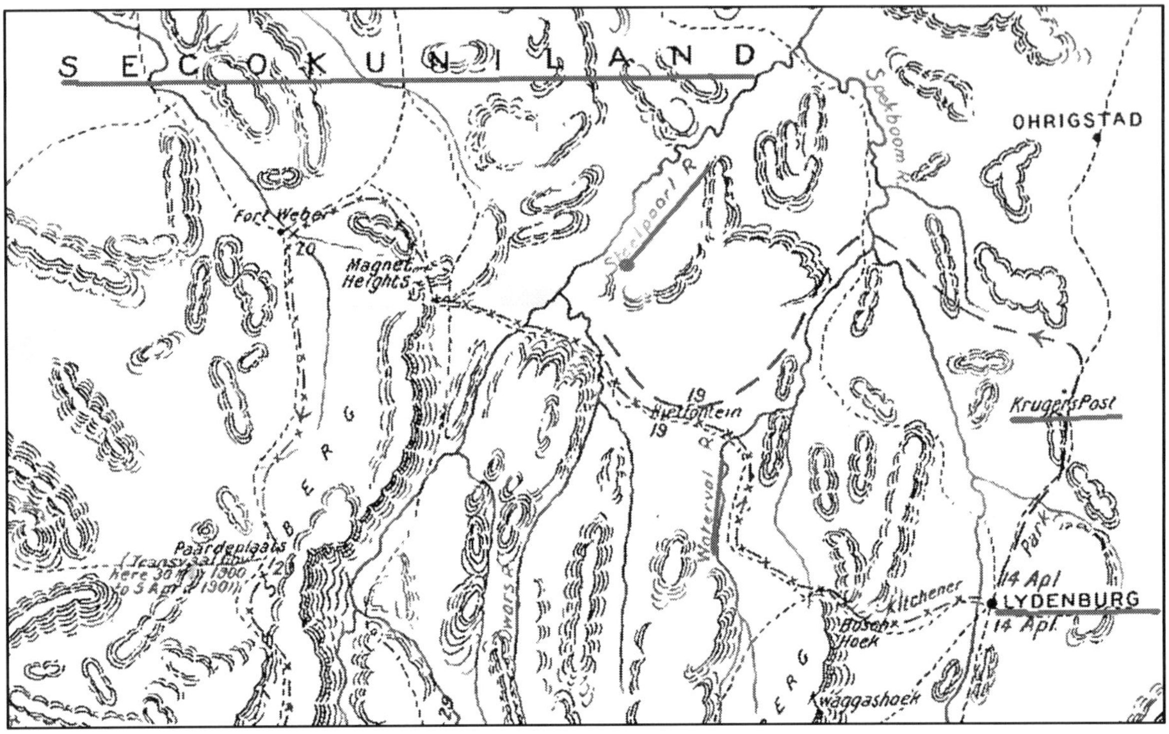

Secokuniland to the North but inside Transvaal as well as Kruger's Post and Lydenburg along with Waterval river and Steelpoort river.

The British around us were now posted as follows: At Diepkloof on the Tautesberg to the north-west of us; at Roodekraal, between Tautesberg and Bothasberg, to the west of us; at Koebold, under Roodehoogte; at Windhoek, to the east of us; at Oshoek, to the north-east; and to the north of us between Magneetshoogte and Klip Spruit. We were positioned on Mapochsberg near Roos Senekal, about midway between Tautesberg and Steenkampsberg. We had carts, waggons, two field-pieces, and a Colt-Maxim.

We speedily discovered that we should have to leave our baggage and guns, and rely mainly on our horses and rifles. We had placed our hospitals as well as we could, one in an empty school-building at Mapochsberg with 10 wounded, under the care of Dr. Manning; the other, our only field-hospital, at Schoonpoort, under the supervision of Dr. H. Neethling. Whether these poor wounded Boers would have to be abandoned to the enemy, was a question which perplexed us considerably. If so, we should have been reduced to only one physician, Dr. Leitz, a young German who might get through with a pack-horse. Many officers and men, however, had lost all hope of escape.

It was about the 20th of April when the British approached so close that we had to fight all day to maintain our positions. I gave orders that same night that we should burn our waggons, destroy our guns with dynamite, and make a dash through the enemy's lines, those burghers who had no horses to mount the mules of the convoy. Hereupon about 100 burghers and an officer coolly informed me that they had had enough fighting, and preferred to surrender. I was at that time powerless to prevent them doing so, so I took away all their horses and ammunition, at which they did not seem very pleased. Before dusk our camp was a scene of wild confusion. Waggons and carts were burning fiercely, dynamite was being exploded, and horseless burghers were attempting to break in the mules which were to serve them as mounts. Meanwhile a skirmish was going on between our outposts and those of the enemy.

Boer Soldiers Burning a Camp or Laager.[284]

It was a strange procession that left Mapochsberg that night in our dash through the British lines. Many Boers rode mules, whilst many more had no saddles, and no small number were trudging along on foot, carrying their rifles and blankets on their shoulders. My scouts had reported that the best way to get through was on the southern side along Steelpoort, about a quarter of a mile from the enemy's camp at Bothasberg. But even should we succeed in breaking through the cordon around us, we still had to cross the [railroad] line at Wondersfontein before daybreak, so as not to get caught between the enemy's troops and the blockhouses.

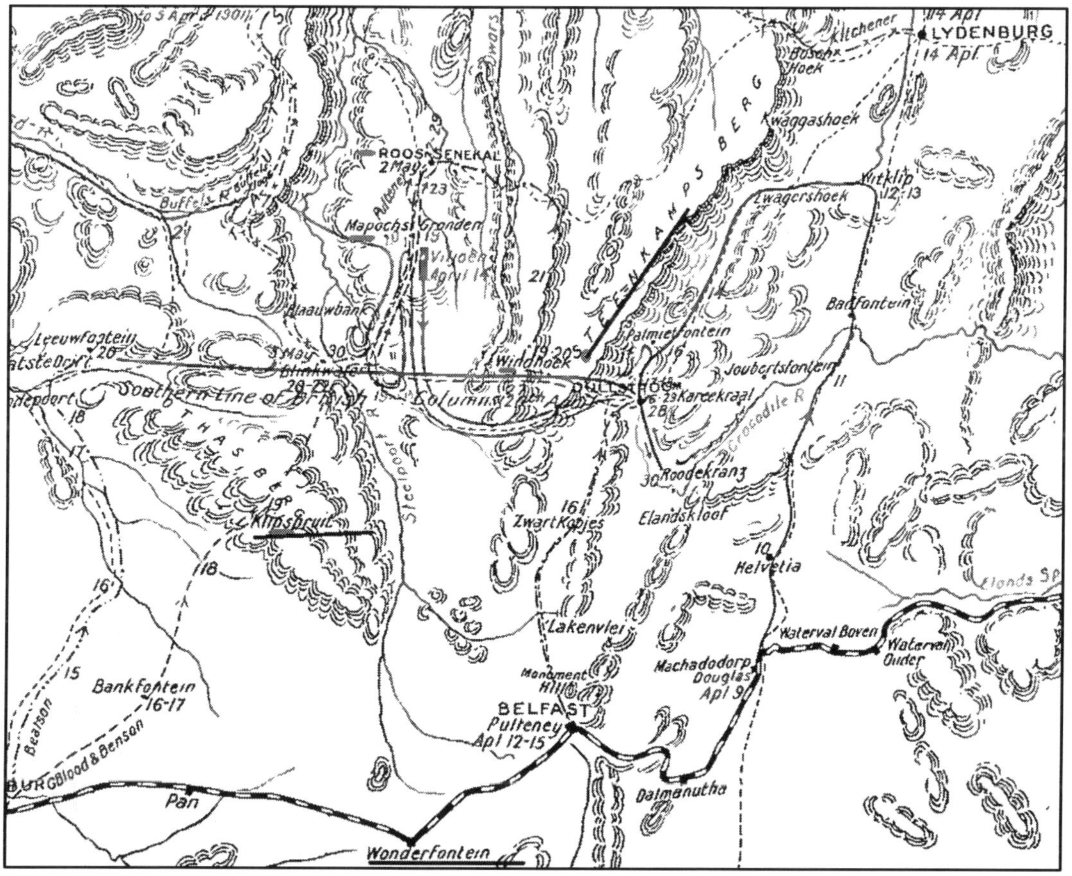

Area of operations. Note Steenkampsberg as a range not a town. Wonderfontein at the bottom.

About 100 scouts, who formed our advance-guard, soon encountered the enemy's sentries. They turned to the right, then turned to the left; but everywhere the inquisitive "Tommies" kept asking: "Who goes there?" Not being over anxious to satisfy their curiosity, they sent round word at once for us to lie low, and we started very carefully exploring the neighbourhood. But there seemed no way out of the mess. We might have attacked some weak point and thus forced our way through, but it was still four or five hours' ride to the railway line, and with our poor mounts we should have been caught and captured. Besides which the enemy might have warned the blockhouse garrisons, in which case we should have been caught between two fires.

No; we wanted to get through without being discovered, and seeing that this was that night hopeless, I consulted my officers and decided to return to our deserted camp, where we could take up our original positions without the enemy being aware of our nocturnal excursion.

Next morning the rising sun found us back in our old positions. We despatched scouts in all directions as usual, so as to make the enemy believe that we intended to remain there permanently, and we put ourselves on our guard, ready to repel an attack at any point on the shortest notice.

But the enemy were much too cautious, and evidently thought they had us safely in their hands. They amused themselves by destroying every living thing, and burned the houses and the crops. The whole veldt all round was black, everything seemed in mourning, the only relief from this dull monotony of colour being that afforded by the innumerable specks of khaki all around us. I believe I said there were 25,000 men there, but it now seemed to me as if there were almost double that number.

We had to wait until darkness set in before making a second attempt at escape. The day seemed interminable. Many burghers were loudly grumbling, and even some officers were openly declaring that all this had been done on purpose. Of course, these offensive remarks were pointed at me. At last the situation became too serious. I could only gather together a few officers to oppose an attack from the enemy on the eastern side, and something had to be done to prevent a general mutiny. I therefore ordered a burgher who seemed loudest in his complaints to receive 15 lashes with a sjambok,[182] and I placed a field-cornet under arrest. After this the grumblers remained sullenly silent.

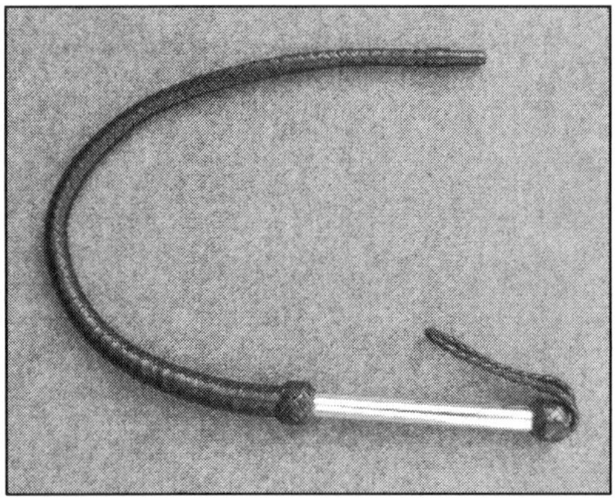

A Sjambok.

The only loophole in the enemy's lines seemed to be in the direction of Pietersburg on the portion held by General Plumer, who seemed far too busy capturing cattle and sheep from the "bush-lancers" to surround us closely. We therefore decided to take our chance there and move away as quickly as possible in that direction, and then to bear to the left, where we expected to find the enemy least watchful. Shortly before sunset I despatched 100 mounted men to ride openly in the opposite direction to that which we intended to take, so as to divert the enemy's attention from our scene of operations, and sat down to wait for darkness.

[182] The sjambok or litupa, is a heavy leather whip. It is traditionally made from adult hippopotamus or rhinoceros hide, but it is also commonly made out of plastic.

Chapter 34. Eluding the British Cordon.

"The shades of eve were falling fast" as we moved cautiously away from Mapochsberg and proceeded through Landdrift, Steelpoort, and the Tautesberg. At 3 o'clock in the morning we halted in a hollow place where we would not be observed, yet we were still a mile and a half from the enemy's cordon. Our position was now more critical than ever; for should the enemy discover our departure, and General Plumer hurry up towards us that morning, we should have little chance of escape.

During the day I was obliged to call all the burghers together, and to earnestly address them concerning the happenings of the previous day. I told them to tell me candidly if they had lost faith in me, or if they had any reason not to trust me implicitly, as I would not tolerate the way in which they had behaved the day before. I added:—

"If you cannot see your way clear to obey implicitly my commands, to be true to me, and to believe that I am true to you, I shall at once leave you, and you can appoint someone else to look after you. We are by no means out of the wood [s] yet, and it is now more than ever necessary that we should be able to trust one another to the fullest extent. Therefore, I ask those who have lost confidence in me, or have any objection to my leading them, to stand out."

No one stirred. Other officers and burghers next rose and spoke, assuring me that all the rebels had deserted the previous night, and that all the men with me would be true and faithful. Then Pastor J. Louw addressed the burghers very earnestly, pointing out to them the offensive way in which some of them had spoken of their superior officers, and that in the present difficult circumstances it was absolutely necessary that there should be no disintegration and discord amongst ourselves. I think all these perorations had a very salutary effect. But such were the difficulties that we officers had to contend with at the hands of undisciplined men who held exaggerated notions of freedom of action and of speech, and I was not the only Boer officer who suffered in this respect.

About two in the afternoon I gave the order to saddle up, as it was necessary to start before sunset in order to be able to cross the Olifant's River before daybreak, so that the enemy should not overtake us should they notice us. We dismounted and led our horses, for we had discovered that the English could not distinguish between a body of men leading their horses and a troop of cattle, so long as the horses were all kept close together. All the hills around us were covered with cattle captured from our "bush-lancers," and therefore our passage was unnoticed.

We followed an old waggon track along the Buffelskloof, where a road leads from Tautesberg to Blood River. The stream runs between Botha's and Tautesbergen, and flows into the Olifant's River near Mazeppa Drift. It is called Blood River on account of the horrible massacre which took place there many years before, when the Swazi kaffirs murdered a whole kaffir tribe without distinction of age or sex, literally turning the river red with blood.[183]

Towards evening we reached the foot of the mountains, and moved in a north-westerly direction past Makleerewskop. We got through the English lines without any difficulty along some footpaths, but our progress was very slow, as we had to proceed in Indian file, and we had to stop frequently to see that no one was left behind. The country was thickly wooded, and frequently the baggage on the pack-horses became entangled with branches of trees, and had to be disentangled and pulled off the horses' backs, which also caused considerable delay.

It was 3 o'clock in the morning before we reached the Olifant's River, at a spot which was once a footpath drift, but was now washed away and overgrown with trees and shrubs, making it very difficult to find the right spot to cross. Our only guide who knew the way had not been there for 15 years, but recognised the place by some high trees which rose above the others. We had considerable difficulty in crossing, the water reaching to our horses' saddles, and the banks being very steep. By the time we had all forded the sun had risen. All the other drifts on the river were occupied by the enemy, our scouts reporting that Mazeppa Drift, three miles down stream, was entrenched by a strong English force, as was the case with Kalkfontein Drift, a little higher up. I suppose this drift was not known to them, and thus had been left unguarded.

[183] I'm not sure he is referring to this battle of 1838. https://en.wikipedia.org/wiki/Battle_of_Blood_River

Having got through we rode in a northerly direction until about 9 o'clock in the morning, and not until then were we sure of being clear of the enemy's clutches. But there was a danger that the English had noticed our absence and had followed us up. I therefore sent out scouts on the high kopjes in the neighbourhood, and not until these had reported all clear did we take the risk of off-saddling. You can imagine how thankful we were after having been in the saddle for over 19 hours, and I believe our poor animals were no less thankful for a rest.

We had not slept for three consecutive nights, and soon the whole commando, with the exception of the sentries, were fast asleep. Few of us thought of food, for our fatigue and drowsiness were greater than our hunger. But we could only sleep for two hours, for we were much too close to the enemy, and we wished to make them lose scent of us entirely.

The burghers grumbled a good deal at being awakened and ordered to saddle up, but we moved on nevertheless. I sent some men to enquire at a kaffir kraal for the way to Pietersburg, and although I had no intention of going in that direction, I knew that the kaffirs, so soon as we had gone, would report to the nearest British camp that they had met a commando of Boers going there. Kaffirs would do this with the hope of reward, which they often received in the shape of spirituous liquor.

We proceeded all that day in the direction of Pietersburg until just before sunset we came to a small stream. Here we stopped for an hour and then went on again, this time, however, to the left in a southerly direction through the bush to Poortjesnek near Rhenosterkop, where a little time before the fight with General Paget's force had taken place. We had to hurry through the bush, as horse-sickness was prevalent here and we still had a long way before us. It was midnight before we reached the foot of the Poortjesnek.

Here my officers informed me that two young burghers had become insane through fatigue and want of sleep, and that several, while asleep in their saddles had been pulled off their horses by low branches and severely injured. Yet we had to get through the Nek and get to the plateau before I could allow any rest. I went and had a look at the demented men. They looked as if intoxicated and were very violent. All our men and horses were utterly exhausted, but we pushed on and at last reached the plateau, where, to everybody's great delight, we rested for the whole day. The demented men would not sleep, but I had luckily some opium pills with me and I gave each man one of them, so that they got calmer, and, dropping off to sleep, afterwards recovered.

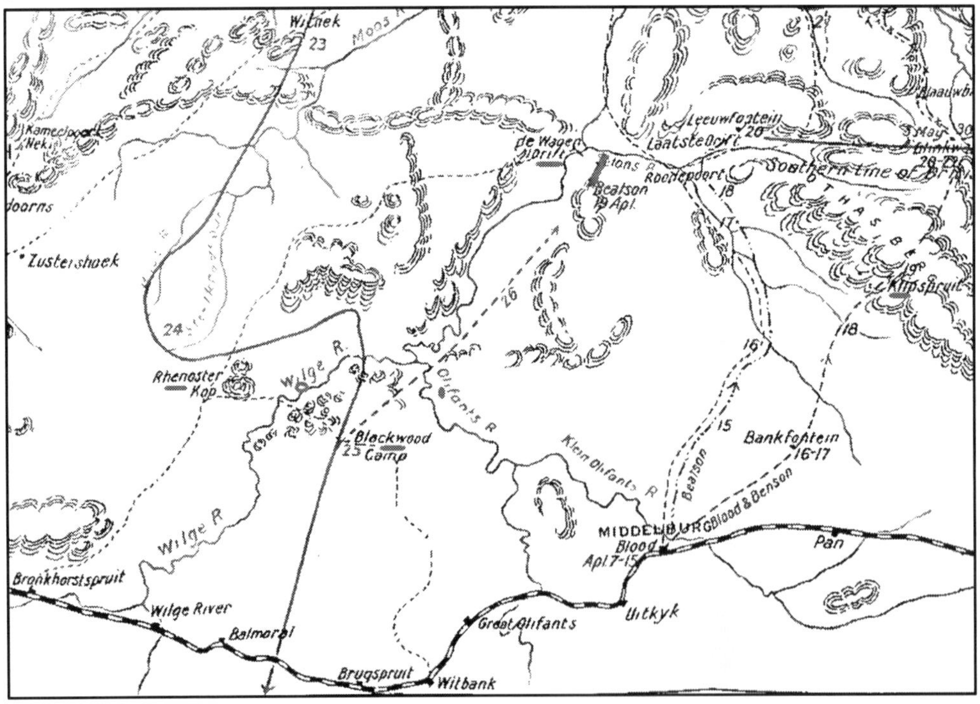

Rhenoster Kop and Blackwood Camp and de Wagen Drift.

My scouts reported next day that a strong English patrol had followed us up, but that otherwise it was "all serene." We pushed on through Langkloof over our old fighting ground near Rhenosterkop, then through the Wilge River near Gousdenberg up to Blackwood Camp, about nine miles north of Balmoral Station. Here we stayed a few days to allow our animals to rest and recover from their hardships, and then moved on across the railway to the Bethel and Ermelo districts. Here the enemy was much less active, and we should have an opportunity of being left undisturbed for a little time. But we lost 40 of our horses, who had caught the dreaded horse-sickness whilst passing through the bush country.

On the second day of our stay at Blackwood Camp I sent 150 men under Commandants Groenwald and Viljoen [His brother Wynand] through the Banks, via Staghoek, to attack the enemy's camp near Wagendrift on the Olifant's River. This was a detachment of the force which had been surrounding us. We discovered that they were still trying to find us, and that the patrol which had followed us were not aware of our having got away. It appears that they only discovered this several days afterwards, and great must have been the good general's surprise when they found that the birds had flown and their great laid schemes had failed.

My 150 men approached the enemy's camp early in the morning, and when at a short range began pouring in a deadly rifle fire on the western side. The British soldiers, who were not dreaming of an attack, ran to and fro in wild disorder. Our burghers, however, ceased firing when they saw that there were many women and children in the camp, but the enemy began soon to pour out a rifle and gun fire, and our men were obliged to carry on the fight.

After a few days' absence they returned to our camp and reported to me that "they had frightened the English out of their wits, for they thought we were to the east at Roos Senekal, whereas we turned up from the west."

Of course the British speedily discovered where we were, and came marching up from Poortjesnek in great force. But we sent out a patrol to meet them, and the latter by passing them west of Rhenosterkop effectually misled them, and we were left undisturbed at Blackwood Camp.

This left us time to prepare for crossing the railway; so I despatched scouts south to see how matters stood, and bade them return the next day. We knew that a number of small commandos were located on the south side of the railway, but to effect a junction was a difficult matter, and we would risk getting trapped between the columns if we moved at random. The railway and all the roads were closely guarded, and great care was being taken to prevent any communication between the burghers on either side of the line.[184]

Chapter 35. Boer Government's Narrow Escape.

During the first week of May, 1901,[185] we split up into two sections, and left Blackwood Camp early in the evening. General Muller took one section over the railway line near Brugspruit, whilst I took the other section across near Balmoral Station. We naturally kept as far from the blockhouses as possible, quietly cut the barbed-wire fences stretched all along the line, and succeeded in crossing it without a shot being fired. To split up into two sections was a necessary precaution, first because it would have taken the whole commando too long to cross the line at one point, and secondly, we made more sure of getting at least one section across. Further, had the enemy encountered one of the sections they would probably have concluded that that was our whole force.

We halted about six miles [South] from the railway-line, as it was now 2 o'clock in the morning. I ordered a general dismount, and we were at last able to light up our pipes, which we had been afraid of doing in the neighbourhood of the railway for fear of the lights being seen by the enemy. The men sat round in groups, and smoked and chatted cheerfully. We passed the rest of the night here, and with the exception of the sentinels on duty, all were able to enjoy a refreshing sleep, lying down, however, with their unsaddled horses by their side, and the bridles in their hands—a most necessary and useful precaution. Together with my adjutant, Nel, I made the round of the sentries, sitting a few moments with each to cheer them up and keep them awake; for there is nothing to which I object more than to be surprised by the enemy, when asleep.

The few hours of rest afforded us passed very quickly, and at the first glimmer of dawn I ordered the men to be called. This is simply done by the officers calling "Opzâal, opzâal" (saddle-up) in loud tones. When it was light

[184] Victory by General Babington over Delarey at Ventersdorp. Nine Boer guns captured.
4/6/1901 General French, in his sweeping operations in the Eastern Transvaal, captured all the enemy's guns in that district.
4/8/1901 Colonel Plumer captured Pietersburg, the terminus of the railway running due north from Pretoria.
4/10/1901 Civil administration resumed in the Transvaal.
4/15/1901 Smuts' commando defeated near Klerksdorp. Two guns captured.
4/19/1901 Generals Plumer and Walter Kitchener co-operated with General French in clearing the Eastern Transvaal and Lydenburg district.

[185] 4/30/1901 General Blood discovered documents and banknotes of Transvaal (Government at Roosenekal, from which place Mr. Schalk Burger fled.

enough to look round us we had the satisfaction of seeing that all was quiet and that no troops were in the immediate neighbourhood. We made for a place called Kroomdraai, about halfway between Heidelberg and Middelburg, where we knew there were some mealies left; and although we should be between the enemy's camps there, I felt there would be no danger of being disturbed or surprised.

I also sent a report to the Commandant-General, who was at that time with the Government near Ermelo, and described to him all that had happened. I received a reply some days later, requesting me to leave my commando at Kroomdraai and proceed to see him, as an important Council of War was to be held between the various generals and the Government.

Four days later I arrived at Begin der Lijn ("beginning of the line") on the Vaal River, south-east of Ermelo, accompanied by three of my adjutants, and reported myself to the Commandant-General.

Crossing Railway Line Northward (Between Balmoral, and Brugspruit Stations). Commandant Ben Viljoen and staff.[285]

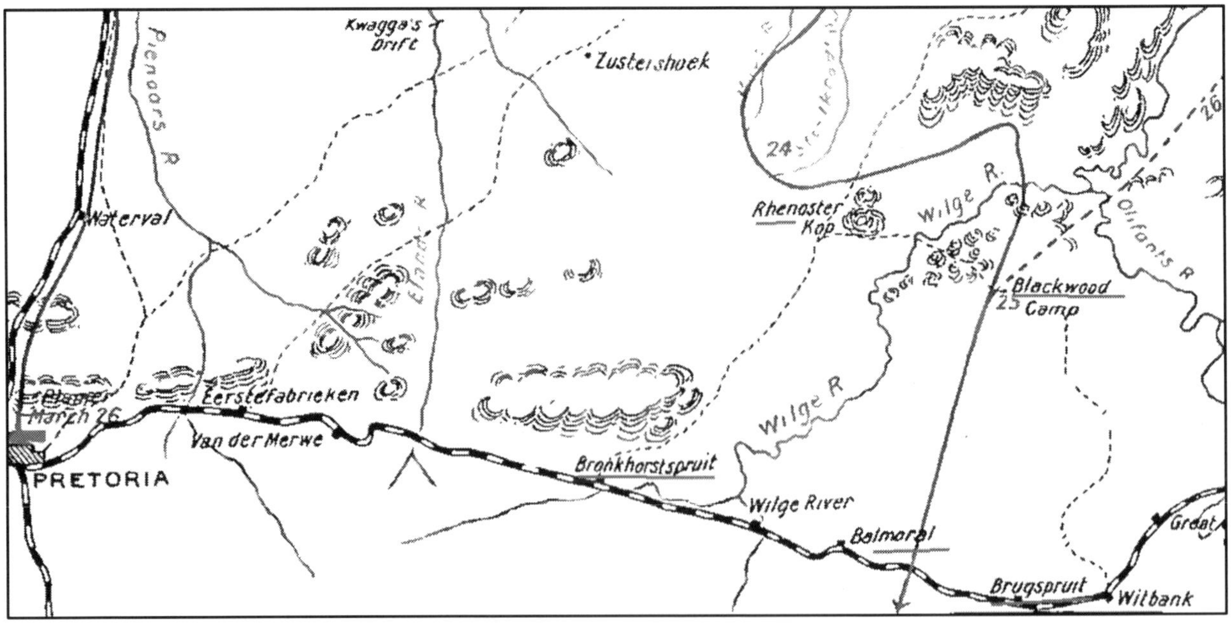

Balmoral South of Blackwood Camp.

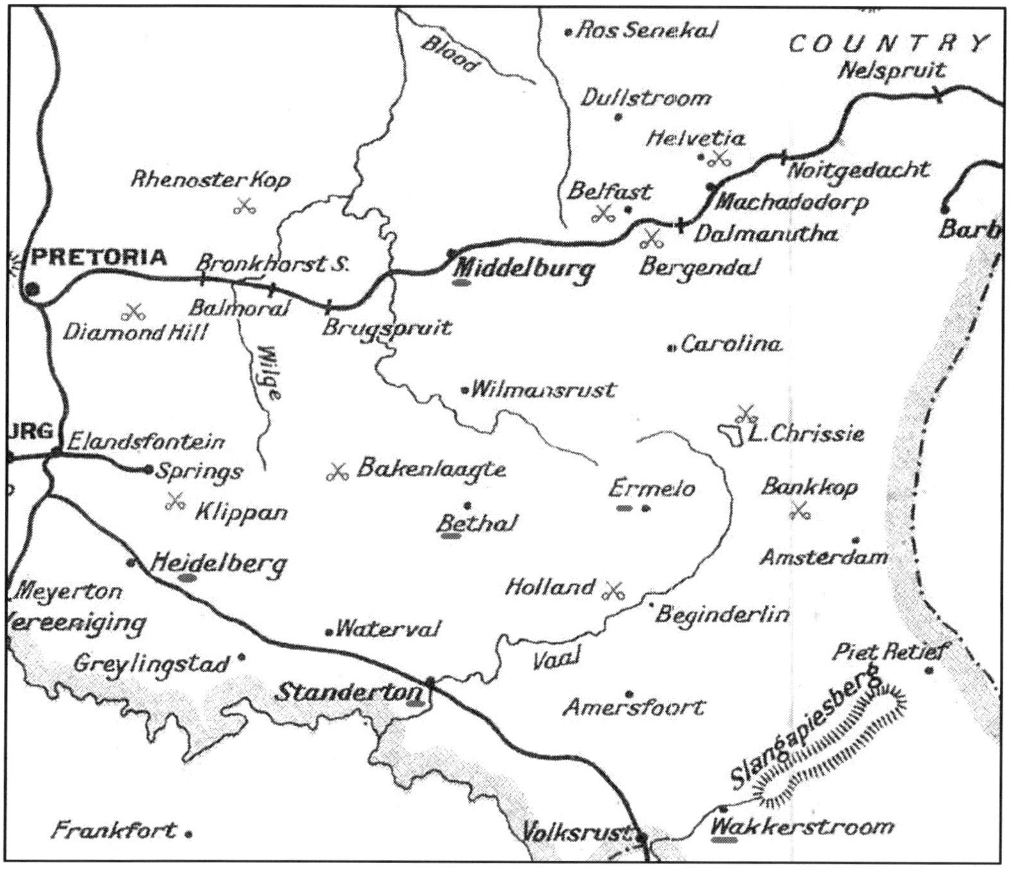

Heidleberg South and Middleburg on rail line.

Simultaneously with my arrival there came two British columns, commanded by our old friend Colonel Bullock, whose acquaintance we had previously made at Colenso. They came apparently with the idea of chasing us, possibly thinking to catch us. This was far from pleasant for me. I had been riding post-haste for four days, and I and my horse were very tired and worn out. However, there was no help for it. I had barely time to salute the members of the Government, and to exchange a few words with General Botha, when we had to "quit."

For eight days we wandered round with Colonel Bullock at our heels, always remaining, however, in the same neighbourhood. This officer's tactics in trying to capture us were childishly simple. During the day there would be skirmishes between the enemy and General Botha's men, but each evening the former would, by retiring, attempt to lull us into a sense of security. But as soon as the sun had set, they would turn right about face, return full speed to where they had left us, and there would surround us carefully during the night, gallantly attacking us in the morning and fully expecting to capture the whole Boer Government and at least half a dozen generals. This was a distinct nuisance, but the tactics of this worthy officer were so simple that we very soon discovered them. Accordingly, every evening we would make a fine pretence of pitching our camp for the night; but so soon as darkness had set in, we would take the precaution of moving some 10 or 15 miles further on.

Next morning Colonel Bullock, who had been carefully "surrounding" us all night, would find that we were unaccountably absent. Much annoyed at this, he would then send his "flying" columns running after us. This went on for several days, until finally, as we expected, his horses were tired out, and I believe he was then removed to some other garrison, having been considered a failure as a "Boer-stalker." No doubt he did his best, but he nevertheless managed his business very clumsily.

Not until nine days after my arrival at this perambulating seat of Government did we have an opportunity of snatching a few hours' rest. We were now at a spot called Immegratie, between Ermelo and Wakkerstroom. Here a meeting was held by the Executive Council, and attended by the Commandant-General, General Jan Smuts, General C. Botha, and myself. General T. [Tobias] Smuts could not be present, as he was busy keeping Colonel Bullock amused.

At this meeting we discussed the general situation, and decided to send a letter to President Steyn, but our communication afterwards fell into the enemy's hands. In accordance with this letter, President Steyn and Generals De Wet and De la Rey joined our Government, and a meeting was held later on.[186]

The day after this meeting at Immegratie I took leave of my friends and began the journey in a more leisurely fashion back to my commando at Kroomdraai, via Ermelo and Bethel. The Acting-President [S. Burger] had made me a present of a cart and four mules, as they pitied us for having had to burn all our vehicles in escaping from Roos Senekal. We were thus once more seated in a cart, which added considerably to the dignity of our staff. How long I should continue to be possessed of this means of transport depended, of course, entirely on the enemy. My old coloured groom "Mooiroos," who followed behind leading my horse, evidently thought the same, for he remarked naïvely: "Baas, the English will soon fix us in another corner; had we not better throw the cart away?"

Cart on a lonely road.[286]

A mixed race commando unit.[287]

We drove into Ermelo that afternoon. The dread east wind was blowing hard and raising great clouds of dust around us. The village had been occupied about half a dozen times by the enemy and each time looted, plundered, and evacuated, and was now again in our possession. At least, the English had left it the day before, and a Landdrost had placed himself in charge; a little Hollander with a pointed nose and small, glittering eyes, who between each sentence that he spoke rolled round those little eyes of his, carefully scanning the neighbouring hills for any sign of the English. The only other person of importance in the town was a worthy predicant,[187] who evidently had not had his hair cut since the commencement of the War, and who had great difficulty in keeping his little black wide-awake on his head. He seemed very proud of his abundant locks.

There were also a few families in the place belonging to the Red Cross staff and in charge of the local hospitals. One of my adjutants was seriously indisposed, and it was whilst hunting for a chemist in order to obtain medicine that I came into contact with the town's sparse population. I found the dispensary closed, the proprietor having departed with the English, and the Landdrost, fearing to get himself into trouble, was not inclined to open it. He grew very excited when we liberally helped ourselves to the medicines, and made himself unpleasant. So we gave him clearly to understand that his presence was not required in that immediate neighbourhood.

Our cart was standing waiting for us in the High Street, and during our absence a lady had appeared on the verandah of a house and had sent a servant to enquire who we were. When we reappeared laden with our booty she graciously invited us to come in. She was a Mrs. P. de Jager and belonged to the Red Cross Society. She asked us to stay and have some dinner, which was then being prepared. Imagine what a luxury for us to be once more in a house, to be addressed by a lady and to be served with a bountiful repast! Our clothes were in a ragged and dilapidated condition and we presented a very unkempt appearance, which did not make us feel quite at our ease. Still the good lady with great tact soon put us quite at home.

We partook of a delicious meal, which we shall not easily forget. I cannot remember what the menu was, and I am not quite sure whether it would compare favourably with a first-class café dinner, but I never enjoyed a meal more in my existence, and possibly never shall.

[186] This is the meeting where the SAR made a proposal to the OFS to stop the war and sent the letter. It was received and that's why Steyn and De Wet traveled to the Transvaal to put a stop to the idea of ending the war at that time. See Chapter 36 infra.

[187] A Preacher or Minister.

After dinner the lady related to us how on the previous day, when the British entered the village, there were in her house three convalescent burghers, who could, however, neither ride nor walk. With tears in her eyes she told us how an English doctor and an officer had come there, and kicking open the doors of her neatly-kept house, had entered it, followed by a crowd of soldiers, who had helped themselves to most of the knives, forks, and other utensils. She tried to explain to the doctor that she had wounded men in the house, but he was too conceited and arrogant to listen to her protestations. Fortunately for them the men were not discovered, for the English, on leaving the village, took with them all our wounded, and even our doctor. With a proud smile she now produced this trio, who, not knowing whether we were friend or foe, were at first very much frightened.

I sympathised with the lady with respect to the harsh treatment she had received the previous day, and thanking her for her great kindness, warned her not to keep armed burghers in her house, as this was against the Geneva Convention.

We told her what great pleasure it was for us to meet a lady, as all our women having been placed in Concentration Camps, we had only had the society of our fellow-burghers. Before leaving she grasped our hands, and with tears in her eyes wished us God speed:—"Good-bye, my friends! May God reward your efforts on behalf of your country. General, be of good cheer; for however dark the future may seem, be sure that the Almighty will provide for you!" I can scarcely be dubbed sentimental, yet the genuine expressions of this good lady, coupled perhaps with her excellent dinner, did much to put us into better spirits, and somehow the future did not seem now quite so dark and terrible as we were previously inclined to believe.

We soon resumed our journey, and that night arrived at a farm belonging to a certain Venter. We knew that here some houses had escaped the general destruction and we found that a dwelling house was still standing and that the Venter family were occupying it. It was not our practice to pass the night near inhabited houses, as that might have got the people in trouble with the enemy, but having off-saddled, I sent up an adjutant to the house to see if he could purchase a few eggs and milk for our sick companions. He speedily returned followed by the lady of the house in a very excited condition:—

"Are you the General?" she asked.

"I have that honour," I replied. "What is the matter?"

"There is much the matter," she retorted loudly. "I will have nothing to do with you or your people. You are nothing but a band of brigands and scoundrels, and you must leave my farm immediately. All respectable people have long since surrendered, and it is only such people as you who continue the War, while you personally are one of the ringleaders of these rebels."

"Tut, tut," I said, "where is your husband?"

"My husband is where all respectable people ought to be; with the English, of course."

"'Hands-uppers,' is that it?" answered my men in chorus, even Mooiroos the native joining in. "You deserve the D.S.O.," I said, "and if we meet the English we will mention it to them. Now go back to your house before these rebels and brigands give you your deserts."

She continued to pour out a flood of insults and imprecations on myself, the other generals, and the Government, and finally went away still muttering to herself. I could scarcely help comparing this patriotic lady to the one in Ermelo who had treated us so kindly. I encountered many more such incidents, and only mention these two in order to show the different views held at that time by our women on these matters, but in justice to our women-folk I should add that this kind were only a small minority.

It was a bitterly cold night. Our blankets were very thin, and the wind continually scattered our fire and gave us little opportunity of warming ourselves. There was no food for the horses except the grass. We haltered them close together, and each of us took it in turn to keep a watch, as we ran the risk at any moment of being surprised by the enemy, and as many in that district had turned traitors, we had to redouble our precautions. During the whole cold night I slept but little, and I fervently wished for the day to come, and felt exceedingly thankful when the sun arose and it got a little warmer.

Proceeding, we crossed the ridges east of Bethel, and as this village came in sight my groom Mooiroos exclaimed: "There are a lot of Khakis there, Baas."

I halted, and with my field-glasses could see distinctly the enemy's force, which was coming from Bethel in our direction, their scouts being visible everywhere to the right and left of the ridges. While we were still discussing what to do, the field-cornet of the district, a certain Jan Davel, dashed up with a score of burghers between us and the British. He informed me that the enemy's forces were coming from Brugspruit, and that he had scattered his burghers in all directions to prevent them organizing any resistance. The enemy's guns were now firing at us, and although the range was a long one the ridges in which we found ourselves were quite bare, and afforded us no cover.

We were therefore obliged to wheel to our right, and, proceeding to Klein Spionkop, we passed round the enemy along Vaalkop and Wilmansrust.

At Steenkoolspruit I met some burghers, who told me that the enemy had marched from Springs, near Boksburg, and were making straight for our commando at Kroomdraai. We managed to reach that place in the evening just in time to warn our men and be off. I left a section of my men behind to obstruct the advance of the enemy, whom they met the following day, but finding the force too strong were obliged to retire, and I do not know exactly where they got to. At this time there were no less than nine of the enemy's columns in that district, and they all tried their level best to catch the Boers, but as the Boers also tried their best not to get caught, I am afraid the English were often disappointed. Here the reader will, perhaps, remark that it was not very brave to run away in this fashion, but one should also take our circumstances into consideration.

considerable difficulty in effecting our escape. The enemy, moreover, had every advantage of us. They had plenty of guns, and could cut our ranks to pieces before we could approach sufficiently near to do any damage with our rifles; they far surpassed us in numerical strength; they had a constant supply of fresh horses—some of us had no horses at all; they had continual reinforcements; their troops were well fed, better equipped, and altogether in better condition. Small wonder, therefore, that the War had become a one-sided affair.

On the 20th of May, 1901, I seized an opportunity of attacking General Plumer on his way from Bethel to Standerton.

No sooner did we attack one column than we were attacked in our turn by a couple more, and had then we had effected a junction with Commandant Mears and charged the enemy, and but for their having with them a number of Boer families we would have succeeded in capturing their whole laager. We had already succeeded in driving their infantry away from the waggons containing these families, when their infantry rushed in between and opened fire on us at 200 paces. We could do nothing else but return this fire, although it was quite possible that in doing so we wounded one or two of our own women and children. These kept waving their handkerchiefs to warn us not to fire, but it was impossible to resist the infantry's volleys without shooting. Meanwhile the cavalry replaced their guns behind the women's waggons and fired on us from that coign of vantage.

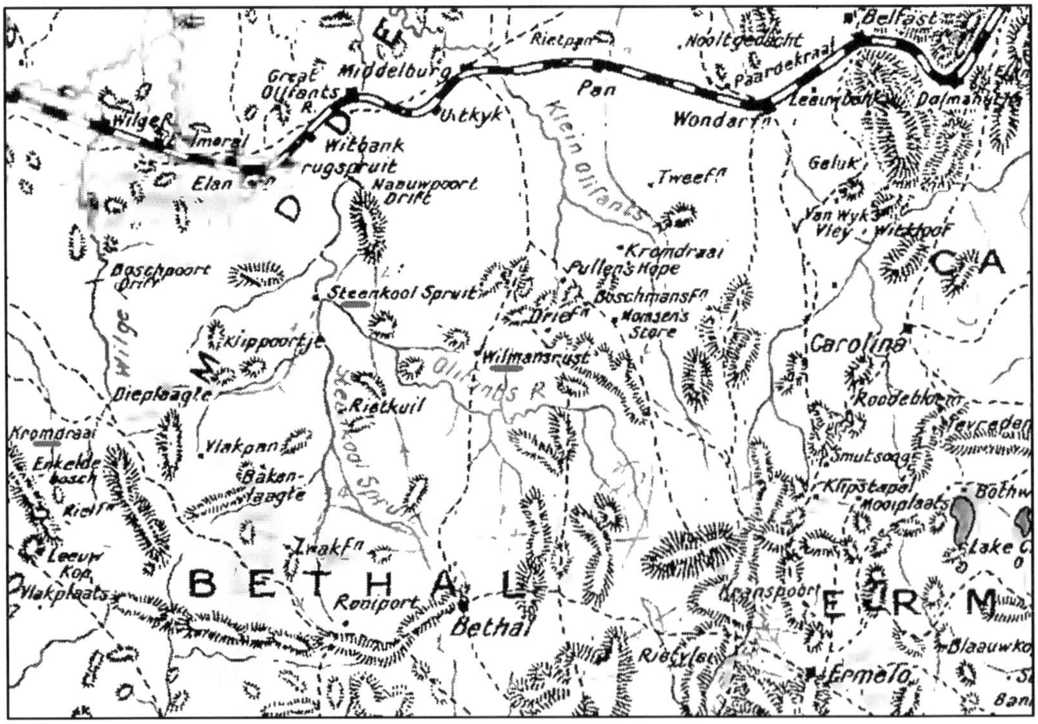

Kromdraai Steenskool and Wilmamsrust.

Here we took 25 prisoners, 4,000 sheep and 10 horses. Our losses were two killed and nine wounded. The enemy left several dead and wounded on the field, as well as two doctors and an ambulance belonging to the Queensland Imperial Bushmen, which we sent back together with the prisoners we had taken.

On this occasion the English were spared a great defeat by having women and children in their laager, and no doubt for the sake of safety they kept these with them as long as possible. I do not insinuate that this was generally the case, and I am sure that Lord Kitchener or any other responsible commanding officer would loudly have condemned such tactics; but the fact remains that these unpleasant incidents occasionally took place.

About the beginning of June, 1901 (I find it difficult to be accurate without the aid of my notes) another violent effort was made to capture the members of the Government and the Commandant-General. Colonel Benson now appeared as the new "Boer-stalker," and after making several unsuccessful attempts to surround them almost captured the Government in the mountains between Piet Retief and Spitskop. Just as Colonel Benson thought he had them safe and was slowly but surely weaving his net around them—I believe this was at Halhangapase—the members of the Government left their carriages, and packing the most necessary articles and documents on their horses escaped in the night along a footpath which the enemy had kindly left unguarded and passed right through the British lines in the direction of Ermelo. On the following day the English, on closing their cordon, found, as they usually did, naught but the burned remains of some vehicles and a few lame mules.

Together with the late General Spruit,[188] who happened to be in that neighbourhood, I had been asked to march with a small commando to the assistance of the Government and the Commandant-General and we had started at once, only hearing when well on our way that they had succeeded in escaping.

We proceeded as far as the Bankop, not knowing where to find them, and it was no easy matter to look for them amongst the British columns.

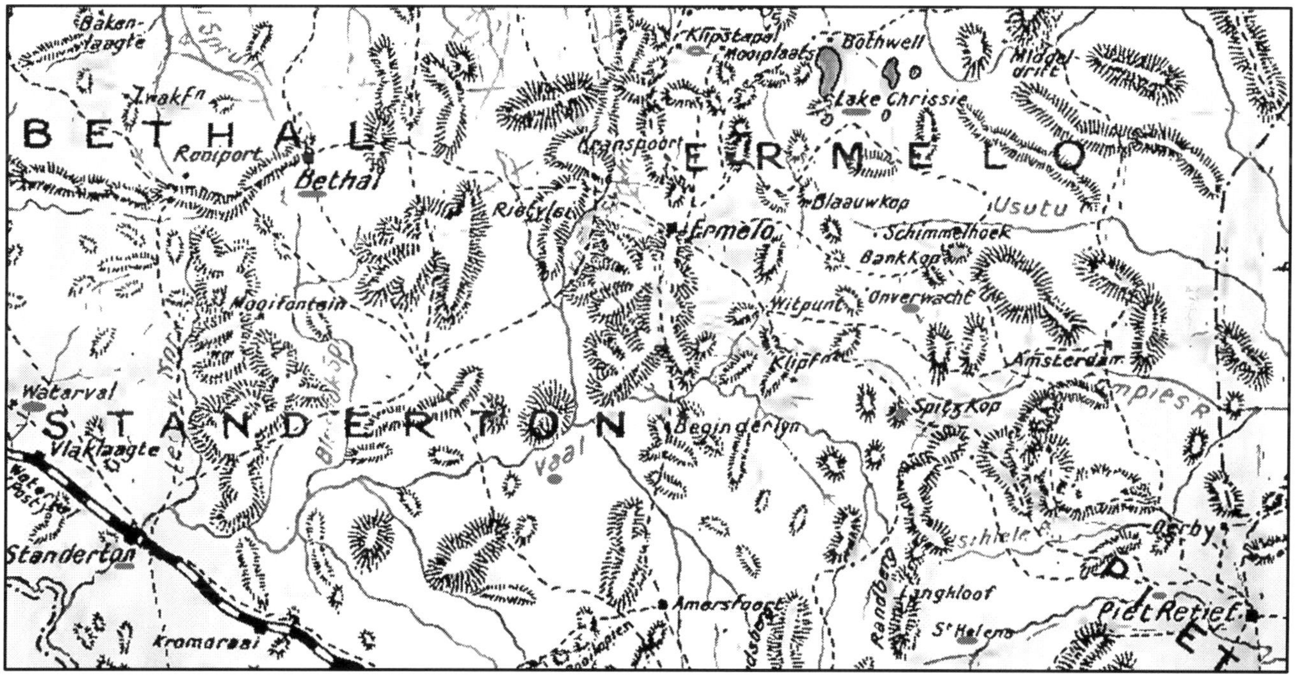

Waterval, Standerton and Bethal to the West and Piet Retief and Spitz Kop to the East.

[188] Cornelis Johannes Spruyt. See MOA.

Chapter 36. A Government on Horseback.

For ten days we searched the neighbourhood, and finally met one of the Commandant-General's despatch-riders, who informed me of their whereabouts, which they were obliged to keep secret for fear of treachery. We met the whole party on William Smeet's farm near the Vaal River, every man on horseback or on a mule, without a solitary cart or waggon. It was a very strange sight to see the whole Transvaal Government on horseback. Some had not yet got used to this method of governing, and they had great trouble with their luggage, which was continually being dropped on the road.[189]

Commando on Horseback.[288]

General Spruit and myself undertook to escort the Executive Council through the Ermelo district, past Bethel to Standerton, where they were to meet the members of the Orange Free State Government. I had now with me only 100 men, under Field-Cornet R. D. Young; the remainder I had left behind near Bethel in charge of General Muller and Commandants Viljoen and Groenwald, with instructions to keep on the alert and to fall on any column that ventured a little ahead of the others.

It was whilst on my way back to them that a burgher brought me a report from General Muller, informing me that the previous night, assisted by Commandants W. Viljoen and Groenwald, he had with 130 men stormed one of the enemy's camps at Wilmansrust, capturing the whole after a short resistance on the enemy's part, but sustaining a loss of six killed and some wounded. The camp had been under the command of Colonel Morris, and its garrison numbered 450 men belonging to the 5th Victorian Mounted Rifles. About 60 of these were killed and wounded, and the remainder were disarmed and released. Our haul consisted of two pom-poms, carts and waggons with teams in harness, and about 300 horses, the most miserable collection of animals I have ever seen. Here we also captured a well-known burgher, whose name, I believe, was **Trotsky**, and who was fighting with the enemy against us. He was brought before a court-martial, tried for high treason, and sentenced to death, which sentence was afterwards carried out.

Our Government received about this time a communication from General Brits, that the members of the Orange Free State Government had reached Blankop, north of Standerton, and would await us at Waterval. We hurried thither, and reached it in the evening of the 20th of June, 1901. Here we found President Steyn and Generals De Wet, De la Rey, and Hertzog, with an escort of 150 men. It was very pleasant to meet these great leaders again, and still more pleasing was the cordiality with which they received us. We sat round our fires all that night relating to each other our various adventures. Some which caused great fun and amusement, and some which brought tears even to the eyes of the hardened warrior. General De Wet was then suffering acutely from rheumatism, but he showed scarcely any trace of his complaint, and was as cheerful as the rest of us.

[189] 5/29/1901 Jan Kemp inflicted significant casualties on a British force led by General Dixon. Vlakfontein is the farm where the Rustenburg/Ventersdorp and Koster/Hekpoort wagon roads cross. General Kemp possessed a force of 400–500 citizens. https://af.wikipedia.org/wiki/Slag_van_Vlakfontein
6/1/1901 Severe engagement between General Dixon and Delarey at Vlakfontein, in the Magaliesberg. Enemy repulsed with heavy loss. Our casualties also heavy. British: 49 killed, 130 wounded. Boers: 41 killed.
6/6/1901 De Wet severely defeated near Reitz. By General Elliot, who made large captures.

Christiaan De. Wet, June 1902. General Jacob Hercules De La Rey.[289] Judge (now Commandant) Brand J. B..M. Hertzog.[290] [190]

1901 SAR Letter

See De Wet Chapter 28 for the circumstances regarding this letter and the real reason the two governments met. The letter Ben refers to in the previous chapter.

Government Offices, In the Field, District Ermelo, South African Republic,

To the Government Secretary, O.F.S.

May 10th, 1901

Sir,—

I have the honour to report to you that to-day the following officers met the Government, namely, the Commandant-General, General B. Viljoen, General J.C. Smuts (Staats-Procureur), the last-named representing the western districts. Our situation was seriously discussed, and, among others, the following facts were pointed out:—

1. That small parties of burghers are still continually laying down their arms, and that the danger arising from this is becoming every day more threatening, namely, that we are exposed to the risk of our campaign ending in disgrace, as the consequence of these surrenders may be that the Government and the officers will be left in the field without any burghers, and that, therefore, heavy responsibility rests upon the Government and War Officers, as they represent the nation and not themselves only.

2. That our ammunition is so exhausted that no battle of any importance can be fought, and that this lack of ammunition will soon bring us to the necessity of flying helplessly before the enemy. And that through this same lack it has become impossible for us to afford adequate protection to our people and their cattle, with the result that the general population is being reduced to poverty and despair, and that even the troops will soon be unable to be supplied with provisions.

3. That through the above-mentioned conditions the authority of the Government is becoming more and more weakened, and that thus the danger arises of the people losing all respect and reverence for lawful authority, and falling into a condition of lawlessness. And that to prolong the war can only lead to hastening the ruin of the people, and making it clear to them that the only authority in the country is that of the enemy.

[190] General James Barry Munnik Hertzog KC (3 April 1866 – 21 November 1942), better known as Barry Hertzog or J. B. M. Hertzog, was a South African politician and soldier. He was a Boer general during the Second Boer War who served as the third prime minister of the Union of South Africa from 1924 to 1939. Hertzog advocated for the development of Afrikaner culture and was determined to prevent Afrikaners from being excessively influenced by British culture. He founded the National Party in 1914.
In 1941, Hertzog, who had resigned after South Africa - against his efforts - entered World War II on the side of the Allies, issued a statement in which he openly praised Nazism and said it needed to be adapted to South African needs under a fascist dictatorship. https://en.wikipedia.org/wiki/J._B._M._Hertzog

4. That not only is our nation becoming disorganized in the manner above referred to, but that it will also most certainly happen that the leaders of the nation, whose personal influence has hitherto kept it together, will fall into utter contempt, and lose that influence which is our only hope for reviving the national spirit in the future.

5. That the people are constantly demanding to be told what hope still exists of successfully prosecuting the war, and that they have the right to expect to be informed in an honest and straightforward manner that their cause is hopeless, whenever this has become evident to the Government and the Leaders.

Up to the present time the Government and the nation have been expecting that, with the co-operation of their Deputation and by the aid of European complications, there would be some hope for the success of their cause, and the Government feels strongly that before taking any decisive step, an attempt should again be made to arrive with certainty at the results of the Deputation and the political situation in Europe.

Having taken all the facts into consideration, the Government, acting in conjunction with the above-mentioned officers, have arrived at the following decision:

Firstly, that a request should be addressed this very day to Lord Kitchener, asking that through the intervention of ambassadors sent by us to Europe, the condition of our country may be allowed to be placed before President Kruger, which ambassadors are to return with all possible speed.

Secondly, that should this request be refused, or lead to no results, an armistice should be asked for, by which the opportunity should be given us of finally deciding in consultation with your Government, and the people of the two States, what we must do.

This second proposal is, however, subject to any solution which your Government, taking into consideration the above-mentioned grievances, may be able to suggest.

The Government feels very keenly that it would no longer be right to allow things to go on as they have been going on, and that the time has arrived for taking some definite steps; it will, therefore, be glad to receive an answer from your Government as soon as possible.

I have the honour to be,

Yours, etc., F.W. Reitz,

Secretary of State. [of the Transvaal]

The answer which the President [Steyn] sent to this letter was formerly in my possession, but has been lost with many of my documents. I am able, however, to give an extract, which I received from the Rev. J.D. Kestell. It was to the following effect:—

The President was much disappointed with the letter of the Transvaal Government; he said that although there had been in the past some surrenders in the Free State, this difficulty had now been overcome. Moreover, although the ammunition had for a long time been scarce, nevertheless, after every fight, there had been enough to begin the next with.

To the question, What probability was there of their being able to continue the struggle? he would reply by asking another question—What hope had the two little Republics, at the beginning of the war, of winning the fight against the might of England? If they had trusted in God at the beginning, why did they not continue to trust in Him?

He also pointed out that if the Boer cause was really quite hopeless, the Deputation would have been sure to send word to that effect. Further, he assured the Transvaal Government that if an armistice were to be obtained, and if during it the people of the Free State were to be asked for their opinion, the decision of the burghers who were still in the field would be to continue the war.

He could not approve of the decision of the Transvaal Government to ask Lord Kitchener to allow ambassadors to be sent to Europe, for, by so doing, the Government would be showing its hand to the enemy; he added that he was very sorry that such a decision had been taken without first consulting the Free State.[191]

[191] The reply to the enquiry, facilitated by Kitchener, had not yet been received. 7/5/1901 In reply to Botha's inquires about ending the war, Kruger telegraphed to Botha to continue fighting.

As to the fear expressed by the Transvaal Government, that the Authorities and the Officers in the field would be left without burghers, the President said, that even if the Government and the Officers of the Free State were to surrender, the nation would not do so.

It would be a great misfortune, he added, if the Orange Free State, which had not only lost its property and the lives of many of its burghers but also even its very independence, in the defence of the sister Republic, should now be abandoned by that Republic; that then all confidence in one another and all co-operation between Afrikanders would come to an end for ever: and that, under such circumstances, it would be too much to expect that the African nation should ever be able to rise again. If then the Boers wished to remain a nation, it was absolutely necessary to continue the war.

After having quoted various appropriate passages from the newspapers, the President went on as follows:—"All these considerations combine to make me believe that we should be committing a National murder if we were to give in now. Brethren! Hold out a little longer. Let not our sufferings and our struggles be in vain; let not our faith in the God of our fathers become a byword. Do all that you can to encourage one another."

The President concluded this very remarkable and powerful letter with the question:—

"Are we again to leave the Colonial burghers in the lurch? God forbid."

We decided to set out for the Transvaal in order to discuss the matter with the Government; and on the evening of the 5th of June we marched four or five miles from Liebenbergsvlei, to a place opposite Verkijkersdorp. We were, all told, between sixty and seventy men, including the staff and part of the bodyguard of President Steyn, the staff of General De la Rey, and eight of my staff officers.

Marthinus Theunis Steyn.[291]

President Steyn [Seated] with his Burghers.[292]

From De Wet's Chapter 29

On the fourth day after this [~June 11th] we met the Transvaal Government and held a conference at once, in accordance with the letter mentioned in my last chapter. It grieved us much that things should have taken this turn, for it nearly always happened that somehow matters of this sort came to the ears of the English. But the Transvaal Government had again taken courage, as they had received an answer to the cable which they had sent to the Deputation, which answer instructed them to hold out; and also because two successful battles had taken place shortly before—one fought by General Kemp, and the other by Commandant Muller.

We remained there for two days, and after it had been settled by the two Governments that the war should be continued with all our might, and also that days of thanksgiving and humiliation should be appointed, we went away accompanied by the genial and friendly Commandant Alberts, of Standerton, who brought us across the Natal-Transvaal railway. Captain Alberts was renowned as a valiant soldier; we now also found him to be a most sociable man. He beguiled the time with agreeable narratives of events in which he had taken part, and almost before we realized it we had reached the railway line. We crossed in safety and took a hearty farewell of our friendly Commandant and his burghers.

Resume Viljoen

Next day we parted, each going separately on our way. We had decided what each of us was to do, and under this agreement I was to return to the Lydenburg and Middelburg districts, where we had already had such a narrow escape. I confess I did not care much about this, but we had to obey the Commandant-General, and there was an end of it. Meanwhile, reports came in that on the other side of the railway the burghers who had been left behind were surrendering day by day, and that a field-cornet was engaged in negotiations with the enemy about a general laying down of arms. I at once despatched General Muller there to put an end to this.

We now prepared once more to cross the railway line, which was guarded more carefully than ever, and no one dared to cross with a conveyance of any description. We had, however, become possessed of a laager—a score of waggons and two pom-poms—and I determined to take these carts and guns across with me, for my men valued them all the more for having been captured. They were, in fact, as sweet to us as stolen kisses, although I have had no very large experience of the latter commodity.

Chapter 37. Blowing Up an Armoured Train.

We approached the line between Balmoral and Brugspruit, coming as close to it as was possible with regard to safety, and we stopped in a "dunk" (hollow place) intending to remain there until dusk before attempting to cross. The blockhouses were only 1,000 yards distant from each other, and in order to take our waggons across there was but one thing to be done, namely, to storm two blockhouses, overpower their garrisons, and take our convoy across between these two. Fortunately there were no obstacles here in the shape of embankments or excavations, the line being level with the veldt.

We moved on in the evening (the 27th of June), the moon shining brightly, which was very unfortunate for us, as the enemy would see us and hear us long before we came within range. I had arranged that Commandant Groenwald was to storm the blockhouse on the right, and Commandant W. Viljoen that to the left, each with 75 men. We halted about 1,000 paces from the line, and here the sections left their horses behind and marched in scattered order towards the blockhouses. The enemy had been warned by telephone that morning of our vicinity, and all the pickets and outposts along the line were on the "qui vive."[192] When 150 yards from the blockhouses the garrison opened fire on our men, and a hail of Lee-Metford bullets spread over a distance of about four miles, the British soldiers firing from within the blockhouses and from behind mounds of earth. The blockhouse attacked by Commandant [Wynand] Viljoen offered the most determined resistance for about twenty minutes, but our men thrust their rifles through the loopholes of the blockhouses and fired within, calling out "hands-up" all the time, whilst the "Tommies" within retorted, "You haven't V.M.R.'s to deal with this time!"[193] However, we soon made it too hot for them and their boasting was exchanged into cries of mercy, but not before three of our men had been killed and several wounded. The "Tommies" now shouted: "We surrender, Sir; for God's sake stop firing." My brave field-cornet, G. Mybergh, who was closest to the blockhouses, answered: "All right then, come out." The "Tommies" answered: "Right, we are coming," and we ceased firing.

Field-Cornet Mybergh now stepped up to the entrance of the fort, but when he reached it a shot was fired from the inside and he fell mortally wounded in the stomach. At the same time the soldiers ran out holding up their hands. Our burghers were enraged beyond measure at this act of treachery, but the sergeant and the men swore by all that was sacred that it had been an accident, and that a gun had gone off spontaneously whilst being thrown down. The soldier who admitted firing the fatal shot was crying like a baby and kissing the hands of his victim. We held a short consultation amongst the officers and decided to accept his explanation of the affair. I was much upset, however, by this loss of one of the bravest officers I have ever known.

Meanwhile the fight at the other blockhouse continued. Commandant Groenwald afterwards informed me that he had approached the blockhouse and found it built of rock; it was, in fact, a fortified ganger's house built by the Netherlands South Africa Railway Company. He did not see any way of taking the place; many of his men had fallen, and an armoured train with a search-light was approaching from Brugspruit.

[192] A state of heightened vigilance, especially prior to battle.

[193] A mounted rifle battalion. Not sure which one but one not highly regarded by their British comrades.

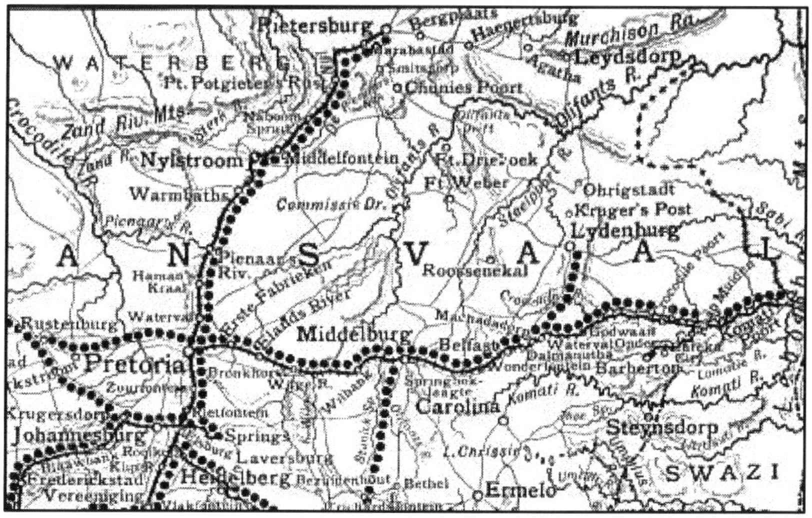

The Nearly Completed Blockhouse System.[194]

Blockhouse Near Wolseley, Western Cape.[293]

Traveling Searchlight designed by Percy Scott.[294] [Noted H.M. S. Terrible on siding.] A gun carrying carriage named "Cock of the North" of No. 3 Artillery Train at the railway station at Bronkhorstspruit, Gauteng Province.[295]

Pattern 3 Gun Train.[296]

A monster gun in the Transvaal.[297] Photo Taken at Belfast.

On the other side of the blockhouse we found a ditch about three feet deep and two feet wide. Hastily filling this up we let the carts go over. As the fifth one had got across and the sixth was standing on the lines, the armoured train came dashing at full speed in our midst. We had had no dynamite to blow up the line, and although we fired on the train, it steamed right up to where we were crossing, smashing a team of mules and splitting us up into two sections. Turning the search-light on us, the enemy opened fire on us with rifles, Maxims and guns firing grape-shot. Commandant Groenwald had to retire along the unconquered blockhouse, and managed somehow to get through.

The majority of the burghers had already crossed and fled, whilst the remainder hurried back with a pom-pom and the other carts. I did not expect that the train would come so close to us, and was seated on my horse close to the surrendered blockhouse when it pulled up abruptly not four paces from me. The search-light made the

[194]The distance between blockhouses was calculated to be able to extent the reach as far as possibly but at the same time to prevent the Boers from passing unnoticed between them. The average distance apart was about 1,000 yards (910 m). Line of sight was vital and this required a greater density of blockhouses in some areas. Anglo Boer War - Blockhouses

surroundings as light as day, and revealed the strange spectacle of the burghers, on foot and on horseback, fleeing in all directions and accompanied by cattle and waggons, whilst many dead lay on the veldt. However, we saved everything with the exception of a waggon and two carts, one of which unfortunately was my own. Thus for the fourth time in the war I lost all my worldly belongings, my clothes, my rugs, my food, my money.

My two commandants were now south of the line with half the men, whilst I was north of it with the other half. We buried our dead next morning and that evening I sent a message to the remainder of the commandos, telling them to cross the line at Uitkijk Station, south-west of Middelburg, whilst Captain Hindon was to lay a mine under the line near the station to blow up any armoured train coming down. Here we managed to get the rest of our laager over without much trouble. The "Tommies" fired furiously from the blockhouses and our friend the armoured train was seen approaching from Middelburg, whistling a friendly warning to us. It came full speed as before, but only got to the spot where the mine had been laid for it. There was a loud explosion; something went up in the air and then the shrill whistle stopped and all was silent.

The next morning we were all once more camped together at Rooihoogte.

Overturned Train.[298]

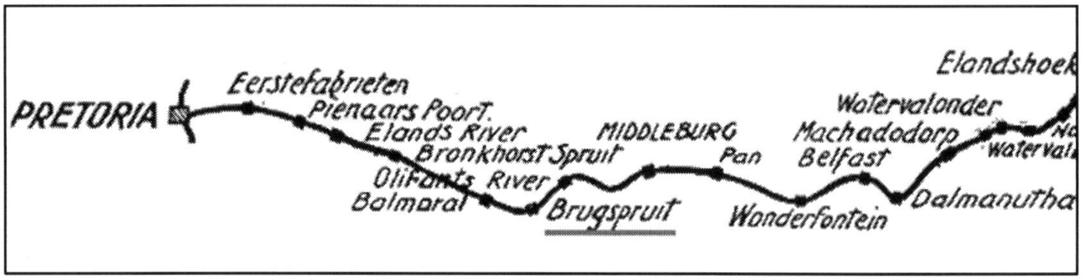

Brugspruit.[299]

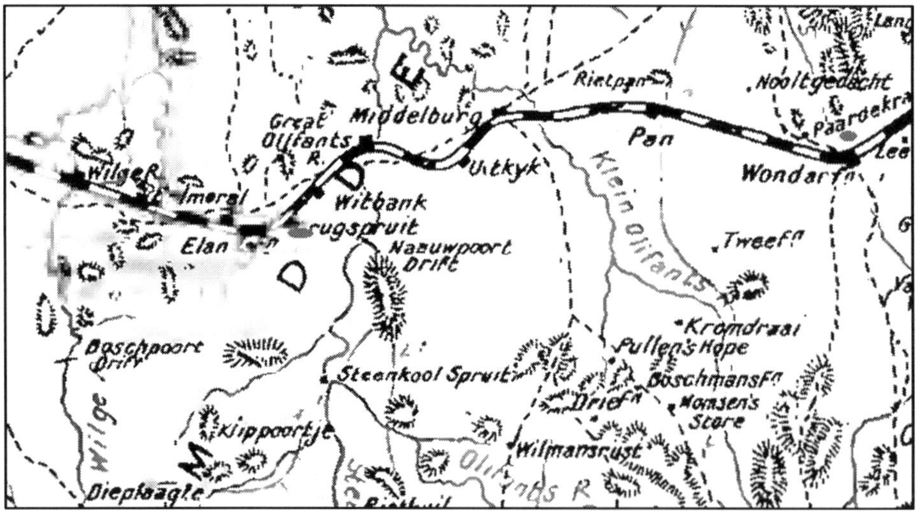

Brugspruit.

Chapter 38. Trapping Pro-British Boers.

In the month of July, 1901, we found ourselves once more on the scene of our former struggles, and were joined here by General Muller, who had completed his mission south of the railway. This district having been scoured for three weeks by thirty thousand English soldiers, who had carefully removed and destroyed everything living or dead, one can imagine the conditions under which we had to exist.

No doubt from a strategical point of view the enemy could not be expected to do otherwise than devastate the country, but what grieved us most was the great amount of suffering this entailed to our women and children. Often the waggons in which these were being carried to imprisonment in the Concentration Camps were upset by the unskillful driving of the soldiers or their kaffir servants, and many women and children were injured in this way.

"Rough Justice in Wartime."[300] "The Penalty of Foul Play. South African Light Horse Burning a Farm which has fired on Our Troops."

Boer family photographed outside their house.[301]

Ruined town-appears to be Frankfort, Orange Free State.[302]

"No Help For It. Pack up….."[303]

British soldier guarding refugees.[304]

Moreover, a certain Mrs. Lindeque was killed by an English bullet near Roos Senekal, the soldiers saying that she had passed through the outposts against instructions. Small wonder, therefore, that many of our women-folk fled with their children at the enemy's approach, leaving all their worldly possessions behind to fall a prey to the general destruction. We often came across such families in the greatest distress, some having taken shelter in caves, and others living in huts roughly constructed of half-burnt corrugated iron amongst the charred ruins of their former happy homes. The sufferings of our half-clad and hungry burghers were small compared to the misery and privations of these poor creatures. Their husbands and other relations, however, made provision for them to the best of their ability, and these families were, in spite of all, comparatively happy, so long as they were able to remain amongst their own people.

A Boer refugee family peering out from their covered wagon.[305] Family of Boer Refugees.[306]

Our commandos were now fairly exhausted, and our horses needed a rest very badly, the wanderings of the previous few weeks having reduced them to a miserable condition. I therefore left General Muller near the cobalt mines on the Upper Olifant's River, just by the waggon drift, whilst I departed with 100 men and a pom-pom to Witpoort and Windhoek, there to collect my scattered burghers and reorganise my diminished commando, as well as to look after our food supplies. At Witpoort the burghers who had been under the late Field-Cornet Kruge, and had escaped the enemy's sweeping movements, had repaired the [flour] mill which the English had blown up, and this was now working as well as before. A good stock of mealies had been buried there, and had remained undiscovered, and we were very thankful to the "bush-lancers" for this bounty.

Still, things were not altogether "honey." Matters were rather in a critical state, as treachery was rampant, and many burghers were riding to and fro to the enemy and arranging to surrender, the faithful division being powerless to prevent them. We had to act with great firmness and determination to put a stop to these tendencies and within a week of our arrival half a dozen persons had been incarcerated in Roos Senekal gaol under a charge of high treason. Moreover we effected a radical change in leadership, discharging old and war-sick officers and placing younger and more energetic men in command.

Several families here were causing considerable trouble. When first the enemy had passed through their district they had had no opportunity of surrendering with their cattle. But when the English returned, they had attempted to go to the enemy's camp at Belfast, taking all their cattle and moveables with them. At this the loyal burghers were furious and threatened to confiscate all their cattle and goods. Seeing this, these families, whom I shall call the Steenkamps, had desisted from their attempt to go over to the enemy and had taken up their abode in a church at Dullstroom, the only building which had not been destroyed, although the windows, doors and pulpit had long disappeared. Here they quietly awaited an opportunity of surrendering to the enemy, whose camp at Belfast was only 10 or 12 miles distant. We were very anxious that their cattle and sheep, of which they had a large number, should not go to the enemy, but we could bring no charge of treachery home to them, as they were very smooth-tongued scoundrels and always swore fealty to us.

I have mentioned this as an example of the dangerous elements with which we had to contend amongst our own people, and to show how low a Boer may sink when once he has decided to forego his most sacred duties and turn against his own countrymen the weapon he had lately used in their defence. Such men were luckily in the minority. Yet I often came across cases where fathers fought against their own sons, and brother against brother. I cannot help considering that it was far from noble on the part of our enemy to employ such traitors to their country and to form such bodies of scoundrels as the National Scouts.

Andries Petrus Johannes Cronjé.[307] [195]

General Peter De Wet.[308] [196]

Amongst all this worry of reorganising our commandos and weeding out the traitors we were allowed little rest by the enemy, and once we suddenly found them marching up from Helvetia in our direction. A smart body of men, chiefly composed of Lydenburg and Middelburg men, and under the command of a newly-appointed officer, Captain Du Toit, went to meet the enemy between Bakendorp and Dullstroom. Here ensued a fierce fight, where we lost some men, but succeeded in arresting the enemy's progress. The fight, however, was renewed the next day, and the British having received strong reinforcements our burghers were forced to retire, the enemy remaining at a place near the "Pannetjes," three miles from Dullstroom.

Imperial Light Horse.[309]

[195] Andries Petrus Johannes Cronjé (A.P.J. Cronjé, 18 June 1849–23 February 1923) was a Second Boer War general, like his older brother Piet Cronjé (1836–1911). After surrendering to the British he became a prominent member of the National Scouts Corps, who assisted the British military against the remaining Boer guerrillas in the last years of the Boer War. A.P.J. Cronjé should not be confused with Boer general Andries Petrus Cronjé (A.P. Cronjé, 1833–1916), who was captured and sent to Bermuda by the British. https://en.wikipedia.org/wiki/Andries_Petrus_Johannes_Cronj%C3%A9

[196] Piet de Wet led his 200 men of the Bethlehem Commando near Nicholson's Nek outside Ladysmith on 5 October 1899 and during the siege of Ladysmith. Then he was tasked to march west and attack Vaalkop near Arundel on 16 December 1899. He impressed president Steyn of the Free State who gave him the command of all troops south of the Orange River with headquarter at Colesberg. British troops continued their advance and occupied Bloemfontein, the capital of Orange Free State, on 13 March 1900. On 28 May Orange Free State was annexed by Lord Roberts as the Orange River Colony. But in the mean time the brothers Christiaan and Piet de Wet won skirmishes with the British at Abrahamkraal (10 March), Sanna's Post (Sannaspost, 31 March) and Dewetsdorp (20 April). On 31 May 1900 Piet de Wet with General Marthinus Prinsloo captured 13th Battalion Imperial Yeomanry near Lindley. Piet de Wet bagged in total 530 men, including Spragge, Lord Longford, Lords Ennismore, Leitrim and Donoughmore (and the future Lord Craigavon) who were marched off to the eastern Transvaal northwards. However, after the occupation by Roberts of Kroonstad on 12 May 1900, Free State soldiers were disheartened and Piet de Wet had started to doubt the use of fighting on.
On 18 May 1900, British brigade general Robert Broadwood received a message that Piet de Wet was willing to surrender on the condition that he and his men could return to their farms. However, Commander-in-chief Roberts demanded an unconditional surrender so that the fighting continued. But in July 1900, Piet de Wet and several of his staff surrendered to the British at Kroonstad, becoming "joiners," "hensoppers" (hands-upper), and "wapenneerlêers" (who put down their weapons). His unavailing letter of 11 January 1901 to his brother Christiaan de Wet to stop his guerrilla war, was published in the Bloemfontein Post and separately as a pamphlet "Broeder tot broeder" (Brother to brother). In March 1902 Piet de Wet established the Orange River Colony Volunteers on the British side. and had joined the National Scouts headed by Andries Cronjé (1849–1923), who served the British troops as Boer auxiliaries.
https://en.wikipedia.org/wiki/Piet_de_Wet

The English camp was now close to our friends, the Steenkamps, who were anxiously waiting an opportunity to become "hands-uppers." They had, of course, left off fighting long ago, one complaining that he had a disease of the kidneys, another that he suffered from some other complaint. They would sit on the kopjes and watch the fighting and the various maneuvres, congratulating each other when the enemy approached a little nearer to them.

Boers watching the fight at Dundee.[310]

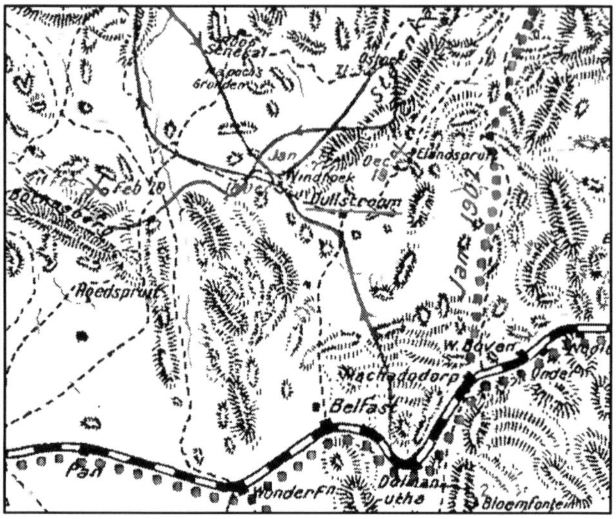

Dullstroom and other familiar landmarks.

I will now ask the reader's indulgence to describe one of our little practical jokes enacted at Dullstroom Church, which was characteristic of many other similar incidents in the Campaign. It will be seen how these would-be "hands-uppers" were caught in a little trap prepared by some officers of my staff.

My three adjutants, Bester, Redelinghuisen, and J. Viljoen, carefully dressed in as much "khaki" as they could collect, and parading respectively as Colonels Bullock, "Jack," and "Cooper," all of His Majesty's forces, proceeded one fine evening to Dullstroom Church, to ascertain if the Steenkamps would agree to surrender and fight under the British flag. They arrived there about 9 p.m., and finding that the inmates had all gone to sleep, loudly knocked at the door. This was opened by a certain youthful Mr. Van der Nest, who was staying in the church for the night with his brother. J. Viljoen, alias "Cooper," and acting as interpreter between the pseudo-English and the renegade Boers, addressed the young man in this fashion:—

"Good evening! Is Mr. Steenkamp in? Here is a British officer who wishes to see him and his brother-in-law."

Van der Nest turned pale, and hurried inside, and stammering, "Oom Jan, there are some people at the door," woke up his brother and both decamped out of the back door. Steenkamp's brother-in-law, however, whom I will call Roux, soon made his appearance and bowing cringingly, said with a smile:—

"Good evening, gentlemen; good evening."

The self-styled Colonel Bullock, addressing "Cooper," the interpreter, said: "Tell Mr. Roux that we have information that he and his brother wish to surrender."

As soon as "Cooper" began to interpret, Roux answered in broken English, "Yes, sir, you are quite right; myself and my brother-in-law have been waiting twelve months for an opportunity to surrender, and we are so thankful now that we are able to do so."

"Colonel Bullock": "Very well, then; call your people out!"

Roux bowed low, and ran back into the church, presently issuing with three comrades, who all threw down their arms and made abeyance.

The "Colonel": "Are these men able to speak English?"

Roux: "No, sir."

The "Colonel": "Ask them if they are willing to surrender voluntarily to His Majesty the King of Great Britain?"

The burghers, in chorus: "Yes, sir; thank you very much. We are so pleased that you have come at last. We have wished to surrender for a long time, but the Boers would not let us get through. We have not fought against you, sir."

The "Colonel": "Very well; now deliver up all your arms."

And whilst the pseudo-colonel pretended to be busy making notes the burghers brought out their Mausers and cartridge-belts, handing them over to the masquerading "Tommies."

Roux next said to the "Colonel": "Please, sir, may I keep this revolver? There are a few Hollanders in the hut yonder who said they would shoot me if I surrendered; and you know, sir, that it is these Hollanders who urge the Boers to fight and prolong the War. Why don't you go and catch them? I will show you where they are."

Resisting an impulse to put a bullet through the traitor's head, the "Colonel" answered briefly: "Very well, keep your revolver. I will catch the Hollanders early to-morrow."

Roux: "Be careful, sir; Ben Viljoen is over there with a commando and a pom-pom."

The "Colonel" (haughtily): "Be at ease; my column will soon be round him and he will not escape this time."

The women-folk now came out to join the party. They clapped their hands in joy and invited the "Colonel" and his men to come in and have some coffee.

The "Colonel" graciously returned thanks. Meanwhile a woman had whispered to Roux: "I hope these are not Ben Viljoen's people making fools of us."

"Nonsense," he answered, "Can't you see that this is a very superior British officer?" Whereat the whole company further expressed their delight at seeing them.

The "Colonel" now spoke: "Mr. Roux, we will take your cattle and sheep with us for safety. Kindly lend us a servant to help drive them along. Will you show us to-morrow where the Boers are?"

Mr. Roux: "Certainly, sir, but you must not take me into dangerous places, please."

The "Colonel": "Very well; I will send the waggons to fetch your women-folk in the morning."

Roux gathered together his cattle and said: "I hope you and I shall have a whiskey together in your camp to-morrow."

The "Colonel" answered: "I shall be pleased to see you," and asked them if they had any money or valuables they wished taken care of. But the Boers, true to the saying, "Touch a Boer's heart rather than his purse," answered in chorus: "Thank you, but we have put all that carefully away where no Boer will find it."

They all bid the "Colonel" good-bye, the "Tommies" exchanging some familiarities with the women till these screamed with laughter, and then the "Colonel" and his commando of two men remounted their big clumsy English horses and rode proudly away. But pride comes before a fall, and they had not proceeded many yards when the "Colonel's" horse, stumbling over a bundle of barbed wire, fell, and threw his rider to the ground. Just as he had nearly exhausted the Dutch vocabulary of imprecations, the Steenkamps, who fortunately had not heard him, came to his assistance and with many expressions of sympathy helped him on his horse, Roux carefully wiping his leggings clean with his handkerchief. After proceeding a little further the "Tommies" asked their "Colonel" what he meant by that acrobatic performance. Whereat the "Colonel" answered: "That was a very fortunate accident; the Steenkamps are now convinced that we are English by the clumsy manner I rode."

The next morning my three adjutants arrived in camp carrying four new Mausers and 100 cartridges each, and driving about 300 sheep and a nice pony. The same morning I sent Field-Cornet [Docs] Young to arrest the brave quartette of burghers. He found everything packed in readiness to depart to the English camp, and they were anxiously awaiting Colonel Bullock's promised waggons.

It was, of course, a fine "tableau" when the curtain rose on the farce, disclosing in the place of the expected English rescuers a burgher officer with a broad smile on his face. They were, of course, profuse in their apologies and excuses. They declared that they had been surrounded by hundreds of the enemy who had placed their rifles to their breasts, forcing them to surrender. One of them was now in so pitiable a condition of fear that he showed the field-cornet a score of certificates from doctors and quacks of all sorts, declaring him to be suffering from every imaginable disease, and the field-cornet was moved to leave him behind. The other three were placed under arrest, court-martialed and sentenced to three months' hard labour, and to have all their goods confiscated.

Two days later the English occupied Dullstroom, and the pseudo-invalid and the women, minus their belongings, were taken care of by the enemy, as they had wished.

Chapter 39. Brutal Kaffirs' Murder Trail.

At Windhoek we were again attacked by an English column. The reader will probably be getting weary of these continual attacks, and I hasten to assure him that we were far more weary than he can ever grow. On the first day of the fight we succeeded in forcing back the enemy, but on the second day, the fortunes of war were changed and after a fierce fight, in which I had the misfortune to lose a brave young burgher named Botha, we gave up arguing the matter with our foes and retired.

The enemy followed us up very closely, and although I used the sjambok freely amongst my men I could not persuade them, not even by this ungentle method, to make a stand against their foes, and as we passed Witpoort the enemy's cavalry with two guns was close at our heels.

Not until the burghers had reached Maagschuur, between the Bothas[berg] and Tautesbergen, would they condescend to make a stand and check the enemy's advance. Here after a short but sharp engagement, we forced them to return to Witpoort, where they pitched camp.

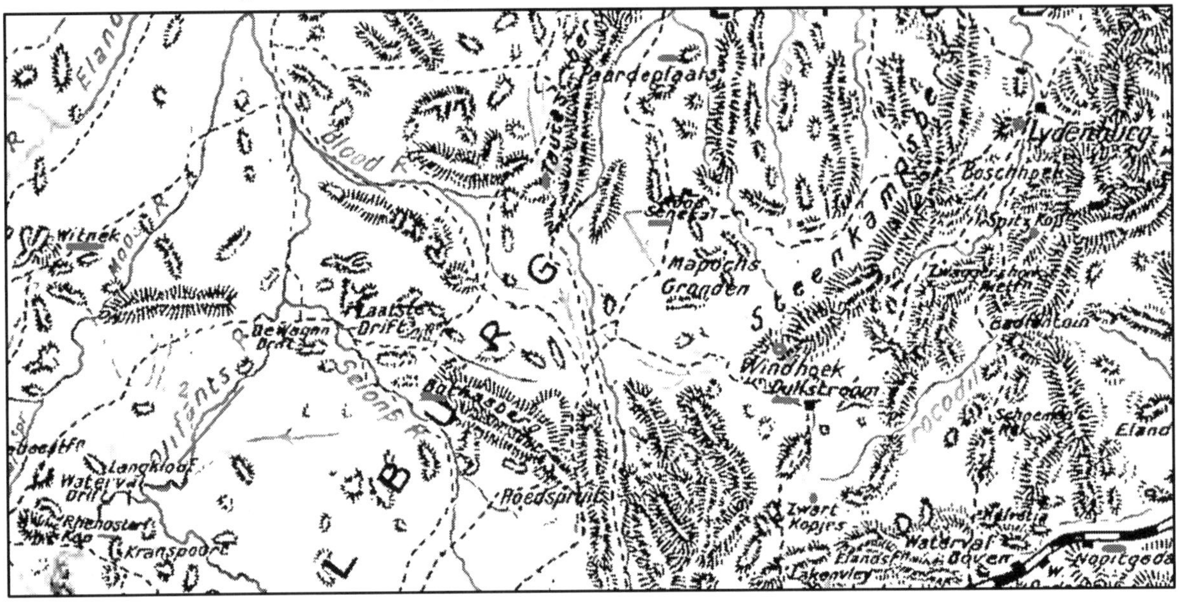

Olifants River south and Blood River to the north. The 'Cobalt Mines' mentioned must be in this area.

Our mill, which I have previously mentioned as being an important source of our food supply, was again burned o the ground.

Our commandos returned to Olifant's River and at the cobalt mine near there joined those who had remained behind under General Muller. The enemy, however, who seemed determined, if possible, to obliterate us from the earth's surface, discovered our whereabouts about the middle of July,[197] and attacked us in overwhelming numbers. We had taken up a position on the "Randts," and offered as much resistance as we could. The enemy poured into us a heavy shell fire from their howitzers and 15-pounders, while their infantry charged both our extreme flanks. After losing many men, a battalion of Highlanders succeeded in turning our left flank, and once having gained this advantage, and aided by their superior numbers, the enemy were able to take up position after position, and finally rendered it impossible to offer any further resistance.

Late in the afternoon, with a loss of five wounded and one man killed—an Irish-American, named Wilson—we retired through the Olifant's River, near Mazeppa Drift, the enemy staying the night at Wagendrift, about three miles further up the stream. The following morning they forded the river, and proceeded through Poortjesnek and Donkerhoek, to Pretoria, thus allowing us a little breathing space. I now despatched some reliable burghers to report our various movements to the Commandant-General, and to bring news of the other commandos. It was three weeks before these men returned, for they had on several occasions been prevented from crossing the

[197] 7/5/1901 In reply to Botha's inquires about ending the war, Kruger telegraphed to Botha to continue fighting.
7/6/1901 A train wrecked on the Pretoria- Pietersburg line.
7/15/1901 Capture of the so-called "Orange Free State Government " at Reitz. Important Boer papers seized. Steyn alone of the members of his "Government" escaped — in his shirt.
7/16/1901 Important success by General French in Cape Colony.
7/19/1901 Publication of Kitchener's despatch embodying contents of important documents seized at Reitz. Death of Mrs. Kruger.

railway line, and they finally only succeeded in doing so under great difficulties. They reported that the English on the high veldt were very active and numerous.

About the middle of July I left General Muller to take a rest with the commando, and accompanied by half a score of adjutants and despatch riders, proceeded to Pilgrimsrust in the Lydenburg district to visit the commandos there, and allay as much as I could the dissatisfaction caused by my reorganisation.

At Zwagerhoek, a kloof some 12 miles south of Lydenburg, through which the waggon track leads from Lydenburg to Dullstroom, I found a field-cornet with about 57 men. Having discussed the situation with them and explained matters, they were all satisfied.

Zwaggershoek.

Commandant Piet Moll (Standing Second from the Right) His trek wagon and staff.[311]

Here I appointed as field-cornet a young man of 23 years of age, a certain J. S. Schoenman, who distinguished himself subsequently by his gallant behaviour.

We had barely completed our arrangements when we were again attacked by one of the enemy's columns from Lydenburg. At first we successfully defended ourselves, but at last were compelled to give way.

I do not believe we caused the enemy any considerable losses, but we had no casualties. The same night we proceeded through the enemy's line to Houtboschloop, five miles east of Lydenburg, where a small commando was situated, and having to proceed a very roundabout way, we covered that night no less than 40 miles.

Another meeting of all burghers north of Lydenburg was now convened, to be held at a ruined hotel some 12 miles west of Nelspruit Station, which might have been considered the centre of all the commandos in that district. I found that these were divided into two parties, one of which was dissatisfied with the new order of things I had arranged and desired to re-instate their old officers, while the other was quite pleased with my arrangements. The latter party was commanded by Mr. Piet Moll, whom I had appointed commandant instead of Mr. D. Schoeman, who formerly used to occupy that position. At the gathering I explained matters to them and tried to persuade the burghers to be content with their new commandants. It was evident, however, that many were not to be satisfied and that they were not to be expected to work harmoniously together. I therefore decided to let both commandants keep their positions and to let the men follow whichever one they chose, and I took the first opportunity of making an attack on the enemy so as to test the efficiency of these two bodies.

Taking the two commandos with their respective two commandants in an easterly direction to Wit River, [White River[198]] we camped there for a few days and scouted for the enemy on the Delagoa Bay Railway, so as to find

[198] White River (Afrikaans: Witrivier) is a small holiday and farming town situated just north of Mbombela in Mpumalanga, South Africa. The farms in the region produce tropical fruits, macadamia nuts, vegetables, flowers and timber. As of 2011, White River had a population of 16,639 https://en.wikipedia.org/wiki/White_River,_Mpumalanga

out the best spot to attack. We had just decided to attack Crocodilpoort Station in the evening of the 1st August, when our scouts reported that the English, who had held the fort at M'pisana's Stad, between our laager in Wit River and Leydsdorp, were moving in the direction of Komati Poort with a great quantity of captured cattle.

Our first plan was therefore abandoned and I ordered 50 burghers of each commando to attack this column at M'pisana's fort at once, as they had done far too much harm to be allowed to get away unmolested. They were a group of men called "Steinacker's Horse,"[199] a corps formed of all the desperadoes and vagabonds to be scraped together from isolated places in the north, including kaffir storekeepers, smugglers, spies, and scoundrels of every description, the whole commanded by a character of the name of——. Who or what this gentleman was I have never been able to discover, but judging by his work and by the men under him, he must have been a second Musolino.[200] This corps had its headquarters at Komati Poort, under Major Steinacker, to whom was probably entrusted the task of guarding the Portuguese frontier, and he must have been given carte blanche as regards his mode of operation.

From all accounts the primary occupation of this corps appeared to be looting, and the kaffirs attached to it were used for scouting, fighting, and worse. Many families in the northern part of Lydenburg had been attacked in lonely spots, and on one occasion the white men on one of these marauding expeditions had allowed the kaffirs to murder ten defenceless people with their assegais and hatchets, capturing their cattle and other property.

In like manner were massacred the relatives of Commandants Lombard, Vermaak, Rudolf and Stoltz, and doubtless many others who were not reported to me. The reader will now understand my anxiety to put some check on these lawless brigands. The instructions to the commando which I had sent out, and which would reach M'pisana's in two days, were briefly to take the fort and afterwards do as circumstances dictated. If my men failed they would have the desperadoes pursue them on their swift horses, and all the kaffir tribes would conspire against us, so that none would escape on our side.

A kaffir was generally understood to be a neutral person in this War, and unless found armed within our lines, with no reasonable excuse for his presence, we generally left him alone. They were, however, largely used as spies against us, keeping to their kraals in the daytime and issuing forth at night to ascertain our position and strength. They also made good guides for the English troops, who often had not the faintest idea of the country in which they were. It must not be forgotten that when a kaffir is given a rifle he at once falls a prey to his brutal instincts, and his only amusement henceforth becomes to kill without distinction of age, colour, or sex. Several hundreds of such natives, led by white men, were roaming about in this district, and all that was captured, plundered or stolen was equally divided among them, 25 per cent. being first deducted for the British Government.

I have indulged in this digression in order to describe another phase with which we had to contend in our struggle for existence. I have reason to believe, however, that the British Commander-in-Chief, [Kitchener] for whom I have always had the greatest respect, was not at that time aware of the remarkable character of these operations, carried on as they were in the most remote parts of the country; and there is no doubt that had he been aware of their true character he would have speedily brought these miscreants to justice.

[199] Labotsibeni Mdluli (c. 1859–15 December 1925), also known as Gwamile, was the queen mother and queen regent of Swaziland. She was largely successful in keeping Swaziland out of the war. Exceptional incidents were Thinthitha Dlamini's attack on a party from the Piet Retief commando near Hlatikhulu in February 1901, and General Tobias Smuts's attack on a small unit of the British irregular force Steinacker's Horse at Bremersdorp in July 1901. Labotsibeni had apparently called in the Boers to remove this troublesome group of freebooters and to release Prince Mancibane, a member of the royal family whom the British had detained on suspicion of spying, but she regretted the Boers' destruction of the small town. https://en.wikipedia.org/wiki/Labotsibeni_Mdluli

[200] Giuseppe Musolino (September 24, 1876–January 22, 1956), also known as the "Brigante Musolino" or the "King of Aspromonte," was an Italian brigand and folk hero. Musolino received great notoriety and admiration in Calabria for escaping prison and committing a string of murders in retaliation for false testimony delivered against him while on trial; his later, second, trial after his recapture was subject to extensive international media coverage and attention. https://en.wikipedia.org/wiki/Giuseppe_Musolino

Chapter 40. Capturing a Freebooter's Lair.

Early in the morning of the 6th of August, as the breaking dawn was tinting the tops of the Lebombo Mountains with its purple dye and the first rays of the rising sun shed its golden rays over the sombre bushveldt, the commando under Commandants Moll and Schoeman were slowly approaching the dreaded M'pisana's fort.

When within a few hundred paces of it they left the horses behind and slowly crept up to it in scattered order; for as none of us knew the arrangement or construction of the place, it had been arranged to advance very cautiously and to charge suddenly on the blowing of a whistle. Nothing was stirring in the fort as we approached, and we began to think that the garrison had departed; but when barely 70 yards from it the officers noticed some forms moving about in the trenches, which encompassed it. The whistle was blown and the burghers charged, a cheer rising from a hundred throats. Volley after volley was discharged from the trenches, but our burghers rushed steadily on, jumped into the trenches themselves and drove the defenders into the fort through secret passages.

The English now began firing on us through loopholes in the walls and several of our men had fallen, when Commandant [Piet] Moll shouted, "Jump over the wall!" A group of burghers rushed at the 12-foot wall, and attempted to scale it; but a heavy fire was directed on them and seven burghers, including the valiant Commandant Moll, fell severely wounded. Nothing daunted, Captain Malan, who was next in command of the division, urged his men to go on, and most of them succeeded in jumping into the fort, where, after a desperate resistance, in which Captain ——, their leader, fell mortally wounded, the whole band surrendered to us. Our losses were six burghers killed, whilst Commandant Moll and 12 others were severely wounded. The burghers found one white man killed in the fort, and two wounded, whilst a score of kaffirs lay wounded and dead. We took 24 white prisoners and about 50 kaffirs. I repeat that the whites were the lowest specimens of humanity that one can possibly imagine.

Hardly was the fight over and our prisoners disarmed when a sentry we had posted on the wall called out:

"Look out, there is a kaffir commando coming!" It was, in fact, a strong kaffir commando, headed by the chief M'pisana himself,[201] who had come to the rescue of his friends of Steinacker's Horse. They opened fire on us at

[201] Labotsibeni Mdluli (c. 1859–15 December 1925), also known as Gwamile, was the queen mother and queen regent of Swaziland and the wife of King Mbandzeni.
For most of the three-year crisis of the South African War, Labotsibeni was, with the support of a co-regent, Prince Logcogco (a son of King Mswati II), and her council, the last independent ruler in Africa south of the Zambezi. During this period she adopted the habitual stance of a Swazi monarch. While leaning towards the British, she sought to preserve Swaziland as a neutral space and maintained a diplomatic relationship with the South African Republic's forces. She was largely successful in keeping Swaziland out of the war. Exceptional incidents were Thinthitha Dlamini's attack on a party from the Piet Retief commando near Hlatikhulu in February 1901, and General Tobias Smuts's attack on a small unit of the British irregular force Steinacker's Horse at Bremersdorp in July 1901. Labotsibeni had apparently called in the Boers to remove this troublesome group of freebooters and to release Prince Mancibane, a member of the royal family whom the British had detained on suspicion of spying, but she regretted the Boers' destruction of the small town.
On the outbreak of the South African War in October 1899 the Transvaal's special commissioner, J. C. Krogh, and the British consul, Johannes Smuts, withdrew from Swaziland. General Piet Joubert wrote to Bhunu, indicating that the South African Republic was leaving Swaziland in his hands. He resumed full authority over his kingdom, but he did not live long to enjoy untrammelled power: he died two months later at Zombodze on 10 December. Labotsibeni now became queen regent as well as queen mother and acted in the name of Bhunu's son, Mona, also known as Nkhotfotjeni, who was chosen to succeed at the age of six months; he eventually became paramount chief, and later King Sobhuza II. It was widely believed that Labotsibeni would have preferred her second son, Malunge, a handsome, intelligent, eloquent, and able young man, to succeed Bhunu. His succession would have avoided a very long minority, but it would have been an unacceptable break with Swazi custom.
Although Bhunu had been installed as ngwenyama or king, with the title Ngwane V, in February 1895, Labotsibeni retained considerable authority. As queen mother she was, in terms of the unwritten constitution of the country, a dual monarch with political influence equal to that of the king, and with the supernatural power to make rain. Her position was strengthened by the reckless behaviour of Bhunu, who established his own base at Mampondweni in the Mdzimba mountains above Zombodze. When he was implicated in the murder of Labotsibeni's senior induna, Mbhabha Nsibandze, and two other indunas, at Zombodze in April 1898, the Transvaal administration sought to bring him to trial. Accompanied by his brother, Malunge, he fled across the border into the British colony of Natal. He was saved from deposition by the intervention of the British high commissioner in South Africa, Lord Milner, who held that the Transvaal's attempt to try him was ultra vires. He returned to Swaziland under British protection and a commission of inquiry imposed a fine on him, holding that he had allowed disorderly behaviour within his kingdom. Britain and the Transvaal then combined to add a protocol to the Swaziland convention that purported to reduce his status from king to paramount chief, and removed his powers of criminal jurisdiction. https://en.wikipedia.org/wiki/Labotsibeni_Mdluli
And: Mbandzeni (also known as Dlamini IV, Umbandine,) (c. 1855–1889) was the King of Swaziland (now Eswatini) from 1872 until 1889. Ingwenyama Mbandzeni was the son of Mswati II and Nandzi Nkambule. https://en.wikipedia.org/wiki/King_Mbandzeni

about 100 yards, and the burghers promptly returned their greeting, bowling over a fair number of them, at which the remainder retired.

Alongside the fort were about 20 small huts, in which we found a number of kaffir girls. On being asked who they were, they repeated that they were the "missuses" of the white soldiers. Inside the captured fort we found many useful articles, and the official books of this band. They contained systematic entries of what had been plundered, looted and stolen on their marauding expeditions and showed how they had been divided amongst themselves, deducting 25 per cent. for the British Government.

A long and extensive correspondence now took place about this matter between myself and Lord Kitchener. I wished first to know whether the gang was a recognised part of the British Army, as otherwise I should have to treat them as ordinary brigands. After some delay Lord Kitchener answered that they were a part of His Majesty's Army. I then wished to know if he would undertake to try the men for their misdeeds, but this was refused. This correspondence ultimately led to a meeting between General Bindon Blood and myself, which was held at Lydenburg on the 27th August, 1901.

The captured kaffirs were tried by court-martial and each punished according to his deserts. The 24 Englishmen were handed over to the enemy, after having given their word of honour not to return to their barbarous life. How far this promise was kept I do not know; but from the impression they made upon me I do not think they had much idea of what honour meant. The captured cattle which we had hoped to find at the fort had been sent away to Komati Poort a few days before our attack and according to their "books" it must have numbered about 4,000 heads. Another section of this notorious corps met with a like fate about this time at Bremersdorp in Swaziland.[202] They did not there offer such a determined resistance, and the Ermelo burghers captured two good Colt-Maxims and two loads of ammunition probably intended for Swaziland natives.

[202] Eswatini, and the Kingdom of Swaziland, is a landlocked country in Southern Africa. It is bordered by South Africa on all sides except the northeast, where it shares a border with Mozambique. At no more than 200 km (120 mi) north to south and 130 km (81 mi) east to west, Eswatini is one of the smallest countries in Africa; despite this, its climate and topography are diverse, ranging from a cool and mountainous highveld to a hot and dry lowveld. https://en.wikipedia.org/wiki/Eswatini

Chapter 41. Ambushing the Hussars.

On August 10th,[203] shortly after our arrival with the prisoners-of-war at Sabi, and while I was still discussing with Lord Kitchener the incident related in the previous chapter, General Muller sent word to me from Olifant's River, where I had left him with my men, that he had been attacked by General Walter Kitchener three days after I had left him. It appears that his sentries were surprised and cut off from the commandos, these being divided into different camps.

The burghers who were farthest away, the Middelburg and Johannesburg men, had, contrary to my instructions, pitched camp on the Blood River, near Rooikraal, and were suddenly and unexpectedly attacked by the enemy at about two o'clock in the afternoon, whilst their horses were grazing in the veldt.

Some horses were caught in time and some burghers offered a little resistance, firing at a short range, several men being killed on both sides. The confusion, however, was indescribable, horses, cattle, burghers and soldiers being all mixed up together. A pom-pom, together with its team of mules and harness, and most of the carts and saddles, were captured by the enemy. Our officers could not induce the men to make a determined stand until they had retired to the Mazeppa Drift, on the Olifant's River.

Here General Muller arrived in the night with some reinforcements and awaited the enemy, who duly appeared next morning with a division of the 18th and 19th Hussars, and, encouraged by the previous day's success, charged our men with a well-directed fire which wrought havoc in their ranks. The gallant Hussars were repulsed in one place, and, at another, Major Davies (or Davis) and 20 men were made prisoners. At last some guns and reinforcements reached the enemy, and our burghers wisely retired, going as far as Eland's River, near the "Double Drifts," where they rested.

De Wagan Drift on Olifants River. Elands River to the N.W.

[203] 8/2/1901 More murders by Boers officially announced. One of the murdered men was an Imperial Yeoman.
8/8/1901 Commandant de Villiers and two Field Cornets surrendered at Warmbaths.
8/10/1901 Kitchener by proclamation called upon the Boer leaders to surrender on or before the 15th of September.
8/13/1901 Kitchener reported the largest return of Boer losses yet sustained in a week. More than 800 prisoners, 700 waggons, and 33,000 cattle.
8/27/1901 Kitchener received letters from Steyn and De Wet protesting against his proclamation. See De Wet volume.

On the third day General W. Kitchener had discovered our whereabouts, and our sentries gave us warning that the enemy was approaching through the bushes, raising great clouds of dust. While the waggons were being got ready the burghers marched out, and awaited the English in a convenient spot between two kopjes. The latter rode on unsuspectingly two by two, and when about 100 had been allowed to pass, our men rushed out, calling, "Hands up!" and, catching hold of their horses' bridles, disarmed about 30 men.

This caused an immediate panic, and most of the Hussars fled (closely pursued by our burghers, who shot 10 or 12 of them). The Hussars left behind a Colt-Maxim and a heliograph for our usage. The ground was overgrown here with a prickly, thorny bush, which made it difficult for our foes to escape, and about 20 more were overtaken and caught, several having been dragged from their horses by protruding branches, and with their face and hands badly injured by thorns, whilst their clothes were half torn off their bodies.

Meanwhile the enemy continued to fire on us whilst retreating, and thus succeeded in wounding several of their own people. This running fight lasted until late in the evening, when the burghers slackened off their pursuit and returned, their losses being only one killed, Lieut. D. Smit, of the Johannesburg Police. The enemy's losses were considerable, although one could not estimate the exact number, as the dead were scattered over a large tract of ground and hidden amongst the bushes, rendering it difficult to find them. Weeks afterwards, when we returned over the same ground, we still found some bodies lying about the bush, and gave them decent burial.

Our burghers were now once more in possession of 100 fresh horses and saddles, whilst their pom-pom was replaced by a Colt-Maxim. General W. Kitchener now left us alone for a while, for which relief we were very thankful, and fell back on the railway line. The respite, however, was short-lived; soon fresh columns were seen coming up from Middelburg and Pretoria, and we were again attacked, some fighting taking place mostly on our old battlefields. General Muller repeatedly succeeded in tearing up the railway line and destroying trains with provisions, whilst I had the good fortune of capturing a commissariat train, near Modelane, on the Delagoa Bay line; but, as I could not remove the goods, I was forced to burn the whole lot. A train, apparently with reinforcements, was also blown up, the engine and carriages going up in the air with fine effect.

Destruction of a train of supplies near Greylingstad, February 13th, 1901.[312] Ganna, circa 1902. Damage to rolling stock after a train was captured by Boer forces at Ganna siding near Nelspoort.[313]

Train tracks. Note block house top left.[314]

Highlanders Repairing the Line.[315] Note, they are wearing kilts.

Chapter 42. I Talk with General Blood.

About the end of August, 1901, I met General Sir Bindon Blood at Lydenburg by appointment.

We had arranged to discuss several momentous questions there, as we made little progress by correspondence. In the first place, we accused the English of employing barbarous kaffir tribes against us; in the second place, of abusing the usage of the white flag by repeatedly sending officers through our lines with seditious proclamations which we would not recognise, and we could only obey our own Government and not theirs; in the third place, we complained of their sending our women with similar proclamations to us from the Concentration Camps and making them solemnly promise to do all that they could to induce their husbands to surrender and thus regain their liberty. This we considered was a rather mean device on the part of our powerful enemy. There was also other minor questions to discuss with regard to the Red Cross.

I went into the English line accompanied by my adjutants, Nel and Bedeluighuis, and my secretary, Lieutenant W. Malan. At Potloodspruit, four miles from Lydenburg, I met General Blood's chief staff officer, who conducted us to him. At the entrance of the village a guard of honour had been placed and received us with military honours. I could not understand the meaning of all this fuss, especially as the streets through which we passed were lined with all sorts of spectators, and to my great discomfort I found myself the chief object of this interest. On every side I heard the question asked, "Which is Viljoen?" and, on my being pointed out, I often caught the disappointed answer, "Is that him?" "By Jove, he looks just like other people." They had evidently expected to see a new specimen of mankind.

In the middle of the village we halted before a small, neat house, which I was told was General Blood's headquarters. The General himself met us on the threshold; a well-proportioned, kindly-looking man about 50 years of age, evidently a genuine soldier and an Irishman, as I soon detected by his speech. He received us very courteously, and as I had little time at my disposal, we at once entered into our discussion.

It would serve little purpose to set down all the details of our interview, especially as nothing final was decided, since whatever the General said was subject to Lord Kitchener's approval, whilst I myself had to submit everything to my Commandant-General.

Going Under the White Flag to a Conference With General Blood at Lyndburg. Sir Bindon Blood [204] General Ben Viljoen.[316]

General Blood promised, however, to stop sending out the women with their proclamations, and also the officers on similar missions, and the Red Cross question was also satisfactorily settled. The kaffir question, however, was left unsettled, although General Blood promised to warn the kaffir tribes round Lydenburg not to

[204] General Sir Bindon Blood GCB GCVO (7 November 1842–16 May 1940) was a British Army commander who served in Egypt, Afghanistan, India, and Southern Africa. Bindon Blood - Wikipedia Winston Churchill, who served under Blood on the North-West Frontier in 1897, dedicated his first non-fiction book, The Story of the Malakand Field Force (1898), to "Major-General Sir Bindon Blood, K.C.B., under whose command the operations therein recorded were carried out; by whose generalship they were brought to a successful conclusion; and to whose kindness the author is indebted for the most valuable and fascinating experience of his life."

interfere in the War and not to leave the immediate vicinity of their kraals. (Only the night before two burghers named Swart had been murdered at Doorukoek by some kaffirs, who pretended to have done this by order of the English). The interview lasted about an hour, and besides us two, Colonel Curran and my secretary, Lieutenant Malan, were present. General Blood and his staff conducted us as far as Potloodspruit, where we took leave. The white flag was replaced by the rifle, and we returned to our respective duties.

Chapter 43. Mrs. Botha's Baby and the "Tommy."

In September, 1901, after having organized the commandos north of Lydenburg, I went back with my suite to join my burghers at Olifant's River, which I reached at the beginning of September. The enemy had left General Muller alone after the affair with the Hussars. Reports were coming in from across the railway informing us that much fighting was going on in the Orange Free State and Cape Colony, and that the burghers were holding their own. This was very satisfactory news to us, especially as we had not received any tidings for over a month. I again sent in a report to our Commandant-General relating my adventures.

We had much difficulty in getting the necessary food for the commandos, the enemy having repeatedly crossed the country between Roos Senekal, Middelburg, and Rhenosterkop, destroying and ravaging everything.

I therefore resolved to split up my forces, the corps known by the name of the "Rond Commando" taking one portion through the enemy's lines to Pilgrimsrust, North of Lydenburg, where food was still abundant. Fighting-General Muller was left behind with the Boksburg Police and the Middelburg Commando, the Johannesburg corps going with me to Pilgrim's Rest, where I had my temporary headquarters. We had plenty of mealies in this district and also enough cattle to kill, so that we could manage to subsist on these provisions. We had long since dispensed with tents, but the rains in the mountain regions of Pilgrim's Rest and the Sabi had compelled us to find the burghers shelter. At the alluvial diggings at Pilgrim's Rest we found a great quantity of galvanized iron plates and deals, which, when cut into smaller pieces, could be used for building. We found a convenient spot in the mountains between Pilgrim's Rest and Kruger's Post, where some hundreds of iron or zinc huts were soon erected, affording excellent cover for the burghers.

Ben Viljoen with some of his soldiers and baboon with a rifle, in the Pilgrim's Rest district in 1901.[205]

[205] General Ben Viljoen with armed Boer soldiers and a baboon (Papio ursinus) with a rifle. Gen. Ben Viljoen (4th from the right back row) and some of his commandos in the Pilgrim's Rest district in 1901. Note the medic with a Red Cross headband; also the Baboon with 'toy' gun.
https://commons.wikimedia.org/wiki/File:General_Ben_Viljoen_with_armed_Boer_soldiers_and_a_baboon_with_a_rifle,_1901.jpg

Patrols were continually sent out round Lydenburg, and whenever possible we attacked the enemy, keeping him well occupied. We succeeded in getting near his outposts from time to time and occasionally capturing some cattle. This seemed to be very galling to the English, and towards the end of September we found they were receiving reinforcements at Lydenburg.[206]

This had soon become a considerable force, in fact in November they crossed the Spekboom River in great numbers, and at Kruger's Post came upon our outposts, when there was some fighting. The enemy did not go any further that night. The following day we had to leave these positions and the other side took them and camped there. Next day they moved along Ohrigstad River with a strong mounted force and a good many empty waggons, evidently to collect the women-folk in that place. I had to proceed by a circuitous route in order to get ahead of the enemy. The road led across a steep mountain and through thickly grown kloofs, which prevented us from reaching the enemy until they had burnt all the houses, destroyed the seed plants, and loaded the families on their carts, after which they withdrew to the camp at Kruger's Post. We at once charged the enemy's rearguard, and a heavy fight followed, which, however, was of short duration. The English fled, leaving some dead and wounded behind, also some dozens of helmets and "putties" which had got entangled in the trees.

[206] 9/2/1901 Another case of train-wrecking on the Pretoria-Pietersburg railway.
9/4/1901 Smuts invades the Cape.
9/7/1901 J. C. Lotter and his entire commando captured in Cape Colony. https://en.wikipedia.org/wiki/Battle_of_Groenkloof
9/17/1901 Battle of Elands River. (Modderfontein) https://en.wikipedia.org/wiki/Battle_of_Elands_River_(1901)
9/17/1901 General Louis Botha overran the Mounted Infantry under the command of Lieutenant-Colonel H de la P Gough.
In the Battle of Blood River Poort or Scheeper's Nek on 17 September 1901 a Boer commando led by Louis Botha crushed a British force commanded by Major Hubert Gough during the Second Boer War.
In August 1901, the Boer leaders determined to send forces south into Natal and the Cape Colony hoping to cause an uprising in the Dutch-majority Cape Colony or at least to gain recruits for their armies. Accordingly, a commando under Botha moved southeast toward Natal while another commando under Jan Smuts raided south into the Cape Colony.
British Intelligence detected the plan, but Botha evaded the British intercepting columns. The cold spring rains made the march especially difficult for the Boers' horses. On 14 September, Botha let his 1,000-man commando camp near Utrecht to permit the horses to recover. Meanwhile, Gough's 24th Mounted Infantry (MI) made a 500-mile (800 km) move by train from Kroonstad in the Orange Free State to Dundee in Natal. Gough received intelligence that Botha and 700 Boers were nearby.
Gough led his MI from Dundee to De Jaeger's Drift, a ford on the Buffalo River. Dismissing the intelligence report as exaggerated, he led three companies on a reconnaissance across the river. Through his field glasses, he spotted 300 Boers who dismounted at a farm near Blood River Poort. Leaving his colleague Lieutenant-Colonel H. K. Stewart with 450 MI in the rear, Gough moved forward into a plain in the early afternoon, planning to surprise the Boers at the farm. Unknown to Gough, Botha was moving around his right flank with 700 men.
Botha's mounted attack completely swamped Gough's outnumbered force. Lieutenant Llewellyn Price-Davies of the King's Royal Rifle Corps won the Victoria Cross for valiantly defending the field guns. Gough was captured, escaped, captured again and finally escaped on foot in the darkness. On the British side, four officers and 19 other ranks were killed or mortally wounded, 2 officers and 19 men wounded, and 6 officers and 235 men captured. According to Boer policy, the captured were stripped of their weapons and any useful gear, and most of their clothing, and were allowed to walk to the nearest British post. The Boers seized two field guns, 180 rifles and a large quantity of small arms ammunition. The 200 captured horses turned out to be in poor condition and of little use to the raiders. Boer losses were light.
Botha was unable to exploit his victory because he found all the crossings of the Buffalo River blocked by the British. The Boers moved to the southeast, hoping to find a place to cross into Natal. On the Zululand border, Botha attacked a British camp named Fort Itala, believing it to be weakly defended. Instead, the Boers received a bloody nose when 56 of their men were killed or wounded. When Botha realized that British forces were approaching in overwhelming strength, he turned back into the Transvaal, his raid a failure.
British Casualties: https://en.wikipedia.org/wiki/Battle_of_Blood_River_Poort And:
https://af.wikipedia.org/wiki/Slag_van_Bloedrivierpoort
9/21/1901 Reverse at Vlakfontein, near Sanna's Post. Two guns lost, (Afterwards recovered.)
9/23/1901 The camp of Lovat's Scouts rushed by Kruitzinger near Herschell. Koch's commando captured near Edenburg. The Carolina commando captured by Colonel Benson.
9/26/1901 Attacks on Fort Itala and Fort Prospect. Boers repulsed with very heavy losses at both places. The attempt of Botha and De Wet to invade Natal foiled.
9/29/1901 Proclamation issued in Pretoria providing for the sale of the properties of Boers still in the field, in accordance with Kitchener's proclamation.
9/30/1901 Great attack by Delarey and Kemp on Colonel Kekewich's camp near Magato Nek, in the Magaliesberg. Boers repulsed. Severe losses on both sides. The Scottish Horse distinguished themselves and sustained severe loss.

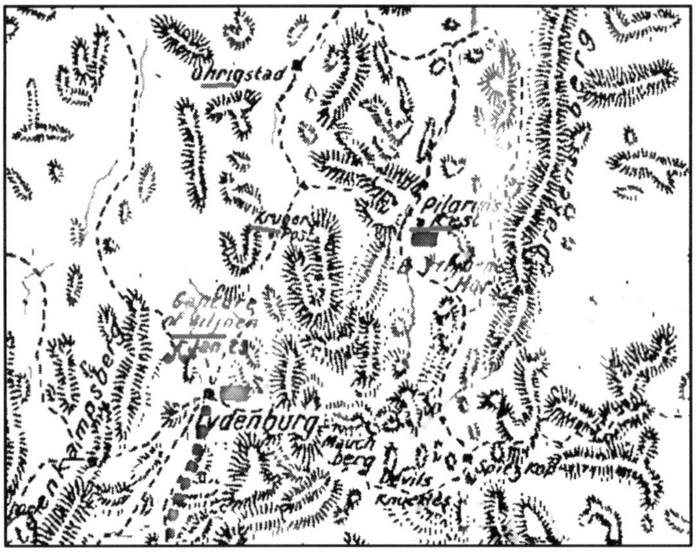

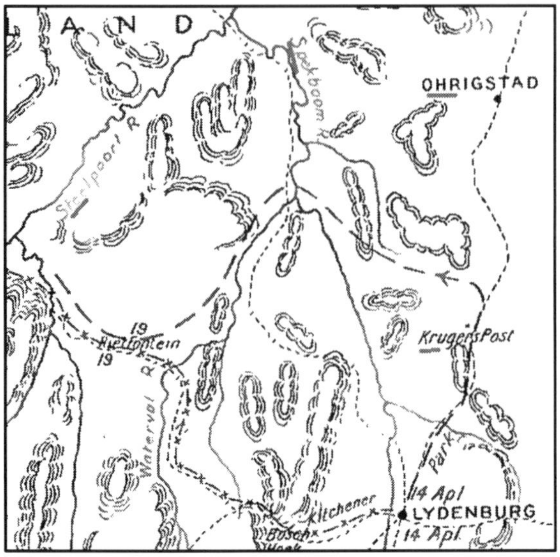

Ohrigstad, Kruger's Post, Pilgrims Rest and Capture of Viljoen Jan, 23. Second map: That's as far as NE TV map goes.

We also captured a waggon loaded with provisions and things that had been looted, such as women's clothes and rugs, a case of Lee-Metford ammunition and a number of uniforms. Some days after the enemy tried to get through to Pilgrim's Rest, but had to retire before our rifle fire.

British supply column on the move.[317]

Supply wagon being pulled out mud by troops and oxen.[318]

They managed, however, to get to Roosenkrans, where a fight of only some minutes ensued, when they retired to Kruger's Post. They only stopped there for a few days, marching back to Lydenburg at night time just when we had carefully planned a night attack. We destroyed the Spekboom River bridge shortly after, thus preventing the enemy's return from Lydenburg to Kruger's Post in a single night. Although there is a drift through the river it cannot be passed in the dark without danger, especially with guns and carts, without which no English column will march. Every fortnight I personally proceeded with my adjutants through the enemy's lines near Lydenburg to see how the commando in the South were getting on and to arrange matters.

Column of troops of one of the British mounted regiments on the move.[319]

Field artillery crossing the Klip River.[320]

The Royal Artillery gun crew transporting their gun across a river with the help of horses.[321]

The Royal Artillery gun crew transporting their gun across a river with the help of horses.[322]

Taaibosch Spruit Bridge, from the north bank.[323] A more complete wreck was made of this bridge, than of any other throughout the State. It is about five miles from the Transvaal border. On the opposite bank can be seen the Rail Head or construction train which meant the line was repaired up to that point. Taaibosch Spruit Bridge, from the south bank. [324] A large quantity of dynamite must have been used to destroy this bridge for great pieces of steel were torn from the body of the structure and were to be seen afterwards as much as 150 yards away. The explosion was heard many miles off.

The month of November, 1901, passed without any remarkable incidents. We organized some expeditions to the Delagoa Bay Railway, but without much success, and during one of these the burghers succeeded in laying a mine near Hector's Spruit Station during the night. They were lying in ambush next day waiting for a train to come along when a "Tommy" went down the line and noticed some traces of the ground having been disturbed which roused his suspicions. He saw the mine and took the dynamite out. Two burghers who were lying in the long grass shouted "Hands up." Tommy threw his rifle down and with his hands up in the air ran up to the burghers saying, before they could speak, "I say, did you hear the news that Mrs. Botha gave birth to a son in Europe?"

They could not help laughing, and the "Tommy," looking very innocent, answered:

"I am not telling you a fib."

One of the burghers coaxed him by telling him they did not doubt his word, only the family news had come so prematurely.

"Well," returned "Tommy," "Oi thought you blokes would be interested in your boss's family, that's why I spoke."

The courteous soldier was sent back with instructions to get some better clothes, for those he had on his back were all torn and dirty and they were not worth taking.

The expedition was now a failure, for the enemy had been warned and the sentries were doubled along the line.[207]

[207] 10/6/1901 General Walter Kitchener and General Bruce-Hamilton engaged Botha's forces in the south-east of the Transvaal. Botha escaped to the north.
10/9/1901 Martial law extended to the whole of Cape Colony.
10/11/1901 Commandant Lotter sentenced to death. Death sentence on five members of his commando was commuted to penal servitude for life.
10/13/1901 Lieut.-Colonel Julian Byng attacked laager at Jackfontein and captured eighteen prisoners.
10/15/1901 Major Damant took prisoner Adjutant Theron. Colonel de Lisle surprised laager at Wilge River and captured fifteen prisoners.
10/16/1901 Colonel Rawlinson returned to Standerton with twenty prisoners.
10/21/1901 Colonel Lukin surprised Vander Venter's laager near New Bethesda.
10/22/1901 Colonel Benson captured laager at Klippoortje.
10/23/1901 Gallant attack on laager in Pongola Bosch.
10/24/1901 Colonel von Donop's brilliant defeat of 1,000 Boers at Kleenfontein.
10/25/1901 Botha's farm surrounded at Schimmelhoek. His papers captured.
10/26/1901 Colonel Benson repulsed attack on his rearguard on the Steenkool Spruit.
10/27/1901 Colonel Williams' force occupied the Witnek Pass and routed a strong body of Boers from the position.
10/30/1901 Attack on Colonel Benson's force at Bakenlaagte. Colonel Benson and Colonel Guinness killed. Colonel Kekewich captured a laager at Beestekraal.
The Battle of Bakenlaagte in Eastern Transvaal, South Africa, occurred on 30 October 1901 during the guerrilla phase of Anglo-Boer war of 1899–1902. The battle saw the Eastern Transvaal Boer commandos of Generals Johan Grobler, Coen Brits, Piet Viljoen and Louis Botha attack the rearguard of Colonel Benson's much feared No. 3 Flying Column while it was in marching formation to its base camp. On the Boer side general Daniel Opperman was killed in the engagement. The Bakenlaagte battlefield is located on the Kriel-Kinross road at the intersection of the R547 and R580 roads in Mpumalanga Province, just south of Matla Power Station.
General Botha arrived with about 800 reinforcements after riding about 40 kilometres (25 mi) without stopping, on arrival, Botha observed that the strung-out column provided an ideal opportunity for an overwhelming force to roll up the isolated and spread-out groups of Commonwealth troops piecemeal and immediately ordered a large Boer force of mounted men to attack the small isolated rearguard of the column.
Outnumbered four to one, the column's rearguard of 210 Commonwealth troops set up a defensive position on Gun Hill and fought about 900 Boers in a close-quarter twenty-minute gun fight that ended only when the column rearguard was annihilated.
Great bravery was demonstrated by the men on both sides, with combined casualties numbering approximately 87 killed with 182 wounded. Colonel Benson (a veteran of the Battle of Magersfontein, 11 December 1899) was to die the next morning from wounds received on the field of battle.
This rearguard action allowed the main column time to deploy and set up a defensive perimeter under Lt Colonel Wools-Sampson. This deployment prevented the attacking Boer forces from riding on and capturing the main column as originally planned. The Boers left the field with whatever spoils they could carry, and the British carried in the wounded to the entrenched camp during the night.
The 73 dead Commonwealth troops were buried on Gun Hill but later reinterred in Primrose Cemetery, corner of Cemetery Road and Beaconsfield Road, Germiston, in the 1960s. Lt Colonel Benson's grave is located at 26°11′45.20″S 28°8′46.05″E. Some of the white granite graves have been vandalised and the brass letters on Colonel Benson's granite grave have been stolen.
https://en.wikipedia.org/wiki/Battle_of_Bakenlaagte
11/2/1901 Patrol under Captain Walker captured twenty-one prisoners near Wolvekop.
11/7/1901 Attack on Piquetberg repulsed by garrison under Major Wilson and Town Guard. General B. Hamilton commenced operations against Botha in the Eastern Transvaal.
11/8/1901 Major Wiggin (26th Mounted Infantry) surrounded laager near Mahamba. Fourteen prisoners secured.
11/9/1901 Line blown up at Myburg Siding by Fouche.
11/11/1901 Major Pack Beresford and detachment of South African Constabulary captured laager at Doornhock.
11/13/1901 Squadron Imperial Yeomanry detached from Hickie's force surprised and surrounded. Rescued by reinforcements.
11/14/1901 Rearguard of Colonel Byng's column attacked near Heilbron by 400 of the enemy under De Wet. Boers repulsed. British loss considerable.
11/16/1901 Further captures by Major Wiggin within Swaziland border.

In December, 1901, we tried an attack on a British convoy between Lydenburg and Machadodorp. I took a mounted commando and arrived at Schvemones Cleft after four days' marching through the Sabinek via Cham Sham, an arduous task, as we had to go over the mountains and through some rivers. Some of my officers went out scouting in order to find the best place for an attack on the convoy. The enemy's blockhouses were found to be so close together on the road along which the convoy had to pass as to make it very difficult to get at it. But having come such a long way nobody liked to go back without having at least made an effort.

We therefore marched during the night and found some hiding places along the road where we waited, ready to charge anything coming along. At dawn next day I found the locality to be very little suitable for the purpose we had in view, but if we were now to move the enemy would notice our presence from the blockhouses. We would, therefore, either have to lie low till dusk or make an attack after all. We had already captured several of the enemy's spies, whom we kept prisoners so as not to be betrayed. Towards the afternoon the convoy came by and we charged on horseback.

The English, who must have seen us coming, were ready to receive our charge and poured a heavy fire into us from ditches and trenches and holes in the ground. We managed to dislodge the enemy's outer flanks and to make several prisoners, but could not reach the carts on account of the heavy fire from a regiment of infantry escorting the waggons. I thought the taking of the convoy would cost more lives than it was worth, and gave orders to cease firing. We lost my brave adjutant, Jaapie Oliver, while Captain Giel Joubert and another burgher were wounded. On the other side Captain Merriman and ten men were wounded. I do not know how many killed he had.

British Army supply column crossing a river, possibly the Molopo River.[325]

We went back to Schoeman's Kloof the same day, where we buried our comrades and attended to the wounded. The blockhouses and garrisons along the convoy road were now fortified with entrenchments and guns, and we had to abandon our plan of further attacks. It was raining fast all the time we were out on this expedition, which caused us serious discomfort. We had very few waterproofs, and, all the houses in the district having been burnt down, there was no shelter for man or beast. We slowly retired on Pilgrim's Rest, having to cross several swollen rivers.

11/20/1901 Engagement with Buys near Villiersdorp. Major Fisher killed. Buys captured by Colonel Rimington. Captain Elliot successfully engaged Boers in Grinqualand. Captain Elliot killed. Three officers wounded.
11/24/1901 General Dartnell, with Highland Light Infantry, engaged Boers near Harrismith. Captured twelve and killed two. Offer of Canadian Government to raise 600 more troops for service in South Africa accepted.
11/25/1901 General Dartnell's force surprised Boers near Bethlehem and took twelve prisoners.
11/26/1901 Basing engaged Joubert in Orange River Colony. Joubert wounded and captured. Major Pack Beresford attacked convoy near Paardeberg.
11/27/1901 Attack on Colonel Rimington's rearguard by De Wet repulsed. Many prisoners taken. Separately, Imperial Light Horse under Colonel Mackenzie took twenty four prisoners, &c.
11/28/1901 Van Rensburg and thirteen burghers captured by Colonel Lowry Cole in Wepener district.

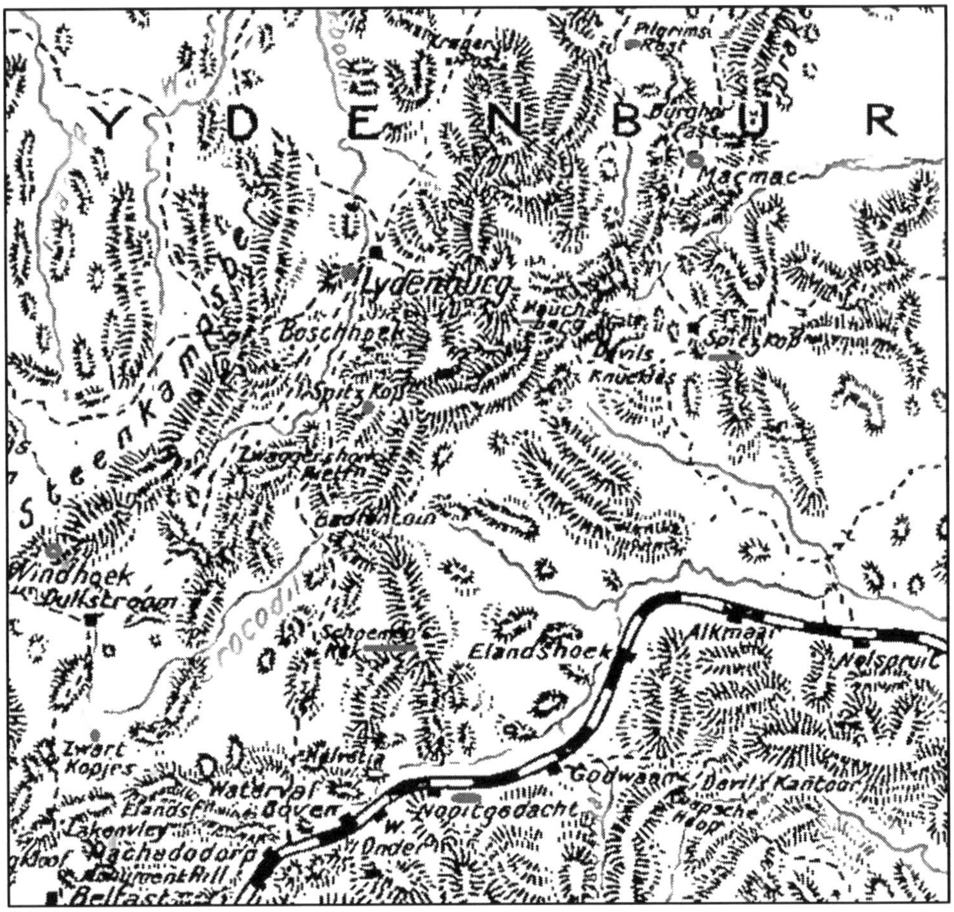

Schoeman's Nek, lower center.

On our arrival at Sabi I received the sad tidings that four burghers named Stoltz had been cruelly murdered by kaffirs at Witriver. Commandant Du Toit had gone there with a patrol and found the bodies in a shocking condition, plundered and cut to pieces with assegais, and, according to the trace, the murderers had come from Nelspruit Station.

Another report came from General Muller at Steenkampsberg. He informed me that he had stormed a camp during the night of the 16th December, but had been forced to retire after a fierce fight, losing 25 killed and wounded, amongst whom was the valiant Field-Cornet J. J. Kriege. The enemy's losses were also very heavy, being 31 killed and wounded, including Major Hudson.[208]

It should not be imagined that we had to put up with very primitive arrangements in every respect. Where we were now stationed, to the north of Lydenburg, we even had telephonic communication between Spitskop and Doornhoek, with call-offices at Sabi and Pilgrim's Rest. The latter place is in the centre of the diggers' population here, and a moderate-sized village.[209] There are a few hundred houses in it, and it is situated 30 miles

[208] 12/1/1901 General Elliot reached Kroonstad with 15 prisoners, 114 waggons, 89 carts, 2470 cattle, and 1280 horses.
12/3/1901 Colonel Colenbrander broke up Badenhorst's commando, and took fifteen prisoners and all the waggons.
12/4/1901 Laager surprised at Oshoek (twenty miles from Ermelo) by Spens' and Rawlinson's columns. Ninety three prisoners taken.
12/7/1901 Colonel C. Mackenzie, in night march towards Watervaal (Eastern Transvaal), took sixteen prisoners. Colonel Holland surprised Brand's laager and took six Boers.
12/11/1901 Badenhorst and twenty-two burghers secured by Colonels Colenbrander and Dawkins, near Zandriverspoort.
12/13/1901 Brilliant surprise of Boers by General B. Hamilton at Witkraus, Laager broken up. One of Benson's guns recovered.
12/16/1901 Haasbrock killed in encounter with Colonel Barker's men in the Doornberg. Capture of Kruitzinger by Colonel Dorans' and Charles Benlinck's columns.

[209] Pilgrim's Rest (Afrikaans: Pelgrimsrus) is a small museum town in the Mpumalanga province of South Africa which is protected as a provincial heritage site. It was the second of the Transvaal gold fields, attracting a rush of prospectors in 1873, soon after the MacMac diggings started some 5 kilometres (3.1 mi) away. Alluvial panning eventually gave way to deeper ore mining.
The alluvial gold was discovered by prospector Alec Patterson. He panned Pilgrim's Creek, as it became known, when the nearby MacMac diggings became too crowded. He kept his find a secret, but a gold rush resulted when fellow prospector William Trafford registered his claim with the Gold Commissioner at MacMac. After it was officially declared a gold field in September 1873, the town suddenly grew to 1,500 inhabitants searching for alluvial gold.

north-east of Lydenburg. Here are the oldest goldfields known in South Africa, having been discovered in 1876. This village had so far been permanently in our possession. General Buller had been there with his force in 1900 but had not caused any damage, and the enemy had not returned since.

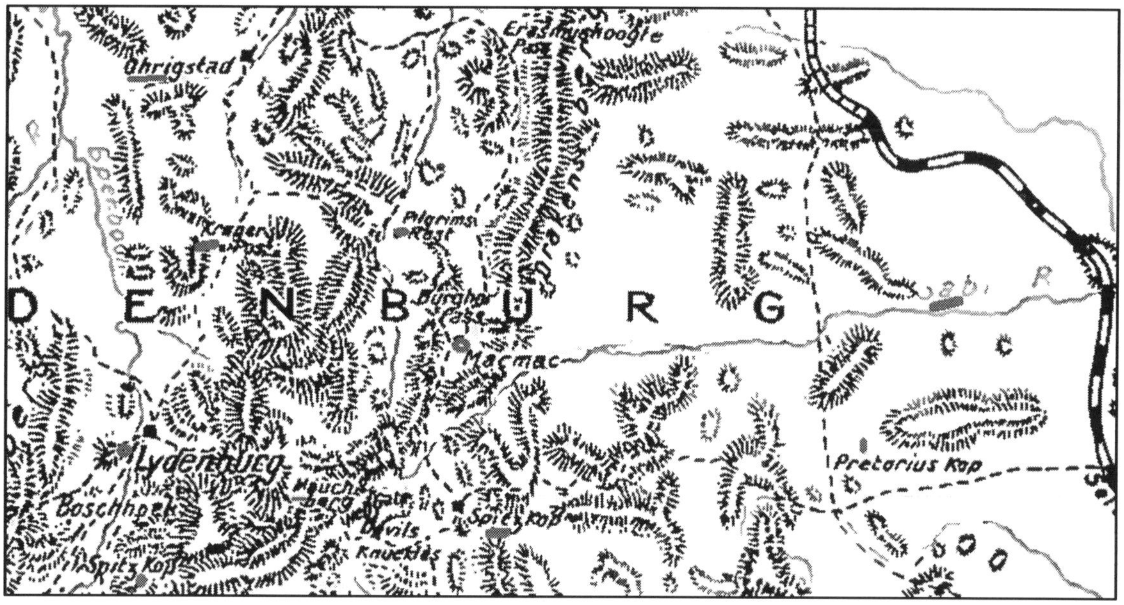

Sabi River. Spitz Kop(2) and Pilgrim's Rest. Sabi is half way between Macmac and Spitz Kop.

The mines and big stamp-batteries were protected by us and kept in order by neutral persons under the management of Mr. Alex. Marshall. We established a hospital there under the supervision of Dr. A. Neethling. About forty families were still in residence and there was enough food, although it was only simple fare and not of great variety. Yet people seemed to be very happy and contented so long as they were allowed to live among their own people.

Eight head stamps powered by waterwheel.[326] Note the large lifting cams. [210] Bisbee, Arizona ca. 1900.[327] A Five Stamp Mill.[328]

As the volumes of gold ore increased, the engineers constructed small, local hydro-electric plants to generate electricity for the electric tramway and the ore crushers at the reduction works, built in 1897.
Pilgrim's Rest was the location of an emergency mint during the Second Boer War. This mint struck the famous and extremely rare Veld Pond.

[210] A stamp mill consists of a set of heavy steel (iron-shod wood in some cases) stamps, loosely held vertically in a frame, in which the stamps can slide up and down. They are lifted by cams on a horizontal rotating shaft. As the cam moves from under the stamp, the stamp falls onto the ore below, crushing the rock, and the lifting process is repeated at the next pass of the cam. Each one frame and stamp set is sometimes called a "battery" or, confusingly, a "stamp" and mills are sometimes categorised by how many stamps they have, i.e. a "10 stamp mill" has 10 sets. https://en.wikipedia.org/wiki/Stamp_mill

PF: Examples of alluvial deposit mines. In these cases for diamonds. The Lace Diamond Mines. Near Kroonstad. [329] A Kimberley Mine 1900[330].

Acting Treasurer with The Boer Forces.[331] Will Barter,[211] John Hygelsen, Klopper.[332] The Treasury Department in the Field. William Barter, chief of the Money Printing Division, on the left.

Coins of the SAR.[333] The Government Veldt Mint in the Last Ditch.[334]

[211] Military Commissioner William Joseph Henry Barter: Literal translation from the Dutch original. Military Commissioner William Joseph Henry Barter, of Irish descent and born in 1865, was born in Sept. 1900 by Komm. Genl. Botha was appointed Military Commissioner in Lijdenburg. As such he accompanied Genl. Gravett who sent him from the Highveld to Pilgrim's Rest at to gather provisions. Then I got the blank Government Notes in my hands had fallen to the enemy, Barter was appointed one of the members of the supervisory board State Printing House in the Field, and when the State Mint in the Field was created he was appointed Treasurer; afterwards he established a military tannery and tailoring. From: Our [Transvaal] Military Officers: Dutch: Onze krijgs-officieren: album van portretten met levens-schetsen der. by Gustav Schoeman Preller, Frans Vredenrijk Engelenburg 1904 Catalogue Record: https://archive.org/search?query=title%3A%28Onze%20Krijgs-Officieren%29

Chapter 44. The Last Christmas of the War.

December, 1901, passed without any important incident. We only had a few insignificant outpost skirmishes with the British garrison at Witklip to the south of Lydenburg. Both belligerents in this district attempted to annoy each other as much as possible by blowing up each other's mills and storehouses.

Two of the more adventurous spirits amongst my scouts, by name [L] Jordaan and Mellema, succeeded in blowing up a mill in the Lydenburg district used by the British for grinding corn, and the enemy very soon retaliated by blowing up one of our mills at Pilgrim's Rest. As the Germans say, "Alle gute dingen sind drei."[212] Several such experiences and the occasional capture of small droves of British cattle were all the incidents worth mentioning. It was in this comparatively quiet manner that the third year of our campaign came to a termination. The War was still raging and our lot was hard, but we did not murmur. We decided rather to extract as much pleasure and amusement out of the Christmas festivities as the extraordinary circumstances in which we found ourselves rendered possible.

The British for the time being desisted from troubling us, and our stock and horses being in excellent condition, we arranged to hold a sort of gymkhana[213] on Christmas Day. In the sportive festivities of the day many interesting events took place. Perhaps the most noteworthy of these were a mule race, for which nine competitors entered, and a ladies' race, in which six fair pedestrians took part. The spectacle of nine burly, bearded Boers urging their asinine steeds to top speed by shout and spur provoked quite as much honest laughter as any theatrical farce ever excited. We on the grand stand were but a shaggy and shabby audience, but we were in excellent spirits and cheered with tremendous gusto the enterprising jockey who won this remarkable "Derby." Shabby as we were, we subscribed £115 in prizes. After the sports I have just described the company retired to a little tin church at Pilgrim's Rest, and there made merry by singing hymns and songs round a little Christmas tree.

Later in the evening a magic-lantern, which we had captured from the British, was brought into play, and with this we regaled 90 of our juvenile guests. The building was crowded and the utmost enthusiasm reigned. The ceremony was opened by the singing of hymns and the making of speeches, a harmonium adding largely to the enjoyment of the evening.

A Phoenix Stereopticon.[335]

Mason & Hamlin Harmonium Pump Organ.[336 214]

I felt somewhat nervous when called upon to address the gathering, for the children were accompanied by their mothers, and these stared at me with expectant eyes as if they would say, "See, the General is about to speak; his words are sure to be full of wisdom." I endeavoured to display great coolness, and I do not think I failed very markedly as an extemporaneous orator. I was helped very considerably in the speechmaking part of the programme by my good friends the Rev. Neethling and Mr. W. Barter, of Lydenburg. I have not now the slightest idea of what I spoke about except that I congratulated the little ones and their mothers on being preserved from the Concentration Camps, where so many of their friends were confined.

[212] All good things come in threes.

[213] Gymkhana is a British Raj term which originally referred to a place of assembly. The meaning then altered to denote a place where skill-based contests were held and as a term coined during the British Raj for gentlemen's club. https://en.wikipedia.org/wiki/Gymkhana

[214] A pump organ. Pump organ-Wikipedia

I have mentioned that there were young ladies with us who participated in the races. These were some whom the British had kindly omitted to place in the Concentration Camps, and it was remarkable to see how soon certain youthful and handsome burghers entered into amorous relations with these young ladies, and matters developed so quickly that I was soon confronted with a very curious problem.

We had no marriage officers handy, and I, as General, had not been armed with any special authority to act as such. Two blushing heroes came to me one morning accompanied by clinging, timorous young ladies, and declared that they had decided that since I was their General I had full authority to marry them. I was taken aback by this request, and asked, "Don't you think, young fellows, that under the circumstances you had better wait a little till after the termination of the war?" "Yes," they admitted, "perhaps it would be more prudent, General, but we have been waiting three years already!"

In General De la Rey's Commando, which comprised burghers from eight large districts, it had been found necessary to appoint marriage officers, and quite a large number of marriages were contracted. I mention this to show how diversified are the duties of the Boer general in war-time, and what sort of strange offices he is sometimes called upon to perform.

It will be seen from what I have said that occasionally the dark horizon of our veldt life was lit up by the bright sunshine of the lighter elements of life. At most times our outlook was gloomy enough, and our hearts were heavily weighed down by cares. I often found my thoughts involuntarily turning to those who had so long and so faithfully stood shoulder to shoulder with me through all the vicissitudes of war, fighting for what we regarded as our holy right, to obtain which we were prepared to sacrifice our lives and our all. Unconsciously I recalled on this Christmas Day the words of General Joubert addressed to us outside Ladysmith in 1899: "Happy the Africander who shall not survive the termination of this War." Time will show, if it have not already shown, the wisdom of General Joubert's words.

Just about this time rumours of various kinds were spread abroad. From several sources we heard daily that the War was about to end, that the English had evacuated the country because their funds were exhausted, that Russia and France had intervened, and that Lord Kitchener had been captured by De Wet and liberated on condition that he and his troops left South Africa immediately. It was even said that General Botha had received an invitation from the British Government to come and arrange a Peace on "independence" lines.

Nobody will doubt that we on the veldt were desperately anxious to hear the glad tidings of Peace. We were weary of the fierce struggle, and we impatiently awaited the time when the Commandant-General and the Government should order us to sheathe the sword.

But the night of the Old Year left us engaged in the fierce conflict of hostilities, and the dawn of the New Year found us still enveloped in the clouds of war—clouds whose blackness was relieved by no silver lining.[215]

[215] 12/1/1901 General Elliot reached Kroonstad with 15 prisoners, 114 waggons, 89 carts, 2470 cattle, and 1280 horses.
12/3/1901 Colonel Colenbrander broke up Badenhorst's commando, and took fifteen prisoners and all the waggons.
12/4/1901 Laager surprised at Oshoek (twenty miles from Ermelo) by Spens' and Rawlinson's columns. Ninety three prisoners taken.
12/7/1901 Colonel C. Mackenzie, in night march towards Watervaal (Eastern Transvaal), took sixteen prisoners. Colonel Holland surprised Brand's laager and took six Boers.
12/11/1901 Badenhorst and twenty-two burghers secured by Colonels Colenbrander and Dawkins, near Zandriverspoort.
12/13/1901 Brilliant surprise of Boers by General B. Hamilton at Witkraus, Laager broken up. One of Benson's guns recovered.
12/16/1901 Haasbrock killed in encounter with Colonel Barker's men in the Doornberg. Capture of Kruitzinger by Colonel Dorans' and Charles Benlinck's columns.
12/18/1901 Colonel Steele, with South African Constabulary, captured thirty-six Boers in the region of the Magaliesberg. Four hours' fighting between De Wet and General Dartnell. Boers driven off. Methuen reported capture of thirty-two Boers.
12/19/1901 Colonel Allenby captured thirty-two of the enemy near Heidelberg.
12/20/1901 Colonel Damant attacked by 800 Boers. Two officers killed, three wounded. Boers repulsed.
12/21/1901 Capture of Smuts' convoy, near Bothwell, by Colonel Mackenzie.
12/22/1901 Seven hundred Cape raiders attacked columns of Colonels Wyndham and Crabbe. Were driven off with loss of five killed and twenty wounded.
12/23/1901 Successful attack on Grobelaar's laager by General B. Hamilton.
12/24/1901 Colonel Du Moulin surprised laager near Jagersfontein. Captured two Field-Cornets and twenty other Boers.
12/25/1901 Colonel Firman's camp at Tweefontein rushed by huge force under De Wet.
In the Battle of Groenkop (Battle of Tweefontein) on 25 December 1901, Christiaan de Wet's Boer commando surprised and defeated a force of Imperial Yeomanry under the command of Major Williams. https://en.wikipedia.org/wiki/Battle_of_Groenkop British: 68 killed 77 wounded 206 captured. Boers 11 killed 30 wounded.
12/28/1901 Successful engagement near Burghersdorp by Colonel Price. Field-Cornet Jan Venter killed.

Chapter 45. My Last Days on the Veldt.

The first month of 1902 found the storm of death and destruction still unabated, and the prospect appeared as dark as at the commencement of the previous year. Our hand, however, was on the plough, and there was no looking back. My instructions were, "Go forward and persevere."

To the south of Lydenburg, where a section of my commando under General Muller was operating, the enemy kept us very busy, for they had one or more columns engaged. We, to the north of Lydenburg, had a much calmer time of it than our brethren to the south of that place, for there the British were pursuing their policy of exhausting our people with unsparing hand. I attribute the fact that we in the north were left comparatively undisturbed to the mountainous nature of the country. It would have been impossible for the British to have captured us or to have invaded our mountain recesses successfully without a tremendous force, and, obviously, the British had no such force at their disposal.

Probably also the British had some respect for the prowess of my commando. An English officer afterwards told me in all seriousness that the British Intelligence Department had information that I was prowling round to the north of Lydenburg with 4,000 men and two cannons, and that my men were so splendidly fortified that our position was unconquerable. Of course, it was not in my interest to enlighten him upon the point. I was a prisoner-of-war when this amusing information was given me, and I simply answered: "Yes, your intelligence officers are very smart fellows."

The officer then inquired, with an assumption of candour and innocence, whether it was really a fact that we had still cannon in the field. To this I retorted: "What would you think if I put a similar question to a British officer who had fallen into my hands?" At this he bit his thumb and stammered: "I beg your pardon; I did not mean to—er—insult you." He was quite a young chap this, a conceited puppy, affecting the "haw-haw," which seems to be epidemic in the British Army. His hair was parted down the centre, in the manner so popular among certain British officers, and this style of hair-dressing came to be described by the Boers as "middel-paadje" (middle-path). As a matter of fact, my men only numbered as many hundreds as the thousands attributed to me by the British. As for cannons, they simply existed in the imagination of the British Intelligence Department.

Affairs were daily growing more critical. Since the beginning of the year we had made several attempts at destroying the Delagoa Bay Railway, but the British had constructed so formidable a network of barbed wire, and their blockhouses were so close together and strongly garrisoned, that hitherto our attempts had been abortive. The line was also protected by a large number of armoured trains.

Canadian Troops in Armoured Train with Signal Light on top.[337]

Armoured Train.[338]

6-Inch Rail Truck Gun.[339]

Waterval-Boven, circa 1901 with Armoured train.[340]

In consequence of our ill-success in this enterprise, we turned our attention to other directions. We reconnoitred the British garrisons in the Lydenburg district with the object of striking at their weakest point. A number of my officers and men proceeded under cover of darkness right through the British outposts, and gained the Lydenburg village by crawling on their hands and knees. On their return journey they were challenged and fired on several times, and managed only with difficulty to return to camp unhurt.

The object of the reconnaissance was, however, accomplished. They reported to me that the village was encompassed with barbed wire, and that a number of blockhouses had been built round it, and also that various large houses of the village had been barricaded and were strongly occupied. My two professional scouts, Jordaan and Mellema, had also reconnoitred the village from another direction, and had brought back confirmatory information and the news that Lydenburg was occupied by about 2,000 British soldiers, consisting of the Manchester Regiment and the First Royal Irish, together with a corps of "hands-uppers" under the notorious Harber. Three other Boer spies scouting about the forts on the Crocodile Heights also brought in discouraging reports.

At the Council of War which then took place, and over which I presided, these reports were discussed, and we agreed to attack the two blockhouses nearest the village, and thereafter to storm the village itself. I should mention that it was necessary for us to capture the blockhouses before attempting to take the village itself, for had we left them intact we should have run the danger of having our retreat cut off.

The attack was to take place next night, and as we approached the British lines on horseback, between Spekboom River and Potloodspruit, we dismounted, and proceeded cautiously on foot. One of the objective blockhouses was on the waggon path to the north of the village, and the other was 1,000 yards to the east of Potloodspruit. Field-Cornet Young, accompanied by Jordaan and Mellema, crept up to within 10 feet of one of these blockhouses, and brought me a report that the barbed wire network which surrounded it rendered an assault an impossible task in the darkness. Separating my commando of 150 men into two bodies, I placed them on either side of the blockhouse, sending, in the meanwhile, four men to cut down the wire fences. These men had instructions to give us a signal when they had achieved this object, so that we could then proceed to storm the fort. It would have been sacrificing many in vain to have attempted to proceed without effecting the preliminary operation of fence cutting, since, if we had stormed a blockhouse without first removing the wire, we should have become entangled in the fences and have offered splendid targets to the enemy at a very short range, and our losses would, without doubt, have been considerable.

My fence-cutters stuck doggedly to their task despite the fact that they were being fired upon by the sentries on guard. It was a long and weary business, but we patiently waited, lying on the ground. Towards 2 o'clock in the morning the officer in command of the wire-cutters returned to us, stating that they had accomplished their object in cutting the first wire barrier, but had come across another which it would require several hours to cut through. The sentries had, in the meantime, grown unpleasantly vigilant, and were now frequently firing on our men. They were often so close that at one time, in the darkness, they might have knocked up against the Boers who were cutting their fences.

It being very nearly 3 o'clock, it appeared to me that the attempt would be ineffectual owing to the approach of daylight, and we were forced to retire before the rays of the rising sun lit the heavens and exposed us to the well-aimed fire of the British. I therefore resolved, after consulting my officers, to retire quietly, and to renew my attempt a week later at another point. We returned to camp much disappointed, but consoled ourselves with the hope that success would attend our next efforts.

Chapter 46. I Am Ambushed and Captured.

I may say that the barbed wire fences by which the blockhouses were encompassed, constituted very formidable obstacles to our attacks. Our men were comparatively few, and we could not afford to lose any of them in futile attempts to capture strongly garrisoned British forts. Moreover, there were many other ways of inflicting damage on the enemy that did not lay us open to so much danger.

Heavy and continuous rains had been experienced for some time, and the rivers and spruits were greatly swollen. The whole of the Lydenburg district, in which we were operating, was besides enveloped in a thick mist, and both these causes rendered reconnoitring very difficult and perilous, as we never knew how near the enemy's patrols might be.

About the 15th of January, 1902, I obtained information that our Government were being chased all over the country, and had now encamped at Windhoek near Dullstroom, to the south of Lydenburg. At the same time I received an order from Acting-President Schalk Burger, stating that he wished to see me. This latter intelligence was very acceptable, for I was anxious to renew acquaintance with the President, and with a personal friend of mine, Mr. J. C. Krojk, who was attached to the Field Government. Therefore, on receiving this instruction, I set out from Pilgrim's Rest accompanied by Adjutants Nel, Coetzee, Bester, and Potgieter, for the place where the Government were encamped. I little expected as I rode along that this would be my last and most fateful expedition.[216]

I calculated that I should be away eight days, and, wishing to be present at any active operations that might be conducted, I instructed my brother, [Wynand J, Viljoen] whom I left in charge of my forces, to make no attack during my absence.

After leaving Pilgrim's Rest, I and my companions rode briskly forth along the path past Dornbock, Roodekrans and Kruger's Post. We encamped at the latter place at night-fall. Next day we again set out, and having succeeded in passing the British forts and blockhouses to the north of Lydenburg, we came upon the Spekboom River. This river was so swollen by the recent rains that no fording was possible, and we were only able to cross by making our horses swim. At one o'clock we reached Koodekraus, and off-saddled there. This place is about 15 miles to the west of Lydenburg. At dawn the next day, after having reconnoitred the country in the neighbourhood, we proceeded cautiously in the direction of Steenkampsberg [Mountain Range] until we were meet by messengers, who told us precisely where our Government was to be found. That evening we found our locomotive Administration encamped at Mopochsburgen, to which place they had retreated before a hostile column, which was operating from Belfast.

The greetings that were exchanged were of the heartiest character, and we sat chatting round the camp fires far into the night. That we had much to talk about and many stories to relate of the vicissitudes of war needs no saying. I personally received the very lamentable tidings that my sister, her husband, and three of their children had died in the Concentration Camp at Pietersburg.

[216] 1/3/1902 Capture of General Erasmus by General Bruce-Hamilton.
1/10/1902 Surprise of laager near Ermelo by Colonel Wing and capture of forty-two prisoners.
1/12/1902 More captures by General B. Hamilton.
1/13/1902 Fight for a convoy by De Villiers. Gallant charge of Munster Fusiliers.
1/16/1902 Capture of laager and twenty-four prisoners by Methuen.
1/18/1902 Execution of Scheepers on various charges of murder at Graaff Reinet.
1/21/1902 Colonels Park and Urmston engaged parly of Boers under Muller and Trichardt, occasioning stampede of Boer Government from Houtenbek.
1/24/1902 Important captures by General Plumer's troops. Thirty burghers secured by Colonel Fry, West Yorkshire Regiment. Attack on Pietersburg repulsed. Volunteer Town Guard distinguished itself.

Kruger's Post, Lydenburg and between them the notation, 'Capture of Viljoen Jan, 25'

Two days after we arrived, the Government received a report from General Muller stating that two hostile columns were approaching. We had not long to wait. The enemy attacked us in the afternoon, but did not succeed in driving us from our position. We were not, however, in a position to sustain a long battle, owing to scarcity of ammunition. Many of our burghers had only five cartridges left and some had not even one. Therefore, that same night—I think it was the 21st of January although I had lost count of dates—the Government, whom I accompanied, departed and proceeded to the Kloof Oshoek, between Dullstroom and Lydenburg. The weather was very unpropitious, rain falling in torrents, and as may be understood, we were in a sad plight. We were protected by nothing except our mackintoshes, and greatly envied a member of the party who was the proud possessor of a small piece of canvas.[217]

It had been decided that the Government should proceed on the 25th of January from Oshoek to Pilgrim's Rest, but the information that the British were not pressing their pursuit, caused them to give up this project, for it was thought advisable to await the enemy's next move. I should here mention that the further the Government were chased, the more difficult they found it to keep up communications with the Commandant-General and the Orange Free State Government.

With the latter, however, despatches were being exchanged concerning very important matters which I consider as still improper to disclose. The Government having determined not to proceed, I decided to bid farewell, and to proceed with my attendants on the way to Pilgrim's Rest.

Accordingly, on the 25th of January, we left the Government at Oshoek and rode along to Zwagerhoek, where we remained till sundown. We were now nearing the enemy's country, and so, having carefully reconnoitred the ground, we set forth cautiously at dusk. Two young Boers, who were also on the road to Pilgrim's Rest, had meanwhile joined us, and, including my kaffir servant, [Name?] our party comprised eight persons. We soon passed the fateful spot where Commandant Schoenman[218] had been captured in the early part of the War, and forded the Spekboom River.

I am not superstitious, but I must confess that somehow or other I experienced considerable disquietude about this time, and felt cold shivers running down my back. We were just approaching Bloomplaats, which is about two and half miles to the west of Lydenburg, when we observed something moving. A deadly silence enveloped

[217] 1/18/1902 Execution of Scheepers on various charges of murder at Graaff Reinet.
1/21/1902 Colonels Park and Urmston engaged parly of Boers under Muller and Trichardt, occasioning stampede of Boer Government from Houtenbek.
1/24/1902 Important captures by General Plumer's troops. Thirty burghers secured by Colonel Fry, West Yorkshire Regiment. Attack on Pietersburg repulsed. Volunteer Town Guard distinguished itself.
[218] A David Johannes Schoeman. See MOA.

the country, and the brightly-shining moon gave a weird appearance to the moving objects in the distance which had attracted our attention. Our suspicions were aroused and we went in pursuit, but soon lost sight of the object of our quest.

We discovered afterwards that our suspicions were well-founded, and that the moving objects were kaffir spies, who returned to the British lines and reported our approach. Having failed in this enterprise we returned to the road, I riding in advance with Adjutant Bester, the others following. Presently we approached a deep spruit, and having dismounted, we were cautiously leading our horses down the steep bank, when suddenly we found ourselves the centre of a perfect storm of bullets. We were completely taken by surprise, and almost before we realised what had happened, we found ourselves confronted by two rows of British soldiery, who shouted "Hands up," and fired simultaneously.

Bullets whistled in every direction. The first volley laid my horse low, and I found myself on the ground half stunned. When I recovered somewhat and lifted my head, I discovered myself surrounded, but the dust and the flash of firing prevented me from seeing much of what occurred. It seemed hopeless to attempt escape, and I cried excitedly that I was ready to surrender. So loud, however, was the noise of shouting that my cries were drowned. One soldier viciously pressed his gun against my breast as if about to shoot me, but thrusting the barrel away, I said in English that I saw no chance of escape, that I did not defend myself, and there was no reason therefore why he should kill me. While I was talking he again drove his rifle against me, and I, having grasped it firmly, a very animated argument took place, for he strongly resented my grasping his gun. Outstretching my hand I asked "Tommy" to help me up, and this he did. I afterwards learned that the name of my assailant was Patrick, and that he belonged to the Irish Rifles.

Four or five soldiers now took charge of me, and at my request consented to conduct me to an officer. Just as they were about to lead me away, however, they all fell flat upon their chests, and directed their fire at an object, which turned out later to be a bush. I very soon discovered that the "Tommies" were not very circumspect in their fire, and I sought safety by lying on the ground. Having discovered the innocent nature of their target, my guards conducted me before one of their officers, a young man named Walsh, who seemed to belong to the British Intelligence Department.

This officer enquired, "Well, what is it?" I answered him in his own language, "My name is Viljoen, and not wishing to be plundered by your soldiers, I desire to place myself under the protection of an officer." He was quite a minor officer this Mr. Walsh, but he said kindly, "All right, it is rather a lucky haul, sir; you look quite cool, are you hurt?" I replied that I was not hurt, though it was a miracle that I was still alive, for a bullet had struck my chest, and would have penetrated had my pocket-book not stopped it. The fact was, that my pocket-book had served the providential service of the proverbial bible or pack of cards. Bester was with me, and not seeing my other adjutants, I enquired what had become of them. Walsh did not reply at once, and one of the "Tommies" standing close by said, "Both killed, sor." This information was a terrible blow to me.

Major Orr, of the Royal Irish Regiment, was in charge of the force that had captured me, and presently I was taken before him. He greeted me most courteously and said, "I believe we are old friends, General Viljoen; at least you captured some of my comrades in that regrettable affair at Belfast." I was greatly touched by Major Orr's kindness, and asked that I might see those of my men who had been killed. He immediately consented, and led me a few paces aside. My gaze was soon arrested by a heartrending spectacle. There on the ground lay the two lifeless forms of my brave and faithful adjutants, Jacobus Nel and L. Jordaan. As I bent over their prostrate bodies my eyes grew dim with the sad tears of my great bereavement. Major Orr stood uncovered by my side, touched by my deep emotion and paying homage to the brave dead. "These men were heroes," I said to him with broken voice. "They followed me because they loved me, and they fearlessly risked their lives for me several times." The good Major was full of sympathy, and made provision for the decent burial of my poor comrades at Lydenburg.

Bester and I were now conducted under an escort of 150 soldiers with fixed bayonets to the village, which was two and a half miles off. We reached Lydenburg very wet and gloomy, after having waded through a drift whose waters reached up to our armpits. Major Orr did his best to console us both with refreshment and kind words.

Our procession was presently joined by an officer of the British Intelligence Department, and this gentleman told me that he knew of the approach of my party, and that the chief object of the British in attacking us was to capture our itinerant Government, who they learned were to accompany us. He was very anxious to know where the Government was, and whether it was intended that they should pass that way. But I answered his queries by telling him that it was quite unworthy of a gentleman to put such questions to me, and to attempt to exploit my most unfortunate position.

Arriving at the village, I was treated with great courtesy, and was introduced by Major Orr to Colonel Guinness, the commanding officer. Colonel Guinness declared that he regarded it as an honour to have a man of my rank as a prisoner-of-war, and that we had fought so frequently that we were quite old friends. I thanked him for his compliment, expressing, however, my regret that we had renewed acquaintance under such unfortunate circumstances.

"That is the fortune of war," said the Colonel. "You have nothing to be ashamed of, General." We were treated very well by our captors, and were given accommodation in the apartments of my old friend Captain Milner, who now filled the office of Provost-Marshal. My meeting with this gentleman was very cordial, and we sat up till nearly daybreak relating our different adventures since we had last met at Roos Senekal, where the worthy Captain was made prisoner by me. He assured me that his regiment entertained the highest respect for me and my burghers, and that they appreciated the fact that we had fought fairly and gallantly and had well-treated our prisoners-of-war. Bester and I remained under Milner's care throughout our stay at Lydenburg, and I shall always remember with gratitude the kindness extended me by the officers of the Royal Irish Regiment.

My Capture. General Ben Viljoen and his secretary [Bester] after their capture by the British in front of a tent, January 1902. Handwritten caption: "Gen. Ben Viljeon & Secretary being brought into line after capture. Jan./02."[341] [219]

[219] 1/25/1902 Capture of Viljoen near Kruger's Post by detachment of Royal Irish under Major Orr.

Chapter 47. Shipped to St. Helena.

We were kept at Lydenburg until about the 30th of January, 1902, and during our stay there I obtained leave to write a letter to my burghers. In this I acquainted them and my brother with what had occurred, and exhorted them to keep up their hearts and persevere. Although kindly treated at Lydenberg, I cannot adequately describe the feeling of disappointment and sorrow which my enforced inaction caused me. I would have given anything to have been able to return to my commando, and felt that I would rather have been killed than have fallen into the enemy's hands. Being thus rendered impotent I could but curse my fate.[220]

Friendships which are formed on the veldt are strong indeed, and the men who have lived together through all the vicissitudes of war for twenty-eight months—through sunshine and rain, happiness and sorrow, prosperity and adversity—become attached one to another with lasting affections. My sufferings hit me very keenly. Besides the sadness which separation from my companions caused me, I acutely felt my position as, having been before in the habit of commanding and of being obeyed by others, I was now subject to the humiliation of having to obey the orders of British privates.

We prisoners were conveyed from Lydenburg to Machadodorp under the charge of Colonel Urenston, of the Argyll and Sutherland Highlanders, with an escort of 2,000 men. I was at a loss to know why so large a force should have been sent to guard me, but this seemingly exaggerated precaution was soon explained when I was told that Lord Kitchener had given special orders that great care was to be taken to prevent my commando from rescuing me. I must say that there was not much chance of that occurring. Colonel Urenston was a very courteous soldier, and treated me as well as could be expected.

Reaching Machadodorp four days later, I was handed over at Dalmanutha Station to Captain Pearson, a staff officer, who subsequently conducted me and my fellow prisoners to Pretoria. Some days after my arrival there I was taken before Lord Kitchener, and was received very courteously by him at his office. My interview with this great General lasted about half an hour. The Commander-in-Chief of the British Army in South Africa impressed me as being a real soldier, a man possessed of a strong will not marred by arrogance.

I did not know what the British military authorities proposed to do with me, and felt quite indifferent as to the matter. At dawn on the third day after my arrival I was awakened by a soldier and informed that I was to be taken to the station. The train was in readiness when I arrived, and the officer in charge invited me to take a seat in his compartment. I was then told that we were to proceed to Durban, but no information was given me as to my ultimate destination.

On the train we prisoners were treated with great courtesy, but on reaching Durban a different experience awaited us. Here I was placed under the charge of Colonel Ellet, a very irascible person. This Colonel greeted me with the information that he was quite delighted that I had been captured. He repeated this gratuitous insult three times, and, my patience being exhausted, I asked him to be kind enough to tell me where he was instructed to convey me, and not to cause me unnecessary pain by his taunts.

He apologised lamely and told me that I was to proceed on board ship. This very much surprised me, and I remarked that I had already been taken from home and hearth 500 miles. This ill-tempered creature then lent back arrogantly in his armchair, puffing at his cigar, and said: "Well, ah, you are banished, don't you know. You are to be sent to St. Helena, or as we call it, 'The Rock.' You will shortly embark. It is a large ship you are going in; it is called—ah, let me see, oh, yes, the Britannica. I will proceed to the station and order your kit, and in the meantime you must sign this parole and report yourself forthwith at the docks." I said in Dutch, which the Colonel did not understand, "Lord deliver me from this evil person."

On arriving on board ship I found several other Boer prisoners-of-war, amongst them my old friend Erasmus, who masqueraded as a general in the early stages of the War. Never having been before upon the sea I was soon in the throes of mal de mer,[221] and the prospect was certainly not encouraging. There was no help for it, however. Colonel Curtis, of the Royal Artillery, who was in charge of the troops on board, was a very polite and

[220] 1/26/1902 Successful engagement on the Modder by Major Driscoll's column. Huge laager at Nelspan dispersed by General Bruce-Hamilton.
1/27/1902 Colonel Du Moulin killed in a night attack on his camp. Enemy repulsed by Major Gilbert (Sussex Regiment).
1/30/1902 Colonel Rawlinson's troops after tremendous march surprised Manie Botha's laager and made valuable captures.
1/31/1902 Capture of convoy at Groothoop by Colonel Rimington.

[221] Motion Sickness.

pleasant person, and very welcome after that extraordinary creature, Ellet. We were provided with good cabins and the food was excellent. Before leaving the Bay General Lyttelton visited me and showed himself very friendly. I soon found out that Mrs. Lyttelton was proceeding on the same boat to England. My company must have been rather unattractive, seeing that I was only well for one day during the whole voyage.

The steamer was ordered to call at Cape Town, and when we neared this port the guard kept over us was strengthened. An officer remained with us continually and counted us every two hours to make sure that none of us had escaped. One day two young Boers conspired to make a fool of the officer, and concealed themselves in the lavatory. Their absence was discovered the next time we were counted, and the officer in charge, in a great state of perturbation, demanded of us what had become of them. We took up the joke at once, and replied that they had gone on shore to be shaved and would return at 7 o'clock. This entirely took his breath away. But the absurdity of the situation so got the better of us that we burst out into ironical laughter, and finally set our custodian at ease by producing the two fugitives. We were punished for our little joke, however, by having our paroles withdrawn.

On the 19th of February the ship, with its sorrowful freight, steamed away from Cape Town. We prisoners, assembled on the upper deck, bade a very sorrowful farewell to the shores of our dear Fatherland. Long and sadly did we gaze upon the fast receding land from which we expected to be alienated for ever. Notwithstanding our depressing circumstances, however, we attempted pluckily to keep up our spirits, and with laughter and frivolity to cheer each other. Most of us had never been on a ship before, and only one of our number had ever voyaged away from South Africa. Ours was a very cheerless prospect, for, although we did not know our exact fate, banishment for life loomed over us. The ship's officers were urbanity itself, and did everything in their power for our comfort. I shall always remember their kindness, but it would have required much more than human effort to have made our voyage enjoyable owing to the fact that we suffered so intensely from sea-sickness.

After a very cheerless and discomforting voyage, we dropped anchor on the 24th of February in St. Helena Harbour. "The Rock" rose out of the ocean, bare and rugged, and imprisonment upon it offered a gloomy prospect. No animal was visible, and foliage was wanting, I never saw a less attractive place than Jamestown, the port at which we landed. The houses seemed to be tumbling over one another in a "kloof." We were all gloomily impressed, and somebody near me said, "This will be our living graves." I answered, "No wonder that Napoleon broke his heart upon this God-forsaken rock." I must confess that the feeling grew upon us that we were to be treated as ordinary criminals, since only murderers and dangerous people are banished to such places to be forgotten by mankind.[222]

An English officer came to me and asked what I thought of the Island. My feelings got the better of me, and I replied—"It seems a suitable place for England's felons, but it is very spiteful of England to deport here men whose only crime has been to fight for their country. It would have been much more merciful to have killed us at once than to make us drag out an existence in a manner so dreary."

We were soon taken ashore by boats to Jamestown, and there learned to our great disgust that we were all to be put in quarantine for bubonic plague, and to be isolated at Lemon Valley, a valley in which I afterwards found that lemons were conspicuous by their absence. No greenery was to be seen in this desolate place. While our

[222] 2/2/1902 De Wet's commando gallantly charged by New Zealanders, Queensland Imperial Bushmen, and South African Light Horse.
2/4/1902 Capture and destruction of British convoy by Boers in Cape Colony. Major Crofton killed.
2/5/1902 Surprise and capture of Commandant S. Alberts' laager by Scottish Horse under Major Leader.
2/6/1902 Major Vallancey dispersed Beyers commando. Movements to entrap De Wet started.
2/7/1902 De Wet, by brilliant manoeuvre, ruptured the British cordon and escaped.
2/8/1902 Big capture from Potgieter's laager by Colonel von Donop's force.
2/13/1902 Bouvers' laager in Cape Colony rushed by Colonel Kavanagh's men.
2/18/1902 Capture of Judge Hugo in Cape Colony. Boers cut off and surrounded a portion of squadron of Scots Greys south-east of Springs.
2/20/1902 Two laagers surprised by Colonel Park's troops ; 164 prisoners taken.
2/21/1902 Capture of laager at Buffelskloof by Colonel E. Williams' column.
2/24/1902 Some East Griqualand rebels surrendered to Colonel Stanford.
2/24/1902 De la Rey pounced on a wagon convoy commanded by Lieutenant Colonel S. B. Von Donop. For the loss of 51 Boers, De la Rey killed, wounded or captured 12 officers and 369 men. In response, Methuen tried to track the Boer leader down. Pakenham, p. 583. https://en.wikipedia.org/wiki/Battle_of_Tweebosch

debarkation was proceeding one of the boats capsized, but, happily, everybody escaped with nothing worse than a ducking.

Quarantine regulations were enforced for six days at Lemon Valley. The accommodation was very inadequate, and our culinary utensils, though not primitive, were very bad, the food being such as might have been the portion of criminals.

Luckily for us a British Censor named Baron von Ahlenfeldt, and a doctor named Casey had accompanied us, and owing to their instrumentality we were allowed better food and treatment. At the end of our detention in the quarantine camp some of our number were removed to Broadbottom Camp, while the others were quartered at Deadwood Camp. Lieutenant Bathurst, who now assumed the position of our custodian, was a good prototype of friend Ellet at Durban, and he was at pains to treat us as felons rather than as prisoners-of-war.

Chapter 48. Life in Bonaparte's Prison.

In order to reach Broadbottom Camp we had to ascend a remarkably rocky cliff named "Jacob's Ladder," the face of which was cut into a multitudinous series of steps. Having reached the summit we found a pleasing view of the Island opened before us. We now discovered that St. Helena was not the totally-barren rock we had at first been led to suppose. Patches of trees and greenery met our gaze, and in the midst of a carefully-cultivated plantation we espied a beautiful house, the habitation of the Governor of the Island. On our way we encountered a party of our fellow-prisoners, who, having been guilty of insubordination, were being taken to the dreary fort at High Knoll for punishment. Amongst these unfortunates we recognised several friends, but were not permitted to talk to them.

'Turbulant' Boers at High Knoll Fort.

At sundown our destination was reached at Broadbottom Camp, which is situated under High Peak. Before us stretched a large space enclosed by four encirclements of barbed wire containing the tents and houses which formed the temporary homes of the prisoners-of-war. Sentries were posted at every hundred paces. There were 2,000 prisoners stationed here, and as they wandered aimlessly round they forcibly reminded me of the Israelites in exile.

On entering the camp I was received by the commandant, Colonel Wright, a typical Briton, who made no pleasant impression upon me. I shall not be querulous, although the Colonel very bluntly notified to me that he had no instructions but to treat me in the same manner as the ordinary prisoners, and added that as my name had appeared in the list of Boer officers who were sentenced to banishment, he doubted whether I was entitled even to the treatment accorded to the ordinary prisoners-of-war.

However, a tent was erected for me, and I and my companions in adversity were given beds and culinary utensils. My bed consisted of two khaki blankets and a waterproof sheet, and my kitchen utensils comprised a pot, a washing basin, a pail, two enamelled plates, two large mugs, and a spoon. This is a complete inventory of the articles with which I was provided. I and the prisoners who had accompanied me had not tasted food throughout the whole day, and we would have gone supperless to bed had it not been that some compassionate brother prisoners ministered to our inner needs by providing us with some bully beef and bread, which, though but a frugal meal, was very welcome to us.

Camp life of the kind I now experienced was wearisome indeed. There was nothing to do, and we tried to while away the time by singing psalms and songs.

Church Service on St. Helena.[342]

At night the camp and its environments were rendered almost as bright as day by the glaring light of huge naphtha flares and by large search-lights which played round, making attempts at escape hopeless. It appeared to me that the search-lights were continually being turned in my direction, and I can assure you that I wished these glaring abominations at Hades. The buzzing and roaring noise given forth by the naphtha lamps, the monotonous chanting of the prisoners, the perpetual "All's well" of the sentries, and the intermingling notes of the bugle calls suffused the air with their distracting sounds and made me feel as if my head were in a maelstrom. The bugler was so amiable a person that he always made it a point of standing close to my tent when launching forth to the world his shrieking calls. Happily I became acclimatised to my distasteful surroundings, or I fear I should have soon graduated as a patient for a lunatic asylum.

I unhappily became at an early date acquainted with Colonel Price, commanding the troops on the Island. I shall never forget his demeanour towards me, for from the first his attitude was arrogant, cruel, and generally unbearable. He refused me parole, and declined to give me a pass beyond the confines of the camp. The unreasonableness of this hard treatment will be seen when it is remembered that not the slightest possibility of escape from the Island existed. The close confinement began to play havoc with my health, and I was in the fair way to the hospital, when a friendly doctor intervened and restored me to health once more. The rigid discipline and the stern regulations that were enforced can only be likened to what is experienced in monastic life. The "red-tape" curse prevailed everywhere.

Subsequently Colonel Price modified his tone towards me and allowed me parole. He was also gracious enough to permit me and some companions to occupy a little house 400 paces from the camp. This was a very agreeable change, for now we were no longer subjected to the harsh treatment of the "Tommies." Our little residence rejoiced in the pleasantly-floral name of the "Myrtle Grove," and was rented by us from an old coloured lady who vigorously insisted upon the punctual payment of the rent, and drew our special attention to the fact that plucking pears in the garden was strictly prohibited.

We had been told that the "Myrtle Grove" was haunted by ghosts, but the ghosts, if any there were, must have been pro-Boers, since they never disturbed us. But though we had no ghostly visitors we certainly had some of another kind. The house was perfectly infested by particularly large and bold rats. These thieving rodents, not satisfied with robbing our larder, had the audacity to sup off our fingers and ears while we were asleep. We waged vigorous war against the vermin, and after considerable difficulty managed to get the residence exclusively to ourselves. With the addition of some furniture, with which Colonel Wright was good enough to provide us, we made our house so comfortable that we felt ourselves almost in a position to invite the Governor to dinner.

Our landlady, Mrs. Joshua, was the proud possessor of several donkeys, which were turned loose in our garden, and a large number of fowls. I may say that Mrs. Joshua was very ill-advised in keeping her fowls so near our house, for our cook, who had been trained in commando, was unable to resist the temptation of appropriating eggs. It did not, however, take our landlady long to find out what was happening, and we were informed that it was very much more Christianlike to purchase eggs. We took the hint, and adopted as far as we could Christianlike methods, though we found it extremely difficult to subscribe to all the principles of Christianity practised by the Islanders.

We whiled away the time by taking daily walks, and, by making excursions to the house at Longwood tenanted by Napoleon Bonaparte for six and a half years, and to the grave where his remains were interred for 19 years. I noticed that both places were being preserved and kept in order by the French Government. We used to sit by the little fountain, where the great French warrior so frequently sat, and read. We were permitted to drink a glass of water from this historical spring.

Longwood House in 19th Century.[343] [223]

Longwood House in 2008.[344]

Prisoners visiting Napoleon's Tomb.[345]

General Cronje and his officers on St. Helena.[346]

[223] Longwood House is a mansion on the British overseas territory island of St. Helena. It was the final residence of Napoleon Bonaparte, the former Emperor of the French, during his exile on the island of Saint Helena, from 10 December 1815 until his death on 5 May 1821. In 1858 it was bought by the French government, along with the Valley of the Tomb for a sum of £7,100. Since then they have been under the ownership of the French Foreign Ministry and a French government representative has lived on the island and has been responsible for managing both properties. https://en.wikipedia.org/wiki/Longwood_House

At Deadwood Camp 4,000 of my compatriots were confined. Some had been there for over two years, and I could not help admiring their discipline. It is not for me to criticise the entirely unnecessary restrictions to which these unfortunate prisoners were subjected, but I will point out that the severity practised towards helpless prisoners by armed soldiers created feelings of great bitterness. It was a stupid policy to pursue and perhaps fateful.

Deadwood Camp from Flagstaff Direction.[347]

The military authorities were entirely unacquainted with the character and mannerisms of the Boers, and were advised in this connection by so-called "Cape" or "English" Afrikanders, who bear an ineradicable hatred to the Boers, and who always did their utmost to cause the prisoners to be treated with humiliation and contempt. Happily a number of English officers whom I met on the Island saw that we were not so black as we had been painted. Most of the officers who acted as our custodians here had come direct from England and knew nothing of South Africa. One of these gentlemen confessed to me that when he left London for St. Helena he had a sort of idea that he was to be placed in charge of a troop of wild barbarians, and that he had been quite agreeably disappointed. He declared, indeed, that he had found that the Afrikander in some respects was superior to men of his own nation.

It was undoubtedly a sad error for England to send officers to look after us, who, not having had any experience of South African warfare, were entirely ignorant of our idiosyncrasies and manners. The result of placing these inexperienced men as our guards was that one misunderstanding followed upon another, and that unnecessarily rigorous regulations were promulgated to preserve discipline and order. This treatment had the effect of nourishing within our bosoms hatred and bitterness.

Not being desirous of having to undergo incarceration with my insubordinate fellow-prisoners at High Knoll Fort, I carefully refrained from being unruly, and practised an orderly and amiable demeanour.

On one occasion I ventured to approach Colonel Price with a view to obtaining some amelioration in our treatment, and some remission of the rigorous regulations meted out to us. After keeping me waiting half an hour he came out of his office to meet me, but instead of extending a greeting he stared at me with ill-concealed amazement, probably expecting that I should jump up and salute him. I, however, merely rose and nodded, and enquired if I had the honour of addressing Colonel Price. He answered stiffly, "Yes, what do you want?" It was greatly disconcerting to be thus unceremoniously and discourteously greeted, and having explained my mission, I withdrew and took care to fight shy of this arrogant soldier in future.

I may say that our little party at "Myrtle Grove" was a few weeks later augmented by the arrival of Vaal Piet Uys and Landdrost T. Kelly.

We had in the meantime improved our acquaintance with Colonel Wright, who always treated us with cordiality and kindness, and allowed us frequently the privilege of spending pleasant afternoons at his house. Mrs. Wright was a charming hostess, and did everything in her power to lessen the feeling of humiliation with which we regarded our sad plight.

I should perhaps mention that St. Helena boasts of some elegant society. A few years before our confinement the Zulu chief, Dinizulu, was banished within the rocky bounds of this island prison. This son of Cain had during his detention here been invited to all the fashionable parties and dances, and had been honoured with an invitation to the Governor's house. He was feted at dinners and public festivities—but of course it must be remembered that Dinizulu was a kaffir and we were only Boers. Fancy, my Afrikander brothers, a self-respecting English young lady consenting to dance with this uncivilised kaffir! Imagine, they allowed him to dine at the same table, and to drive in the same carriage with them! I do not know how this information strikes my readers, but I must say that when the Governor of the Island, an elderly gentleman named Sterndale, with 35 years of the Indian Civil Service behind him, informed me that such had been the case, I was rendered speechless.

Robert A. Sterndale.[348] Picture of Governor Sterndale, upon the departure of another exile, King Dinuzulu.[349]

I would not have it supposed, however, that we prisoners had any special ambition to attend balls and dinners, for we were not in the mood for festivities, and even had we desired we could hardly with propriety have appeared at these elegant boards and gatherings dressed in our shabby apparel.

A number of the prisoners received permission from the authorities to pursue the various crafts and employments with which they were conversant, at the small daily wage of between sixpence and a shilling. This pay was a ridiculously small remuneration for the large amount of work which the men executed. A great diversity of trades were represented by us prisoners. One was a mason, another a farmer, a third an apothecary, while a fourth was a goldsmith, and so far did we go that one man was appointed caterer for the St. Helena Club.

Months had now passed since I had been first brought a captive to this island prison, and it approached the middle of May. Persistent though rather vague reports about Peace continually reached us, but owing to the strictness of the censors, who had an exaggerated idea of their duties, any news from outside came to our anxious ears in very small pieces, and gave us a very meagre idea of what was happening in South Africa and other places outside. That we were all praying earnestly for Peace needs no telling, especially if I may mention that some of my comrades had been incarcerated on the island for two years and eight months. I cannot adequately tell how wearisome their long exile was to them.

Just before I was liberated from confinement, our old antagonists, the 3rd Battalion of "Buffs," under Colonel Brinckman, were detailed to the Island. This regiment had seen two years of active service in South Africa, and they were, therefore, soldiers who did not hold their enemies in contempt.[224]

I do not feel at this time, in view of the present tension of affairs, able to pursue my account further; but if encouraged by a sympathetic public to supplement this effort by a more detailed description of my imprisonment at St. Helena, I may in the near future again seek their indulgence.

Meanwhile, I take what I hope will prove but a temporary leave of my readers, with the following explanatory details and critical comments on the general characteristics of the War.

[224] PF: Col. Theodore Brinckman, C.B. wrote the preface to this volume.

Chapter 49. How We Blew Up and Captured Trains.

Looking at the matter superficially it seems a very barbarous thing to derail and destroy trains with dynamite, but this was the only course left open to us, since large military stores were being continually brought in by the British from the coast. We honestly regretted that, owing to the derailment and destruction of trains, drivers, stokers, and often innocent passengers were launched into eternity. War is at best a cruel and illogical way of settling disputes, and the measures which the belligerent parties are sometimes compelled to take are of such a character that sentimentality does not enter into any of the calculations of the contending parties.

It should not be necessary to assure my readers that we acted entirely within our rights in derailing and destroying trains. This was the only means we had of breaking the British lines of communication and of interrupting the conveyance of British troops and food.

Moreover, we were more than justified in any act of train-derailment that we committed, by the instructions of Lord Wolseley as expressed in his handbook. In that well-known publication this distinguished soldier actually prescribes the use of dynamite, and even suggests the manner in which it may be employed to the best advantage. But although this train-wrecking was in every degree justifiable, I can assure the reader that we regarded it as a very unpalatable duty. I remember that when Lord Kitchener complained to me about the destruction of a certain train, I sent him a reply to the following effect:—

"That the blowing up and destroying of trains was as distasteful to me as I hoped the burning of our houses was to his Excellency; and that when we derailed trains we entered upon the task with hearts quite as heavy as those which I presumed weighed down his troops when they deported our women and children from their homes to the Concentration Camps."

I shall now describe how we went to work in the matter of capturing trains. That this is not so easy a task as appears to be supposed I shall endeavour to show. Perhaps the best way to exemplify our method of procedure would be to describe a particular instance which occurred in March, 1901, between Belfast and Wonderfontein on the Delagoa Bay Railway. The two stations are approximately 12 miles apart. At either station a garrison had been established, and these were provided with two or three cannons and two armoured trains, which latter were held in readiness to proceed to any place within their immediate sphere of action when anything irregular occurred on the line.

They were used besides to carry reinforcements and stores when needed. The armoured train was indeed a very important factor in the British military tactics, and one we had to take fully into account. The railway between these two stations was also guarded by blockhouses. Every morning the British soldiers carefully inspected their particular section of the railway before trains were despatched in any direction. The peril of running trains at night was speedily recognised, and of those that attempted the journey very few indeed escaped capture. On the particular occasion when the incident I am about to relate took place, we were encamped at Steenkampsbergen, enjoying a little remission from the arduous work in which we had been engaged. But we were not idle, and a field-cornetcy of approximately a hundred men was detailed to attempt the capture of a train. I personally reconnoitred the line, and sent a field-cornet with instructions to lay a mine at the most favourable spot for the distasteful operation we were about to perform.

Our modus operandi was to take a Martini-Henri rifle and saw off four inches before and behind the magazine, and then to so file the trigger guard that the trigger was left exposed. Two of the most intelligent burghers were despatched over night with this mutilated rifle and a packet of dynamite to the spot chosen for the mine, while two other burghers kept guard.

Special precautions were taken to prevent footmarks being traced by the British patrols, the burghers walking for a considerable distance on the rails. The mine was prepared by carefully removing the stones from underneath the rails and as cautiously replacing them to again fill up the hole after the instruments of destruction had been adjusted. The trigger was placed in contact with the dynamite, and just enough above ground to be affected by the weight of the locomotive, but so little exposed as to be passed unnoticed. All surplus stones were carried off in a bag and great care was taken to conceal all traces of the mine. Gingerly and cautiously and without leaving any trace of their visit, the burghers now returned to their field-cornet and reported that all was in order. The field-cornetcy took up its position behind a small hill about a mile from the railway, and the men concealed themselves and their horses so ingeniously that their presence was not even suspected by the occupants of the blockhouse close by. According to our information the first train that was to pass next morning was the mail train carrying the European mails, and the prospect of capturing some newspapers and thus obtaining news of the outside world, from which we had been isolated for several months, filled us with pleasant expectation. I especially instructed the field-cornet to obtain newspapers, and to capture as much food and clothing as possible. It being the custom of the British garrisons to send scouts along the railway each day to examine the line, the next morning the track was as usual microscopically inspected, but the scouts failed to discover the trap which we had laid.

Two outpost burghers lay at the top of the hill in the grass, and from their coign of vantage they had a clear view of the railway line.

Ten o'clock in the morning arriving without a train appearing, my men began to grumble. In the excitement of this adventure they had omitted to prepare any food, and they were not now allowed to make fires, because the smoke evolved in culinary operations would have been immediately noticed by the enemy's outpost. We had therefore to remain hungry, or our well-laid plans would have been frustrated. Time passed on, and at 2 o'clock in the afternoon there were still no traces of the expected train. Our horses were saddled up and had been without food since the previous afternoon, and the poor animals also began to show their displeasure by whinnying and stamping their hoofs on the ground. The enemy's scouts had already inspected the line three or four times either by going over it on foot or by using a trolley.

The afternoon was well advanced, and fears were growing in our minds that the mine had been discovered. I should say that it was Sunday afternoon, and that the mine had been laid on Saturday night. This train-wrecking scheme of ours was contrary to the practices of our nation, who regard all such acts on Sunday as a desecration of the Sabbath, but here I will again apply an English precept, "The better the day the better the deed."

About four o'clock my outposts notified to me the approach of smoke, and shortly afterwards we beheld a train coming along. Every man of us mounted his horse, and we sat calmly in the saddle to observe the execution of our plan. We held our breaths. Perhaps the British had detected the mine and removed it, with the result that all our travail would be in vain; or they might possibly have sent a large force of soldiers with cannon on the train to give us a "good hiding" to boot. We watched breathlessly the progress of the train as it rapidly approached the fatal spot, and our hearts thumped wildly as we waited to see the success or failure of our enterprise. We had not long to wait, for with a tremendous shock the mine exploded, overturning the engine, and bringing the train to a standstill.

We now proceeded to storm the train, but I saw the danger of advancing in a mass and shouted to my men to go carefully and spread out. When we were about 500 feet from the train the British fired a volley at us, but in so doing they merely displayed by their firing that there were not many riflemen on the train, and that those that there were shot badly and at random. Thus shown the weakness of the enemy, we stormed with renewed vigour, and on arriving at about a hundred yards distance we dismounted. The defenders did not face our fire long before displaying the white flag. I stopped fire at once and the train was ours.

It was Lieutenant Crossby, of the Remount Department, who waved the white flag, and he now surrendered with about 20 "Tommies."

Among the occupants of the train was an old major, and on his saying that he was very sick, and was on his way to the hospital, we immediately apologised for having disturbed him and for the delay which our little operation had caused him. There were eight sacks of European mail in the train and these we seized. We liberated the "Tommies" after disarming them. The Lieutenant in charge was the sole person detained as a prisoner-of-war, and he was added to six other British officers who were vegetating under our charge. Only a part of the train could be destroyed by us, as one section was occupied by women and children who were being transported to the Concentration Camps.

On the following morning the field-cornet brought me the papers and said with a smile, "You see I have brought you what you required, General." I was overjoyed to obtain tidings from the outside world. The letters were distributed about the laager, and there was abundance of reading matter. I felt rather sorry for the "Tommies" who were being thus mercilessly robbed of their letters, but I consoled myself with the thought that our plight was quite as bad as theirs, for we Boers had had no communication from any members of our families for twelve months, and we felt justified in making the "Tommies" share our misfortune. The Boers did not, however, get much satisfaction out of other men's epistles, and even those who could read English gave up the operation after having perused one or two, and threw away the sackfuls of letters with disappointed faces.

The capture of this train was our second success. Shortly before we had seized a train near Pan Station and had obtained a splendid haul. This particular train was carrying Christmas presents for the British soldiers, and we found a miscellaneous assortment of cakes, puddings and other delicacies. It was very amusing that we should be celebrating Christmas with cakes and puddings which had been intended for our opponents.

A few weeks after we had captured the train carrying the European mails we made another attempt at train wrecking, this time at Wonderfontein Station. All, too, went well on this occasion until we charged, and the British opened fire upon us with cannon. We were not favoured this time by any sort of cover, but had to attack over open ground, exposing ourselves to the heavy fire of the guns and the fusillade of a hundred British riflemen. We had chanced this time upon an armoured train, and the trucks which bore the cannon had remained uninjured. The nut was rather too hard for us to crack, and failing to take the train by storm, we were compelled to retire, after having sustained the loss of three men, of whom one was my brave adjutant, Vivian Cogell. From what I have said I think my readers will agree that the capturing of a train is not always a "cake and ale" operation.

Chapter 50. How We Fed and Clothed Commandos.

As early as March, 1901, we experienced the difficulty of adequately providing our commandos with the necessities of life. So far back as September, 1900, we had said good-bye at Hector's Spruit to our commissariat, and thence, no organized supplies existing, it may very well be imagined that the task of feeding the Boers was one of the most serious, and I may say disquieting, questions with which we had to deal. We were cut off from the world, and there was no means of importing stores. Of course the men who had been previously engaged on commissariat duty were enlisted in the fighting ranks so soon as they became available.

From this date we had to feed ourselves on quite a different system. Each commandant looked after his own men and appointed two or three Boers whose special duty it was to ride round for provisions. It must not be supposed that we commandeered stores without signing receipts, and the storekeeper who supplied us was provided with an acknowledgment, countersigned by field-cornet, commandant, and general. On producing this document to our Government the holder received probably one-third of the amount in cash and the balance in Government notes, better known as "blue-backs." By this time a large portion of the Republic had been occupied by the British, all food-stuffs had been removed or destroyed, and most of the cattle had been captured.

In consequence, everything in the shape of food became very scarce. Flour, coffee, sugar, &c., were now regarded as delicacies remembered from the far-away past. The salt supplies were especially low, and we feared that without salt we would not be able to live, or if we did manage to exist, that we might bring upon ourselves an epidemic of disease. Our fears in this respect were increased by the opinions expressed by our doctors, and we viewed our situation with considerable disquietude. Happily, as experience proved, our apprehensions were not in the least justified, for during the ten months that preceded my capture my burghers lived entirely without salt, and were at the time that I fell into the hands of the British as healthy as could be desired.

Existing as we did solely on mealies and meat, potatoes and other vegetables which we might chance upon were regarded as luxuries indeed. Though it may appear strange it is nevertheless a fact that we were always fortunate enough to obtain adequate supplies of mealies and meat. We ground our mealies in coffee mills if no other mills were available. Mealie pap is cooked in a simple fashion, and occasionally boiling hot pots of it have fallen into the hands of the British. The British soldiers were not much better off than we were, for they were limited to bully-beef and "clinkers," though they frequently supplemented their larder by stores from Boer farms, such as fowls, pigs, &c., and had salt, sugar, and coffee in abundance. Their culinary utensils were not nearly so primitive as circumstances had reduced ours to.

Many Boers did nothing but roam round with their cattle, and I confess that on many occasions they excited my admiration by the "slim" manner in which they evaded capture. Boers of this description were dubbed "bush-lancers," because they always sought the thickest bushes for sanctuary. These "bush-lancers" were of three kinds: There were some who sought by running away with their cattle to escape commando duty, others who hoped by retaining their cattle to obtain a large profit on them after the War was over, while others were so attached to their cattle that they would as lief[225] have lost their own lives as have suffered their cattle to be taken. All three classes of "bush-lancers" contrived to supply us with adequate stores of food. Often, however, it

[225] PF: As Printed in Book. Possibly as Dutch for 'Stomach.'

was a difficult task to get the supplies out of them. When we asked them to sell us cattle we were frequently met by the reply that we had already taken their best cattle, that the British had taken some, and that the little they had left they could not do without. Of course we were not hindered in our purpose of obtaining food by such a reply, and we had sometimes to resort to force. We frequently gave these "bush-lancers" notice when danger threatened, but in most instances they were the first to discover danger, and gave us information as to the movements of the British.

Everybody knows that it is a sore trial for the Boer to live without coffee, but this national beverage disappeared entirely from our menu, and its loss was only partly replaced by the "mealie coffee" which we set about preparing. The process was a very simple one. As soon as we off-saddled a hundred coffee mills were set to work. The mealie was roasted over a fire and afterwards treated in a similar manner to that by which the coffee bean is prepared. This "mealie coffee" made a very palatable drink, especially as we were frequently able to obtain milk to mix with it.

We generally roasted our meat on the coals, as we found that without salt meat was most palatable when treated in this way. This is explained by the fact that the ashes of the fire contain a certain saline quality. We obtained mealies in all sorts of extraordinary ways. Sometimes we harvested it ourselves, but more often we found quantities hidden in caves or kraals. Mealies were also purchased from the natives. Every general did all that was possible to sow in the district in which he was operating, for the soil is very fruitful. We very seldom lacked mealies, although the British frequently destroyed the crops we had been growing. There can be no doubt that when an Afrikander feels hungry he will find something to eat.

I have already mentioned that sometimes when the British swooped down upon us they carried away our culinary utensils, and a question may arise in the minds of my readers as to how we obtained others to replace them. Well, we were not particular in this connection. We found empty tea cans and empty bully-beef tins, and by manipulating barbed wire we speedily converted these crude materials into serviceable culinary implements. We preferred the tar cans because the beef tins often came to pieces after the solder with which they are fastened had been subjected to the heat of the fire. I remember that one day our parson gave as much as five shillings for an empty tar can.

Several British convoys fell into our hands, but the food we found on them consisted usually of bully-beef and "clinkers," things which only dire necessity drove us Boers to eat. Sometimes to our great chagrin we discovered that all our fighting to capture a convoy was only rewarded by the sight of empty trucks or ones loaded with hay and fodder. If perchance we were fortunate enough to capture a camp or a fort we contented ourselves with removing such coffee and sugar as we could carry away on our pack mules.

The clothing question was very perplexing. Whenever we were able to obtain it we bought canvas and converted it into trousers. Sheep skins we tanned and employed either for the purpose of making clothes or for patching. The hides of cattle and of horses that had died of disease were also tanned and employed for the making of boots.

I may point out that no horse was specially slaughtered for this purpose or for the purpose of food. It was only General Baden-Powell and General White who slaughtered their horses to make sausages. Our best clothing supply, however, came from the British Army. Forgive me for saying so; I do not intend to be sarcastic. When we captured a convoy or a fort we always obtained a supply of clothes. At the beginning of the War we Boers had a strong prejudice against any garment which even faintly resembled khaki, but afterwards we grew indifferent and accepted khaki quite as readily as any other material. We generally compelled our prisoners to exchange clothes with us, and often derived much amusement from the disgusted look of the sensitive Briton as he walked away in the clothes of a ragged Boer. Imagine the spectacle! A dandy English soldier, clean shaven, with a monocle adorning one eye, his head covered with an old war-worn slouch hat of broad brim, and his body with ragged jacket and trousers patched with sheep-skin or yarn.

I may say that none of this systematic plundering occurred in my presence. But such things were certainly done, and, after all, who can blame a ragged burgher for resorting to this means, however much to be deprecated, of clothing himself. Remember that the poor Boers were prepared to pay double the value of a suit of clothes, and were, so to speak, cut off from the world, while the British soldier had simply to go back to camp to obtain a new outfit. "Necessity knows no law."

In concluding this chapter I must mention that the lack of matches was very sensibly felt. And when our stock of matches was exhausted we had to resort to the old-fashioned tinder-box and flint and steel. We found this expedient a very poor substitute for the lucifer match, but it was certainly better than nothing at all. Personally I experienced the greatest difficulty in getting fire from a flint and steel, and to do it generally took me quite twice as long as it took anybody else, and I bruised my hands considerably. This latter, however, is an experience to which every amateur is liable, and I was never much more than an amateur at anything.

Tinder Box with flint and sparking (carbon) firesteel.[350]

Chapter 51. Our Friend the Enemy.

In venturing on a judgment of the British soldier, from a military point of view, I may be told that only the man who has had a military training is competent to express an opinion upon the individual capacity of a soldier, be he Boer or Briton. That may be true, as long as people only go theoretically to work; but after my two and a half years of practical experience, my military friends may be gracious enough to allow me to express my simple opinion concerning this important factor, which is undoubtedly fundamental to the efficiency of any army. At the same time I promise to be as impartial in my judgment of the Boer as of the Briton as a fighter, or, at least, as impartial as can be expected from a fallible Boer.

As an officer in the Boer army I encountered the British soldier in many capacities and in many circumstances. The officer of the regular British troops was always prepared to notify that he had no high opinion of the officers of the irregular troops. At the same time the volunteer officer was equally ready to heartily reciprocate the compliment when it was passed upon him by the regular. To be honest, I must say that I specifically give preference to the regular officer, whom I regard as having more initiative, and as being more practical and less artificial than his colleague, the irregular Imperial officer. As regards courage I saw little to choose between them. I certainly can draw no great distinction, since I have never been in a position to fight on the same side as they.

Generally speaking, I consider the British officer a very brave man, though I do think he sometimes is guilty of excess in that respect—that is to say, that he goes impractically to work, and, the young officer especially, is driven by ambition to do desperate and stupid things. To this foolhardiness may be largely attributed the heavy losses in officers suffered by the British Army in the War.

Since I fell into British hands I have found the officers to whom I had been opposed on the battlefield treat me with the utmost magnanimity. After having been in personal contact with a considerable number of officers of various regiments I must plainly say that the British officer is to be encountered in only two species: He is either a gentleman or—the other. The officer of the first species is prepared to be charitable to his antagonists, and generally assumes an attitude of dignity and humanity; whereas the latter possesses all the attributes of the idiot, and is not only detestable in the eyes of his antagonists, but is also despised by his own entourage.

There have been unfortunate British officers in this War, and there have been occasions when a disaster to the British has been immediately attributed to the acts or the tactics of the commanding officer. In this connection I will cite the regrettable instance of General Gatacre at Stormberg. I do not think this reverse is to be attributable to stupidity, or indiscretion, or cowardice.

There is a great deal of luck attached to any adventure in the field, and ill-luck had pursued General Gatacre persistently. But undoubtedly where bad luck pursues a commander on more than one occasion it is not only expedient but necessary to dismiss such an officer, because his troops lose confidence in him, and their spirit is undermined. It has occurred in this War that incapable officers with good men and much luck have performed wonders.

The British soldier, or "Tommy," who draws a very poor daily pay, for which he has to perform a tremendous lot of work, is, if not the most capable fighter, the most willing in all circumstances to offer himself as a sacrifice at the altar of duty, or of what he considers his duty, to his country. But if "Tommy" by any accident be asked to deviate from the usual routine in which he has been trained, he is a thoroughly helpless creature. This helplessness, in my opinion, is caused by exaggerated discipline, and by the system under which "Tommy" is not allowed to think for himself or to take care of himself, and this individual helplessness has undoubtedly been one of the shortcomings of the British soldier during the War. As regards the fortitude of the ordinary British soldier, I must repeat what I have already said—that he is a courageous, willing and faithful warrior, and that it is to his fidelity and patriotism that the British Army may attribute its success. I believe this to be a truism which will defy even criticism.

There are, of course, exceptions to the courageous "Tommy." If I were to draw any comparison between the nationalities, I would say that of the soldiers with whom I was brought into contact on the battlefield, the Irishmen and the Scotsmen were better fighting men than the others. In regard to British soldiers generally, I would remark that, if they could add good shooting and ability to judge distances to their courage, then they would be perhaps perfect soldiers, and certainly be doubly dangerous to their foes.

Taken as a whole "Tommy" is a very warm-hearted fellow, though as regards humanity some distinction must be drawn between the regular soldier and the enlisted volunteer, for the latter is less humane than the former. This was too clearly shown by his conduct in the transporting of women and children and in the plundering of prisoners-of-war. But nevertheless "Tommy," generally speaking, whether regular or irregular, was sympathetic with regard to our wounded, and showed great kindness of heart to a maimed opponent.

I consider that the British infantry bore the brunt of the fighting of this War, especially in its earlier stages. Where the cavalryman failed to break through our lines the infantryman stepped in and paved the way for him. We found we could always better stand an attack from cavalry than from infantry, for this latter, advancing as it did in scattered formation, was much less visible to our marksmen. When advancing to the attack the British foot soldiers were wont to crawl along on their faces, seeking cover whenever that was available; thus advancing, and especially when they were supported by artillery, these men proved very difficult indeed to repulse.

In my opinion a cavalryman has no chance against a good marksman when this latter occupies a good position and is able to await attack. The British cavalry horses are such stupendous creatures that given a good rifle and a keen eye it is difficult for one to miss them. They certainly make most excellent targets. It is my firm opinion that for usefulness the cavalryman cannot be compared to the mounted infantryman. Indeed, my experience during the last 14 months of my active participation in the War taught me that the British mounted infantry was a very hard nut to crack. Of course everything depended upon the quality of the man and the horse.

A good rifleman and a horseman, especially if he were able to fire when mounted, was a very formidable foe. As for horses, I may say that I do not wonder that the great unwieldy horses for which the British cavalrymen have such a predilection cannot be compared to the Basuto ponies with which we went to work. The African pony has, in fact proved itself to be the only useful horse during the campaign. The British cavalryman might have used elephants with almost as much advantage as their colossal horses. Further, in my opinion, the cavalrymen might just as well be discontinued as a branch of an army, for there can be no doubt that the infantry, artillery, and mounted infantry will be the only really useful and, indeed, practicable soldiers of the future.

Mounted Yeomanry at rest.[351]

Troops of one of the British mounted regiments collecting forage for their horses.[352]

Entraining horses at Cape Town.[353]

A shove and up and in.[354]

While I was writing the above a book was placed in my hand written by Count Sternberg, with an introduction from the pen of Lieut-Colonel Henderson. I doubt very much whether Colonel Henderson read the manuscript of the Count's book before penning his introduction, for I cannot suppose that he holds such small-minded and fantastic ideas regarding South Africa as the Count expresses. In this memorable work some extraordinary tales are told of the galloping and trotting feats of the Basuto ponies. The confession that the Count makes that he did not care upon which side he fought so long as he fought is indeed extraordinary. That he ever fought at all the Boer officers who knew him strongly doubt, and none of them will wonder that the Count's bitterest experience in South Africa was that on one occasion some naughty German ambulance people deprived him of a box of lager-beer. This and other amateurs have already overwhelmed the reading public with so much so-called criticism about this War, that I venture upon delicate ground in offering my opinion. I will confine myself to commenting upon what I saw and I know personally, for I know nothing about the topography of Europe and I am not acquainted either with the composition of the European armies or with their manner of fighting.

A Basuto-Pony.[355 226]

Basuto Pony & Rider.[356]

Chapter 52. The Fighting Boer and His Officer.

There is great difference between the relations of a Boer officer to his following and the relation of a European officer to his men, for while in the former case no social distinction between the two exists, in the latter the officers and men are drawn from two distinct branches of society. The Boers in their normal state are independent farmers differing only in wealth. One Boer might be the possessor of perhaps ten farms and be worth a quarter of a million, while another might be but a poor "bywoner" and not worth a hundred pence, yet the two men would occupy the same rank in time of war.

Immediately [when] martial law is promulgated the entire Boer adult male population is amenable for military service. In the ranks of a commando one finds men of every profession, from the advocate and doctor to the blacksmith and plumber. From these ranks the officers are chosen, and a man who one day is but an ordinary soldier might be the next promoted to the rank of field-cornet or commandant, and might possibly in a few days attain the position of a General.

The officer and the men that follow him have in most cases been drawn from the same district, and they know one another personally. If, therefore, a Boer falls in battle, whatever be his rank, his loss is keenly felt by his comrades in arms, for they, having known him of old, lose a personal friend by his death.

The Boer officers can be divided into two classes—the brave and the cowardly. The brave officer fights whenever he gets the chance, whereas his chicken-hearted brother always waits for orders and makes elaborate plans to escape fighting. It is quite easy in the Boer Army to succeed in the course adopted by the latter class, and it not infrequently occurred that the Boers preferred this class of officer to his more reckless comrade, for they argued—"We like to serve under him because he will keep us out of danger." And just as the officers could be divided so could the men.

In this campaign it was noticeable that during the last stages of the struggle the younger officers replaced the older ones. Many of these latter got tired of the War and surrendered to the British, others were removed from their commands as being too old-fashioned in their methods and incapable of adapting themselves to the altered circumstances. Moreover, we found that the younger officers were more industrious, more mischievous, and more reckless. Of course, when I speak of the young Boer officers I do not intend to convey the idea of children of seventeen to twenty years of age, such as I have sometimes encountered among the junior officers of the British Army.

The life training of the burghers in horsemanship and musketry stood them in good stead. I may say that a Boer even early in life is a good horseman and marksman. He does not shoot without purpose for he can generally estimate at a glance the distance at which he is shooting, and he has been taught economy in the use of ammunition. The burgher knows perfectly well how valuable to him is his horse, and he is thus constrained to

[226] The Basuto is a pony breed from Lesotho and South Africa. The Basuto is considered a small horse, since it possesses horse-like characteristics, such as an exceptionally long stride. Basutos have, long necks and long, straight backs, and straight shoulders. They have very tough legs and sound, very hard hooves. They can be up to 14.2 hands high (56.8 inches or ~142 cm). Basutos can be chestnut, brown, bay, gray or black. They are sure-footed, fast, fearless and are known for their stamina. Basuto pony - Wikipedia

use his knowledge in carefully tending it; moreover, considerable affection exists, in many instances, between the master and his beast.

"A party of Boers at target practice."[357]

Practicing Running Shots.[358]

Taken all round the Boer is a brave man, but his attitude on the battlefield is influenced very largely by the character of his officer. And being brave, the Boer is, in the main, sympathetic towards prisoners-of-war, and especially towards such as are wounded. Possessing bravery and humanity the Boer has besides what the British "Tommy Atkins" lacks, the power of initiative. The death of an officer does not throw the ranks of a Boer commando into chaos, for everybody knows how to proceed. It must not be supposed, however, that the death of an officer does not exercise a certain amount of demoralising influence. What I wish to impress is that the members of a commando can act independently of the officer and can exercise their own judgment.

As regards the fortitude of the Boers, I can best illustrate it by pointing to the fact that it frequently happened that having been repulsed with loss one day we attacked our conqueror with better success the next. We often assumed the aggressive when a favourable opportunity offered itself, and did not always wait to be shot at. Frequently we held out for hours notwithstanding severe punishment.

I think even the bitterest of our enemies will allow that the Boers who remained faithful to their country to the last were animated with noble principles. Were it not that so many of my compatriots lacked that which is so largely characteristic of the British soldier, the quality of patriotism and the intense desire to uphold the traditions of his nationality, I would ask what people in the world would have been able to conquer the Afrikander? I say this with great deliberation, and I do not believe that any impartial compatriot will attempt to deny the truth of the statement.

The question suggests itself how would the English have fared had they been placed in a plight similar to that to which we found ourselves reduced? Supposing that we Boers had taken London and other large towns, and had driven the English people before us and compelled them to hide in the mountains with nothing upon which to subsist but mealie pap and meat without salt, with only worn and rent clothes as a covering, their houses burnt, and their women and children placed in Concentration Camps in the hands of the enemy. How would the English have acted under such circumstances? Would they not have surrendered to the conqueror? However that may be, one thing is certain, that the patriotism of a nation is only to be learned when put to such a severe test as this.

In his book, "The Great Boer War," Dr. Conan Doyle has, on the whole, gained the admiration of the Afrikanders by his moderate language. But here and there, where he has been carried away by his English sympathies to use bitter and libelous language with respect to the Boers, that admiration has been changed into contempt. Dr. Conan Doyle attempts to defend the British Army by abusing the Boers. Abuse is not argument. To prove that Van der Merwe is a thief does not exonerate Brown from the crime of theft if he have been stealing.

The author describes the shooting of Lieutenant Neumeyer, for refusing to surrender and for attempting to escape from his captors as murder, and the shooting of kaffir spies it also glibly described as murder; whereas, the incident at Frederickstad, where a number of Boers were shot dead by the British because they continued firing after hoisting the white flag, is justified by him.

Of course, the execution of Scheepers is also justified by the author. I object to such things appearing in a book, because they must tend to sow anew the seeds of dissension, hate and bitterness, and these have been planted sufficiently deep without being nurtured by Dr. Conan Doyle. Neither Boer nor Briton can speak impartially on this question, and both would be better employed in attempting to find out the virtues rather than the vices in one another's characters.

Whoever in the future governs South Africa, the two races must live together, and when the day of Peace arrives and the sword is sheathed, let us hold out our hands to each other like men, forgetting the past and remembering the motto—

"Both Nations have Done their Duty."

[Transvaal] State Seal.[359] A mixture of Dutch and Latin: "Freedom, Guilds, in Health - Immigration."

General Ben Viljoen delivering his lecture on the Boer War at Queen's Hall, October 16 1902.[360]

Viljoen Epilogue

St. Helena Camps

From 1900 to 1902 St Helena had to house nearly 6,000 Boer POW's. See: Boer Prisoner of War Camp on St. Helena.[361] And [362] Knowing that the Boers were coming, and concerned that, as captured enemy soldiers in a war that was still ongoing there was a risk of abuse and ill-treatment by the islanders, Governor Sterndale,[363] published the following advance proclamation:

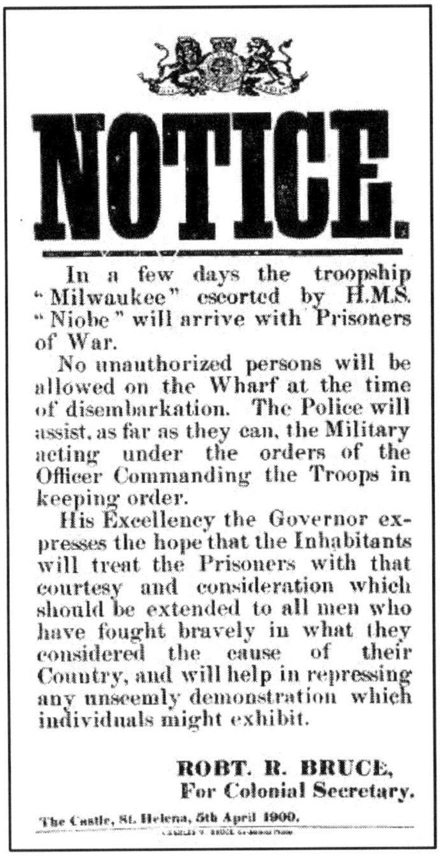

NOTICE.

In a few days the troopship "Milwaukee" escorted by H.M.S. "Niobe" will arrive with Prisoners of War.

No unauthorized persons will be allowed on the Wharf at the time of disembarkation. The Police will assist, as far as they can, the Military acting under the orders of the Officer Commanding the Troops in keeping order.

His Excellency the Governor expresses the hope that the Inhabitants will treat the Prisoners with that courtesy and consideration which should be extended to all men who have fought bravely in what they considered the cause of their Country, and will help in repressing any unseemly demonstration which individuals might exhibit.

ROBT. R. BRUCE,
For Colonial Secretary.

The Castle, St. Helena, 5th April 1900.

Notice.[364]

The first shipment of 514 Pow's arrived on the 10th April 1900, including General Cronjé and his wife, Colonel Schiel and 21 other officers. General Cronje and his officers.[365]

General and Mrs. Cronjé were taken to Kent Cottage in Half Tree Hollow where they were to stay for the duration of their time on the island. The remaining Pow's were marched via Napoleon Street to Deadwood Plain, where they encountered a fence of barbed wire surrounding several hundred square metres containing of canvas tents in which they were to be housed for the duration of their stay on the island.

When the prisoners arrived, there was no jeering or rude remarks from the crowd of islanders who had congregated to see the Pow's land and pass on their way to Deadwood Camp which had been prepared for them.

Arriving at the wharf.[366] The steamship in the background may be "Milwaukee."

Marching from the wharf to the town.[367]

Boers in Jamestown heading to main square.[368]

Prisoners addressed by Governor.[369]

Prisoners on main street in Jamestown.[370]

Deadwood Camp The Boer Camp on Deadwood Plain.[371]

Another instalment of Pow's arrived two weeks later on 26th April. Between April 1900 and February 1902 around five and a half thousand Boer Pow's arrived on the island.

The Regiment sent out from England to guard the prisoners was the Gloucester (Gloster) Regiment, led by Lt Colonel Earl Bathurst.[227]

In addition to General Cronjé the island also hosted another important Boer General: Ben Viljoen who had been ambushed and captured towards the end of the war. Viljoen arrived at St Helena on 25th February 1902 and resided in a small house outside the Deadwood Plain Camp.

It was there that he wrote his autobiography, "My Reminiscences of the Anglo-Boer War." The book is a realistic description of the war from a Boer perspective. Unlike other Boer generals he was not a property owner and thus hesitated to return to South Africa. He returned to South Africa as a pauper but refused to take British

[227] Excerpt from "Boer Prisoners on St Helena" compiled by Barbara B George. http://libcom.org/library/pow-camps-st-helena-during-boer-war

citizenship, thus greatly reducing his chances of resuming a public career. He was also disillusioned by rumours claiming that he had collaborated with the British, and he felt that his contribution to the struggle was not recognised.

Some of the Pow's expressed a desire to become British Subjects, but this caused issues between them and others who were bitterly against the British. To prevent conflict, the authorities were compelled to form a separate camp located apart from the general camp, housing the Pow's desiring to become British Subjects the 'Deadwood No. 2' or 'Peace Camp'. Friction also developed between the 'Freestaters' and the 'Transvaalers' and in early 1901 the authorities decided to open another site at Broad Bottom Camp, in what is now Blue Hill district, housing the burghers of the Orange Free State.

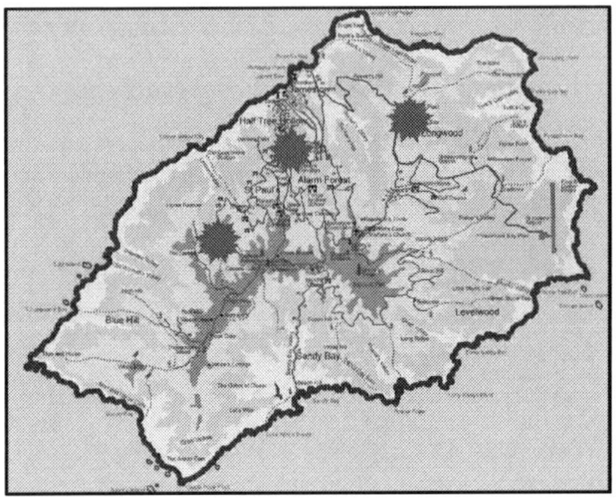

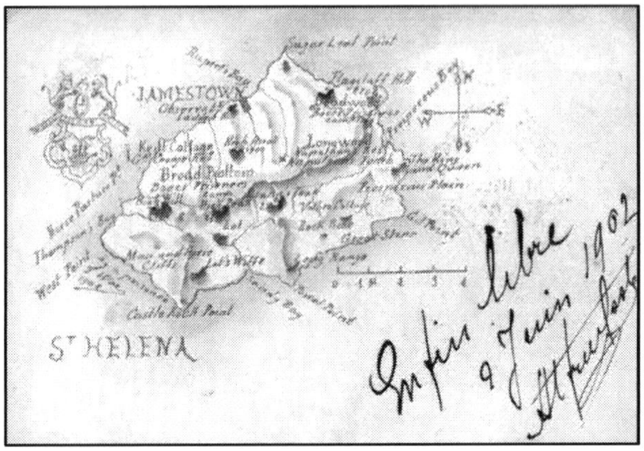

St. Helena map.[372] Deadwood camp is in the upper right. Broad Bottom camp is in the center and Blue Hill is on the left. Also: St. Helena postcard depicting the island.[373]

The bulk of the Pow's were housed in canvas tents or in huts which they later built themselves from biscuit tins. It should be noted that Deadwood Plain is one of the windiest and most exposed spots on the island (which is why the island's windmills have been sited there); and that Broad Bottom is one of the dampest places.

Among the Pow's were many skilled craftsmen and these were employed on various building projects around the island.

Carpenters outside their shop.[374]

NOTICE.

Prisoners of War and others are hereby warned not to cut or remove Trees or Timber of any description from the adjoining Government Forests.

Offenders after this date will be prosecuted.

By order,

C. COTTRELL,
Inspector of Police.

Police Office,
St. Helena 13th May 1901.

Notice regarding timber felling.[375]

Others were given permission to work in and around Jamestown in jobs such as household servants, cooks and grooms. Some were allowed to live in the homes of their British employers provided that they remained well behaved; others were housed in a two-section camp in the Government Gardens and the Botanical Gardens in Jamestown.

The more enterprising of the Pow's involved themselves in establishing a Coffee House, a Brewery, a pawnbrokers and another set himself up as an auctioneer. Some found employment with the Eastern Telegraph Company. The Pow's even started an Afrikaans newspaper, Kamp Kruimels.

Eastern Telegraph Company workers.[376]

The Pow's were allowed to write letters home (some even drew postcards), but only under strict censorship including 'No POW may write about any of the following subjects: the political situation; the British Government or troops; occurrences in the camps; complaints about the food; or anything of a kindred kind and all photographs showing the coats of arms of the late republics, pictures of the leader, and any article whatsoever with the coats of arms aforementioned will be confiscated.' Incoming letters were also censored.

Picture postcard of Deadwood camp written in Dutch, dated September 20th, 1901.[377]

The Pow's interacted well with the local people and two were allowed to marry local women. They established a string quartet, a piano trio, a brass band and a male choir.

Camp Band.[378]

The camps boasted a debating society, a German club, an anti-smoking society and many sports teams. In 1900 they even held a crafts exhibition.

Craft Exhibition.[379]

Industrial Exhibition.[380] PF: Those look like Boer and Orange Free State flags.

Among the Pow's was accomplished artist Erich Mayer; the paintings and sketches he produced provide an intriguing glimpse into the life of the Boer Pow's.

Erich Mayer painting of camp life.[381]

Erich Mayer painting of camp entertainment.[382]

"Cronje's captive warriors: A group of Boer prisoners at Deadwood Camp, St. Helena."[383] Children Prisoners on St Helena.[384] PF: It looks like the boys made slates for practicing writing. The writing on the one that can be resolved says: "Piet Willem Pieterse" One boy has made a monkey on a trapeze and another, kneeling, has made what appears to be a very large tobacco pipe. The reason that there are children included in the population is that when Cronje surrendered his laager and camp train included many women and children who had joined his forces as refugees and to support their family members who were fighting.

Local Residents employed as rubbish removers.[385]

St. Helena Prisoners 1902.[386]

Despite the fact that the Boer Pow's were generally well cared for and accepted by the local people, there were still those among them who plotted to escape from the island. In February 1901 five of the prisoners tried to escape in a boat which they seized from fishermen at Sandy Bay. The fishermen took away the oars and after a struggle the prisoners got into the boat and tore up the bottom boards to make paddles. When they found that this did not work, they then tried to bribe the fishermen, offering them money for the oars. In the meantime one of the fishermen had gone on to report the event and eventually a guard arrived and the Boers were taken into custody.

Possibly the most enterprising escape attempt was that of Andries Smorenburg, who fashioned a crate marked 'Boer Curios', in which he hid with clothing, matches, and food and water for 20 days and posted himself from St Helena on a passing ship. But although the crate was marked 'With Care' and 'This Side Up' it was tossed about and overturned on board and as a result Smorenburg suffered concussion and lost most of his water. In the meantime back on the island, Smorenburg's absence had been discovered when he did not appear for roll call. The authorities on St Helena contacted Ascension Island and Smorenburg was recaptured there and returned to St Helena after only five days at sea.

Andries Smorenburg and his crate.[387]

Andries Smorenburg outside Jamestown Gaol.[388]

Escapes St. Helena and Elopes with Heiress

This story is way more than likely to be an example of Yellow Journalism[228] There is no Fritz Everhardt listed on Anglo Boer War website as a prisoner or part of the Boer Forces..[229] Without any coroberation it is included for entertainment purposes only. See the volume on Blake for more articles about the Boer War Spectacle at Brighton Beach in 1905.

Dashing Boer Lieutenant Elopes With an Heiress

Beautiful Girl Admired Him in Paris Seven Years Ago—Then He Went to War. Daring Escape From St. Helena.

NEW YORK, July 15.—While the guests of the Brighton Beach Hotel are wondering what has become of the beautiful young English heiress who has been attracting attention there for several weeks, Capt. A. W. Lewis, general manager of the Boer War Spectacle, is mourning the disappearance of the blond and dashing Fritz Everhardt, lieutenant of his artillery.

The lieutenant, in a letter left for his old commander, explained that he would never appear in the show again, as he and the English girl had eloped and would be married by the time the letter was received.

No One to Interfere.

Of course there is no one who could interfere with the young couple here, though the father of the girl, Arthur Edward Lawson, chief of the branch of customs in Windsor, Ontario, had opposed her marriage to the young Boer, but only because of financial reasons, Miss Dorothy Lawson having inherited a large fortune from her aunt, while the young Boer had lost all his money before becoming a soldier of fortune indeed.

The story leading up to the elopement of yesterday is a romantic one, in that the girl, who is only 22, had been faithful to the memory of the young Boer ever since they first met in Paris, seven years ago, even though she at one time believed him dead. He had been captured by the English and imprisoned on the island of St. Helena, but escaped. When Miss Lawson heard this she came to this country, despite the opposition of her father, as her mother had called her to her dying bedside and given consent to the marriage.

First Met as Children.

Years ago when both were almost children, Everhardt and Dorothy Lawson met in India, where their parents were partners in the tea business. Previous to the breaking out of the Boer war young Everhardt, who had been educated in a military school in India, went to Paris and Brussels as an attache of the Boer legation.

When hostilities were about to open the elder Everhardt begged his son to return to India and take a half interest in the rich tea business. But the youth was disinclined to enter the mercantile trade, and, encouraged by letters from his mother, who was a loyal Boer, went to the Transvaal as an assistant to Dr. Leyds.

Miss Lawson was in Paris at the time of his departure, and, although they then were deeply in love, the manly young Boer refused to pledge the beautiful girl to an engagement, as he never expected to return alive.

The services of the young assistant to Leyds are a part of the military record of that war. He was made a lieutenant and his education proved of great value, having been trained as an engineer. Several medals which he received for bravery in breaking up the railroad service installed by the British he still wears. Then came an unfortunate raid in which almost all his comrades were shot and the rest captured, and news reached Miss Lawson in Paris that her sweetheart was dead.

The fact was he had only been slightly wounded, and, with the rest of the prisoners, had been taken to St. Helena, with twenty-nine indictments against him by the English hanging over his head. The lieutenant was taken down with a dangerous fever and lay in the hospital for seven months. During his convalesence he planned a scheme of escape, and the better to carry it out pretended to be slower in recovering than he really was.

Escaped by Swimming.

One night, with other daring spirits, he added a chapter to the bloody history of St. Helena by leaping into the sea and trying to swim to a friendly tramp American steamer, which had been signaling far off shore just before dark. The British used searchlights on the fugitives whenever a head showed, and shot the men to death in the water. Young Everhardt, however, was one of the few to escape, and he managed to reach the steamer by diving whenever the searchlight promised to find him out.

From St. Louis he began a search by letter for Miss Lawson, finally finding her in Ontario. Leaving her father, who still was opposed to the match, she went to St. Louis, and from there followed the Boer war party to this city. Young Everhardt saw her every day, and as he said in his letter to his old commander:

"Miss Lawson's wealth always had inclined me to keep apart from her in view of her father's opposition, but having her dying mother's consent, I believe the fact she is an heiress is no longer an obstacle."

Image 6 of The Washington times (Washington [D.C.]), July 15, 1905, (Evening).[230]

[228] In journalism, yellow journalism is the use of eye-catching headlines and sensationalized exaggerations for increased sales, while the yellow press are American newspapers which do so. This term is chiefly used in American English, whereas in the United Kingdom, the similar term tabloid journalism is more common. https://en.wikipedia.org/wiki/Yellow_journalism

[229] https://www.angloboerwar.com/

[230] https://www.loc.gov/resource/sn84026749/1905-07-15/ed-1/?sp=6&q=Boer+War&r=0.596,0.185,0.335,0.21,0 https://tile.loc.gov/storage-services/service/ndnp/dlc/batch_dlc_alpha_ver01/data/sn84026749/100492192/1905071501/0645.pdf

A total of 5,865 prisoners arrived here over the almost two years up to February 1902. Several were on parole early on, and helped in extending the wharf, setting up a new crane (this was a Swedish engineer who fought for the Boers), building a new house and repairing older ones, including adding verandas. There was a small camp in the Castle Gardens and Botanical gardens for those working in the town area, and they set up their own Tea shops in camp – one prisoner actually opened one in town, while another opened a Bakery!

7th May 1900: 180 prisoners died on the island from various epidemics, in may of 1900 A severe epidemic of influenza causes 75 deaths (8 funerals are held in one day).

In 1902 an outbreak of enteric fever in 1902, probably brought to the island by the last contingent of POWs to arrive killed 31 individuals. Only thee shootings occurred, it is not known if all were fatal.

For example, one Saturday night early in 1901 a POW was shot by a sentry. At the Military Court which followed it emerged that for some time the POWs had been hurling stones, sticks, tin cans and other missiles at the sentries, and that on this occasion the sentry in question had been struck in the face by one of these missiles.

The first prisoners were interred in St Paul's Cathedral graveyard, then later in the graveyard at the Baptist Chapel, Knollcombes, where all their names are recorded on two pillars.[231]

First Monument 1913.[389]

The peace treaty was signed on 31st May 1902. The St Helena Guardian of 5th June 1902 carried the headline 'Peace, Perfect Peace' and expressed the hope that the peace would be a lasting one. On 26th June 1902 the first batch of Pow's embarked for Cape Town; the last batch left on the 21st October.

The island suffered from deforestation due to the huge increase in requirement for fuel, and the roads were cut to pieces by all the mule waggon traffic, but poverty was eased for a short time for the majority of the population. Five prisoners stayed behind to live and raise families on the island.

Chronology

To provide context, here is everything we have in our Events Database for 10th April 1900 to 21st October 1902:

10th April 1900: The first shipment of Boer Pow's arrives in James Bay. They disembark three days later.

7th May 1900: A severe epidemic of influenza causes 75 deaths (8 funerals are held in one day).

8th October 1900: New road round to West Rocks is completed, constructed with the help of Boer Pow's.

5th November 1900: The Boer Pow's hold craft exhibition at the Deadwood Camp.

10th November 1900: An 'Industrial Exhibition' is held.

[231] Excerpt from "Boer Prisoners on St Helena" compiled by Barbara B George. http://libcom.org/library/pow-camps-st-helena-during-boer-war

20th December 1900: First issue of the Boer newspaper Kamp Kruimels is published (later Die Krijgsgevangene).

4th February 1901: Five Boer Pow's try to escape in a boat, which they seize from fishermen at Sandy Bay. They are quickly recaptured.

16th February 1901: Bubonic plague in South Africa causes restrictions on ships coming from the Cape to be brought into force in St Helena.

2nd April 1901: Completion of new crane on the wharf, erected by Boer Pow's.

21st June 1901: A water condensing plant is completed in Ruperts Bay, to provide fresh water for the Boer Pow's encamped on Deadwood Plain.

27th June 1901: The St Helena Guardian reports that two ships have brought in a cargo of 200 tons of firewood, with more to follow. This is to feed the Desalination Plant in Ruperts.

25th July 1901: Bubonic plague in South Africa causes extra quarantine regulations to be brought into force in St Helena.

24th August 1901: The Condensing Works (desalination plant) in Ruperts is operated for the first time.

25th February 1902: General Viljoen arrives on Britannica and moves into a small house outside the Deadwood Plain.

31st May 1902: The peace treaty is signed, ending the (Second) Boer War.

1st June 1902: A message is received via the 'Victorian Internet' with one word: PEACE, signalling the end of the (Second) Boer War.

5th June 1902: The St Helena Guardian leads with the headline 'Peace, Perfect Peace', following the end of the (Second) Boer War.

8th June 1902: Churches across the island hold special services to celebrate the end of the (Second) Boer War.

26th June 1902: The first Boer Pow's leave St Helena for Cape Town aboard the Canada.

3rd October 1902: 95 Boer Pow's are reported to have died while on St Helena, mostly from enteric fever.

21st October 1902: The last of the pardoned Boer Pow's leave St Helena for Cape Town aboard the Golconda.

Prisoners Departing.[390]

Graves of Prisoners.[391] The original monuments can be seen at the bottom.

Cronje's Marriage at the St. Louis World's Fair in 1904.

Image 9 of The St. Louis Republic (St. Louis, Mo.), July 5, 1904.[232]

Advertisements for the Boer War Libretto and other shows at the fair.

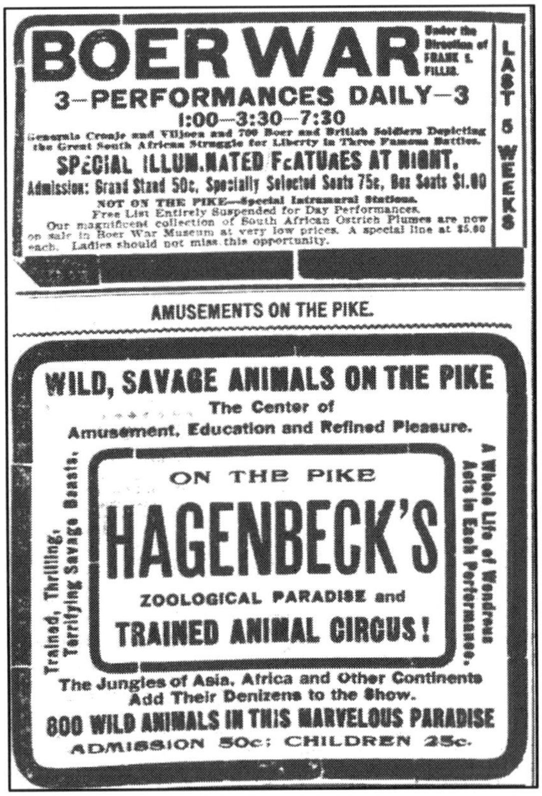

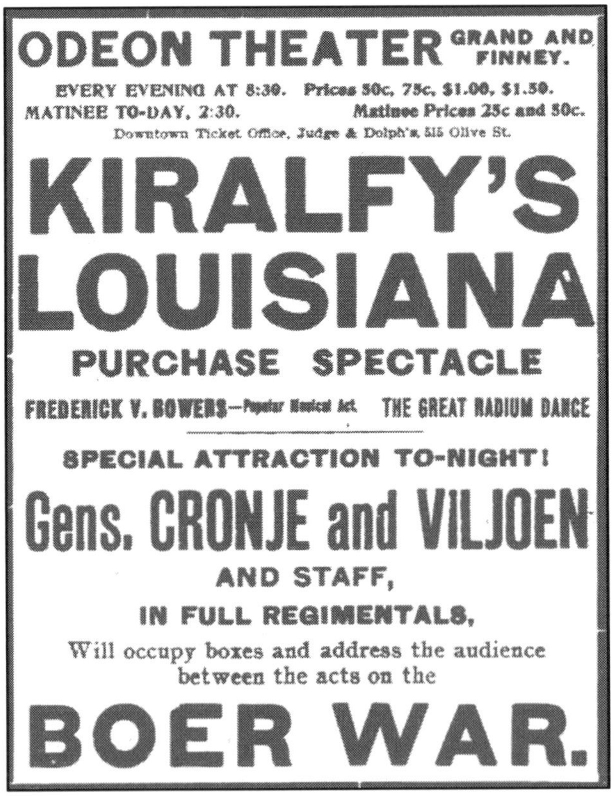

The St. Louis Republic Sunday, October 23. 1904 Page 3.[233]

[232] https://www.loc.gov/resource/sn84020274/1904-07-05/ed-1/?sp=9&q=Anglo-Boer+War&r=0.498,0.081,0.436,0.27,0 https://tile.loc.gov/storage-services/service/ndnp/mohi/batch_mohi_ingalls_ver01/data/sn84020274/00294559747/1904070501/0102.pdf

Boer War Libretto

At the conclusion of the war, the Orange River Colony and Transvaal needed to do everything they could to rebuild their country and recover from their losses. They had debts to pay, veterans and widows to care for and numerous other responsibilities to address in order to restore their respective countries.

There was also a political dimension to try and maintain their relations with the rest of the world and to solicit what good will and assistance they could obtain; as well as to habilitate their valiant if unsuccessful efforts at independence.

To serve both of those ends the South African Exhibition Company embarked on what can only be called and expedition to the 1904 Louisiana Purchase Exposition. This was an opportunity to generate good will and mingle on the world stage with many other counties, exchange ideas, generate new business and well as income to assist themselves and their countrymen.[234] [235]

Description

The Louisiana Purchase Exposition, informally known as the was an international exposition held in St. Louis, Missouri, United States, from April 30 to December 1, 1904. Local, state, and federal funds totaling $15 million were used to finance the event. More than 60 countries and almost all of the American states maintained exhibition spaces at the fair, which was attended by nearly 19.7 million people.

Program sold at the exhibition.[392]

Parade[393] (probably at beginning of performance).

[233] The St. Louis Republic. [volume] (St. Louis, Mo.), 23 Oct. 1904. Chronicling America: Historic American Newspapers. Lib. of Congress. https://chroniclingamerica.loc.gov/lccn/sn84020274/1904-10-23/ed-1/seq-17/

[234] Direct Link: https://en.wikipedia.org/wiki/Louisiana_Purchase_Exposition
Permanent Link: https://en.wikipedia.org/w/index.php?title=Louisiana_Purchase_Exposition&oldid=1299869696 Over 900 pictures and books: Available Online, Louisiana Purchase Exposition | Library of Congress (loc.gov) Comprehensive program Guide: World's fair authentic guide; complete reference book to St. Louis and the Louisiana purchase exposition... | Library of Congress (loc.gov)

[235] One might be tempted to regard this extravaganza as a exploitive of a genuinely tragic war that caused untold misery Or even to recall Karl Marx's remark that history repeats itself first as a tragedy and then as a farce; but in this case I don't think it was so. It was an astute effort on both political and business grounds. Also, when Marx made that statement he was referring to the tragedy of the fall of Napoleon I and the later farce of the reign of his nephew Napoleon III.

Arrival of the President and Mrs. Roosevelt.[394]

Overview of the grounds.[395]

Festival Hall.[396]

Pavilion and Floral Garden.[397]

Grand Basin.[398]

Palace of Electricity & Machinery.[399]

Frank E. Fillis produced what was supposedly "the greatest and most realistic military spectacle known in the history of the world." Different portions of the concession featured a British Army encampment, several South African native villages (including Zulu, San, Swazi, and Ndebele) and a 15-acre (61,000 m2) arena in which soldiers paraded, sporting events and horse races were held and major battles from the Second Boer War were re-enacted twice a day. Battle recreations took 2–3 hours and included several Generals and 600 veteran soldiers from both sides of the war.

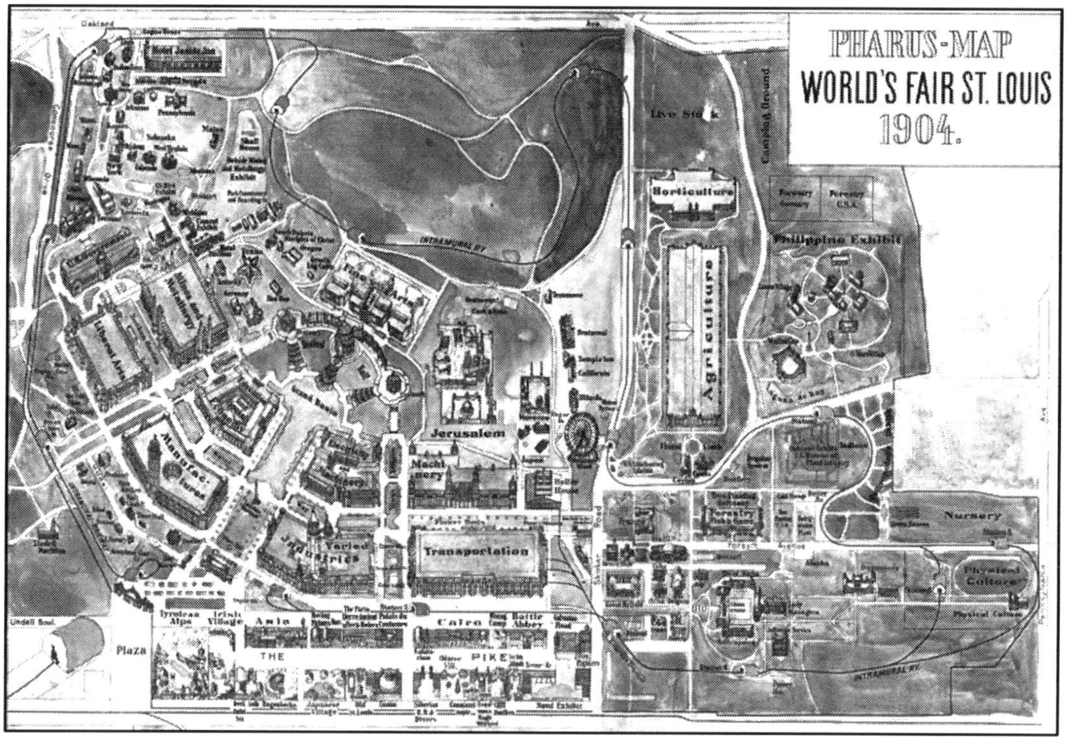

Pharus-map World's Fair St. Louis, 1904.[400]

From left to right: Commander JN Boshoff, General Piet Cronjé, General Ben Viljoen and Commander GM van Dam during a performance of the St. Louis Boer War Spectacle of 1904.[236]

A contemporary description of the performance.[237]

Though located some distance from The Pike proper, the South African Boer war is not to be overlooked by the visitor. It introduces scenes of conflict which recently engaged the attention of the world.

The audience is brought face to face with heroes who played conspicuous roles on Transvaal battle-fields. In a rugged amphitheater of wooded hills on Skinker road, surrounded by a perspective of African scenery, is reproduced the battle of Colenso, fought on the Tugela river December 15, 1899, when eleven field guns were captured from General Long and where Buller's British forces were held in check for many days.

236 From left to right: Commander JN Boshoff, General Piet Cronjé, General Ben Viljoen and Commander GM van Dam during a performance of the St. Louis Boer War Spectacle of 1904. Louisiana Purchase Exposition (St. Louis World's Fair). https://commons.wikimedia.org/wiki/File:Commander_Boshoff,_Generals_Cronj%C3%A9_and_Viljoen,_St._Louis_1904.jpg

237 Article by: By Carlos J. Frank. From: On the Pike, the first complete booklet of the Pike and amusement features of the Louisiana purchase exposition ... | Library of Congress (loc.gov) Page 41.

L to R. Boshoff, Cronje, and Viljoen.[401] General Ben Viljoen commander and president of the South African Boer War Exhibit Company.[402]

At the conclusion of the show, an actor in the role of Boer General Christiaan de Wet would reenact one of his escapes on horseback by leaping from a height into a pool of water.

Showing De Wet's escape scene from the Boer War attraction.[403] A second performance.[404]

Reenactment of British Troops trying to cross the Tugela during the battle of Colenso.[405] Second Image.[406]

1904 World Fair title: "Boer Cavalry ready for Mimic War."[407]

The conclusion of the battle with the Boers crossing to capture the British artillery pieces.[408] Taking possession of the guns.[409]

The battle of Paardeburg, where Gen. Cronje surrendered on February 27, 1900, to Lord Roberts, is also presented. The battles are reproduced twice a day, with uniforms and guns worn in the actual conflict; a battery of galloping colts, rapid-firing guns, a battery of Maxims, all used on the fields of South Africa. Enthusiasm is heightened by the appearance o, the old Transvaal Staatz artillery band, dressed in the same uniforms worn during the engagements in South Africa.

The Boer forces are in charge of Gen. Viljoen, a gallant Boer leader. An added interest is gained by the presence of Gen Piet A. Cronje and Gen. Kemp, who commanded the Boers in the engagement which resulted in the capture of Gen. Lord Methuen. Maj. Scott Harden, who commanded the contingent of Boers which volunteered for service against the Mad Mullah, and who fought in Somaliland, together with Maj. W. P. Steward of the 3d hussars, lead detachments of Boer soldiers.

British shells explode the ammunition in the Boer laager and destroy the Red Cross wagon train. The British then storm the laager. Gen. Cronje surrenders to Lord Roberts and staff, who appear on the scene.

Exploding ammunition wagon.[410]

Another performance.[411]

Battle Scene.[412]

Mock Prisoners reenacting Battle of Paardeberg.[413]

Hand to Hand Fight.[414]

Gordon Highlanders.[415]

In connection with the show is an exhibit of South African metallurgy and curios and a village of Zulus, Swazies and other South African tribes.

Admission ranged from 25 cents for bleacher seats to one dollar for box seats, and admission to the villages was another 25 cents. The concession cost $48,000 to construct, grossed over $630,000, and netted about $113,000 to the fair—the highest-grossing military concession of the fair.

Boer Colony in Mexico

Boer Colony in Mexico

General Viljoen Sails for United States To Complete Plans for Aid of His Countrymen.

London, Dec. 10.—General Benjamin Viljoen, the former Boer commander, sailed for New York on the steam ship Umbria for the purpose of completing arrangements for the establishment of a Boer colony in Mexico. Hs Says that he has already secured 85,000 acres [Recited as 5,000 acres in another article.] acres and expects to get 2,000, 000 more later on, in northern Mexico, where the climate closely resembles that of the Transvaal

General Viljoen gives a discouraging account of the situation of the Boers in South Africa at the present time. He says that it was the policy of the leaders after the war to encourage the hope of securing local self government, but the conditions render it Impossible for them to remain longer. The country has been devastated and cattle and horse diseases seem to make its Immediate recovery impossible.

Arrangements have been completed for His reproduction on a large scale at the St, Louis exposition next year of. the principal battles In the South African war. An attempt will be made to preserve the absolute historical accuracy of the famous .engagements. Artists are now painting the scenery for representation of the South African country. There will be from 700 to 1,000 actual participants on both sides.

Boer Colony in Mexico.[238] Dec. 10, 1903,

Boer Colony 2

Boer Colony Prospers. Chihuahua, In Northern Mexico, the Rise-of a Flourishing Settlement.

A private letter to the East Oregonian says the colony of Boers, which was established In a valley on Conch as river . In the state of Chihuahua. Mexico, two years ago is prospering. The promoter of this colonization scheme is General Snyman. He directs the affairs of the colony.

The land which the colonists own Is very rich, being brought to a high state of cultivation. A largo crop of wheat and corn has been grown this season and the colonists will lay up money. Their farms are well stocked with cattle, horses and hogs.

Unlike other farmers In Mexico, the Boers are doing without Mexican labor as much as possible, the experience with Mexicans in the first year having proved unreliable and unsatisfactory. The colony is constantly receiving recruits from South Africa. The latest to arrive are W. J. Viljoen and wife, the parents of General Benjamin Viljoen. General Viljoen has four grown sons in the colony

Boer Colony Prospers.[239]
August 30, 1903,[240]

Ben Viljoen was one of the South African refugee officers who formed a farm colony in Mexico with the assistance of Theodore Roosevelt. Help with selecting and negotiating for the property were provided by two men hired by Roosevelt family friend Marshall Latham Bond and the husband of a Roosevelt relative Edward Reeve Merritt. It was located at Hacienda Humboldt in the municipality of Julimes, Chihuahua..

[238] Boer Colony in Mexico. Las Vegas daily optic. [volume] (Las Vegas, N.M.), 10 Dec. 1903. Chronicling America: Historic American Newspapers. Lib. of Congress. Las Vegas daily optic. [volume] (Las Vegas, N.M.) 1880-1908, December 10, 1903, Image 1 « Chronicling America « Library of Congress

[239] Boer Colony Prospers. East Oregonian : E.O., August 30, 1905, Daily Evening Edition, Page Six, Image 6
East Oregonian : E.O. (Pendleton, OR) 1888-current, August 30, 1905, DAILY EVENING EDITION, Page PAGE SIX, Image 6 « Chronicling America « Library of Congress

[240] The Brandon news., January 12, 1905, Image 3. 309 families departing SA for the Colony. The Brandon news. (Brandon, Miss.) 1892-1961, January 12, 1905, Image 3 « Chronicling America « Library of Congress

Ben's Newspaper Coverage

Loves Labours Lost and Found

Following is an abstract of a longer article that may appear in full in the Appendix.

Ben at White House

GENERAL VILJOEN AT THE WHITE HOUSE

He Predicts That the Transvaal Republic Will Be Restored Some Day.

WASHINGTON, Feb. 3.—The Boer general Viljoen took luncheon with President Roosevelt today. Viljoen asserts that British rule is distasteful to both Englishmen and the Dutch and predicts that the republic will some day be re-established. He leaves tomorrow for St. Louis, where he is working on a Boer exhibit.

General Viljoen At The White House

He Predicts That the Transvaal Republic Will Be Restored Some Day.

Washington. Feb. 3 [1904] The Boer general Viljoen took luncheon with President Roosevelt today. Viljoen asserts that British rule is distasteful to both Englishmen and the Dutch and predicts that the republic will some day be reestablished. He leaves tomorrow for St. Louis, where he is working on a Boer exhibit.

General Viljoen at The White House.[241] Feb 3rd 1904.

Novel Review

A REMARKABLE BOER NOVEL

General Viljoen Substitutes the Pen for the Sword. Portrays South African Life and Boer War.

General Ben J. Viljoen, whose novel of South African life and adventure during the Boer War, entitled Under the Vierkleur, has just been published by Messrs. Small, Maynard & Company, is one of the most interesting personalities in America today. Although only thirty six years of age, his career has been full of mark and public service both in civil and in military life. During this short period he has witnessed the opening up of a savage country by a few sturdy patriots, and the growth of a perfected civilization out of a native barbarism.

In 1093, General Viljoen made his first visit to the United States, and. with the exception of a brief return to England and South Africa, he has remained in this country ever since. He is at present in St. Louis, as President of the South African Boer War Exhibition Company, which is presenting at the Louisiana Purchase Exposition a wonderfully graphic and complete reproduction of scenes of Boer life and the battles of the recent war.

A Remarkable Boer Novel! General Viljoen Substitutes the Pen for the Sword.

Portrays South African Life and Boer War.

General Ben J. Viljoen, whose novel of South African life and adventure during the Boer War, entitled Under the Vierkleur, has just been published by Messrs. Small, Maynard & Company, is one of the most interesting personalities in America today. Although only thirty six years of age, his career has been full of mark and public service both in civil and in military life, During this short period he has witnessed the opening up of a savage country by a few sturdy patriots, and the growth of a perfected civilization out of a native barbarism.

In 1903, General Viljoen made his first visit to the United States, and. with the exception of a brief return to England and South Africa, he has remained in this country ever since, He is at present in St. Louis, as President of the South African Boer War Exhibition Company, which is presenting at the Louisiana Purchase Exposition a wonderfully graphic and complete re production of scenes of Boer life and the battles of the recent war.

Novel Review.[242] June 15th, 1904.

[241] General Viljoen at The White House. The Evening statesman. [volume] (Walla Walla, Wash.), 03 Feb. 1904. Chronicling America: Historic American Newspapers. Lib. of Congress. https://chroniclingamerica.loc.gov/lccn/sn88085421/1904-02-03/ed-1/seq-1/

[242] Novel Review. The Denison review. [volume] (Denison, Iowa) 1867-current, June 15, 1904, Image 3 « Chronicling America « Library of Congress And https://tile.loc.gov/storage-services/service/ndnp/iahi/batch_iahi_gallup_ver01/data/sn84038095/00415622934/1904061501/0266.pdf

Ben's Novel Cover, etc.

UNDER
THE VIERKLEUR

A Romance of a Lost Cause

By

BEN J. VILJOEN

Late Assistant Commandant-General of the Transvaal Burgher Forces, and Author of "My Reminiscences of the Anglo-Boer War."

SCIRE QVOD SCIENDVM

Boston
Small, Maynard & Company
1904

Copyright 1904 by
Small, Maynard & Company
(Incorporated)

Entered at Stationers' Hall

TO THE DEAREST WOMAN ON EARTH,

MY MOTHER.

First Edition (7,000 copies) April, 1904

Press of
George H. Ellis Co., Boston, U.S.A.

Imposter Article

LECTURER JOUBERT SAID TO BE AN IMPOSTOR

General Viljoen Says the Man Touring the Northwest Was Not an Officer of Boer Army.

PORTLAND, Or., May 17.—Mayor Williams this morning received a telegram from General B. J. Viljoen of the Boer camp at the St. Louis World's Fair stating that Gideon D. Joubert and W. S. O'Donnell, who are lecturing throughout the Northwest, were not known as officers of the Boer army. Joubert, who says he is a nephew of the General Joubert killed at Mafeking, is touring with O'Donnell, prospecting for suitable sites for an immigration of Boers to the Pacific Coast.

As the lecturers had been recognized by all the Irish and German societies and had been introduced by A. E. Kern, editor of the German paper, Nachrichten, the news came like a thunder clap upon the Boer sympathizers here. Joubert, over the long-distance telephone at Forest Grove tonight, declared that Viljoen, whom he knew personally, was trying to injure him. He spoke native Dutch to several Hollanders here, and was everywhere accepted as genuine. Joubert declared that Viljoen was trying to injure him that he might throw sunspicion from the manner in which the Boer camp at the World's Fair is being conducted. Joubert and O'Donnell have lectured all over the Middle West without being challenged. Before that they traveled through Mexico. Joubert exhibited a number of wounds. A civil engineer who had been in the Transvaal talked with him on the country without suspecting that he was an impostor.

Lecturer Joubert Said To Be An Impostor

General Viljoen Says the Man Touring the Northwest Was Not an Officer of Boer Army.

Portland, Or., May 17. — Mayor Williams this morning received a telegram from General B. J. Viljoen of the Boer camp at the St. Louis World's Fair stating that Gideon D. Joubert and W. S. O'Donnell, who are lecturing throughout the Northwest, were not known as officers of the Boer army. Joubert, who says he is a nephew of the General Joubert killed at Mafeking, is touring with O'Donnell, prospecting for suitable sites for an immigration of Boers to the Pacific Coast.

As the lecturers had been recognized by all the Irish and German societies and had been introduced by A. E. Kern editor of the German paper, Nachrichten, the news came like a thunder clap upon the Boer sympathizers here. Joubert. over the long distance telephone at Forest Grove to night declared that Viljoen, whom he knew personally. was trying to injure him. He spoke native Dutch to several Hollanders here, and was everywhere accepted as genuine. Joubert declared that Viljoen was trying to injure him that he might throw suspicion from the manner in which the Boer camp at the World's Fair is being conducted. Joubert and O'Donnell have lectured all over the Middle West without being challenged. Before that they traveled through Mexico. Joubert exhibited a number of wounds. A civil engineer who had been in the Transvaal talked with him on the country without suspecting that he was an impostor.

Imposter Article.[243] May 18th, 1904

Divorce Filed

DESERTED BY GENERAL

Mrs. Viljoen Asks Divorce from the Boer Leader.

New York Sun Special Service.

St. Louis, Mo., Sept. 8.—News of the divorce suit instituted in Johannesburg by his wife against General Viljoen of the Boer war, has caused great surprise in the Boer camp at the world's fair. General Viljoen reluctantly confirmed this report. Mrs. Viljoen states that General Viljoen deserted her in July, 1903, and has never since that time contributed to her support or to the maintenance of her two boys. General Viljoen denied the lack of support.

Deserted By General

Mrs. Viljoen Asks Divorce from the Boer Leader.

New York Sun Special Service.

St. Louis, Mo., Sept. 8. [1904] News of the divorce suit instituted in Johannesburg by his wife against General Viljoen of the Boer war, has caused great surprise in the Boer camp at the world's fair. General Viljoen reluctantly confirmed this report. Mrs. Viljoen states. That General Viljoen deserted her in July, 1903, and has never since that time contributed to her support or to the maintenance of her two boys. General Viljoen denied the lack of support.

Divorce Filed.[244] Sept. 8, 1904.

His wife, Lenie (née Els), did not want to leave South Africa and emigrate to the United States.

[243] Imposter Article. The San Francisco call. [volume] (San Francisco [Calif.]), 18 May 1904. Page 7 Col. 4 Chronicling America: Historic American Newspapers. Lib. of Congress. https://chroniclingamerica.loc.gov/lccn/sn85066387/1904-05-18/ed-1/seq-7/

[244] The Minneapolis journal. [volume], September 08, 1904, Page 3, Image 3 https://tile.loc.gov/storage-services/service/ndnp/mnhi/batch_mnhi_angus_ver02/data/sn83045366/00206537528/1904090801/0117.pdf

Assault & Battery

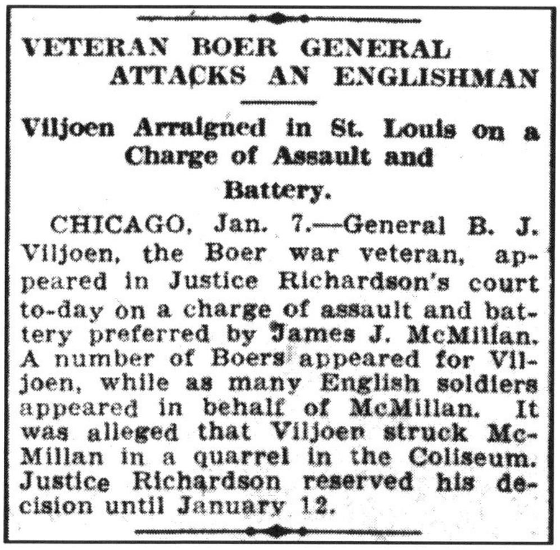
VETERAN BOER GENERAL ATTACKS AN ENGLISHMAN

Viljoen Arraigned in St. Louis on a Charge of Assault and Battery.

CHICAGO, Jan. 7.—General B. J. Viljoen, the Boer war veteran, appeared in Justice Richardson's court to-day on a charge of assault and battery preferred by James J. McMillan. A number of Boers appeared for Viljoen, while as many English soldiers appeared in behalf of McMillan. It was alleged that Viljoen struck McMillan in a quarrel in the Coliseum. Justice Richardson reserved his decision until January 12.

Veteran Boer General Attacks An Englishman

Viljoen Arraigned in St. Louis on a Charge of Assault and Battery.

Chicago, Jan. 7.—General B. J. Viljoen, the Boer war veteran, appeared in Justice Richardson's court today on a charge of assault and battery preferred by "James J. McMillan.

A number of Boers appeared for Viljoen, while as may English soldiers appeared in behalf of McMillan. It was alleged that Viljoen struck McMillan in a quarrel in the Coliseum. Justice Richardson reserved his decision until January 12.

Veteran Boer General Attacks An Englishman.[245] January 8th, 1905

Ben Viljoen in front of his tent in St. Louis in 1904.[416] The Exposition ran from April 30, 1904 to December 1, 1904 Commander Boshoff, and Generals Cronjé and Viljoen, at the St. Louis Exposition in 1904.[417] See the Boer War Libretto in Appendix for description of St. Louis Exposition Boer War Extravaganza.

[245] Veteran Boer General Attacks An Englishman The San Francisco call. [volume] (San Francisco [Calif.]) 1895-1913, January 08, 1905, Page 35, Image 35 « Chronicling America « Library of Congress

May Belfort

May Belfort [246] (née Egan; c. 1872–1929) was an Irish singer, actress and comedian who was famous across Europe and America.

Born in Ireland in about 1872, she was a comedian and singer on the London music halls and then in Paris where she performed her trademark nonsensical songs at the café des Décadents and the Petit Casino. She commissioned the famous 1895 Henri de Toulouse-Lautrec poster showing her wearing the red dress with her ever present black cat. She became a favourite of the artist who painted her more than once. She performed in Russia, South Africa and the United States during her career.

Belfort was friends with Jane Avril.[247] The British born dancer May Milton[248] was her lover, as was Boer General Ben Viljoen. In fact Belfort expected to marry Viljoen. When it turned out he was already married and unwilling to continue the relationship with her, she traveled to Chicago and horsewhipped him in the street. He later achieved a divorce, but he then married another woman.

Belfort retired from the stage due to ill health. Due to bad investments in mining, Belfort lost her money and ended working as a rug weaver and living in poverty. She died after a lingering illness in the United States in 1929.

Portrait of May Belfort.[249]

May Belfort, 1895 by Henri de Toulouse-Lautrec.[250]

[246] May Belfort. https://en.wikipedia.org/wiki/May_Belfort

[247] Jane Avril (9 June 1868–17 January 1943) was a French can-can dancer at the Moulin Rouge in Paris and a frequent subject of painter Henri de Toulouse-Lautrec's art. She was nicknamed La Mélinite, An explosive containing picric acid and guncotton. https://en.wikipedia.org/wiki/Jane_Avril

[248] May Milton was a British cabaret dancer and performer active in late 19th-century Paris, France, and a muse of the painter Henri de Toulouse-Lautrec. She is best known today through his striking lithograph poster May Milton (1895), originally intended to advertise her planned tour of the United States. https://en.wikipedia.org/wiki/May_Milton

[249] Portrait of May Belfort. The Sketch, Vol. XXIII, No. 293, September 7, 1898, page 312 https://commons.wikimedia.org/wiki/File:May_Belfort_(c._1872_%E2%80%93_1929).png

[250] Henri de Toulouse-Lautrec: May Belfort, 1895 https://commons.wikimedia.org/wiki/File:Henri_de_Toulouse-Lautrec_-_May_Belfort_-_Google_Art_Project.jpg

General Viljoen is Horsewhipped

GENERAL VILJOEN IS HORSEWHIPPED

FORMER BOER FIGHTER ROUTED BY AN ACTRESS.

CHARGED WITH JILTING HER

South African Hero Beats Hasty Retreat Before Young Woman—Latter Tells of Romantic Adventures.

Chicago.—Gen. Benjamin Johannes Viljoen, Boer patriot and hero of half a hundred South African battlefields, was horsewhipped at the Coliseum here the other night by Mies May Belfort, an English vaudeville actress, who claims in defense of her act that he jilted her after his wife secured a divorce from him in order to give him his freedom.

The sensational attack upon the famous soldier occurred at seven o'clock, when Miss Belfort appeared before the general and demanded:

"What do you mean by your scurrilous letter to me? Have you acted toward me as a man would toward a woman?"

STRUCK HIM A CUTTING BLOW UPON THE ARM.

Struck Him a Cutting Blow Upon the Arm.[251] Feb. 18, 1905.

General Viljoen Is Horsewhipped

Former Boer Fighter Routed by An Actress.

Charged with Jilting Her

South African Hero Beats Hasty Retreat Before Young Woman—Latter Tells of Romantic Adventures.

Chicago.—Gen. Benjamin Johannes Viljoen, Boer patriot and hero of half a hundred South African battlefields, was horsewhipped at the Coliseum here the ether night by Miss May Belfort, an English vaudeville actress, who claims in defense of her act that he jilted her after his wife secured a divorce from him in order to give him his freedom.

The sensational attack upon the famous soldier occurred at seven o'clock, when Miss Belfort appeared before the general and demanded:

"What do you mean by your scurrilous letter to me? Have you acted toward me as a man would toward a woman?"

"I know of no wrong deeds on my part—" began Viljoen. when the enraged actress snatched a heavy rawhide whip from the folds of her skirt and struck him a cutting blow upon the arm. Gen. Viljoen retreated before a shower of blows, the woman following him. With a last vicious cut, she desisted from her attack and then left the place.

In her room at the Sherman house Miss Belfort sat on the edge of her bed and told in picturesque style of her romantic adventures with Viljoen.

"I'm clean flabbergasted over this affair," she said. "Do you know that I met the Boer in 1903, while he was on a lecture tour in Lunnon.[252] We were very fond of each other, but it was quite by accident, you know, that we left England for Africa on the same boat. Then we had our pictures taken together in Johannesburg."

[251] The commonwealth., February 18, 1905, Image [page] 6 (Greenwood, Miss.) 1896–1923.
The commonwealth. (Greenwood, Miss.) 1896-1923, February 18, 1905, Image 6 « Chronicling America « Library of Congress
And: https://tile.loc.gov/storage-services/service/ndnp/msar/batch_msar_cotton_ver01/data/sn89065008/0019991652A/1905021801/0026.pdf
And: Richmond planet. [volume] (Richmond, Va.), 25 Feb. 1905. Page 2 Chronicling America: Historic American Newspapers. Lib. of Congress. https://chroniclingamerica.loc.gov/lccn/sn84025841/1905-02-25/ed-1/seq-2/ And https://tile.loc.gov/storage-services/service/ndnp/vi/batch_vi_creed_ver01/data/sn84025841/00175047189/1905022501/0071.pdf

[252] Sic? A small town in England or a mis-typesetting of London. See: Lunnon in Swansea.

Miss Belfort then obligingly loaned the picture, a heart-shaped card, showing herself and the general together.

"Do you know," she continued, "he wanted to know if I would marry him if his wife got a divorce through a barrister. I said 'Yes,' but my friends wanted to know what I wanted to marry a Boer general for. It was because I loved him.

"We returned to England together. About a twelvemonth ago the Boer went to St. Louis. I have real 'swads' of letters from him. After Mrs. Viljoen's barrister got her a divorce last fall the tone of his letters changed. He wrote to me that 'he was through with married life,' and that he could not afford to get married.

"He often took the trouble to cable to the London Daily Express and deny that he was engaged to marry me.

"I am sure I hurt him. If his coat had only been thinner I would have thrashed him even more soundly than I did. I did my best, anyway."

Miss Belfort sings French and English songs like vaudeville.

Gen. Viljoen refused to discuss Miss Belfort, saying that the notoriety was distasteful to him. He declared that Miss Belfort was seeking notoriety for herself.

Gen. Viljoen is certain that a man accompanied the woman and made an appointment by telephone for her. Miss Belfort denies emphatically that any other person was connected with the affair.

Boer General Horsewhipped

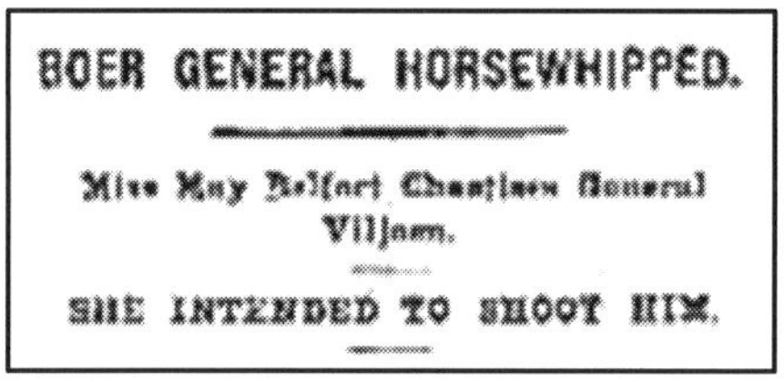
BOER GENERAL HORSEWHIPPED.

Miss May Belfort Chastises General Viljoen.

SHE INTENDED TO SHOOT HIM.

General Viljoen Horsewhipped.[253] May 27th, 1905.[254]

Boer General Horsewhipped.

Miss May Belfort. Chastises General Viljoen.

She Intended to Shoot Him.

Miss May Belfort, the English comedienne, who declares that General Benjamin J. Viljoen, hero of many Boer wars, made love to her during a period covering several years, has made public for the first time many of the love missives she says she received from him. She declares that General Viljoen, although he already had a wife in South Africa, proposed marriage to her. She accepted him, she says, after he had promised to obtain a divorce. Subsequently, she declares, he scorned her.

Miss. Belfort's reasons for giving out the letters are best told by herself.

"I can no longer withstand," said Miss Bel fort, "the oft-repeated denials of General Viljoen that he had paid court to me. He has said he merely knew me as an acquaintance—nothing more. I feel that in justice to myself I should take this step to disabuse the public mind of any erroneous impression the attentions of General Viljoen may have caused."

Death Too Good for Him.

"It is true I horsewhipped General Viljoen publicly in Chicago, after he had made it plain that he had no serious intentions toward me. Indeed, upon that occasion I carried with me a loaded revolver, with which I intended to kill him. But I did not use the pistol because, at the last moment, I concluded that death was too good for the man. So I horsewhipped him."

"I am a woman of sudden impulses," she said, as she handled the gleaming weapon, "and there is no telling what a woman scorned will do."

[253] The World's News, Sydney, NSW: 1901-1955 Sat. May 27th, 1905. Page 25 "Boer General Horsewhipped. - Miss May Belfort Chastises General Viljoen. She Intended to Shoot Him." The World's News. Sydney. 27 May 1905. p. 18. Retrieved 9 October 2023–via Trove. Trove Article Link: https://trove.nla.gov.au/newspaper/article/136667192

[254] All letters prior to October 27th, 1904. May Belfort arrives in New York in January, 1905 and confronts Ben on February 2nd in Chicago.

"I am only a woman. I have no one to defend me, and I must, therefore, defend myself. I have no friends, nor relatives, in this country."

Miss Belfort showed the "American" reporter several large bundles of letters, all of which, she said, were written to her by General Viljoen. Extracts from some of them follow:—

Always My Sweetheart.
Washington, D.C.

My Dearest May,

I am here from New York, after spending some days there getting the stock of our company taken up. I have been invited by the President to dine with him at the White House to-morrow.[255]

Are you going to Australia, sweetheart? Are you going into the world alone again? Ah, May, I will always call you sweetheart, and always hope for the day that will bring us together. As things stand, I cannot say, "Come, love, be patient," so or so long, for it would be wrong for me to hold out any false hopes. I am unable, according to best advice, to proceed with my divorce suit until this company is fixed up and the newspaper story has blown over.

I never loved any woman before as I love you, and I cannot bear the thought of your saying that you are unhappy or doubting me. I am sad, my sweetheart. Heaven alone knows how I suffer. If I live through all this I am coming to you yet. Farewell for the present, sweetheart. God bless and keep you. Ever yours, Ben.

I Can't Provide for You.

My Cherie May,

To-day I received your three-line note. Well, I can't blame you, my dear. I know how you have been neglected. But the reason, I trust you understand from my former letters. Well, my dear, you say I am forgetting you. No, my love, such is not the case, but, alas, my hopes of having you as my wife seem to dwindle away. I fear to hold out any hopes, and cannot deceive you. My love and esteem for you is too high to take you to a desolate home, where I will not be able to provide for you. Do not forget me altogether, love. The girl on horseback in my book is a reproduction of your face. Your devoted Ben.

Bettie had been riding ahead of the wagon (see page 192). Before him stood Bettie, … her arms outstretched. (see page 385)

[255] February 2nd 1904.

Au Revoir for the Present.
On board the Umbria, Irish Channel.

Ma Cherie,

The passengers show little Inclination to be sociable. I am the object of considerable scrutiny from those who, I suppose, never saw a live Boer before. Well, my own darling May, my thoughts are always of you. Your kind face is before me, and I fancy I hear you say, "My Ben." Oh. May, my sweetheart, how I love you Heaven alone can tell. I am sad, and my heart is heavy. The tears rush out of my eyes when I think of the happy days I spent in your dear home.

Now, my heart's love, I must close for the night, and again I bid you au revoir for the present. Good-night, sweetheart. God bless and keep His guiding hand over you. Ever and always your own Ben

Says He Cannot Marry.
St. Louis.

Ma Cherie May,

I received your cable last night, but what can I answer? I am sorry, yes, doubly so, that you should have been taken in by the newspaper man. I fear I am ruined. The divorce lawyers refuse to take my case unless I deny in toto that I contemplate marriage.

My dear, I am unhappy, and hardly know what to say. Now, May, dearest, I will write you again. At present I am in a miserable mood, so forgive. Au revoir, love, Ben.

"Day of Marriage Is Passed."
St. Louis, Oct. 27, 1904.

My Dear May,

Your reproachful letter received. I wrote to you several times to London. As to our relations, my dear May, the position I am in is of such a nature that I cannot look to any prospect of making you a home. I can't afford it. And as my obligations are simply outrageous, it would be folly for any sane person to make promises and drag a woman into ties which would never bring happiness. I will, of course, be glad to see you again. But, my dear girl, the day of marriage with me is past forever. Yours very sincerely. B. J. Viljoen.

American Incident

Ben Viljoen
The American Incident
Letter From May Belfort

The following appears in the 'Rand Daily Mail':—

Sir,—It has just been brought to my notice that General Ben Viljoen had sent a contradiction to the English papers re the fact of his having been horsewhipped and that he intends suing the American papers who published the (as he calls it) unfounded statement.

Now, I cannot understand how he dares to send such a denial, as in the face of it all his interviews, after the occurrence, are in themselves sufficient proof that he was horsewhipped, and as I did the deed I can verify the statement made by the several American papers. However, this man will deny anything, as he denied he asked me to marry him as soon as he got a decree of divorce from his wife in Africa, and the divorce he said he would get as soon as he arrived in America.

He kept up an affectionate correspondence with me until I arrived in this country, late October (22nd 1904), and as there was no letter of welcome awaiting me when I got to New York I wrote to him how surprised I was, etc.

That letter he sent me the one published in the 'Chicago Examiner' of January 19, 1905, and as there was no reason by any action of mine to call for such a letter, so lacking, (in my humble opinion) in decency and good feeling, I sent him several letters asking for a more manly explanation. He treated my letters with undue contempt.

I have been shamefully treated by this man, and as I am earning my living on the stage people are too ready, perhaps- to condemn me and to think, for the same reason, that …..(No further text in this paragraph.)

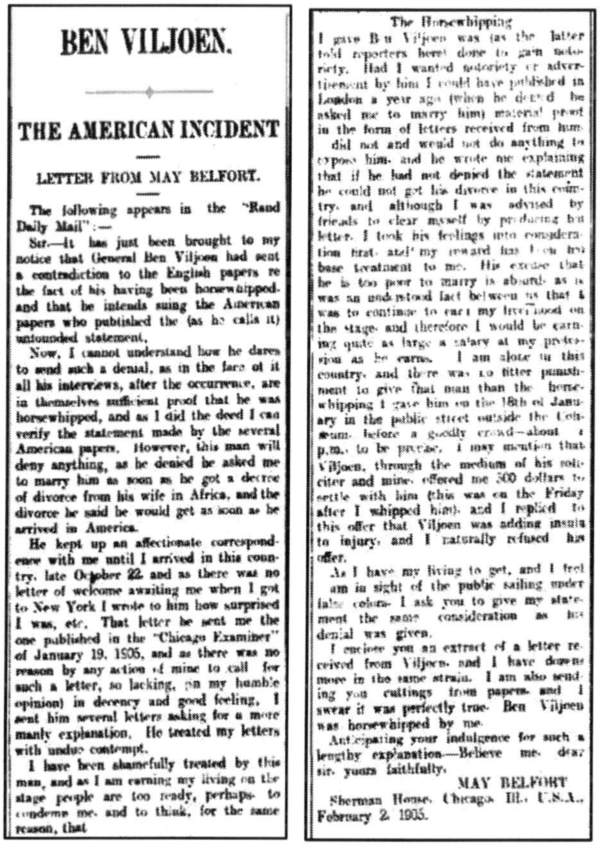

BEN VILJOEN.

THE AMERICAN INCIDENT

LETTER FROM MAY BELFORT.

The following appears in the "Rand Daily Mail":—

Sir,—It has just been brought to my notice that General Ben Viljoen had sent a contradiction to the English papers re the fact of his having been horsewhipped, and that he intends suing the American papers who published the (as he calls it) unfounded statement.

Now, I cannot understand how he dares to send such a denial, as in the face of it all his interviews, after the occurrence, are in themselves sufficient proof that he was horsewhipped, and as I did the deed I can verify the statement made by the several American papers. However, this man will deny anything, as he denied he asked me to marry him as soon as he got a decree of divorce from his wife in Africa, and the divorce he said he would get as soon as he arrived in America.

He kept up an affectionate correspondence with me until I arrived in this country, late October 22, and as there was no letter of welcome awaiting me when I got to New York I wrote to him how surprised I was, etc. That letter he sent me the one published in the "Chicago Examiner" of January 19, 1905, and as there was no reason by any action of mine to call for such a letter, so lacking, in my humble opinion) in decency and good feeling, I sent him several letters asking for a more manly explanation. He treated my letters with undue contempt.

I have been shamefully treated by this man, and as I am earning my living on the stage people are too ready, perhaps, to condemn me, and to think, for the same reason, that

The Horsewhipping

I gave Ben Viljoen was (as the latter told reporters here) done to gain notoriety. Had I wanted notoriety or advertisement by him I could have published in London a year ago (when he denied he asked me to marry him) material proof in the form of letters received from him.

did not and would not do anything to expose him, and he wrote me explaining that if he had not denied the statement he could not get his divorce in this country, and although I was advised by friends to clear myself by producing his letter, I took his feelings into consideration first, and my reward has been his base treatment to me. His excuse that he is too poor to marry is absurd, as it was an understood fact between us that I was to continue to earn my livelihood on the stage, and therefore I would be earning quite as large a salary at my profession as he earns. I am alone in this country, and there was no fitter punishment to give that man than the horsewhipping I gave him on the 18th of January in the public street outside the Coliseum, before a goodly crowd—about 4 p.m., to be precise. I may mention that Viljoen, through the medium of his solicitor and mine, offered me 300 dollars to settle with him (this was on the Friday after I whipped him), and I replied to this offer that Viljoen was adding insult to injury, and I naturally refused his offer.

As I have my living to get, and I feel am in sight of the public sailing under false colors, I ask you to give my statement the same consideration as his denial was given.

I enclose you an extract of a letter received from Viljoen, and I have dozens more in the same strain. I am also sending you cuttings from papers, and I swear it was perfectly true, Ben Viljoen was horsewhipped by me.

Anticipating your indulgence for such a lengthy explanation—Believe me, dear sir, yours faithfully,

MAY BELFORT

Sherman House, Chicago, Ill., U.S.A., February 2, 1905.

American Incident.[256] Penned February 2nd 1905. Published April 28, 1905.

[256] "Ben Viljoen – The American Incident Letter From May Belfort. The Daily News. Perth. 28 April 1905. p. 7. Retrieved 9 October 2023 – via Trove. https://trove.nla.gov.au/newspaper/article/80017966

The Horsewhipping

I gave Ben Viljoen was (as the latter told reporters here) done to gain notoriety. Had I wanted notoriety or advertisement by him I could have published in London a year ago (when he denied he asked me to marry him) material proof in the form of letters received from him.

I did not and would not do anything to expose him, and he wrote me explaining that if he had not denied the statement he could not got his divorce in this country, and although I was advised by friends to clear myself by producing the letter. I took his feelings into consideration first, and my reward has been his base treatment to me.

His excuse that he is too poor to marry is absurd, as it was an understood fact between us that I was to continue to carry my livelihood on the stage and therefore I would be earning quite as large a salary at my profession as he earns. I am alone in this country, and there was no fitter punishment to give that man than the horsewhipping I gave him on the 18th of January in the public street outside the Coliseum, before a goodly crowd—about 1 p.m., to be precise. I may mention that Viljoen, through the medium of his solicitor and mine, offered me 300 dollars to' settle with him (this was on the Friday after I whipped him), and I replied to this offer that Viljoen was adding insult to injury, and I naturally refused his offer.

As I have my living to get, and I feel I'm in sight of the public sailing under false colors, I ask you to give my statement the same consideration as his denial was given.

I enclose you an extract of a letter received from Viljoen, and I have dozens more in the same strain. I am also sending you cuttings from papers, and I swear it was perfectly true. Ben Viljoen was horsewhipped by me

Anticipating your indulgence for such a lengthy explanation—Believe me, dear sir, yours faithfully.

May Belfort

Sherman House, Chicago, Ill., U.S.A., February 2, 1905.

Married and Sued

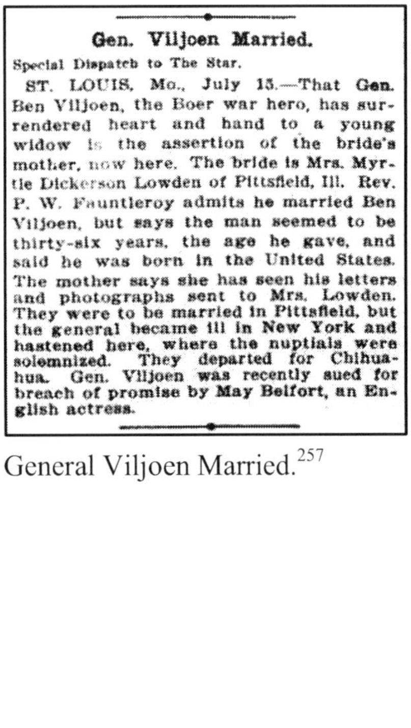

Gen. Viljoen Married.

Special Dispatch to The Star.

ST. LOUIS, Mo., July 15.—That Gen. Ben Viljoen, the Boer war hero, has surrendered heart and hand to a young widow is the assertion of the bride's mother, now here. The bride is Mrs. Myrtle Dickerson Lowden of Pittsfield, Ill. Rev. P. W. Fauntleroy admits he married Ben Viljoen, but says the man seemed to be thirty-six years, the age he gave, and said he was born in the United States. The mother says she has seen his letters and photographs sent to Mrs. Lowden. They were to be married in Pittsfield, but the general became ill in New York and hastened here, where the nuptials were solemnized. They departed for Chihuahua. Gen. Viljoen was recently sued for breach of promise by May Belfort, an English actress.

General Viljoen Married.[257]

Gen. Viljoen Married.

Special Dispatch to The Star. St. Louis, Mo.. July 13. [1905] That Gen. Ben Viljoen, the Boer war hero, has surrendered heart and hand to a young widow the assertion of the bride's mother, now here. The bride Is Mrs. Myrtle Dicker son Lowden of Pittsfield, Illinois.

Rev. P. W. Fauntleroy admits he married Ben Viljoen, but says the man seemed to be thirty-six years, the age he gave, and said he was born In the United States. The mother says she has seen his letters and photographs sent to Mrs. Lowden. They were to be married In Pittsfield. but the general became ill in New York and hastened here, where the nuptials were solemnized. They departed for Chihuahua. Gen. Viljoen was recently sued for breach of promise by May Belfort, an English actress.

[257] Washington, Evening star. [volume], July 17, 1905, Page 2, Image 2.
Evening star. [volume] (Washington, D.C.) 1854-1972, July 17, 1905, Page 2, Image 2 « Chronicling America « Library of Congress

Boer Colony Failure

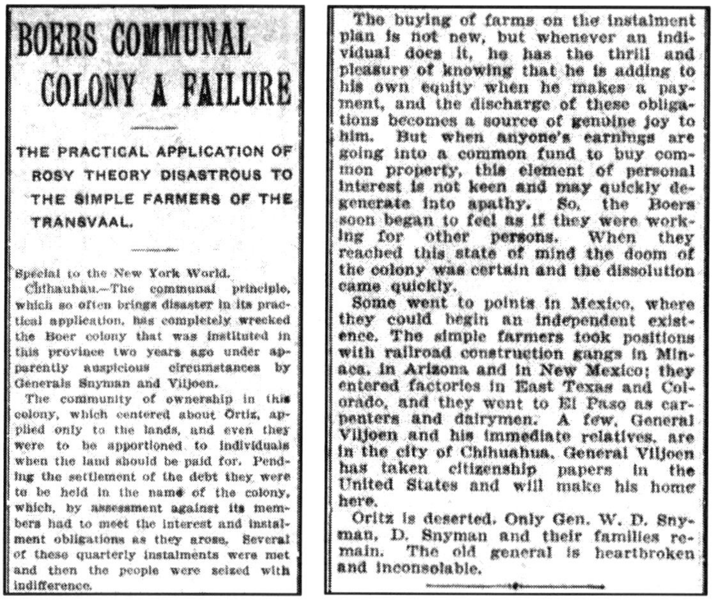

BOERS COMMUNAL COLONY A FAILURE

THE PRACTICAL APPLICATION OF ROSY THEORY DISASTROUS TO THE SIMPLE FARMERS OF THE TRANSVAAL.

Special to the New York World.

Chihuahau.—The communal principle, which so often brings disaster in its practical application, has completely wrecked the Boer colony that was instituted in this province two years ago under apparently auspicious circumstances by Generals Snyman and Viljoen.

The community of ownership in this colony, which centered about Ortiz, applied only to the lands, and even they were to be apportioned to individuals when the land should be paid for. Pending the settlement of the debt they were to be held in the name of the colony, which, by assessment against its members had to meet the interest and instalment obligations as they arose. Several of these quarterly instalments were met and then the people were seized with indifference.

The buying of farms on the instalment plan is not new, but whenever an individual does it, he has the thrill and pleasure of knowing that he is adding to his own equity when he makes a payment, and the discharge of these obligations becomes a source of genuine joy to him. But when anyone's earnings are going into a common fund to buy common property, this element of personal interest is not keen and may quickly degenerate into apathy. So, the Boers soon began to feel as if they were working for other persons. When they reached this state of mind the doom of the colony was certain and the dissolution came quickly.

Some went to points in Mexico, where they could begin an independent existence. The simple farmers took positions with railroad construction gangs in Minaca, in Arizona and in New Mexico; they entered factories in East Texas and Colorado, and they went to El Paso as carpenters and dairymen. A few, General Viljoen and his immediate relatives, are in the city of Chihuahua. General Viljoen has taken citizenship papers in the United States and will make his home here.

Oritz is deserted. Only Gen. W. D. Snyman, D. Snyman and their families remain. The old general is heartbroken and inconsolable.

Boer Colony Failure.[258]

Boers Communal Colony A Failure

The Practical Application of Rosy Theory Disastrous To The Simple Farmers of The Transvaal.

Special to the New York World. Chihuahua.—The communal principle, which so often brings disaster in its practical application, has completely wrecked the Boer colony that was instituted in this province two years ago under apparently auspicious circumstances by Generals Snyman and Viljoen.

The community of ownership in this colony, which centered about Ortiz, applied only to the lands, and even they were to be apportioned to individuals when the land should be paid for. Pending the settlement of the debt they were to be held In the name of the colony, which, by assessment against its members had to meet the interest and instalment obligations as they arose. Several of these quarterly instalments were met and then the people were seized with indifference.

The buying of farms on the instalment plan is not new, but whenever an individual does it, he has the thrill and pleasure of knowing that he is adding to his own equity when he makes a payment, and the discharge of these obligations becomes a source of genuine joy to him. But when anyone's earnings are going into a common fund to buy common property, this element of personal interest is not keen and may quickly degenerate into apathy. So, the Boers soon began to feel as if they were working for other persons. When they reached this state of mind the doom of the colony was certain and the dissolution came quickly.

Some went to points in Mexico, where they could begin an independent existence. The simple farmers took positions with railroad construction gangs in Minaca, In Arizona and in New Mexico; they entered factories in East Texas and Colorado, and they went to El Paso as carpenters and dairymen. A few. General Viljoen and his immediate relatives, are in the city of Chihuahua. General Viljoen has taken citizenship papers in the United States and will make his home here.

Oritz is deserted. Only Gen. W. D. Snyman, D. Snyman and their families remain. The old general is heartbroken and inconsolable.

Boer Colony Failure. Nov, 20, 1905. See Also: Frontiers in the Gilded Age. [259] And. [260]

[258] Boer Colony Failure. The Age-Herald, Monday November 20, 1905. Page 7 Col 2 https://tile.loc.gov/storage-services/service/ndnp/au/batch_au_davis_ver01/data/sn85038485/00340583590/1905112001/0718.pdf

[259] Offenburger, A. (2019). Frontiers in the Gilded Age: Adventure, Capitalism, and Dispossession from Southern Africa to the U.S.-Mexican Borderlands, 1880-1917. The Lamar Series in Western History. Yale University Press. p. 241. ISBN 978-0-300-24525-7. Retrieved 14 November 2019. https://books.google.com/books?id=tiCbDwAAQBAJ&pg=PA241#v=onepage&q&f=false

Civic Life in New Mexico[261]

After Ben returned from Mexico to the US and was instrumental in organising Boer colonies in Doña Ana County at Berino, Chamberino and La Mesa in New Mexico Territory. He became involved in civic affairs throughout the Mesilla Valley, and in 1909 he was granted US citizenship. Viljoen became familiar with both President Theodore Roosevelt and New Mexico Governor George Curry. Viljoen was commissioned as a Major in the territorial National Guard's First Regiment of Infantry. In 1911, he travelled to Washington, D.C., with Curry and Albert Bacon Fall as part of a delegation promoting statehood for New Mexico. In April and May 1911, he fought with the Mexican Revolutionary Francisco Madero at the Battle of Ciudad Juarez.

Viljoen was also influential in agriculture. He introduced new crops and farming practices to the Mesilla Valley. Viljoen was interested in the creation of the Elephant Butte Dam and developed irrigation systems for the valley.

For a short while, he was US Consul in Germany, and also acted as military advisor to Francisco Madero until Madero's assassination in 1913. He died in 1917, at his farm in La Mesa and is buried at the Masonic Cemetery in La Mesa, New Mexico.

Obituary

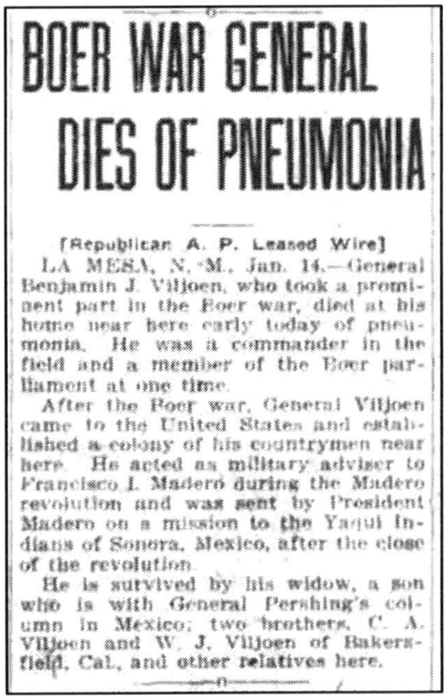
BOER WAR GENERAL DIES OF PNEUMONIA

[Republican A. P. Leased Wire]

LA MESA, N. M., Jan. 14.—General Benjamin J. Viljoen, who took a prominent part in the Boer war, died at his home near here early today of pneumonia. He was a commander in the field and a member of the Boer parliament at one time.

After the Boer war, General Viljoen came to the United States and established a colony of his countrymen near here. He acted as military adviser to Francisco I. Madero during the Madero revolution and was sent by President Madero on a mission to the Yaqui Indians of Sonora, Mexico, after the close of the revolution.

He is survived by his widow, a son who is with General Pershing's column in Mexico; two brothers, C. A. Viljoen and W. J. Viljoen of Bakersfield, Cal., and other relatives here.

Obituary.[262]

Boer War General Dies of Pneumonia.

Republican A.P. Leased Wire.

La Mesa, N.M. January 14. — 4. General I Benjamin J. Viljoen, who took a prominent part In the Boer war, died at his home near here early today of pneumonia. He was a commander in the field and a member of the Boer Parliament at one time. After the war.

General Viljoen came to the United States and established a colony of his countrymen near here. He acted as military adviser to Francisco I. Madero during the Madero revolution and was sent by President Madero on a mission to the Yaqui Indians of Sonora, Mexico, after the close of the revolution.

He is survived by his widow, a son who is with General Pershing's column in Mexico; two brothers, C. A. Viljoen and W. J. Viljoen of Bakersfield, Cal., and other relatives here.

[260] Offenburger, Andrew (2014). "Cultural Imperialism and the Romanticized Frontier: From South Africa and Great Britain to New Mexico's Mesilla Valley." Vol. 59, No. 4, South Africa and the United States in Transnational American Studies (2014), pp. 535-552 (18 pages) JSTOR 44071886. https://www.jstor.org/stable/44071886

[261] Quote from Ben's Wikipedia page Direct Link to Page: https://en.wikipedia.org/wiki/Ben_Viljoen
Permanent Link: https://en.wikipedia.org/w/index.php?title=Ben_Viljoen&oldid=1242327160 (July 10th 2025)

[262] Arizona republican. [volume] (Phoenix, Ariz.), January 15, 1917, Page TWO, Image 2 Arizona republican. [volume] (Phoenix, Ariz.), January 15, 1917, Page PAGE TWO, Image 2

Viljoen Appendix—Correspondence

23rd July, 1901.

23rd July, 1901.

To His Excellency Lord Kitchener,

Commander-in-Chief of His Majesty's Troops in South Africa, Pretoria.

Your Excellency,

I am compelled to emphatically protest against the methods of your officers. Last April your Excellency's brother, General W. Kitchener, took our ambulance veldt-hospital, near Roos Senekal, and only after much trouble were a number of the vehicles restored to us. On that occasion, General W. Kitchener refused to return to me the slaughter oxen belonging to the field-hospital, saying that we could steal such oxen from the kaffirs. In consequence of those acts, my wounded were rendered without food, and robbed of means of transportation.

Now, again, a column of your troops, which was proceeding on the 9th or 10th inst. from Machadodorp across Witpoort, attacked a Red Cross hospital occupied by sick women and children, notwithstanding the patients were in charge of a certificated nurse, named Mrs. W. Botha. One of your officers, misled by a former burgher, who is now treacherously fighting against his own people, declared that the Red Cross was not genuine, and burned all the buildings and food found therein, placed the patients on open trucks, and removed them.

The first night of their deportation the sick patients and nurses slept in a camp at Steelpoortdrift, under the trolley waggons and in the bitter cold, and although the women and children were lamenting and weeping the entire night, their complaints were not listened to. I have declarations testifying to the most inhuman, heartless, and cruel maltreatment committed towards helpless women and children on this occasion.

Probably, your Excellency knows nothing about these incidents, and as regards the bona-fides of our ambulances, I wish to point out to you that British officers depend largely on the assertions of kaffirs, and especially on the allegations of traitors, and on the slightest provocation ignore the rights of the Red Cross.

The column referred to also burned, and plundered and destroyed many houses at Steenkampsberg, Witpoort and many other places, without there being one single shot fired in the neighbourhood by our burghers. And all this was allowed to occur in spite of your Excellency's promises at the meeting of the Commandant-General Botha at Middelburg.

Latterly, it has often occurred that British ambulances have fallen into my hands. At Bethel, three doctors and an ambulance attached to General Plumer's force fell into my hands. Near Vaalkop, Major Morris's ambulance, and near Belfast an ambulance, attached to your brother's forces, were in my power, but I always regarded and treated ambulances flying the Red Cross as neutral and humane institutions, and I even liberated the soldiers employed to attend your wounded.

And not a single one of these doctors or attendants was provided with a certificate, and I have invariably accepted their word that they were legally attached to the Red Cross. But what is the attitude of the British officers towards us?

I trust your Excellency will give me a satisfactory reply to these complaints, and issue orders to remedy them.

I am,

Your Excellency's most obedient servant,

B. J. Viljoen.

Assistant Commandant-General.

20th August, 1901

Some Correspondence between the British and Boer Military Officials.

Lyndenburg,

20th August, 1901.

Assistant Commandant-General

B. J. Viljoen.

Sir,

I have the honour to enclose herewith a copy of a communication received from Lord Kitchener. Begins:—With reference to your letter of the 10th August on the subject of employment of natives, I have the honour to inform you, as I have already informed Commandant-General Botha, that natives are employed by me as scouts and as police in native districts, especially in the low country, where white men, if not by long residence inured to the climate, suffer much from fever.

I would point out to you that in numerous cases armed natives have been employed by the burgher forces, particularly in the commando of General Beyers, and that armed natives have frequently been found in the commandos fighting against us. I do not wish to bring the native population of the country into this quarrel between British and Boers.

I have invariably told the natives that, although I could not forbid their defending themselves if attacked by burghers, they were on no account to attack. I am convinced that but for the strict orders which I have issued on this subject, the hatred engendered by the wholesale slaughter of unarmed natives by the burghers during this War would have led to a native rising, with deplorable results to the Boer race.

It must also be within your knowledge that most of the rifles in possession of M'pisana's natives were sold to them by men of your own commando when moving from Hector's Spruit to Pietersburg last year.

In answer to your questions regarding the British prisoners now in your hands, the persons named are enlisted soldiers in His Majesty's Army, and have been acting under my orders. They should be treated as prisoners-of-war.—Ends.

I have the honour to be, Sir,

Your obedient servant,

A. Curran,

Lieutenant-Colonel

Commanding Lydenburg.

8th September, 1901

District Lydenburg,

To His Excellency, Lord Kitchener, Commanding the British Troops in South Africa, Pretoria.

Your Excellency,

I have the honour to acknowledge receipt of your Excellency's letter to General Blood, dated Pretoria, 31st of August, from which I understand that your Excellency essays to justify the use of the white flag for the dissemination of proclamations through our lines, in connection with which your Excellency offers arguments which I do not hesitate to say are utterly untenable.

Firstly, it is asserted by your Excellency that the sending of these documents addressed to individuals is justified under the white flag; secondly, that your Excellency considers it your Excellency's duty to render us conversant with the contents of your Excellency's proclamations in order that we shall be informed what our fate shall be after the 15th September next, &c., &c., &c.[263]

With regard to the first argument introduced, I regret that I must dispute your Excellency's contention that this is legal, and I am assured that an impartial court would declare it as illegal. I enclose herewith the copy of a letter from General W. Kitchener, dated 1st September last, in answer to a complaint of my locum tenens, "Fighting"-General Muller, with respect to the taking and removing by the said General W. Kitchener's troops of our ambulance and hospital attendants, from which letter it will appear that General Kitchener considers the sending of a white flag despatch concerning important and serious irregularities as "trivial communications." How am I to understand British officers?

Your Excellency thinks that it is permissible to employ the white flag to send pernicious and misleading proclamations within our lines, whereas General W. Kitchener warns us not to employ the white flag when we are compelled to complain concerning the British Army where the latter removes and robs us of our ambulances, as occurred in connection with the ambulance of Dr. Neethling, which was removed to Middelburg, and after being relieved of food, medical instruments, a number of vehicles, eight mules, and 10 oxen, was sent back.

With regard to the second matter, your Excellency, I should say, appears to display as keen an interest in our ultimate fate as Messrs. Dillon and Labouchere, and, if I possessed any prophetic faculty, I should probably be better able to appreciate your Excellency's interest in ourselves.

In the letter referred to above, your Excellency mentions a letter sent to his Honour, Commandant-General Botha, in which your Excellency asserts that certain murders committed by us filled the British public with horror, and that these murders provoked Mr. Chamberlain's remark "that the acts of the Boers justified the description of marauding ruffianisms." I cannot believe such acts have been committed by us or ours with the knowledge of our officers, or that any such acts will be committed. It is, of course, impossible for me to discuss this matter further, as I am ignorant of the circumstances.

With reference to your Excellency's contention that the destruction of our enemy's railway lines is unjustifiable, I can only say that such action is not only regarded as legal by all military authorities, but that in a handbook published by Sir Garnet Wolseley circumstantial instructions are given in this connection for interrupting hostile supplies. As your Excellency rightly remarks, we, as soldiers, must take the rough with the smooth, and not complain petulantly when in certain cases a less gentle treatment is dealt out. Military operations, such as the blowing up of railway lines, are as unpleasant to us as I hope the destruction of our houses, the burning of our food, and the deportation of our families may be to your Excellency.

I have the honour to be,

Your Excellency's obedient servant,

B. J. Viljoen,

Assistant Commandant-General, Transvaal Burgher Forces.

[263] See infra for text of the August proclamation.

Kitchener's Banishment Proclamation

Lord Kitchener's Proclamation.
Surrender or Banishment.
Government Determined To End The War.[264] [265]

A Parliamentary paper, issued to-day, contains the proclamation issued by Lord Kitchener on August 6 with the concurrence of the British Government. It orders surrender before September 15, or permanent banishment for all commandants, field cornets, leaders of armed bands, and all members the Boer Governments. The full text of the proclamation is follows:

Whereas the late Orange Free State and the late South African Republic have been annexed to his Majesty's dominions, and

Whereas his Majesty's forces are and have been tor some considerable time in complete possession of the seats of Government of both of the aforesaid territories with their public offices and the whole machinery of administration as well as of all the principal towns and the whole of the railway lines, and

Whereas the great majority of the burghers of the two late Republics to the number of 35,000 exclusive of those who have fallen in the war are now either prisoners or have submitted to his Majesty's Government and are living peaceably in towns or camps under the control of his Majesty s forces, and

Whereas the burghers of the late Republics still in arms against his Majesty are not only few in numbers but have alto lost almost all their guns and munitions of war and devoid of regular military organization and are therefore unable to carry on regular warfare or to offer any organized resistance to his Majesty's forces in any part of the country, and

Whereas those burghers who are still in arms though unable to carry or regular warfare continue to make isolated attacks upon small posts and detachments of his Majesty's forces to plunder or destroy property and to damage the railway and telegraph lines both in the Orange River Colony and Transvaal and in other portions of his Majesty's South African dominions, and

Whereas the country is thus kept in a state of disturbance checking the resumption of agricultural and industrial pursuits, and

Whereas his Majesty's Government is determined to put an end to a state of things which is aimlessly prolonging bloodshed and destruction and inflicting ruin upon the great majority of the inhabitants who are anxious to live in peace and to earn livelihood (or themselves and their families), and

Whereas it is just to proceed against those still resisting and especially against those persons who being in position of authority are responsible for the continuance of the present state of lawlessness and are instigating their fellow burghers to continue their hopeless resistance to his Majesty's Government:

Now therefore I, Lord Kitchener, &c., under instructions from his Majesty's Government proclaim and make known as follows

"All Commandants, Field-Cornets, and leaders of armed bands, being burghers of the late Republics, still engaged in resisting His Majesty's forces, whether the Orange River Colony and the Transvaal or in any other portion of His Majesty's South African Dominions, and all members of the Governments of the late Orange Free State and the late South African Republic, shall, unless they surrender before September 15 next, be permanently banished from South Africa; the cost of the maintenance of the families of all burghers in the field who shall not have surrendered September 15 shall be recoverable from such burghers and shall be charge upon their property moveable and immovable in the two Colonies."

[264] Kitchener's Proclamation: Pall Mall Gazette August 9, 1901 p. 7 col. 1 BNA File BL_0000098_19010809_042_0007.pdf

[265] Fife Free Press - Saturday 10 August 1901 https://www.britishnewspaperarchive.co.uk/viewer/bl/0001062/19010810/032/0002

Resume Correspondence

21st September, 1901.

District Lydenburg,

21st September, 1901.

To His Honour General Sir Bindon Blood, Middelburg.

Your Honour,

I am compelled to protest against the methods of one of your columns, which during the past week has been operating round about Roos Senekal, and which has burnt and destroyed the food of a number of families which it did not deport. This is surely a most inhuman action, inasmuch that the families mentioned are now in a destitute position. The families in question are those of Mr. Hans Grobler of Klip River, and others at Tondeldoos. I should also like to know why Dr. Manning and his ambulance and wounded have been removed from Tondeldoos, notwithstanding former assurances that the Red Cross should be regarded as neutral and left unmolested.

I have the honour to be,
Your Honour's obedient servant,
B. J. Viljoen.
Assistant Commandant-General.

26th October, 1901.

Head Quarters, Pretoria,
26th October, 1901.
To General Ben Viljoen.
Sir,
I have the honour to acknowledge receipt of your letter of the 8th of October, in which you complain of attacks upon your burghers, and the families and the property of your burghers, by kaffirs. You specify two particular incidents in your letter:—

(a) The incident at Wit River on the 22nd September, 1901.

(b) The burning and plundering of homesteads at Ohrigstad.

I have investigated both cases and find that the facts are as follows:—

(a) In the first case a small body of mounted troops in charge of an officer attempted to capture a number of Boer waggons near Wit River on the 22nd September. A fight took place, and during the battle a band of kaffirs, of whose proximity His Majesty's troops had no knowledge, approached from another direction and commenced shooting on the burghers. This being observed, His Majesty's troops were withdrawn in order to avert any appearance of co-operation with the kaffirs, and a report in connection with the incident was immediately sent in.

(b) In the second case Colonel Parke, the commanding officer of His Majesty's troops in the district named, reports that there is no foundation for the report supplied to you. On the 3rd of September all families in Ohrigstad district were removed by him. Harber's burgher commando was present, but took no part in the operation. On this occasion it was reported by a Boer woman that a number of kaffirs had appeared there the day previous and had plundered the village of Ohrigstad, but the kaffirs were acting independently of His Majesty's troops, and no further information as regards the matter is available except the report as stated above.

In conclusion, I think that it is not improbable that kaffirs have made attacks in the districts named by you, but I can only attribute these attacks to the action of your own burghers, i.e., to the shooting and robbing of kaffirs, and the enmity thereby awakened among the kaffirs by such maltreatment. While at the same time they (the burghers) have supplied the kaffirs, by manner of sale, of weapons and ammunition wherewith the attacks were made concerning which you complain. I emphatically deny that they (the kaffirs) were armed or incited by His Majesty's troops.

I have the honour to be,
Your obedient servant,
Kitchener, Commander-in-Chief in South Africa.

6th November, 1901.

District Lydenburg,

6th November, 1901.

To His Excellency Lord Kitchener, Commander-in-Chief of His Majesty's Forces in South Africa.

Your Excellency,—

I have the honour to acknowledge the receipt of your Excellency's letter of the 26th October, containing a denial of certain acts committed by armed kaffirs in the neighbourhood of Wit River and Ohrigstad.

With respect to the first incident, i.e., that at Wit River, I can only say that it appeared to us not only strange, but even improbable that a band of armed kaffirs could attack simultaneously, and in evident harmony with His Majesty's troops, and that neither party should have any cognisance of the other's presence.

If it were the first occasion that His Majesty's troops had acted in conjunction and with the assistance of kaffirs to make raids on the burghers, then His Excellency's explanation would be feasible.

But, alas, our bitter experience in this War is otherwise. I shall, therefore, be causing your Excellency no surprise if I contend that your Excellency's explanation is untenable. As to what occurred at Ohrigstad, I adhere to what I said, and to my letter of the 8th of October, and I regret to observe that Colonel Parke misled your Excellency by giving you an inaccurate account of the true facts.

To assist Colonel Parke's memory I may state that the same night he left Lydenburg on his way to Kruger's Post, the Boer, Harber, with his band of traitors, proceeded through Klipkloof and across Joubertshoogte, accompanied by 100 armed kaffirs, and passed Field-Cornet Zwart's farm at Uitkomst, where the plundering of Boer families and homesteads was commenced. This was done at the explicit instructions and in the presence of the said Harber.

The same afternoon Harber was met by the forces under Colonel Parke, at Rustplaats, whence they conjointly withdrew to Kruger's Post Nek.

The next morning Colonel Parke once more proceeded to Ohrigstad, where our families were again plundered and deported, and the homesteads raided and burned.

Accordingly, only the last paragraph of Colonel Parke's report is correct; and if your Excellency would take the trouble to question and examine the families now in your hands—as requested in my former letter—your Excellency would easily ascertain the true facts.

Since I construe from your Excellency's letter that Harber and his corps are recognised as attached to His Majesty's forces, His Majesty's officers must be held responsible for the acts of the said Harber and his kaffir hordes. It is not to be assumed that Harber and his corps, all armed, and attired in khaki, only accompanied His Majesty's Army as spectators or military attache's.

In conclusion, I observe that your Excellency repeats the allegation that kaffirs are promiscuously shot, robbed, and maltreated by our burghers, and that arms have been sold to the kaffirs by our burghers; and that you trace the hostile attitude of the kaffirs towards us to these causes.

As regards the hostile attitude of the kaffir races I can refer your Excellency to a letter from his Honour, General Louis Botha, on the same subject, wherein it is notified, inter alia, that prior to the arrival of British troops in these districts, and in Swaziland, the kaffir races, without exception, maintained a pacific attitude, a fact which speaks for itself.

I must again repeat that the allegation that burghers sold arms to the kaffirs is, so far as I know, untrue, and that this is merely one of the many baseless accusations which have emanated from traitors and unscrupulous individuals, and are offered by them as "important information" to the British officers.

That kaffirs were provided by His Majesty's officers with arms can be proved by intercepted documents, and I enclose herewith an extract from the diary of Sergeant Buchanan, of Steinacker's Horse, from which your Excellency will perceive that Lieutenant Gray, an officer of His Majesty's Army, did personally supply kaffirs with arms and ammunition.

I have the honour to be,

Your Excellency's obedient servant,

B. J. Viljoen, Assistant Commandant-General.

7th November, 1901.

District of Lydenburg,

7th November, 1901.

The Officer Commanding Lydenburg,

Dear Sir,—

I shall be obliged by your bringing the following to the attention of Lord Kitchener, namely, that on 29th October last the residence of a certain D. Coetzee, on the Vrischgewaard Farm, in this district, was surrounded during the night of that day, or approximately at that time, by His Majesty's troops, assisted by a number of kaffirs and traitors, and that only the youth Abraham Coetzee, occupied the house, and that this youth, while attempting to escape, was shot through the stomach. Coetzee was, furthermore, left in a shed, and robbed of all his personal goods, and even his clothes.

The following day I found him still alive, but he died shortly after. He declared that in the presence of white British troops he had been robbed, knocked about, and kicked by armed kaffirs. I know beforehand that the officer responsible for this noble and civilised act will attempt to pervert the truth, because I am assured that His Excellency cannot sanction this method of warfare. But this case is personally known to me, and in my opinion, the declaration of a dying man is worthy of credit.

I have the honour to be,

Your very obedient servant,

B. J. Viljoen.

11th November, 1901

On the Veldt,

11th November, 1901.

To His Excellency The Marquis of Salisbury,

Prime Minister of His British Majesty's Government.

Your Excellency,

Whereas His Honour the Commandant-General, and other commanding officers, have already more than once, without any result, protested to the Commanding Officer of your Forces in South Africa against the employment of savage aborigines in this War, and notwithstanding that we have repeatedly assured your military authorities here that on our side every effort is being made to keep kaffirs entirely outside this War, this Government is of opinion that it is its duty to earnestly and solemnly protest to your Government, as we hereby do, and at the same time to point out and direct its attention to the horrible and cruel consequences of this manner of warfare.

Former protests sent in to your military authorities here in this connection have met with the reply that such kaffirs were only employed as unarmed scouts, though we have proof that they actually fight against us, and pursue their destructive methods while in the ranks of your forces, and as isolated commandos directed by British officers.

These kaffirs, being ignorant of the rules of civilised warfare, have not hesitated on various occasions and even in the presence of your troops, to kill prisoners-of-war in a barbarous fashion. This is only one of the evil consequences resulting from the employment of barbarians in war, because it has also occurred that defenceless women and children have been made prisoners by these wild ruffians, and removed to kaffir kraals for detention until they were handed over to the British military authorities.

This Government is prepared, in case the above allegations are denied, to send your Excellency a large number of sworn declarations confirming the facts.

We have the honour to be,

Your Excellency's most obedient servants,

S. W. Burger

(Acting State President).

F. W. Reitz (Acting State Secretary).

1st December, 1901.

Army Headquarters, Pretoria, South Africa.

1st December, 1901.

Sir,

I observe from a communication which his Honour Schalk Burger has requested me to forward to Lord Salisbury, and which I have so forwarded, that his Government complains of the treatment of the women and children in the camps which we have established for their reception.

Everything has been done which the conditions of a state of war allowed to provide for the well-being of the women and children; but as you complain of that treatment and must, therefore, be in a position to provide for them, I have the honour to inform you that all women and children at present in our camps who are willing to leave will be sent to your care, and I shall be happy to be informed where you desire that they should be handed over to you.

I have addressed a reply to His Honour Schalk Burger in the above sense.

I have the honour to be, Sir,

Your obedient servant,
Kitchener,
General Commanding-in-Chief, South Africa.
To General C. De Wet.

End Viljoen Appendix

Viljoen Extra Maps

All the maps in Viljoen's book have been placed close to the events described in the narrative. These two maps do not have a place in the book where the engagement is discussed. They are included here for completeness.

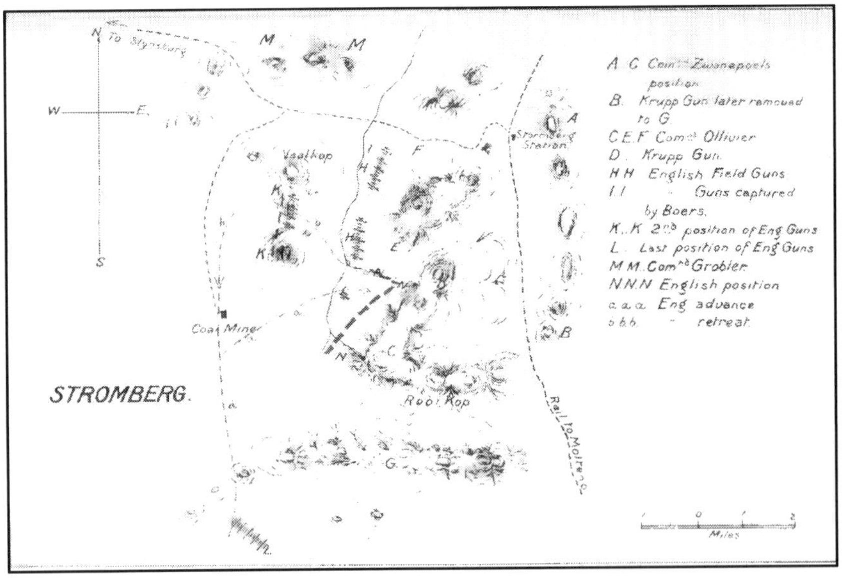

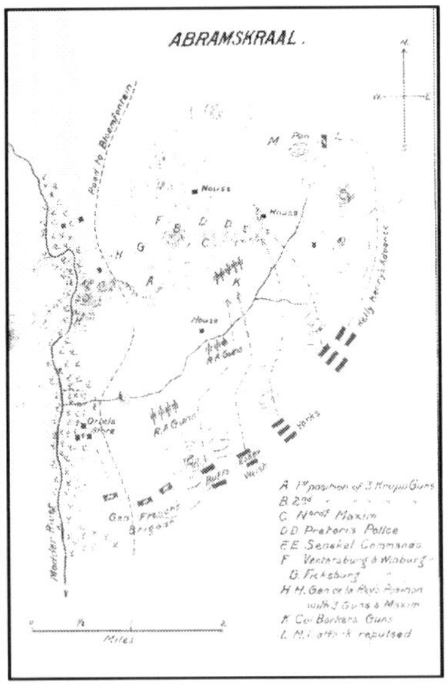

The Battle of Stormberg[266] took place on December 10th 1899 at the Stormberg railway junction, 12 miles (20 km) northwest of Molteno in the Eastern Cape. General William Forbes Gatacre and his Brigade attacked the Boer forces, led by Commandant Jan Hendrik Olivier whose forces were emplaced at the tops of the surrounding hills. Abrahamskrall. 3/7/1900 Roberts dispersed Boers near Poplar Grove. 3/10/1900 Engagement at Driefontein (Abraham's Kraal), fifteen miles east of Poplar Grove.

[266] English Wikipedia: Battle of Stormberg - Wikipedia Afrikaans Wikipedia: Slag van Stormberg - Wikipedia

Our Military Officers Album
Using Portraits with Life Sketches
of
Transvaal Generals and Commandants
E. A.

Published at the "Volksstem"[267] Office, Pretoria. [May] 1904.

ONZE

KRIJGS-OFFICIEREN

ALBUM

VAN

PORTRETTEN MET LEVENS-SCHETSEN

DER

Transvaalse Generaals en Kommandanten

E. A.

UITGEGEVEN TEN „VOLKSSTEM" KANTORE,
PRETORIA.

1904.

Frontispiece of the original publication.

[267] People's Voice.

Military Officer Album Preface

The reproduction of the Military Officer Album requires a short explanation.

I discovered this gold mine of a book by serendipity while researching another topic. It showed up as a reference in Annie Botha's [268] Wikipedia article. I followed the link and found Mr. Pieter van der Merwe's Dutch transcription. Fortunately he provided his email address in his compilation and I was able to contact him and thank him. I also found a copy on the Internet Archive that included an optical character recognition of the Dutch text. I also generated my own OCR using the downloaded PDF from the Archive. See references below.

I lack the skill to translate the Dutch to English so I have taken an alternative approach. Using Google Translate, Bing Translate, and Microsoft Word's translator, I have generated a literal English translation of the original text. For historical and research purposes rather than easy readability the literal translations are sufficient to generate comprehensible text regarding the service history and later careers of the individuals included.

To save space in this combined work I have removed the original Dutch text which had I had previously footnoted for each individual. The original copy most readily available is the New York Public Library's edition donated by the Astor, Lenox and Tilden Foundations.

I have augmented the work with additional pictures and mainly Wikipedia hyperlinks for those individuals whose career intersects with the other narratives in this series. I'm sure that there are additional online materials for other individuals as well but they were outside the scope.

It is interesting how many of these individuals Ben would see when he also arrived at St. Helena.

Payton Fireman

P.S. None of the individuals who surrendered and then enlisted in the National Scouts are mentioned.[269]

[268] Annie Botha (née Emmett; 3 July 1864–20 May 1937) was a South African civic leader and political hostess. She was the wife of Louis Botha, who served as the first Prime Minister of South Africa. She established an orphanage in South Africa and, with Georgiana Solomon, co-founded and chaired the South African Women's Federation. https://en.wikipedia.org/wiki/Annie_Botha

[269] The National Scouts were a military unit in South Africa created by the British authorities in 1900 during the Second Boer War (11 October 1899–31 May 1902). Its membership consisted of former Boer Orange Free State and South African Republic military. They were recruited in significant numbers towards the end of the war from Afrikaner prisoners and defectors, and were commonly known as hensoppers (Afrikaans for Hands-uppers) or joiners among the Boers. Many of their fellow citizens despised them as traitors so that the label of National Scout became a swear word. According to official figures there were 1,480 members in May 1902. After the war, they were largely ostracized by the community and a number of them founded their own church, known as the Kruiskerk (Church of the Cross) in the Transvaal Colony.

After the Boer War, only 10 percent of the total of 1,750 Boers serving on the British side as National Scouts claimed their Queen's South Africa Medals. Ostracized by the Boer Bittereinders and their womenfolk, they had to found their own separate Afrikander church organisation, the Kruiskerk (Church of the Cross) in Pretoria. Towards the end of 1902, General Louis Botha decided to destroy the list of National Scouts compiled by the Boers during the war, to so keep their descendants in ignorance about their actions during the war. Credit Dr. David Biggins. See: https://en.wikipedia.org/wiki/National_Scouts

Alphabetical List of Names

Transvaal Military Officer Album names in Alphabetical Order and indicating their rank.

Brits.—C. J. Brits.—General
Buhrmann.—W. H. Buhrmann.—Commandant
Trichardt.—S. P. E. Trichardt.—Commandant
van Velden.—Reg. Secretary Dirk E. van Velden.—Military Commissioner
Alberts.—H. A. Alberts.—General
Alberts.—J. J. Alberts.—Commandant
Alberts.—S. F. Alberts.—Commandant
Alleman.—F. S. Alleman.—Commandant
Anderssen.—Rev. E. C. Anderssen.—Civilian Authorities
Arnoldi.—Military Commissioner B. H. V. K. Arnoldi.—Civilian Authorities
Badenhorst.—A. D. Badenhorst.—Commandant
Badenhorst.—C. P. S. Badenhorst.—Commandant
Barter—W. J. H. Barter—Military Commissioner
Beyers.—C. F. Beyers.—General
Bierens De Haan.—Dr. J. C Bierens De Haan.—Military Commissioner
Bierman.—J. T. Bierman.—Commandant
Blake.—Colonel J. Y. F. Blake.—Military Chief Officer
Blignaut.—J. S. F. Blignaut.—Commandant
Botha—L. Botha—
Botha.—Chris Botha.—General
Botha.—J. D. L. Botha.—Commandant
Botha.—Captain G. R. Botha.—Military Chief Officer
Bouwer.—B. D. Bouwer.—Commandant
Breytenbach.—D. J. J. Breytenbach.—Commandant
Breytenbach.—B. H. Breytenbach.—Commandant
Burger.—S. W. Burger.—General
Buys.—S. Buys.—Commandant
Celliers.—J. G. Celliers.—General
Coetsee—Rev. M. P. A. Coetsee—Military Commissioner
Coster.—Dr. H. J. Coster.—Military Commissioner
Cronje.—P. A. Cronje.—General
Dames.—Captain P. A. Dames.—Military Chief Officer
De— Adv. N. J. De— Military Commissioner
De Beer—J. F. De Beer—Commandant
De Jager.—Adjudant-General H. De Jager.—Military Chief Officer
De La C Schroeder.—J.P. De La C Schroeder.—Commandant
De La Rey.—J. H. De La Rey.—General
De Witt Hamer.—Capt. B. G. V. De Witt Hamer.—Military Chief Officer
Deventer.—J. L. V. Deventer.—General
Dreyer.—T. F. J. Dreyer.—Commandant
Du Plessis.—C. J. H. Du Plessis.—Commandant
Du Toit.—S. P. Du Toit.—General
Du Toit.—Captain H. Du Toit.—Military Chief Officer
Du Toit.—Captain A. F. Du Toit.—Military Chief Officer
Emmett.—J. C. Emmett.—General
Erasmus.—D. J. E. Erasmus.—General
Erasmus.—J. L. P. Erasmus.—Commandant
Ferreira.—Adv. I. S. Ferreira.—Military Commissioner
Fourie—J. C. Fourie—General
Fourie.—C. E. Fourie.—General
Goddefroy.—Rev. M. J. Goddefroy.—Military Commissioner
Gravett.—G. H. Gravett.—General
Greyling.—P. J. Greyling.—Commandant
Greyling.—I. J. Greyling.—Commandant
Grobler.—F. A. Grobler.—General
Grobler.—J. N. H. Grobler.—General
Grobler.—H. S. Grobler.—Commandant
Grobler.—S. Grobler.—Commandant
Grothaus.—Captain H. C. W. Grothaus.—Military Chief Officer
Hindon.—Kapitein O. J. Hindon.—Military Chief Officer
Hohls.—Dr. J. O. Hohls.—Military Commissioner
Jacobsz.—Adv. L. J. Jacobsz.—Military Commissioner
Joubert.—D. J. Joubert.—General
Joubert.—J. A. Joubert.—Commandant
Jurriaanse—Dr. A. Jurriaanse—Military Commissioner
Kemp.—J. Kemp.—General
Kock—J. H. M. Kock—General
Kriegler.—J. C. Kriegler.—Commandant
Kriel.—Rev. A. P. Kriel.—Military Commissioner
Krogh, L.U.R.—J. C. Krogh, L.U.R.—Military Commissioner
Kruger.—J. A. D. Kruger.—Commandant
Lemmer.—L. A. S. Lemmer.—General
Lemmer.—H. R. Lemmer.—General
Liebenberg.—P. J. Liebenberg.—General
Liebenberg.—D. P. Liebenberg.—Commandant

Lombard.—J. P. La G. Lombard.—Commandant
Louw.—Rev. J. M. Louw.—Military Commissioner
Martins.—J. T. Martins.—Commandant
Meyer—J. G. Meyer—Commandant
Meyer.—L. Meyer.—General
Meyer.—H. F. Meyer.—Commandant
Meyer.—T. C. G. Meyer.—Commandant
Mijburgh.—H. W. Mijburgh.—General
Mouton.—W. J. Mouton.—Commandant
Muller.—C. H. Muller.—General
Naude.—Captain J. J. Naude.—Military Chief Officer
Neethling.—Rev. H. J. Neethling.—Military Commissioner
Oosthuizen.—S. F. Oosthuizen.—General
Opperman.—J.D. Opperman.—General
P. J. Joubert—P. J. Joubert—
Paff.—Captain P. C. Paff.—Military Chief Officer
Pienaar.—J. J. Pienaar.—Commandant
Potgieter.—F. J. Potgieter.—Commandant
Preller.—R. C. Preller.—Commandant
Pretorius.—W. F. Pretorius.—Commandant
Pretorius.—Majoor J. L. Pretorius.—Military Chief Officer
Prinsloo.—H. F. Prinsloo.—Commandant
Prinsloo.—J. P. N. Prinsloo.—Commandant
Reitz.—Staatssekretaris F. W. Reitz.—Military Commissioner
Riekert.—P. J. Riekert.—Commandant
Roos.—J. Roos.—Commandant
Scheepers.—G. J Scheepers.—Commandant
Schoeman.—D. J. Schoeman.—Commandant
Scholtz.—Captain P. Scholtz.—Military Chief Officer
Slegtkamp.—Captain H. P. Slegtkamp.—Military Chief Officer
Smuts.—J. C. Smuts.—General
Smuts.—T. Smuts.—General
Spruyt.—C. J. Spruyt.—General
Steenkamp.—P. Steenkamp.—Commandant
Strasheim.—Rev. J. W. G. Strasheim.—Military Commissioner
Theron.—Captain D. Theron.—Military Chief Officer
Tileman.—Dr. H. Tileman.—Military Commissioner
Trichardt.—P. F. Trichardt.—Commandant
Van Dam.—G. M. J. Van Dam.—Commandant
Van Niekerk.—P. C. Van Niekerk.—Commandant
Van Rensburg.—H. C. J. Van Rensburg.—Commandant
Van Staden.—M. P. Van Staden.—Commandant
Van Zijl.—J.Van Zijl.—General
Vermaas.—H. C. W. Vermaas.—Commandant
Viljoen.—B. Viljoen.—General
Viljoen.—P. R. Viljoen.—General
Viljoen.—W. Viljoen.—Commandant
Villebois-Mareuil.—De Villebois-Mareuil.—General
Von Wichmann.—Captain V. E. Von Wichmann.—Military Chief Officer
Williams.—Captain H. Williams.—Military Chief Officer
Wolmarans—M. J. Wolmarans—Commandant
Wolmarans—Major J. F. Wolmarans—Military Chief Officer

List by Rank

Introduction Louis Botha
P. J. Joubert
L. Botha

Generals.
H. A. Alberts.
C. F. Beyers.
C. J. Brits.
S. W. Burger.
Chris Botha.
J. G. Celliers.
P. A. Cronje.
J. L. V. Deventer.
J. C. Emmett.
D. J. E. Erasmus.
J. C. Fourie
C. E. Fourie.
G. H. Gravett.
F. A. Grobler.
J. N. H. Grobler.
D. J. Joubert.
J. Kemp.
J. H. M. Kock
L. A. S. Lemmer.
H. R. Lemmer.
P. J. Liebenberg.
L. Meyer.
C. H. Muller.
H. W. Mijburgh.
S. F. Oosthuizen.
J.D. Opperman.
J. H. De La Rey.
J. C. Smuts.
T. Smuts.
C. J. Spruyt.
S. P. Du Toit.
B. Viljoen.
P. R. Viljoen.
De Villebois-Mareuil.
J.Van Zijl.

Commandants
J. J. Alberts.
S. F. Alberts.
F. S. Alleman.
A. D. Badenhorst.
C. P. S. Badenhorst.
J. F. De Beer
J. T. Bierman.
J. S. F. Blignaut.
J. D. L. Botha.
B. D. Bouwer.
D. J. J. Breytenbach.
B. H. Breytenbach.
W. H. Buhrmann.
S. Buys.
G. M. J. Van Dam.
T. F. J. Dreyer.
J. L. P. Erasmus.
P. J. Greyling.
I. J. Greyling.
H. S. Grobler.
S. Grobler.
J. A. Joubert.
J. C. Kriegler.
J. A. D. Kruger.
D. P. Liebenberg.
J. P. La G. Lombard.
J. T. Martins.
H. F. Meyer.
J. G. Meyer
T. C. G. Meyer.
W. J. Mouton.
P. C. Van Niekerk.
J. J. Pienaar.
C. J. H. Du Plessis.
F. J. Potgieter.
R. C. Preller.
W. F. Pretorius.
H. F. Prinsloo.
J. P. N. Prinsloo.
H. C. J. Van Rensburg.
P. J. Riekert.
J. Roos.
G. J Scheepers.
D. J. Schoeman.
J.P. De La C Schroeder.
M. P. Van Staden.
P. Steenkamp.
S. P. E. Trichardt.
P. F. Trichardt.
H. C. W. Vermaas.
W. Viljoen.
M. J. Wolmarans

Military Chief Officers
Colonel J. Y. F. Blake.
Captain G. R. Botha.
Captain P. A. Dames.
Captain H. C. W. Grothaus.
Kapitein O. J. Hindon.
Adjudant-General H. De Jager.
Captain J. J. Naude.
Captain P. C. Paff.
Majoor J. L. Pretorius.
Captain P. Scholtz.
Captain H. P. Slegtkamp.
Captain D. Theron.
Captain H. Du Toit.
Captain A. F. Du Toit.
Captain V. E. Von Wichmann.
Captain H. Williams.
Capt. B. G. V. De Witt Hamer.
Major J. F. Wolmarans

Smuts' Commando

The Lijdenburg Kommando

Civilian Authorities
Rev. E. C. Anderssen.
Military Commissioner B. H.
V. K. Arnoldi.

Military Commissioners
W. J. H. Barter
Dr. J. C Bierens De Haan.
Rev. M. P. A. Coetsee
Dr. H. J. Coster.
Adv. I. S. Ferreira.
Rev. M. J. Goddefroy.
Dr. J. O. Hohls.
Adv. L. J. Jacobsz.
Dr. A. Jurriaanse
Rev. A. P. Kriel.
J. C. Krogh, L.U.R.
Rev. J. M. Louw.
Rev. H. J. Neethling.
Staatssekretaris F. W. Reitz.
Rev. J. W. G. Strasheim.
Dr. H. Tileman.
Reg. Secretary Dirk E. van
Velden.
Adv. N. J. De W

Original Foreword.

This Album aims to help perpetuate the reverent memento of the Transvaal Military Officers, who, by their exemplary conduct, made the Afrikaner name awe-inspiring. have. Many of them are known only in a small circle and deserve the permanent appreciation of all his people.

In this Album of Transvaal Generals and Commandants, some persons were also who have behaved creditably on the battlefield. The collection is not completely; some will be missed, especially General P. de Villiers (Griqualand); Commandanten Breytenbach (Western borders), the late Coetzee (Pretoria), J. Coetser, Mol, Möller, van der Merwe (Krugersdorp), Nolte (Boksburg), the late Potgieter (Western borders), Wolmarans (Potchefstroom), J. van Zijl (Krugersdorp); Captain Mike du Toit: Dr. Albrecht, Dr. Neethling, Dr. Rennenkampf, A. Malan, Mears, M. de Jager, S. Eloff and others. It is possible that — when this work will be and a future Album will be devoted to the other Officers, this lack of be restored.

To all those who have been instrumental in the compilation of this People's Book, a word of heartfelt thanks by.

Dr. F. V. Engelenburg.

G. S. Preller.

Pretoria, April 1904.

Introduction by Louis Botha

I am happy to comply with the request of the Volksstem editorial to present this Album to our people to recommend. Any honourable attempt to keep alive the memory of the national Afrikaner virtues that proved so brilliant during the last war, the support of the well-disposed. I appreciate that the Volksstem's editorial has begun to show the world who the men were who fought for three years persisted. What did I, as Commandant-General, what did the Chief-Generals have can do without such men?

How much the compilers of this Album acknowledge its incompleteness, as an attempt as referred to above, their work has been a decidedly successful [undertaking] in which I rejoiced with all my heart.

May this Album satisfy everyone who will leaf through it in many respects and the striving for national unity and self-esteem in every Afrikaner strengthen.

Moreover, may this book be given the favourable reception to which it has received all the so that the compilers will be encouraged to continue their work. and in time to produce a second Album of all Field Cornets and other Military officers.

Louis Botha.

Pretoria, 28 March 1904.

"Intended for Afrikaner Posterity."

P. J. Joubert

Petrus Jacobus Joubert is one of the best-known figures in the early history of the Z.A. Republic and has served its people in numerous important state relations. Already in the First War of Independence he held the rank of Commandant General of the Transvaal Armed Forces and as such took part in all wars which had taken place since then occurred. During the last war he took special charge of the military operations in Natal in his own right, and was present at many a meeting near Ladysmith. But his constitution was severely undermined and shortly after Komdt. General Joubert to Pretoria had returned for the organization of the country's defence, he was suddenly grave, mourned by all the people. This was March 27, 1900.

Louis Botha

Louis Botha, born 27 September 1862 in Greytown, Natal, was the third son of Louis Botha, Lion's Head, district Peace, Orange Free State, which as Field Cornet about sixty in 1860, under punishment and impediment of the Government of Natal, this colony to assist the Free Staters against the Basutos. This L. Botha was second again son of the well-known pioneer Philip Rudolph Botha, who sold all his property in the Left Cape Colony to lead a trek through Grikwaland East, via Kokstad, to Natal, in 1835-6.

When the last war broke out, the subject of this biography was a member of the First People's Council for the district of Freedom, to whose origins and further fortunes he had taken a large share. Although without any war-experience worth mentioning, Louis Botha was two months after the outbreak of the war as a result of his act at Colenso for his people an authoritative general and for the world a became a famous Transvaler. Under General Lukas Meyer, Botha attended the Battle of Dundee at, became 30 October 1899 - the day of the Mud Sprout and Lombard Cup - appointed as General, took part in the Battle of Estcourt, during which the Willow grange battle took place, fought at Colenso, Spioenkop and elsewhere along the Tugela, commanding the western force at Pieters Heights, bombarded the English camp at Elandslaagte, after the relief of Ladysmith, fought from Zandrivier, through Kliprivier as far as Pretoria against Lord Roberts, led the fighting at Donkerhoek, Diamond Hill, Bergendal, and Machadodorp, and led its citizens at Chrissies Lake, Spieshoek, Itala, Onverwacht [Literally 'Unexpected'] (where General Opperman was killed) and Bakenlaagte.

After the conclusion of the Vereniging treaty, Commandant-General Botha, with Generals Christiaan de Wet and de la Key, a visit to England and to several mainland capitals; in the latter he was repulsed with rare enthusiasm by the audience. At the beginning of 1903, he was given a seat in the Transvaal Legislative Council.

Generals.

H. A. Alberts.

Hendrik Abraham Alberts

General Hendrik Abraham Alberts took part, as a private citizen, in several battles, which his Kommando, the Heidelbergers, in Natal. In Spioenkop he acted as acting Field Cornet. After the capture of Pretoria, he and his citizens Standing in Machadodorp, Alberts became Commandant over the Heidelberg district while the Commandant concerned C. J. Spruyt as General instead of the General Gravett was appointed.

In the absence of General Spruyt Alberts usually acted in his place; and when General Spruyt was killed, Alberts has been entrusted with important assignments on several occasions. In November 1901, after the battle of Beacon, he was entrusted by the Commandant General with the supreme command of the in the Free State sent Transvaalers.

In February 1902 Alberts came over the railway line to the Hogeveld and in May 1902 he was born, at the same time as Messrs. P. R. Viljoen and D. J. E. Opperman, delegate to the Vereniging. On 5 June, the Heidelberg Kommando's at Kraalstatie lay down their arms.

C. F. Beyers.

General C. F. Beyers And.[418]

Born in Banhoek, Stellenbos, on September 21, 1869, Christiaan F. Beyers left Victoria College in 1889 and settled in Transvaal, first in Pretoria, where he passed his procureurs exams five years later. Established at Boksburg, his name soon became favourably known;

As a participant in the commandos that resisted the Jameson invasion with such success, he obtained full Transvaal voting rights in 1896. He also distinguished himself on the sports field as a fearless footballer. Deployed to war as a private citizen, he rose to the rank of Ass. Commander-General within twelve months and, with varying degrees of luck, confronted the enemy's superior forces in the Waterberg and Zoutpansberg districts. Delegated by his audience to the Peace Conference, he was elected Chairman and therefore played an important role in the relevant negotiations.

See Also: https://en.wikipedia.org/wiki/Christiaan_Frederik_Beyers

C. J. Brits.

General C. J. Brits.

Coenraad Jacobus Brits was born on 19 April 1868 in the Wakkerstroom district and When the war broke out, he moved as WD. Field cornet with his citizens into Natal, where he took part in the battles of Dundee, Moddersspruit, Colenso and Spioenkop.

Subsequently he took part in several battles in the Free State and the Transvaal. He succeeded at the 700 citizens of the Standerton, who at first seemed inclined to fight the war. to get an active commando. At the beginning of June 1901 C. J. British appointed by General Botha as Fighting General over Standerton, Bethal and a part of Middelburg and in that capacity he participated in several battles, including at Itala, Spieshoek, Bakenlaagte, Begin-der-Lijn and Bankkop. In the beginning In February 1902 he was appointed Assistant Commandant General and on 2 April General British fought his last battle at Standertonkop where he killed 500 fat oxen and over 30 saddled horses. He belonged to the Transvalers who were delegated to Vereeniging.

S. W. Burger.

General S. W. Burger.

Schalk Willem Burger has served his fellow citizens for many years, first as a Field Cornet, then in the war of 1880-1881 as Commandant, later as a member of the People's Council and at least and finally as a member of the Executive Council.

At the start of the war, General Burger commanded the Suffering Citizens at the siege of Ladysmith. And then in the In the second half of 1900 it was decided that President Kruger would go to Europe, General Burger was appointed acting President of the State, in which capacity he was appointed during the remainder of the war he was in charge of the government and at the time of the conclusion of the Peace as one of the chief negotiators to the Unification Tract Occurred. Incidentally, his political career is too well known for a broad description would be appropriate here.

Chris Botha.

Christiaan Botha

Christiaan Botha, a younger brother of Komdt. Genl. Louis Botha, became 6 October Born in Greytown, Natal in 1864, and was Commandant when the last war broke out of Police in Bremersdorp, Swaziland. He took part in the battles of Colenso, Spioenkop and Doornrand and almost lost the while he and 15 of his men were placed in a ditch south of Tugela. an lyddite bomb burst at him, killing eleven men.

At the end of April 1900, Chris Botha, in the place of Genl. Meyer, who was summoned to the Government at Pretoria, appointed as assistant Komdt. General and led all the fights from Pietershoogten against Genl. Buller until his breakthrough. Until the beginning of December 1900 he was commanding officer of all South-Eastern Kommandos, then took part in the attack on Belfast, on Freedom, Itala and other battles. At Tilt height by Genl. Buller approached with a proposal for unconditional surrender, Genl. [Headquarters] replied [to] Chris Botha that he could only accept a proposal of peace. Delegated by his citizens to Vereniging, he came there, as a result of during the war he experienced hardship, fell ill and died not long afterwards in Kokstad, whither he had travelled for private occupation, to the consequences of a heavy operation

J. G. Celliers.

Johan G. Celliers

Johan G. Celliers was born on July 12, 1861 in Frasersburg, Cape Colony. in the Transvaal and on the Diamond Fields of Kimberley. As a youth in a Farmer's business in Zeerust, he went with the Marico Burgers during the Revolutionary War of 1880 and participated in it from beginning to end under the then field cornet J. H. de la Rey, and then then on the western frontiers complications arose, Celliers was one of the volunteers, who exercised the authority of the chief Moshette. In those days, Celliers also served as Postmaster General and Wd. Registrar of Deeds of the territory called Land Goosen. Subsequently, in the Mariko district, he devoted himself to laud construction, but did not when the Witwatersrand goldfields were discovered. When the war broke out, Celliers was part of the Police Department in Krugersdorp. He has a took part in the many brave acts by which the Rand Police first in Natal and then in the North of the Cape Colony. When the war took on a guerrilla character, after the death of General Lemmer, Celliers appointed in his place over the commandos of Lichtenburg and Zeerust, raised to a numerical strength of over 1.OO0 horsemen. In the event of an attack on in the village of Lichtenburg, 4 March 1901, he was severely wounded, but after two months Struggling to flee around, recovered and vigilant as never before. General Celliers managed to hold the field until the end of the war.

P. A. Cronje.

Pieter Arnoldus Cronje

General Pieter Arnoldus Cronje had an honourable martial career behind him, when the last war placed him at the head of the military operations to the west of the Boer republics. His name is inseparable from the siege of Potchefstroom, during the First War of Independence, and the capture of Jameson, commencing 96. After the heroic capitulation at Paardeberg, which followed on 27 February 1900 at the siege of Kimberley, Genl. Cronje as a prisoner of war to St. Helena where he remained until the end of the war.

J. L. V. Deventer.

Jacob Louis van Deventer

General Jacob Louis van Deventer was, when the war broke out, battery chief of the State Artillery in Pretoria, and was assigned to the Genl. Armed Forces. Cronje classified. On the morning of 13 In October he fired the first shot at the armoured train that had run aground at Kraaipan and since a never-resting field life, first along the Western border, then in the Free State, then with Genl. Beyers in various Transvaal districts and then with Genl. Smuts in the Cape Colony. On 20 July 1901 van Deventer left the Zwartruggens, distr. Rustenburg, with the aim of Cape Colony. On 7 August he succeeded between Squadron station and Brandfort to take a blockhouse and the same month he broke through the line of battle between Bloemfontein and Ladybrand, so that he joined Genl. J. C. Smuts, to which On 4 September, the Orange River was crossed near Herschel. The cold rains, with which no less than the enemy Kommandos wrestled with, horses did not do so much Kommando evil. On the 17th of September, the camp of the 17th Lancers, with 2 guns, in the direction of Cradock, to Uitenhage, which was taken on 30 Sept. was reached. At the beginning of October a few were again seen at Groot Zuurberg enemy camps, whereupon the railway line was built at Victoria West station. crossed over, north of which with lasting success against the English commandos was operated on. On 23 January 1902 van Deventer fought at Jakhalsvlei, distr. Sutherland, and on the 5th of February he took a bearing at Middelpas, distr. Calvinia, where Commandant Boshoff and V.K. van den Berg were wounded. On 25 February, Deventer when taking a bearing at Windhoek, distr. Van Rynsdorp, wounded. On 1 In April, Springbok, in Namakwaland, was captured, after which the copper mines at Port Nolloth. Colonel Cooper arrived from Cape Town for reinforcements, beaten back at the end of April. On the 2nd of June the command at Katkop received telegraphy notice that peace had been concluded, and thus ended the military career of Vecht General van Deventer.

J. C. Emmett.

Joseph James Cheere Emmett

General Joseph James Cheere Emmett, son of the well-known John George Cheere Emmett and descendant of Robert Emmett, the Irish patriot, was born 19 January 1866 at Swellendam, Cape Colony, born and was Field Cornet of district II, Freedom, then the last war Broke out. With his citizens marching out, he took part in the battles of Dundee, Moddersspruit and Lombards-kop, was the first officer to go with his men to Colenso came to the ten English guns which the enemy left on the field of battle, fought at Pietershoogten, Willowgrange, Scheepersnek, in the attack on Liberty; at Spieshoek (against Colonel Gouch), Itala and almost all the smaller battles in Freedom, Natal and the Reserve; in March 1902 General Emmett - he was in December 1900 appointed as General - during a nightly trek taken prisoner of war with 20 two of whom were killed and five wounded.

D. J. E. Erasmus.

Daniel J. E. Erasmus

General Daniel J. E. Erasmus was born in 1845 at the place Dorenkloof, then known as Irene, distr. Pretoria, and has taken part in all the military enterprises, of which the Transvaal since 1861. During the First War of Independence, he attended the battle at Bronkhorstspruit and then assisted as a fighter Commandant in the encirclement of Pretoria, during which three attacks by the enemy were each time with success. Foiled. After the restoration of independence, he was a member of the People's Councillor and then Commandant for Pretoria District, in the latter capacity he took part in the kommandos against Malaboch and Mpefoe. In 1899 by a general court-martial in Zandspruit, Natal border, appointed assistant Commandant General, was general Erasmus with his citizens in many a fight nowadays. It is known how a part of the of the Pretoria Kommando had a large share in the military operations at Peony head. In the later part of the war, General Erasmus acted as Chairman of the Military Court for the districts of Heidelberg, Pretoria, Middelburg and Bethal. On 3 In January 1902, however, he was not far from Amsterdam by a troop division of General Bruce-Hamilton from Ermelo was taken prisoner of war and taken to St. Helena, where General Erasmus had to stay for about five months.

See Also: https://en.wikipedia.org/wiki/Daniel_Jacobus_Elardus_Erasmus

J. C. Fourie

Joachim Christoffel Fourie.[270]

General Joachim Christoffel Fourie was born on February 1, 1855 in Grahamsstad, K.K. and grew up in the district of Lijdenburg, Z.A.R. He served in the first war of independence as Veldkornet in the district and afterwards took part in several nature wars. In 1893 he was elected a member of the People's Council. As a private citizen, he joined the Transvaal ranks in 1899 and was ordained in March 00 at Biggarsberg as Assistant Fighting Agent for Carolina, Lijdenburg and Standerton elected. General Fourie took part in several battles, notably at Mud shoot, Spioenkop, Witrand - Dalmanutha and Chrissiesmeer (October 1900). General Fourie was killed on 7 November 1900 in a battle at the town of Witkloof, Carolina and was arrested the next day at his residence Welgevonden, there, in the earth.

See Also: https://en.wikipedia.org/wiki/Joachim_Christoffel_Fourie
https://en.wikipedia.org/wiki/Battle_of_Leliefontein

[270] Joachim Christoffel Fourie (1 February 1845–7 November 1900) was an Anglo-Boer War Combat General (Afrikaans: Veggeneraal) and Assistant Commander-in-Chief (Assistent Kommandant Generaal) for the South African Republic. He should not be confused with his colleagues, the Second Boer War generals Christiaan Ernst Fourie (1858-1943) and Petrus "Piet" Johannes Fourie (1842-1916), and his son and namesake Joachim Christoffel Fourie (1877–1958), military commander of the Carolina Commando in the South West Africa campaign (1914-1915) in the First World War.

C. E. Fourie.

Christian E. Fourie

General Christian E. Fourie was born on 17 June 1858 in the village of Lijdenburg and made several natural wars, including those against Sekoekoenie and Njabel. In 1888 he settled at the town of Blesbokfontein in the district of Middelburg. In 1890 he was a member of a committee which travelled through Mashonaland and lived through many a gripping adventure. In 1896 he was elected Field Cornet for the Olifantsrivier district, he was in the following year to Commandant and from the government to Naturellen kommissaris Appointed. Present at the battles of Dundee and Modderspruit, he was arrested on 13 December appointed as a Fighter General, took part in several battles on the Tugela River, After the relief of Ladysmith, he went to the Free State, where he was in the midst of the Heidelberg civilians fought. Then sent to the Hogeveld for the the designation of positions of the artillery, the threatening attitude of the Sekukunie blackes urgently needed his presence there, and commandant Fourie succeeded, thanks to his predominant influence, tranquillity, and order for the present natural materials. In addition, he managed to send many loads of salt and noodles to the Hogeveldse kommandoes, but fell by treachery in the latter part of March 01 in the hands of the enemy. But he succeeded, in the company of Mr. C. Schutte, at Middelburg, where both had been imprisoned, a couple of riding horses with which they were rescued on June 11, 1901, in a dark thunderstorm night. of the English

G. H. Gravett.

Gerhardus Hendrik Gravett

General Born at Alexandria, Cape Colony, on June 28, 1858, Gerhardus Hendrik Gravett, when the war broke out, living at Elandstfontein (Germiston), where he was born in 1888 and which place its origin was almost chiefly due to his zeal and work force. When our commandos entered the Cape Colony at Colesberg, Gravett served as an ordinary citizen, but was soon dismissed, on account of the peculiar suitability shown by him to do so, was chosen as Field Cornet and took then took an active part in the fighting near Colesberg. At Norval's Pont he was elected Commandant by his citizens and the success of the masterly retreat from the Cape Colony, through the eastern districts of the Free State, past Bloemfontein - already in the hands of the enemy - is becoming proficient in its skilful tactics and unwearied vigilance. At Lijdenburg he was appointed as Fighter general, and when he was killed, he was Assistant Commandant General of the North-eastern districts. At Witpoort, Mapochsgronden, with a handful of bnrgers fighting valiantly, Gravett was knocked next to a Nordenfeldt Maxim by a shrapnel seriously wounded, and died on 26 October 1900 near Roossenekal. After the war, his stoffelik surplus, under the greatest public interest, was buried the cemetery at Germiston for the last time. Commandant General Botha was present on this occasion and testified of Gravett: "He was a man who loved his people dearly; In the darkest days, he was always a source of courage and hope for all who came in contact with him; of the civilian ranks he was ascended to the highest position that a brave man could hold in our country; He was faithful unto death and did what he could.

F. A. Grobler.

Frederick Albertus Grobler

General Frederick Albertus Grobler was arrested by the Government at the outbreak of the last war. appointed Assistant Commandant General on the northern frontier of the Transvaal; with citizens of Waterberg and Zoutpansberg he came into contact with the enemy's posts at Rhodesdrift. He then went with some of his citizens at the beginning of January 1900 to Colesberg, where he led some advantageous battles. When Bloemfontein was Generals Grobler and Lemmer took a detour through the eastern part of the country. of the Orange Free State and achieved, without any loss of human or animal damage. Transvaal territory. This tour is rightly referred to as a masterpiece. Together with a number of Free State officers, he then fought in Tabaksberg and then took up positions south of Johannesburg, at Kliprivier. In the In the later part of the war, General Grobler clashed with the enemy on many occasions. During his defence of the Waterberg district he was seized with a fever and died on 13 May 1901 at his residence in Sterkfontein.

See: https://af.wikipedia.org/wiki/Slag_van_Onderstepoort

J. N. H. Grobler.

Johannes Nicolaas H. Grobler

General Johannes Nicolaas H. Grobler was born in Suffering. Here he saw it was born on February 27, 1864; later his father moved to Ermelo, and in 1891 Grobler was chosen to serve his fellow citizens as an assistant Field Cornet; Three years later he became Veldkornet, and in 1899 in place of Komdt. Tobias Smuts - The One People's councillor was elected as a district Commandant. Under Commandant Grobler the Ermelo kommando was among the first to cross the Natal border, after the outbreak of the last war, covered. In addition to the field cornets of districts 1, 2 and 3, among him Field cornetts Roelof Jansen van Vuren, N J Breijtenbach, F. W. Ernst and a section State artillery. With his commando he took an active part in the battles of Dundee, Moddersspruit, Colenso, Spioenkop, Bosrand, Pietershoogteten and more others. From Natal his commando was sent to the Free State. After the battle of Donkerhoek he was ordered to Ermelo and after Dalmanutha he returned also back to the Hoogeveld. When it was decided to carry out a second raid in Natal, the Ermelo commando was also sent southwards and was now in charge of the Melmoth, Itala, and the other battles that followed. When the Ermolo October 1901 - Grobler was appointed Fighting General and succeeded by Herr Buhrmann as Commandant. Seventeen of the citizens who were Beacons were killed, belonged to General Grobler's kommando. Subsequently, he operated in many smaller battles and skirmishes on the Hoogeveld, had many a narrow escape and was also delegated to Vereniging. On 9 June 1902 his commando laid down his arms near Ermelo.

D. J. Joubert.

David Johannes Joubert

General Joubert was the son of the former Commandant General of the district of Lijdenburg, David Johannes Joubert was born on 7 November 1849. With the storming of Majuba in 1880, Joubert acted as Voldkornet of the Boven Vaalrivier district, distr. Middelburg, under Fighting General Nicholas Smith. At the end of the war, he was Field cornet of the above-mentioned district chosen. In April 1885 he was appointed as Commandant of the district of Ermelo and re-elected as such in 1890. Three years later he thanked him for this position and settled in the distr. Carolina, where he was appointed as acting Commandant the following year, and a year later as to be chosen in such a way. At the same time, Joubert was elected as a member of the First Volksraad, and in 1898 resigned for the Commandantship. Show the war broke out, he was appointed to the staff of Assistant Commandant General S. W. Burger and later elected Commandant of the Carolina kommando. In January 1900 he was appointed as Fighter-General, but shortly afterwards resigned due to illness, lived in May of that year the last session of the Volksraad and, returning to the Natal border, he once again acted as a Fighting General and in November of the same year he was definitively appointed as a Commandant, which he remained until the end of the war. He died of fever on March 13, 1903 in German South East Africa.

J. Kemp.

Jan Kemp

General Jan Kemp, a son of Jurie Kemp, in former days Field cornet of the district Piet Retief, and a grandson of J. C. Greyling and Piet Retief, both by Dingaan murdered, attended the Staatsgymnasium in Pretoria at the age of sixteen, where he took his exam for the First Class Teachers Certificate. After several years at the farm, Kemp entered the state service of the Education Department, and later became responsible clerk of the Mining Commissioner in Krugersdorp. During the Jameson Raid, he became Lieutenant of the Krugersdorp Volunteer Corps and made later the expedition against 'MPefu mead. When the last war broke out, he went out as a civilian but was already in Septeniber 1899 elected assistant Field Cornet and later to Field Cornet. In March 1900 he was elected and re-elected in June 1900 as Commandant over the Krugersdorp kommando. In February of the following year, he was appointed as a Fighter General in the western districts. General Kemp made as well as all the great battles at Natal, including those at Donkerhoek, Dalmanutha, Nooitgedacht, and the principal battles in the western districts. He remained until the end of the war in the field.

J. H. M. Kock

Johannes Hermanus Michiel Kock

General Johannes Hermanus Michiel Kock, during the first War of Independence Landdrost of Potchefstroom and, when the last war broke out, a member of the Executive Council, was appointed to General of a division of the Transvaal Armed Forces, whose task was to was given to seize the railway connection between Dundee and Ladysmith. Bee This object was attained, but attacked by the garrison of Ladysmith, The said troop division suffered a defeat and was driven from its positions. Genl. Kock was - it happened on 21 October 1899 - seriously wounded and on the morning after carried this battle to Ladysmith, where he died on 31 October, the day after the English defeat at Modderspruit. The dusty remains of the brave warrior was then extradited and buried with military honours in Pretoria.

L. A. S. Lemmer.

General L. A. S. Lemmer

General L. A. S. Lemmer [was] Born in the year 1864, the 26th of October, at the place Buffelshoek, district Klein Marico, L. A. S. Lemmer received his education after the first War of Independence in Stellenbos in the Cape Colony, and, after his return to the Transvaal, worked for some years as Goevernement's teacher at an outdoor school. In March 1896 he was appointed as clerk at the Belastinggaarder in Ottoshoop, and in June two years later became a tax collector become. At the outbreak of the last war he went on a commando as an ordinary citizen, then became a corporal and soon had to act as Wd. Field Cornet. After the eviction of Zeerust, and in the general reorganization of the republican judiciary, he was appointed Landdrost for the said district, but soon had to regain his composure. to the kommando's, and after the death of General H. R. Lemmer (19 January 01) appointed Commandant of the Mariko horse command. At the storming the village of Lichtonburg, in March 1901, he was severely wounded, but soon recovered, and in June of June he was appointed Assistant Fighter-General over the Mariko kommando. During the attack on Lord Methuen's kommando, he was again wounded by the leg. During General de la Rey's absence to Klerksdorp, the negotiations that led to peace, he was appointed Vecht General over the Pretoria, Krugersdorp and Rustenburg kommandoes, in place of General Kemp, who replaced General de la Rey as Assistant Commandant General of the western districts.

H. R. Lemmer.

Hermanus Richard Lemmer.

General Hermanus Richard Lemmer was born on 29 June 1849, at the place Hartebeestfontein, distr. Potchefstroom. Not before his thirty-second year was his Calm life as a cattle farmer, disturbed by extraordinary events. It was two years after the first War of Independence, he volunteered to enlist in the war of Moshette vs. Mochua. Here he served as a Field Cornet over a division of the volunteers and was badly injured. Later, Lemmer represented his district in the First People's Council until the last war broke out. He made the battle of Kraaipan as but was later appointed as a Fighting General and in the beginning operated in Colesberg, then on the Natal border and then at various places in the Free State and Transvaal, in numerous battles and smaller battles. In a bloody battle between Lichtenburg and Marico, during the last days of 1900, he lost his life, while he led a storm-attack at the head of only a small number of civilians. Two of his sons were later killed as Field Cornets.

P. J. Liebenberg.

Petrus Johannes Liebenberg

Well before the war, Petrus Johannes Liebenberg represented his district in the second Volksraad, and when the war broke out, together with the Kommando that commanded by General Piet Cronjé, to the western frontier, and took part in the first battle at Kraaipan, where an armoured train from Mafeking was captured. He then took part in the first part of the siege of Mafeking, and went with General Cronjé to the Free State, took an active part in the battle of Modderrivier, while at Magersfontein under Cronjé he served as Commandant over some of the forts. In January 1900 Liebenberg was appointed Fighting General for a komraando destined to enter the Cape Colony at Prieska; Later, he entered the same capacity at Rooidam over Grikwaland West. As captain of the citizens of Bechuanaland and those of Lichtenburg, he fought at Kudurand and fought the relief detachment under Colonel Mahon advancing on Mafeking. In July 1900 at Donkerhoek, and was later definitively appointed general for the Potchefstroom Kommando, in which capacity he remained until the end of the war.

See also: https://en.wikipedia.org/wiki/Petrus_Johannes_Liebenberg

L. Meyer.

General Lukas Meyer

General Lukas Meyer first became widely known for his successful performance as one of the founders of the new Republic, later incorporated into the S. A. Republic and became part of the Natal constituting. From Naturellen Kommisssaris he became a member of the People's Council and later chairman of the First People's Council. As General of Freedom Citizens, he participated in the operations in Natal's field, but had to give up his duties due to illness. General Botha. vomit. Later, General Meyer held the office of Executive Member of the Council and, as such, took part in the final peace negotiations. After the Tract of Vereniging had become a fait accompli, Lukas Meyer went to Europe and died suddenly in Brussels on 8 August 1902 shortly after his arrival there. His dust remains were returned to his native land and rest in the cemetery at Vrijheid.

C. H. Muller.

Christiaan H Muller

General Muller enlisted as a private citizen in battle, Christiaan H Muller was promoted to Chief Corporal in Zandspruit of the Boksburg Kommando; with the main commando he crossed the frontier and took part in all the great battles and most of the smaller battles in the Natalse. At Spioenkop Muller acted for the first time as acting field cornet, at Pietershoogten, where his Field Cornet was wounded and the assistant was killed, he acted as a Field cornet, and then assistant field cornet, and on 26 May 1900 field cornet to be elected. Three days later the kommando left Colley's head of Majuba; some days later they fought at Babsfontein. The battles of Donkerliuk, of Olifansfontein, Dalmanutha and all the smaller battles in that vicinity he witnessed and was killed on 10 Elected Commandant in October, under General Ben Viljoen, He served as such in the protracted battle in November 1900, at Rhenosterkop, against Paget's detachment, for subsequently, on 30 December, to be appointed Fight-General over the Boksburge, Johannesburg and Police Kommando's. On 7 January 1901 he took a trip with his citizens to Helvetie the lyddite gun "Lady Roberts," and during the general attack on the Oosterlijn, who followed shortly afterwards, played a prominent part in the fighting. Monument Hill near Belfast. In April d. a. v., when the commandos north of the Eastern Railway, Muller was wounded in the shoulder, escaped with his commando by rail, and two months later, at Wilmansrust, captured the camp of the Australian volunteers, where two maxims were taken. Weather about it trail, he made two good blows with his commando in the Bosveld; and when General Ben Viljoen was taken prisoner of war, Muller was attacked as assistant Komm. Genl. for the north-eastern districts. In front of the Police, Boxing Burg and a part of the Middelburg kommando, he was delegated to Vereniging.

H. W. Mijburgh.

Marthinus Wilhelmus Mijburgh

General Marthinus Wilhelmus Mijburgh, born in Molenvlei, district Bredasdorp, Cape Colony, came to the Transvaal in 1888, at the age of sixteen. On 4 November, after the outbreak of the war, he was commanded and joined at Moddersspruit Commandant Dirksen. After the reconnaissance trip to experience Estcourt he remained at Colenso and joined the First Reconnaissance Corps there. Subsequently he took part in the battles of Colenso, Spioenkop, Middeldrift, Pontdrift, Pietershoogten, Allemansnek and more others in the Natalse. From Charlestown, the Commandant of this corps back to Krugersdorp; but Mijburgh and captains Thomas and Dames, refused to return with twelve men, and was appointed Lieutenant in the corps, he saw numerous fights, until Dames was wounded when he succeeded him. Afterwards, the corps took part in the battles at 'MTalene (also known as Itala), Spieshoek and Unexpected. On 23 March 1902 Mijburgh was appointed Fighting General over the kommandos of Utrecht and Vrijheid, instead of Genl. Emmett, who was a prisoner of war had been taken. Although as a representative of his Commandants for the Vereniging, Mijburgh, in the confused state in which those commandos in that and the threatening attitude of the blacks, had no opportunity to district. On the 6th of May a detachment of his men, consisting of 56 men, was Holkrans murdered in cold blood by a Zulu impi. On the same day, only three hour late, he came with a commando to the horrible scene of the murder, but the Zulus had already retreated within the English lines. When, after the peace, the southern districts were annexed to Natal, Mijburgh was elected a member of the House of Representatives of Utrecht.

S. F. Oosthuizen.

Sarel François Oosthuizen

The first years of Sarel François Oosthuizen's life passed quietly. Born on the place Sterkfontein, district of Pretoria, on 22 March 1862, he barked his father in his farm helpful, until he later started his own farm in the same place. The First War of Independence He witnessed, and served as a private citizen in the siege of Pretoria. In 1882 he took part in the expedition against Mampoer, then in the war against Mapoch, in 1894 against Malaboch, and in 1896 he was among the citizens who at Waterfall and Thornhead fought against Jameson. Immediately afterwards he was elected to assistant field cornet of district 3, Krugersdorp. In 1898 he took up the leadership of his civilians, legen 'MPefoe, and after his return Oosthuizen became an effective Field Cornet elected. As such, he served at the Battle of Dundee, where he received three wounds, then took part in the march around Estcourt, and on his return, at Colenso, appointed as WD. Commandant in the absence of Commandant Potgieter.

Posted on both sides of the railway at Colenso, he completed with his kommando in battle which was delivered there on 15 Dec. 1899, a distinguished role; in the same capacity he took an active part in the battle of Spioenkop. When Commandant Potgieter Petershoogten was badly wounded, Oosthuizen took his place, and was arrested for two days later appointed Fighting General. After the relief of Ladysmith, he went out with the Krugersdorp kommando to Veertienstromen, fought the battle at Witrand, at which point he ordered to advance on Gatsrand, and then took part in the fighting to Johannesburg and Pretoria. After taking part in the battle in Donkerhoek, he was Commandant-General ordered to return to the Krugersdorp and to remove the kommandos to get back on its feet there. With only fifteen men he undertook the journey of Balmoral, and on the 11th of July he chose at the place Dwarsvlei, Krugersdorp, against a enemy division position with 120 men. In the evening after the battle that followed, stormed Oosthuizen at the head of only 20 of his men, some guns and of which servants and horses had already been shot, and here he received the mortal wound, to which he succumbed on the 14th of August at Zanddrift, Crocodile River.

See Also: https://en.wikipedia.org/wiki/Sarel_Oosthuizen

He was succeeded by Commandants van der Merwe and Coetzee, the latter of whom shortly afterwards at Pienaarsrivier, met a hero's death.

J.D. Opperman.

Jacobus Daniel Opperman

General Jacobus Daniel Opperman was born 16 July 1861 in Cradock, Cape Colony, and came to in 1873 with his parents naair district Kliprivier, distriki Harrismith, Orange Free State. In In 1884 he joined the expedition against Uzebepu as a volunteer, and in 1893 was arrested by the Transvaal Government was appointed Lieutenant of Police and then went to Zwazieland sent. After the outbreak of the war, he went as a Field Cornet with the Zwazieland civilians to the scene of the war. During the war he was made a General. He experienced several battles until he was 20 steps away from the enemy. (Plumer's column) was killed with a bullet in the forehead. This was on January 4, 1901 at the place "Unexpected" in the district of Ermelo. This brave warrior remained faithful to his Land and People to death.

J. H. De La Rey.

Jacobus H. de la Rey

For a long number of years, Jacobus H. de la Rey had been holding his people in public and served on the battlefield and was a member of the First People's Council for Lichtenburg, the district where he lived, when the war broke out, de la Rey was considered Vecthtgenl. appointed to the commandos who went to the Western frontiers. Soon Genl. de la Rey held the rank of Assistant Commandant General. He first led the siege of Kimberley, but surrendered it to Genl. Sarel du Toit, when he was The proposal of the Government left for the Colesberg district and, with Genl. H. R. Lemmer the Transvaal Armed Forces. After the relief of Kimberley, he set out on his way to the from Bloemfontein to the north and, after witnessing the battle of Donkerhoek, he was ordered to defend the seven western districts of the Transvaal. organising, a task in which Genl. de la Rey had succeeded in success, when, shortly after the imprisonment of Gen. Lord Methuen, the peace negotiations began and led to the Vereniging tract. His fine military career has been with Genl. De La Rey had the esteem of friend and foe. After peace was established, Genl. de la Rey several European countries and became, as well as Genl. Louis Botha and Genl. B.C. the Law, wherever he appeared, treated with sincere admiration. Then, at the beginning of 1903, to Genl. the la Rey was offered a seat in the Transvaal Legislative Gouncil, the latter believed that the to refuse the offer. In the autumn of 1903, Genl. de la Rey went to British India, where he succeeded in to induce a number of reluctant prisoners of war to accept the terms of peace, on which their return to the fatherland depended. Mrs. de la Rey has written down in a book worth reading her experiences of her eventful Kommando life. Genl. de la Rey's eldest son was killed on 28 November 1899 at Tweerivier.

J. C. Smuts.

Jan Christiaan Smuts

General Jan Christiaan Smuts was born on 24 May 1870 in the Western Province of the Cape Colony studied at Stellenbos and Cape Town and then at Cambridge, acted as admitted to the London Bar and as such settled first in Cape Town, then in in Johannesburg. In June 1898 he was appointed State Prosecutor of the Z. A. Republic, succeeding Dr. Coster, and, after war was declared, he accompanied the kommandos to Natal and was present at the battles of Dundee, Modderspruit and Colenso. He was then recalled to Pretoria, but left the capital on the eve of the capture by the English troops. After the battle of Donkerhoek, Smuts went with Genl. de la Rey to the western districts, where he took an active part in the military enterprises. In August 1901 he received of Generals de la Rey and Chr. de Wet was appointed assistant Commandant General on the Boer Kommandos in the Cape Colony; and in this capacity he was in charge of the perilous but successful and adventurous journey through the Cape Colony, in which about two hundred Republicans took part, and which began at Herschel, to end in the north-east corner, at Port Nolloth, where the message of peace the brave leader of the brave Kommando. At the beginning of 1903 General Smuts was given a seat in the Transvaal Legislature. Council, which, however, he thought it had to refuse.

T. Smuts.

Tobias Smuts

General Tobias Smuts, born 25 November 1861 in Krugerspost, in the district of Lijdenburg, was educated in Stellenbos, K. Colony. During the Houtbosberg expedition he served as Voldkornet, and with the war against Magato as Kommandiint of Ermelo. At the beginning of 1899 he was elected to the First Volksraad (First People's Council). represent. When the last war broke out, Smuts went out with the department that went to Natal, was present at the battle of Modderspruit, made the reconnaissance to Estcourt and fought on the Tugela as a civilian. Subsequently, he was appointed as assistant Commandant of the Ermelo kommando, and, after the battle of Colenso, acted as Fighter General on the Upper Tugela, in which capacity he was present in the battle of Spioenkop. Some time thereafter he was appointed with the authority of assistant Commandant General of the Transvaal Kommando's, who went from Natal to the Free State and took part in the fighting at Brandfort, whereupon he went to Pretoria returned to be present at the last meeting of the People's Council held there. was held. He fought in Donkerhoek, then went to the Wakkerstroomse and remained in the Eastern districts until the end of the war.

C. J. Spruyt.

Cornelis Johannes Spruyt

At the outbreak of the war, Cornelis Johannes Spruyt was the village's field cornet Heidelberg and as such took part in the siege of Ladysmith; later to Commandant elected, he was immediately prepared to work with his citizens, to the relief of General Cronje. at Paardeberg. Taken prisoner at Kudurand (20 February 1900) he was bound in a railway waggon, under the command of an English guard. sent southwards. When his guards, under the influence of liquor, had fallen asleep, Commandant Spruyt jumped out of the train with success. five days after his escape, surrounded by enemies, he reached Colesberg, where the republican army still held out. After General Gravett's death Spruyt was appointed Asst. Commandant General in his place and as such a lively part in the war operations. On the evening of July 20, 1901, he attempted to between Val and Vlaklaagte stations, located on the Natal line, and was Dodelik wounded by a bullet that penetrated between his shoulders. An hour later, he the spirit and was buried in the place of Mr. Abraham Pretorius. In November In 1903 his surplus was surplus as well as that of nine other Heidelberg citizens solemnly interred at the cemetery in Heidelberg

S. P. Du Toit.

Sarel P. du Toit

General Sarel P. du Toit was, when the last war began, a member of the First People's Council for the Wolmaransstad, to which he was elected in 1896. On 9 October 1899 he closed the joined the western commandos at Polfontein, and was soon arrested by General Piet. Cronje was appointed assistant general for the districts of Wolmaransstad, Bloemhof and Lichtenburg. With General de la Rey he marched from Vrijburg to Kimberley, where he He was entrusted with the siege of the city. In December he was basketed with a horse sent to Tweerivier, and took an active part in the battle, which delivered there. General du Toit then appeared at Veertienstromen; After the capture of Pretoria along the Eastern Line on 'Msilkaatsnek, and then went to the evacuated south-western districts to bring the civilians back under the control of the weapons. Three days after his arrival at Wolmaransstad, du Toit had another Kommando of over 300 men. He then operated under General de la Rey in the districts Wolmaransstad, Lichtenburg, Potchefstroom and Bloemhof; a.o. in the battles to Hartebeestfontein, Lichtenburg, Makwassiebergen, Brakspruit (1 March 1902) and Roodewal (11 April 1902) in wolke last battle his kommando alone had a loss of 75 dead and wounded, including the brave Commandant Potgieter on field cornet G. J. van Wijk. A few days later, du Toit was sent by his citizens to the Vereniging Congress.

B. Viljoen.

Ben Viljoen

General Ben Viljoen was born in the Cape Colony, Ben Viljoen, when he was chief Transvaal officer at the He took part in the last war, already behind him a multifaceted career. Originally a member of the police force of Krugersdorp, he devoted himself to then to the Yurnalist, helped to found a local Volunteer Corps and was elected People's Councillor for the W.W. Rand. Under Genl. Kock to Elandsfontein, Natal, who had penetrated, managed to put the manore, who was placed under the command of Ben Viljoen. Johannesburgers to escape the consequences of defeat. Along the Tugela Ben Viljoen's men distinguished themselves in many battles, as well as in Pieters Heights and Klip River, at the W.W. Rand. Ben Viljoen operated as a General then north of the Delagoa Bay line and was taken prisoner of war in late 1901 and sent to St. Helena. General Viljoen chronicled his adventures in a work published shortly after the peace. was issued.

P. R. Viljoen.

Piet Retief Viljoen

General Piet Retief Viljoen, a grandson of the famous people's leader from the Great Trek, whose name he bears, performed his civic duty in the natural wars before the first Freedom War, served in the latter war as Adjutant to the Commandant- General, and also served in the Mapoch war in the same capacity. In In 1887 he was appointed Mining Commissioner in Heidelberg, and served the State as such until the last war broke out, when he received instructions from the Government to go to Natal in order to assist the Commandant-General there. When General Joubert became ill, the administration was temporarily given to Viljoen and General S.W. Citizen entrusted. After the death of General Joubert, he acted as Landdrost te Potchefstroom, and after the capture of Bloemfontein he functioned for a while as fighting general, until General A. Cronje was appointed, and when he resigned Potchefstroom, Viljoen fled the village with about thirty Grikwalanders. When he arrived in Heidelberg, he was involved in the reorganization of the command unanimously elected Fighting General, and marched to the Hogeveld. After in the he succeeded in having taken an active part in Witrand and Dalmanutha appointed President of the Military Court in Barberton. Then this place was overwhelmed, he escaped with seven men and returned to Bethal and Heidelberg and soon brought another commando of 300 men under arms. At a court-martial elected General, he continued to serve as such until his command was dissolved became part of the Heidelberg commando under General Spruyt. In the At the battle of Chrissiesmeer he was seriously wounded and after his recovery he was sworn in as a member of the Executive Council. When General Spruyt died, he was appointed in his place, remained in the field until the end and represented his command at the conference Vereniging.

See: https://en.wikipedia.org/wiki/Battle_of_Bakenlaagte

De Villebois-Mareuil.

Count de Villebois-Mareuil

Shortly after the outbreak of war, Count de Villebois-Mareuil published in the Transvaal, former chief officer in the French army, driven by sympathy for the cause of Farmers. Taking advantage of the offer of his services, the government sent him to Natal, where the siege of Ladysmith had begun. After the battle of Colenso having attended, he proceeded to the Free State, where General Roberts was inside stocky. Appointed general at Kroonstad, de Villebois collected a small group commando with which it moved in a southwesterly direction to the east of Kimberley to operate. On April 5, 1900, General de Villebois-Mareuil was killed nearby Boshoff, deeply regretted by the Boer commandos.

J.Van Zijl.

General Jacobus van Zijl

General Jacobus van Zijl was a citizen of the Cape Colony by birth. He studied to become a preacher in Stellenbos, but had no strong inclination for teaching feeling that way, van Zijl settled in the Christiana district as a farmer. Then the youngest war broke out, he joined the commandos under Generals de la Rey, Lemmer and Cronje, operated on the western border of the Transvaal. He took one active part in several major battles, and then the western commandos, after capturing the Transvaal capital, under Asst. Commandant-Generaal de la Rey were reorganized, Van Zijl - who had already recruited his fellow citizens before that time had served in various other positions - as a Fighting General in command of the Grikwalanders. With this commando, which has acquired a favorable name, he took over subsequently took part in the main military operations there. General van Zijl remained in the field until the end of the war. He has since published a book about his experiences and also visited Europe. General J. Van Zijl. General Jacobus van Zijl was a citizen of the Cape Colony by birth. He studied to become a preacher in Stellenbos, but had no strong inclination for teaching feeling that way, van Zijl settled in the Christiana district as a farmer. Then the youngest war broke out, he joined the commandos under Generals de la Rey, Lemmer and Cronje, operated on the western border of the Transvaal. He took one active part in several major battles, and then the western commandos, after capturing the Transvaal capital, under Asst. Commandant-Generaal de la Rey were reorganized, Van Zijl - who had already recruited his fellow citizens before that time had served in various other positions - as a Fighting General in command of the Grikwalanders. With this commando, which has acquired a favorable name, he took over subsequently took part in the main military operations there. General van Zijl remained in the field until the end of the war. He has since published a book about his experiences and also visited Europe.

Commandants

J. J. Alberts.

Commandant J. J. Alberts

Commandant J. J. Alberts. Born in the Pretoria district, June 8, 1872, Komm. Alberts raised in the district Standerton. Here he farmed until the war broke out. As a private citizen, he was appointed Vecht corporal in May 1900, he fought at Rhenoster river in the O. V. S. and was located in September, 1900, north of the Delagoa railroad. Over there his commando was surrounded by the English, which resulted in the people became so discouraged that they surrendered. Only Alberts escaped by horse. When he finally returned to the Standerton district, he learned that his Veldkornet had surrendered with his men, but he could not do so gain the heart to lay down his weapon. The Kommando, then by himself was formed, consisted of only 15 men from the Waterval District. This corporalship later grew back into a Kommando and Alberts became Asst in December 1900. Field cornet appointed by Commandant C. Brits. In June 1901 the famous Conference between our Government and that of the O.V.S. held in the Waterval district, whereby Commandant C. Brits became Battle General and Alberts became Commandant of the Standorton district, are appointed. So Alberts replaces Kommdt. C. British. Since that time he has remained Commandant of the District of Standerton until the end of the war, when Alberts was represented by some of the citizens of Wakkerstroom and Standerton was chosen to represent them at the meeting Vereniging, the 15th of May 1902.

S. F. Alberts.

Sarel Francois Alberts

Commandant Sarel Francois Alberts, born May 6, 1872, did military service for the first time Komm. F. J. Potgieter, following the Jameson raid. On December 30 He came in 1895 at the Queens' Battery, Krugersdorp, under fire for the first time and lived the next day the historic battle at the town of Doornkop, during which his horse was injured. In the autumn of 1898 Alberts and his Krugersdorp Kommando was sent against Magato in Zoutpansberg and acted as adjutant to the Commandant. At the outbreak of war against the British Empire, Alberts was reinstated Komm's adjutant. F.J. Potgieter; he took part in the battle at Talana Heuvel and then to all the battles where the Krugersdorp Komm. was involved. On the On the evening of the battle at Pieters Heights, the citizens of district 3 Krugersdorp elected him to Asst. Field cornet and a few months later it was acquired by Genl. S. F. Oosthuizen to Field cornet appointed. On June 24, 1901, the majority of Krugersdorpers voted Alberts as Commandant and as such he served until February 5. 1902, when he met a division of his men at Gruisfontein, distr. Lichtenburg, was made a prisoner of war. Carried away to St. Helena, Komm came. Alberts returned to Krugersdorp at the end of August 1902.

F. S. Alleman.

Frederik S. Alleman

Commandant Frederik S. Alleman was born on May 11, 1861 in Potchefstroom, a grandson of Coenraad Godfried Alleman, who served as First Lieutenant under Governor Van Need. Alleman received his first education at the Gray College in Bloemfontein, and settled there afterwards settled in Klerksdorp, where he met his fellow citizens in various public meetings relations served. Went to war as a private citizen under General Gronje he took part in the first battle at Kraaipan. In Koedoesberg he acted as an assistant Field cornet and was seriously wounded in that battle and, when he still had to go on crutches, too Klerksdorp captured. Transported to the Cape Colony, he jumped off at night December 26 at Drie-Angel station from a train in full motion, succeeded in his escape and shortly afterwards reached General Hertzog's command at Calvinia. With reports sent to General de la Rey, he arrived at Hartebeestfontein in March 1901, where he rejoined his own commaudo and soon became a field cornet was elected. In June of that year, field cornet Alleman accompanied General de la Rey to the Eastern districts and Branddrift, where the famous conference was held became. Returning to the western districts, he was arrested by General Jan Smuts asked to follow him to the Cape Colony, and in December of the same year he was appointed Commandant. At the beginning of 1902, Alleman was removed from Calvinia district, by German South West Africa, sent to Europe with a report of General Smuts to President Kruger, which difficult mission he successfully accomplished. During the war he experienced, among other things, the battles of Tweerivier, Magersfontein, Koedoesberg, Reitz and Tarkastad.

A. D. Badenhorst.

Adriaan Dirk Badenhorst

Commandant Badenhorst was born June 3, 1852, in Natal, Adriaan Dirk Badenhorst came in 1860 with his father Casper Hendrik settled in the Verzamelbergen, distr. Wakkerstroom, establish, where when barely half a dozen Boers lived there. To the wars of 1880, against the English, and 1882 against Mapoch, Badenhorst took part as a private citizen; he also captured in the last war under Commandant J. A. Joubert and Veldkornet H. Sassenberg - the latter died on the mountain near Dundee - entered military service. Until deputy of V.K. H. Sassenberg was chosen beforehand, he resigned major fighting in Natal, Transvaal and the Free State part. Near on February 6, 1901 Ermelo was wounded by a bullet in the leg, Badenhorst had to be unemployed for two months stay, but soon afterwards he found himself again with a crutch in one and the other Mauser in the other hand, into the ranks of his citizens and held with them to the last ready for battle. During this war he lost his wife and one of his eight brothers.

C. P. S. Badenhorst.

Christoffel Petrus Stephanus Badenhorst

Commandant Christoffel Petrus Stephanus Badenhorst, born September 29. 1870 on the site where later the village of Wolmaransttad was established, lost his father Hendrik at the age of 6 Johannes Badenhorst, who died in the Sekoekoenie war. As a private citizen with the Pretoria Kommando Natal, he was withdrawn after battle of Nicholsont's Nek, Corporal and subsequently officiated at every engagement in which its citizens took part, as wd. Field cornet, especially in the battles of Colenso, Tugela, Spioenkop and Helpmekaar. After the latter meeting as Asst. Field cornet elected, he served as such until after the battle of Selekaatsnek, where Komm. Coetzee was killed and Badenhorst was appointed Commandant. In this capacity he took took part in the battles at Nooitgedacht, Hekpoort, Benoni, Boksburg, Vlakfontein and many more nameless skirmishes. On December 13 In 1901 he was taken prisoner of war and sent to prison St. Helena transported.

J. F. De Beer

Jacobus Frederik de Beer

Commandant Jacobus Frederik de Beer, on 18 Nov. 1853 in Vlakfontein, Wolmaransstad, born and trained in the Transvaal, served as a field cornet in the War of Freedom from 1880 - 1881 and was elected as Commandant for the Bloemhof district in 1882. He took part in the Mapoch War and in 1885 was with its citizens in Momoesa on the Western border. In In 1897 he was elected Volksraad member for Bloemhof. The Bear served as Commandant during the last war and took part in several major battles, especially in Boorlaagte and Tweerivier; he was wounded in Paardeberg and took over following events played an active role, including in the fighting of Boekenhoutskloof, Selikaatsnek and Hartebeestfontein.

J. T. Bierman.

J. T. Bierman

Commandant Bierman as a private citizen who went to war at the end of November 1899, J. T. Bierman was captured was briefly elected and served as Adjutant by the citizens of his district of Utrecht several times in the absence of the Field Cornet in its place. On November 18 1901 he was finally chosen as their field cornet by the citizens of District II, Utrecht. Short before the end of the war, General M. W. Myburgh appointed him Commandant of the district.

J. S. F. Blignaut.

J. S. F. Blignaut

Commandant J. S. F. Blignaut eldest son of Pieter G. Blignaut, who died in Elandslaagte on October 21, 1899, prayed J. S. F. Blignaut served with distinction in the Artillery Corps at Pretoria, and afterwards in the Police Force, where he held the rank of sergeant. When taking over Swazieland he was appointed Second Lieutenant and later First Lieutenant promoted. He went to war with the Swazieland commando and became Commandant der Citizen and Police Force appointed and killed on September 26. 1901 in Itala, Natal.

J. D. L. Botha.

Commandant J.D.L. Botha

Commandant J.D.L. Botha. Born October 28, 1837 in distr. Oudtshoorn, Cape Colony. Arrived in Transvaal the year 1849, he was chosen as a field cornet in 1864 for the Klein Marico district, distr. Marico and in 1884 as Kommdt. for the dist. Marico. He served through the last war until the end.

B. D. Bouwer.

Commandant Barend Daniel Bouwer.

About twenty-five years ago a number of Boer families left the Transvaal in order to settle in Damaraland and it was here, during this "Dorstland-Trek," that Barend Daniel Bouwer, on January 31. 1875, born beheld. Returning to Transvaal, he received his education in Pretoria and entered the profession subsequently in state service, initially at the Commandant General's office, then on the Department of the State Attorney. When the war broke out, he was a public servant Prosecutor in Ermelo. Then, during the last days of 1895, it became known Jameson had crossed the border, Bouwer informed him, by order of the Government, to return immediately before any fighting had taken place. As a witness in the known He visited the Parliamentary Inquiry that was subsequently instituted against Jameson next year England. At the outbreak of the latest war he joined it Ermelo commanded, contributed to the main battles in Natal and later joined the Transvaal squadron of Pres. Steyn, under Commandant J. Boshoff, to the west districts. At the end of 1900 he was appointed Commandant of General de la Rey one of the commandos who, under General Jan C. Smuts, went to the Cape Colony. Here about 150 Colonials joined him and he took then, until the end of the war, an active share in military operations there. He was wounded four times: once in Rustenburg, then in Windhoek, Cape Colony, then in the Van Rhijnsdorp district, Cape Colony, and finally, at the siege of O'okiep, Namakwaland, on April 9, 1902. See also: https://af.wikipedia.org/wiki/Ben_Bouwer

D. J. J. Breytenbach.

Daniel Johannes Jacobus Breytenbach

Commandant Daniel Johannes Jacobus Breytenbach was a field cornet at the outbreak of the war of district 1. district Bethal, and was elected Commandant after the battle of Dundee. In the battle of Bakenlaagte, October 30, 1901, his arm was shot off, and on January 5 In 1902 he fell into enemy hands and was transported to St. Helena.

B. H. Breytenbach.

Barend H. Breytenbach

Commandant Barend H. Breytenbach, born in the year 1862, Utrecht district, is a son of a of the Natal pioneers of that name. He grew up in the Utrecht district and took Already at the age of seventeen, he took part in the Volunteer Corps that was founded in 1879, under the late Piet Uijs fought against the Zulus. In the War of Freedom of 1880 - 81, he served from start to finish as a corporal. In the past he acted as a citizen in all the smaller ones natural wars of his time, which was also the case in the natural wars after the first War of Freedom. In 1894 he was chosen as a field cornet, and then the last war broke out, Breytenbach and his department of the Utrecht people moved under General Lukas Meyer entered Natal and took an active part in many of the greatest battles that took place there. During the latter part of the war he was appointed Commandant and served as such until he was finally elected to his commando by the citizens was appointed to represent them at the Peace Conference.

W. H. Buhrmann.

Willem Hendrik Buhrmann

Commandant Willem Hendrik Buhrmann, the son of the well-known Hendrik Theodoor Buhrman, was born on August 1, 1861 in Lijdenburg, from where his parents lived on the Hogeveld, in the Ermelo district. By his parents and by his own inclination towards the Driven to theology, he studied in Stellenbos until 1888, but had to resign from teaching for health reasons. During the Jameson raid he took part in the Ermelo commando, which, however, was not included fire has come. During the Magato War his life was in danger from both enemy lead as from febrile illness. And at the outbreak of the last war, belonged he was among the first to enter enemy soil on October 11. After as usual citizen to have taken part in all the battles with his commando, including in March Having moved to Brandfort in the Free State in 1900, he belonged to them Ermelo people who showed great agility with their commando. From the He was given membership of the Horse Commission on October 23, 1901 by Komm. Genl. Botha, when he appointed him as Commandant of Ermelo in place of J. N. H. Grobler, who was promoted to Fighting General. Seven days after his appointment he took part with his citizens in the great battle of Bakenlaagte and afterwards in the battle in Onverwacht, where General Opperman, as well as his faithful adjutant Adriaan Buhrmann, before the Boer powers were lost. Commandant made it to the end Buhrmann joined the battle and finally led his loyal citizens to the place Twijfelfontein, to lay down their weapons there with them.

S. Buys.

Stephanus S. Buys (Bujis)

Stephanus S. Buys had served the Transvaal Government since 1881 and took over as Commandant der Heidelberg citizens participated in the war. He attended the battles Mud spruit, Colenso, Spioenkop and Pontdrift were added on September 9. 1900 near Lijdenburg was wounded by two bullets, of which only one could be removed. On Nov 20 in 1900 his leg was shot off and he fell into the hands of the enemy having witnessed more than fifty battles, battles and skirmishes. Commandant Buys lost his eldest son at the battle of Moddersspruit. He has tans a wooden leg.

G. M. J. Van Dam.

Gerard Mari Johan van Dam

Commandant Gerard Mari Johan van Dam, born in Delfshaven, Netherlands, October 9, 1855, came to Transvaal with his parents in 1868 and lived in Potchefstroom. When President Pretorius was captured by the English in January 1880, Van Dam was one of the burghers who formed the Kommando at Nauwpoort, near Potchefstroom, and was with Genl. Kock was chosen as representative of the producers to monitor the course of the case. In November 1880 he was with the Schoonsprnit-Kommando for the protection of Messrs. P. A. Cronje, Bezuidenhout and others, and was among those against whom orders for arrest had been issued because of what happened to Bezuidenhout's car in P.C. Stroom. In the meeting of the People's Committee with the English Colonial Secretary Hudson in Kaalfontein, a few days later, Van Dam was appointed Secretary of the Committee, and at the Meeting in Paardekraal on December 12, Secretary of the People's Council. As such, he had the honor of being the first sworn official of the restored Republic. After raising the flag in Heidelberg he left with the first commando under Komdt. Genl. Joubert to the Natal border, and took part in the fighting at Lang's Nek and Ingogo. To P.C. Sent to Stroom due to illness, he later served under Genl. Cronjé, and nowadays at the capitulation of the English camp under Col. Winsloe. In September 88 at the end of Komdt. of Police on the Klerksdorp Goldfields, from there transferred to Johannesburg as second in command. Appointed Komdt in November 1896. of the Witwatersrand Gold Fields Police, which position he held at the outbreak of war. On April 19, 1901, sent with two others to the Swazi captain Umbudula on the border of Transvaal and No Man's Land, they were ambushed in the kraal by a group of English and armed Swazies, during which the Sergt.-Major Lombaard was shot dead and Van Dam was wounded. Taken prisoner and sent to Durban, he went to camp at Cape Town and was sent back to Ladysmith, where he was held until the end of the war.

T. F. J. Dreyer.

T. F. J. Dreyer

Commandant T. F. J. Dreyer, as a civilian of the Potchefstroom commando, carried out the siege General Piet Cronjé's knowledge of Mafeking and all the fighting on the western border to escape lower to Magersfontein, participated in the battle of Abrahamskraal and was defeated the occupation of Pretoria was first elected field cornet and then Commandant. He as such fought several heavy battles in the districts of Rustenburg, Klerksdorp and Potchefstroom and followed with seventeen during the second half of 1901 year-old son of General J.C. Smuts on his journey through the Free State, as well as with a seventy Potchefstromers. However, Commandant Dreyer failed to reach the Cape Colony and lived to the age of 28 September 1901, in the south of the Free State, captured after fierce fighting and sent to India.

J. L. P. Erasmus.

J. L. P. Erasmus

Commandant J. L. P. Erasmus was born in Burgersdorp, Cape Colony, October 13, 1865, J. L. P. Erasmus arrived in 1887 to Transvaal, where two years later he settled as a lawyer in Johannesburg and in was naturalized as a citizen in 1896. He attended most of the major battles in Natal and subsequently served under General Gravett and General Groot Fred. Grobler, wow he joined General Celliers and the Zeerust Kommando. Started in 1901 in the Having entered the Free State, Erasmus managed to persuade 18 Free Staters to move into the to withdraw into Cape Colony, with which he reached the Burgersdorp district and generalized Kritzinger met. There Erasmus was appointed Commandant. In August 1901 Commandant Erasmus was surrounded by superior force in the Steynsburg district captured and sent to India the following October. He was twice injured during the war.

P. J. Greyling.

Pieter Jacobus Greyling

Commandant Pieter Jacobus Greyling, born 5 February 1847 in the district of Ladysmith, Natal, died in 1865 for the first time on a command to help the Free Staters against Masoes. At thirty When he was old, he left the Potchefstroom district for the Hogeveld and settled in the Standerton district. He took part in the battles of Langenek and Ingogo, 1881, and during the Mapoch War and was elected Field Kornet around 1895. Then the new district of Bethal had been proclaimed, it was unanimously adopted by the citizens in 1888 elected Commandant, and as such took part in the Natal operations next year. Due to deafness, Commandant Greyling was forced to return to March 00 to resign his position, but kept the field and was dismissed on April 14, 1902 made prisoner of war and sent to British India, from where he first returned after General de la Rey remained behind there through his personal intervention had induced citizens to sign the declaration in question.

I. J. Greyling.

Izaak J. Greyling

Commandant Izaak J. Greyling began his military career as a private citizen and served with two others brothers took part in the battle of Dundee, where one of the three was killed and two were wounded became. After Colenso he became Commandant of the Wakkerstroom commando chosen with part of which he witnessed the battle of Spioenkop. During the battle van Platrand he lost three men, including the brave Veldkornet Slabberts, who with his son fell on the mountain. Three more of its citizens were killed at Pieterstatie. Subsequently withdrawn from the Free State, this commando lost at Tafelkop nine troops to wounded. During the remainder of the war, Commandant operated Greyling on the Hogeveld and took part in the battle at Spioenkop, Vrijheid, and in the fighting near Itala, where the brave Field Cornet Pieter Badenhorst was killed and a number of Wakkerstromers were injured. After at Melmoth the enemy's convoy had been taken, they came to blows again in the Freedom. Shortly before closing the peace, his eldest brother, Veldkornet A.C. Greyling, was killed in a battle on the place Vijfhoek, and shortly afterwards field cornet Stephanus Moolman from district II. After a To persevere steadfastly, Commandant Greyling's citizens were forced to move to 12 June 1902 to lay down their arms.

H. S. Grobler.

Hendrik Stephanus Grobler

Commandant Hendrik Stephanus Grobler, born in the Rustenburg district, moved in 1870 to age eleven, with his father to a part of distr. Middelburg, now a part forming part of the Bethal district. He experienced the first English war as a civilian, and belonged to the Boer force, which surrounded Pretoria. It next year he took an active part in the war against Mapoch. After the Jameson Grobler was chosen for the first time as assistant field cornet for the district Middelburg, and a year later, when Bethal was declared a separate district, to be chosen as an effective field cornet, while he is a government official received appointment as Natural Commissioner. At the outbreak of the last war, he went to Natal with his commando and the first battle was fought there, namely in Dundee, partly. Later he was too Trichardsdrift and Coienso. In 1900 he was sent to the Netherlands with a commando of 250 men Orange Free State sent; the Free State capital was already in the hands of the enemy, and with the other commandos he fell back to Kliprivier, nearby Johannesburg, where stubborn resistance was offered for two days. He then took part with his people in the fighting in Donkerhoek (Diamondhill), Greylingsstad and Leewkloof, and after the enemy at Bergendal (Dalmanutha) had broken through, he was taken to their district with his commando was sent back, and was present there with General Muller, among others, when the camp of New Zeelanders was captured at Wilmansrust. In 1901 he was appointed by the Commandant General promoted to Commandant, was in the battle of Bakenlaagte and was delegated to the Vereniging by his commando at the peace congress.

Grobler.

Salomon Grobler

Commandant S. Grobler. Commandant Salomon Grobler of the Freedom district, initiated the last war from start

to finish with his three sons and two sons-in-law. At first he became called up as a field cornet from district 3 Vrijheid, to later function as Commandant and later still for a time as acting Fighting General. Made in Natal Grobler took part in all the main battles as well as in the southeastern districts of the Transvaal, such as Dundee, Moddersspruit, Platrand, and more, which are too Spieshoek, Itala, and others.

J. A. Joubert.

Jozua Adriaan Joubert

Commandant Jozua Adriaan Joubert is the son of Dayid Jozua Joubert and was born October 15 53 in the Potchefstroom district. After a short stay in Natal, the family settled Joubert is located in the Wakkerstroom district. He passed away at the age of 21 Commandant C. Joubert (Lang Christiaan) a government expedition, partly for formal reasons recognition of the Swazie chief Bandene. Then he made the first one Freedom War and was defeated in the battle of Lange-nek, on January 28, 1881, in the shoulder injured. The following year he took part in the war operations against Mapoch. In 1892 Joubert was proclaimed Field Kornet for district II of his district, and was found three years later he joined the Wakkerstroom commando that rushed to Johannesburg to Jameson's raid. In 1897 he became Commandant and took part in it commando against Magato. In the most recent war he led his citizens in the fighting Dundee, Modderspruit, and Colenso, in which last battle he was wounded; Like this he had to stay in the hospital in Pretoria for some time. Not before May 22, 1900 recovered to such an extent that he could again take command of his Commandant. Exactly one years later, his arm was treacherously shot in the village of Amersfoort and his horse was so injured under him that it soon gave way, and he had to go on foot to go. However, with the help of his adjutant, Pieter Spaarwater, he succeeded and became admitted to the field hospital of Dr. Jurriaanse, to be in that hospital a month later to be made a prisoner of war. His arm was amputated in Wakkerstroom and he died to the prison camp at Ladysmith, where he remained until the end of the war.

J. C. Kriegler.

Johan Christian Kriegler

Commandant Johan Christian Kriegler, born in the Cape Colony, lived during the last war broke out in Heidelberg and soon went to war under the command of the then Commandant Weilbach. After experiencing most of the major battles in Natal part of the commando of which Kriegler was an officer was sent to the Free State sent. In the battle that was fought soon afterwards at Tafelkop, he received a serious bullet wound, the consequences of which he succumbed to in Brandfort on June 9, 1900. He was also generally respected and loved in church matters and served his community as a deacon for many years. After the war, his remains were recovered, together with those of Mr eight others of the Heidelberg commando, on October 17, 1903, under great public interest was buried in Heidelberg.

J. A. D. Kruger.

Jacobus Andries Diederik Kruger

Commandant Jacobus Andries Diederik Kruger was born in Rietvlei, district Potchefstroom, on May 31, 1871.

In 1891 he enlisted in the State Artillery, later joined the Swazielaind Riding Police and held the rank of sergeant when the war broke out. Under the late General Chris Botha, who was then Commandant of the Swazieland Police, he took part in the battle of Colenso, and then in that too Forest edge, Petershoogten, Helpmekaar, Scheepersnek, Allemansnek, Spieshoek, Talene, Melmoth, and numerous smaller battles. Kruger became a lieutenant from Tweedo August 1901 promoted to First Lieutenant and then after the Battle of Talene, to Commandant of the Swazieland Commando. He became heavy several times wounded and remained in the field until the end of the war.

D. P. Liebenberg.

Daniel Pieter Liebenberg

Commandant Daniel Pieter Liebenberg, born in Graaff Reinet, on June 7, 1869, left the Cape Colony in 1887 and settled in Pretoria. In 1888 he entered government service, and was, when the last war broke out, working as Assistant Chief of Mines. He took part in the military operations in Natal with the first commandos then went with part of the Pretoria commando to the Free State in Brandfort. After the surrender of the Transvaal capital, Liebenberg moved with the commandos the Hogeveld and took an active part in the battles of Donkerhoek and Dalmanutha. By Hectorspruit with the Krugerdorp commando via Leydsdorp to Pietersburg, and thence under General Kemp to the western districts. In February 1901 he became served as Field Kornet of district 1 Krugersdorp, and a year later as Commandant of the Pretoria village command. He stayed in the field and experienced several battles until June 7, 1902, when he arrived at Doornkom, distr. Rustenburg, with his kommando laid down his arms and returned to Pretoria after an absence two years and seven days.

J. P. La G. Lombard.

Johannes Petrus
la Grange Lombard

Commandant Johannes Petrus la Grange Lombard was born on July 24, 1846 in the Riversdale district, Cape Colony, born. At the age of ten he accompanied his parents, who went to the O.V. left the state and settled in the Bloemfontein district. In 1865 he took part in the Basoetu war (Moshesh) and occasionally fulfilled the position of Veldkornet and Commandant. In 1873 he settled in Standerton county. When it was decided in 1880 to war with England, he was elected Commandant of the Standerton district and also functioned as such during the first English war. He accompanied in 1881 as hostage the troops under Buller from New Castle to Potchefstroom and back. From 82 to 1894, he represented Standerton in the First Volksraad, then left to Middelburg, and then also resigned as Commandant since 1880 clothed with him, because the law required a Commandant in the district of which if he was, he had to be resident. In 1896 he was again made a member by Bethal First Volksraad elected and remained so until the war. At the outbreak of the latest war, the Dutch Corps offered it to him Commandantship, which he accepted, but then that command, after Elandslaagte, ceased to exist separately, he served as a private citizen until the Government appointed him first as a member and later as chairman of the military court in the eastern districts, where he remained until the end of the war.

J. T. Martins.

Commandant Jan Thomas Martins

Commandant Jan Thomas Martins was a field cornet in the Gatsrand district for 21 years Potehefstroom and had previously been the site of Commandant taken. He was also appointed Commandant in the last war took part in the military operations under General P. A. Cronjé, at the same time as he Paardeberg was captured. Transferred to St. Helena, he died there on October 7, 1901 at the age of approximately 58 years.

H. F. Meyer.

Commandant Heinrich Friedrich Meyer

Commandant Heinrich Friedrich Meyer was born September 26, 1857 and was born as a field cornet elected by the Citizens of District II, Distrikt Piet Retief. In October 1899 he took took up arms with his department. After having learned a lot in warfare in various large and small battles, he had the misfortune of falling into the hands of the enemy on March 1, 1901. Of he fought off the contempt of death with his faithful corporal F. Martin, until they did not could fight more. He was put before the enemy's council, and by many false ones accusations were made to him that there was no mercy for him and that he had to die. After he had heard all the threats, he took his Bible, which... the enemy had left him and let it fall open, and at the first glance it caught his eye Psalm 41. He read from the 1st to the 3rd verse. By these 3 verses he was completely reassured, and told his faithful Corporal F. Martin: "We have nothing to fear." And it was Like this. After a few days he was informed that he would now be taken as a prisoner of war to be considered as.

J. G. Meyer

Commandant J. George Meyer

Commandant J. George Meyer was a field cornet from the Kliprivier district, Heidelberg district, for many years was promoted to Commandant during the war, in which capacity he served participated in several battles until the end of the war.

T. C. G. Meyer.

Commandant Theodor Christian Gotthard Meyer

Commandant Theodor Christian Gotthard Meyer, from Amalienruh, Luneberg, Piet Retief district, is born in Umpumulo, Natal, October 23, 1863. He served at the beginning of the war as corporal in the Piet Retief commando and was present at the battles of Dundee and that during the siege of Ladysmith. After the capture of his older brother Friedrich he was elected field cornet at the end of February 1901. On 6 Elected Commandant in June 1901, this election was made by the Government of Z.A.R. approved per appointment. He fought until the end of the war, and although he fought countless battles attended, he was never injured

W. J. Mouton.

Commandant Willem Johannes Mouton

Commandant Willem Johannes Mouton was born December 26, 1866, in the Calvinia district, Cape Colony, but was only five years old when his father moved to the Transvaal and located in the district. Middelburg established. On February 8, 1881, his father was killed in action battle of Schuinshoog or Ingogo. When the last war broke out, Mouton was one wealthy cattle farmer in the Middelburg district. As a citizen he already made the most important ones fought in Natal, and was chosen as a non-commissioned officer for the first time after reorganization of the Middelburg commando on the Hogeveld. He was in Middelburg kommando, when it did not act in collaboration with other kommandos, along the railway stood guard and did patrol work. From October 1900 to February 1902, his relatively small corporalship lost 12 men killed and 17 injured, including himself. Ordered from his station to Talene, he was there appointed as field cornet. In the battle of Bakenlaagte he lost a brother and for that one time had also already fallen. On February 13, 1902, Mouton was appointed as Commandant of part of the Middelburg Command south of the railway, in in which capacity he and his commando laid down their arms on June 7, 1902.

P. C. Van Niekerk.

P. C. van Niekerk

Commandant P. C. van Niekerk (later commonly known as Oubaas van Niekerk) came in 1887 from the Humansdorp district, Cape Colony, to the Transvaal, and has since lived in Germiston (Elandsfontein). His first military operations here were related to the commandos who had to deploy so unexpectedly in 1895 against Jameson's invasion, giving himself full citizenship of Transvaal acquired. On October 30, 1899, Van Niekerk joined the commando as a private citizen. to Colesberg, and then in February 1900 his field cornet, later General Gravett, was appointed Commandant, van Niekerk was chosen as field cornet in his place. In September of the same year, while his commando was in Lijdenburg, he was appointed Commandant and subsequently took part in many battles Hogeveld, was wounded twice and was lucky to be in the field until the end to return home with its other citizens after the peace negotiations.

J. J. Pienaar.

Johannes Jacobus Pienaar

Commandant Johannes Jacobus Pienaar, a grandson of the old Free State Commandant same name, was born at Ladybrand, April 16, 1877. His parents moved to the Transvaal when he was a youth; and when the family returned in 1894, Pienaar continued to work in Johannesburg as a civil servant at the Post Office. With the distribution of the Witwatersrand into special Field Kornetships, towards the end of 1896 he clerked in the Central Department office. When war broke out, he was appointed assistant field cornet and moved with the General Schoeman's department went to Colesberg, where he was elected Field Kornet under General Lemmer's command. He took part in the battle of Rensburg Siding and was wounded in another battle there. Then Pienaar fought in Donkerhoek, Dalmanutha, Witpoort, Helvetia, Belfast, Vaalkop and more other places wounded several times and suffered in the latter part of the war appointment as Commandant.

C. J. H. Du Plessis.

Casper Jan Hendrik du Plessis

Commandant Casper Jan Hendrik du Plessis, in 1841 at the place Modderfontein, distr. Rustenburg, born, took part in the Kommandos against Mapoch, Katze-Kats (Zoutpansberg) and Mapela, as well as the first War of Freedom. Since its foundation, he has been a member of the Second Volksraad for Rustenburg. When the last war broke out, he was with a commando along the North Westelike borders, destined to resist the incursions of Linchwe's people. In the Komdt fought for the later settlement of the military operations. du Plessis at Elands River, north of Pretoria, and in Rustenburg and on 7 July 1901, when he attacked the village of Rustenburg, seriously injured. Until Olifantsnek by its citizens released, the latter had to give way to enemy superior power and fell Coming. du Plessis into the hands of the English, who took him to the hospital in Rustenburg and a few weeks later sent to Ceylon, from where he returned after peace returned to his district.

F. J. Potgieter.

Frederik J. Potgieter

From the age of sixteen, when Frederik J. Potgieter was a conscripted citizen became the Transvaal, he performed his civic duty in all the natural wars of his time, and also experienced the first War of Freedom from 1880-81, from start to finish, Potgieter also fought in the later natural wars, until he was promoted to Commandant in 1895. was chosen over the Krugersdorp district. Six days after he was as such sworn in, he received telegraphic instructions from the Commandant-General to call on citizens to oppose Jameson. The operations against it followed Malaboch and other rebellious black chiefs. When the last war broke out, it closed he joined General Meyer with his commando in Zandspruit, and was present in all the battles in which this department took part. During the fortnight's battle the Tugela and Pieters Heights, Commandant Potgieter was seriously wounded in the chest, as a result of which he had to refrain from all activity for eight months. After death of General Oosthuizen, Mr J. L. v. d. Merwe as Commandant over the Krugersdorpers, but always consulted Commandant Potgieter, who, wounded as he was was still with the commando. In August 1900 he resumed his official position duties and operated in the western districts under General de la Rey. In June 1901 Potgieter was appointed Landdrost of the Krugersdorp district and in February of it next year together with Commandant Alberts in the Lichtenburg district taken prisoner and sent to St. Helena.

See Also: https://en.wikipedia.org/wiki/Frederik_Jacobus_Potgieter

R. C. Preller.

Commandant Robert C. Preller

Commandant Robert C. Preller, born May 5, 1846 in Pietermaritzburg, moved to the Transvaal, was in the Orange five years later, when the Basoetu war broke out Free State; and under Veldkornet F. van Niekerk and Commandant Theunis Botha he made the battles at Nauwpoort, Thaba Bosigo and Koraanderberg. In the war against Sekoekoeni, he acted as an assistant Field Kornet and in the battle against Mapoch as a civilian. In In 1893 he was appointed member of the Zoutpansbergse Okkupatie Komvisie, in which he remained in this capacity until 1895. When the last war broke out, he was appointed Commandant of a Commando, consisting of citizens of Johannesburg and Pretoria, moved to the southern borders of the Free State and is elected there by the public in the same position. below generals H. R. Lemmer, De la Rey, and the Free State Chief Commandant, he operated there against the English section, which advanced to the relief of Kimberley, made, among other things, the battle of Rensburg Siding and several other battles in the vicinity of Colesberg co. He was wounded in Rustenburg and became a prisoner of war on September 7, 1901 taken and transported to the Shajahanpur camp, India.

W. F. Pretorius.

Commandant Willem Francois Pretorius

Commandant Willem Francois Pretorius, born May 22, 1852, was at the time of the last war broke out, Veldkornet from the Hogeveld district of Heidelberg, shared the fate of this komraando during the first part of the war in Natal, and was in the promoted to Commandant in further course of military operations. After the capture of Pretoria and the battle of Dalmanutha, Pretorius operated with his kommando Heidelbergers north of the Delagoa Bay track, in the vicinity of Rhenosterkop, to switch to the eastern side during the last part. He was wounded shortly after the battle of Bakenlaagte, but quickly recovered and returned to duty to stand at the head of his commando until the end.

H. F. Prinsloo.

Hendrik F Prinsloo

At the outbreak of the war, Hendrik F Prinsloo was a field cornet of one of the three departed from Carolina, and after the resignation of Commandant David Joubert, was placed in his position place chosen. He distinguished himself at the various battles in Natal and also later, and was exceedingly beloved by the citizens who served under him. And when he got up November 7, 1901, together with General J. C Fourie, in Witkloof, Carolina district, was killed, the civilians were so defeated that they could not take a step towards the enemy who was already fleeing continued further. Hendrik Frederik Prinsloo (1861–1900)
https://en.wikipedia.org/wiki/Hendrik_Frederik_Prinsloo_(1861%E2%80%931900)

Later he was killed in action together with general Joachim Christoffel Fourie by the British at Witkloof in the Battle of Leliefontein (Battle of Witkloof), district Carolina, on 7 November 1900. His burgher soldiers felt so dejected by the loss of Prinsloo that they did not pursue the fleeing enemy.
https://en.wikipedia.org/wiki/Battle_of_Leliefontein

J. P. N. Prinsloo.

Joachim Petrus Nikolaas Prinsloo

Commandant Joachim Petrus Nikolaas Prinsloo, born in the Pretoria district around 1863, lived at the place Rietfontein, Bronkhorstspruit district, when the last war started. In When he went to the front in November, he attended the main battles in Natal, was wounded near Standerton, fought at Donkerhoek and then went with Genl. Erasmus to 't Bosveld north of the Oosterlijn. Returning to his neighborhood, Prinsloo was chosen as a field cornet and shortly afterwards as Commandant. While the peace negotiations were already underway, Komdt was killed. Prinsloo as a brave man that he was, in a fierce battle at Verkleum-kop, district Heidelberg, on April 31, 1902, in which same battle Veldkornet van Niekerk, from the Bethal Kommando, assistant Veldkornet J. Herman, from Pretoria, and thirty other citizens, while, among others, Lt. Farrel, of Naudé's Scouts, was wounded.

H. C. J. Van Rensburg.

Hendrik Christoffel
Janse van Rensburg

Commandant Hendrik Christoffel Janse van Rensburg was born July 21, 1841 in the district Rustenburg, but moved to Zoutpansberg in 1857 and settled in Schoemansdal. In 1879 he was chosen as Commandant for the district Zoutpansberg. He took an active part in the first English war, as well as in the natural wars against Malabochi, Magoeba, Majaadje and Magato. When the last war broke out, he moved with his citizens to the northern borders, but was later ordered to Natal, where he took part in the battle of Colenso. After taking Pretoria, he withdrew with his commando to Zoutpansberge and was there Pienaarsrivier, Warmbad and Buiskop in many battles these days. Then the enemy When he advanced into Zoutpansberg, the commando was completely dispersed and before he retreated could join General Beyers, he was forced to do so by a fever, which was minor number of its citizens, the presence of a large number of women lower and the threatening attitude of the blacks, to surrender his command in May 1901, and was prisoner of war sent away.

P. J. Riekert.

Commandant P.J. Riekert

Before the start of the first War of Freedom as a field cornet of the Elandsrivier district, appointed in the Rustenburg district, P.J. Riekert served the State for twelve years relation. In February 1882 he was charged with a mission with only three others to the rebellious Calafin and Gatsisibia natural tribes on the Western border, and If Riekert, because of his good acquaintance with the natives, would be successful to find a peaceful solution to these problems. From 1892 to 1896 he served as Tax collector for the South-Western border, and was then appointed Commandant for the Border Police, also Special Justice of the Peace there. Lost in the last war he already on November 27, 1899, in Derdepoort, his entire property and escaped with only his riding horse; notwithstanding this, he had six sons, two of whom were heavy and two were slightly wounded and one was killed, the battle was full, and was repeated several times during during the war by generals de la Rey and Botha with important missions.

J. Roos.

Francois Johannes Roos

Commandant Francois Johannes Roos took part in the siege of Mafeking as a private citizen, during which he became Asst. Field cornet was chosen. Then he became a tot Appointed Commandant over the Klerksdorp citizens and assumed as such battles at Tweerivier, Magersfontein and Paardeberg, on which a two-year-old captivity at St. Helena followed.

G. J Scheepers.

Gideon James. Scheepers

Among the impressive and truly sympathetic Afrikaner figures, which is the last war has produced, that of Gideon James. Scheepers undoubtedly to one distant posterity will continue to live on in sincere admiration! In the Transvaal district Born in Middelburg, he worked for several years at the Veld-telegraaf in Pretoria before he moved - shortly before the war - at the request of the Free State Government Bloemfontein had passed over there, under Major Albrecht of the Artillery To set up Free State Field Telegraphy. When the odds of war turned, they turned Scheepers with the Republican force from the Cape Colony back to the Free State, where he was adjutant to General Chr. became the Law and then joined one of the small sections that re-entered the Cape Colony at the end of 1900. Soon climbed he rose to the rank of Commandant and at the end of August 1901 had approximately one several hundred men, who seriously damaged the English commissariat and were formed into a select corps due to their military qualities. Confined to bed due to illness, Scheepers became a prisoner of war on October 10, 1901 and brought before a court-martial for various alleged offenses against it law of war. He personally led his defense, but became ill as he was was shot dead on January 18, 1902. His heart-touching diary, which was published until recently his death was recorded provides moving evidence of the extraordinary soul power of this young hero.

D. J. Schoeman.

David Johannes Schoeman

Commandant David Johannes Schoeman was born on January 25, 1853 in Krugerspost, Lijdenburg district. Shortly before the first War of Freedom, his fellow citizens chose him as a delegate to Paardekraal, but before that place was reached, his commander Frans Joubert received orders to turn the English section marching from Lijdenburg to Pretoria; During the ensuing battle at Bronkhorst-spruit, Schoeman's horse was shot dead under him. Ordered back to Lijdenburg, he was chosen as a field cornet for the Krokodil-rivier district, in which capacity he experienced the entire war, as well as the wars against Mapoch, Madjaadje Magoeba (as deputy Commandant) and against Magato. In 1899 he became Commandant van 't dist. Lijdenburg chosen. At the outbreak of the last war, having entered Natal via Swaziland with 50 civilians, he took part in the first battles at Ladysmith, while his entire command subsequently took part in the siege. He was present at the battles of Allemans-nek and Dalmanutha, and when the enemy occupied Lijdenburg, he retreated to Spitskop and from there via Pelgrimsrust and Orighstad to Boshoek, where he again organized his commando. Surrounded by great force majeure in April 1901, Komdt succeeded. Schoeman, with 30 of his men, to push through the enemy and then escape the armed blacks. He then made the railway unsafe between Waterval and Nelspruit and prevented an attempt by the English to take control of Peigrimsrust and to destroy the seed near Orighstad. During a nighttime journey to one of his field troops, Komdt fell. Schoeman in an ambush, during which one of his sons was seriously injured. Komdt was voted almost unanimously. Schoeman was elected as a delegate to the Association for the Lijdenburg district.

J.P. De La C Schroeder.

Jacobus
Philip de la Cour Schroeder

Commandant Jacobus Philip de la Cour Schroeder was born in Cape Town on October 1, 1863 and enjoyed military education in Germany. He fully experienced the first War of Freedom and took part in the expedition against Ikalefyn in 1881. Then he served the commandos against Mapoch, Massouw, Maleboch, Seleboel, Magoeba and other chiefs and distinguished himself particularly at Blauwberg. In 1895 he accompanied Generals Joubert and Smit to Swaziland to establish Bunu as head captain; next he was present at Jameson's defeat and at the capture of Mpefu's city. In the latter During the war he led a commando, was taken prisoner of war and sent to Bermuda sent. Since 1884 he had been in the service of the State Artillery, then the Johbg. Police and was promoted to Police Commandant in 1900. On many occasions Schroeder worked as a journalist, including in 1895 for the Volksstem during the Swaziland expedition, and of numerous important events of historical importance he eyewitness.

M. P. Van Staden.

Marthinus P. van Staden

Commandant Marthinus P. van Staden was born in the year 1853, in Beaufort West, Cape Colony. In 1860 he moved with his parents to the Waterberg district, Z.A.R., and has since 76 lived in Nijlsoog in that district. Before the war, Van Staden may have been best known as an industrious elder of the Reformed Church. During the first War of Freedom he took over as private citizen participated in the siege of Potchefstroom and Pretoria. Then the latest war broke out, he went to Krokodil River together with the Waterberg Commando and submitted December of the same year in the Marico district, and in early 1900 at Bastardsnek near Colesberg. He then took part in the fighting at Tabaksberg, in the Free State, at Kliprivier, near Johannesburg, and was in September of that year with the declared Commandant by unanimous vote of its citizens. Then he served General Beyers at Nooitgedacht (Clement's camp), Hekpoort, in January 1901 at Krugersdorp, Kaalfontein and Chrissies Lake. Returned to the Waterberg district, he fought most of the battles that were fought there - including at Naboomspruit, Prussia and Malipspoort - co and remained in the field until the end of the war.

P. Steenkamp.

Commandant P.S. Steenkamp.

When the last war broke out, Commandant P.S. Steenkamp left at the behest of the Commandant-General, with his commando to Mafeking, where he lay for three months, when he was ordered to move to Natal with part of the commando. After here After having served for a while, he had to return to Mafeking; then this place was relieved, he fell back on Johannesburg. In a battle at Kliprivier, May 29, 1900, he was seriously wounded and remained on the battlefield. He was still struggling at night the battle from here to Roodepoort on foot, and had not yet fully recovered, he joined General Lemmer's commando rallied and then attacked Rustenburg tried and tested. At Slijpsteenkop, Steenkamp was again seriously injured, later taking one took an active part in the battle of Moedwil (30 September 1901, when the camp of Colonel Kekewich was attacked), at the battle at Driefontein, and later at Marico, where he lost 60 of his men in one battle killed and wounded, including the non-commissioned officers Kritzinger, Van Tonder and Heldinger. On May 29, 1901 he was nowadays during the storming of the camp at Vlakfontein, and become a short distance away of the enemy guns - which had already been put out of action - once again injured. Commandant Steenkamp remained under arms until the end of the war.

S. P. E. Trichardt.

Stephanus P. E. Trichardt

Commandant Stephanus P. E. Trichardt was born in Ohrigstad, Lijdenburg district, in 1847. In recent years he has taken an active part in numerous historians events that preceded the last war. As a youngster he already took part in... first war against Mapoch part under Veldkornet Malan; about a year after that he was under Field Kornet P. de Villiers and Commandant Dirk Coetzee, on a expedition against the Swazies; in the war against Sekoekoeni led by President Burger, he served as wd. He made the field cornet, the first War of Freedom also as Veldkornet from Olifantsrivier district, distr. Middelburg, and also the second war against Mapoch, 1882 - 3, as a field cornet, while at the capture of "Hell," at north of Middelburg, acted as koonmandant. In April, 1889, he was appointed as Commandant of the Middelburg district and acted as such in the Malaboch, Madjaadje, and Magoeba wars, while at the same time he was a Vredjejudge and Natural Commissioner of his district served. After the death of Commandant Henning Pretorius, Trichardt was taken over by the government appointed as Lieutenant Colonel of the State Artillery and served as such in the expedition against Magato. When the commandos were dismissed here, he was given as assistant general, ordered to stay there with an artillery department and build forts and to keep the blacks going until the next winter, but achieved the complete subjugation of the tribe concerned before winter arrived. Tn the youngest war, he spent most of the war as Commandant of the State Artillery and then that corps that played such an important role in republican military operations fulfilled, dissolved or due to lack of ammunition for the field pieces still present was assigned to the various commandos, Trichardt received an appointment as Commandant of that part of the Middelburg commando that is north of the railroad operated, and as such he served until the close of the war.

P. F. Trichardt.

Commandant Piet F. Trichardt

Commandant Piet F. Trichardt, son of the well-known pioneer C. J. Trichardt, was born on September 3 Born in 1856 at Rozekrans, Lijdenburg district. He served in the war against Sekoekoeni, as well as in the first War of Freedom and with the Kommandos against Njabel and Jameson. He was a field cornet and, at the outbreak of the last war, Commandant of Middelburg district. As such he led his citizens in Natal and carried out further military operations.

H. C. W. Vermaas.

Commandant H.C.W. Vermaas.

When the war broke out, Commandant H. C. W. Vermaas moved under the department of General Cronje to the south-western border and took part in the first battle Crowpan. After remaining for some time in Mafeking, he moved to Kimberley and into the Free State. He contributed to almost all the important battles there, as well as those at Scholtznek and Magersfontein, where his brave field cornet Johannes Coetzee and other citizens had to lose their lives. Later he took an active share in the fighting at Kimberley, Kalmberg, Roodepoort, Donkerhoek, 'Mslkaats Nek, Grootplaats, Marieo, Wonderfontein, Witspruit, Hartsrivier, Brakspruit, Roodewal, where that brave hero, Commandant Potgieter, was killed. Vermaas won twice wounded, and once in the hand, by the explosion of a cartridge, by the enemy with dinamite filled; four horses were shot dead under him. He continued to take an active part in military operations until the end of the war about the western districts.

W. Viljoen.

Wynand Viljoen

Commandant Wynand Viljoen, brother of General Ben Viljoen, was born in the Cape Colony on February 29, 1876 and saw the battle at Rensburg Siding, Cape Colony, on February 16, 1900, as a member of the Johannesburg Police Force. At the end of 1900 he became one of its commanders and took the place of General B. Viljoen when he was taken prisoner of war. Coming. Viljoen represented his department at the peace negotiations at the Association.

M. J. Wolmarans

Marthinus Johannes Wolmarans

Commandant Marthinus Johannes Wolmarans, Commander of the Potchefstroom district, took part in the siege of Mafeking and was wounded there. Subsequently, he and his citizens belonged to the army of General P. A. Cronje and shared his fate in Paardeberg, as well as his captivity on St. Helena. The eight sons of Commandant Wolmarans all took part in the war; one of them later served as Commandant under General Liebenberg.

Military Chief Officers

Colonel J. Y. F. Blake.

Colonel John Y. Filmore Blake

Colonel John Y. Filmore Blake, Irish American by nationality and warrior by inclination, resigned shortly after the outbreak of war as Colonel of an Irish army formed by him Brigade, which served in Natal. At the battle of Moddersspruit, October 31 In 1899, he was wounded in the hand by a grenade and had to rest his hand since use miss After his recovery he joined the Artillery and served until the end of the participate in war operations.

Captain G. R. Botha.

Captain Gerrit Reinier Botha

Captain Gerrit Reinier Botha, an older brother of Komdt-General Louis Botha, turned 20 September 1858 in Greytown, Natal, born and lived in Goedehoop, district Freedom, when the war started. He took part in the fighting at Dundee, Moddersspruit, Colenso, Pieters Heights, Willowgrange, as well as the attack on Vrijheid and to the battles at Bloedrivier, Itala and others, which took place in the second half of the war was fought in the districts of Utrecht and Vrijheid. He was usually at the command of General Emmett and he, like General Emmett, succeeded in everything his people to come out when General French made his foray there. Until the end he did excellent work as captain of a scout corps that bore his name.

Captain P. A. Dames.

Captain P. A. Dames.

Captain P. A. Dames was born in the Cape Colony in 1876, P. Alphonse Dames took part as a volunteer in the Middelburg Kommando on November 4, 1899, from where he transferred to the Volunteer Corps under the command of Commander Edwards in December. Appointed in the course of 1900 as an officer in place of Commander Edwards, Dames served as such until May 4, 1901, when he was involved in a battle with Gen. Bullock was wounded close to Amersfoort and, in the Red Cross hospital of Dr. Jurriaanse was taken into custody on June 4 of that year, at the same time as Commander Joubert van Wakkerstroom, by Genl. Plumer's forces were taken to Natal.

Captain H. C. W. Grothaus.

Heinrich Christian Wilhelm Grothaus

Captain Heinrich Christian Wilhelm Grothaus wad born March 9, 1867, in Dortmund, Germany. He served in conscript age as a non-commissioned officer in the 133rd Field Artillery regiment in Metz. Coming to the Transvaal in 1895, he acquired the full potential with the Jameson raid civil rights and was appointed as an Instructor in the Artillery at Fort in 1897 Johannesburg, and in Dec. promoted to Second Lieutenant the same year. When the war broke out, he pulled the two 7.5 cM. Maxim Nordonfoldt guns - those of Jameson were conquered - to Kliprivier under the late General Jan Kok, but was ordered back to Zandspruit and assigned to the First Battery Field Artillery. On 12 In October Grothaus entered Natal with the main command and took a active role in most of the battles that followed. When we withdrew commandos he fought against the division that advanced from Natal under Bullor, at Grasnek and Amersfoort. He subsequently joined the Ermelo and Carolina commandos Bergendal and then stayed with reported commandos on the Hogeveld. On June 7, 1902, Grothaus, together with the Carolina commando, placed Doubter downed his weapons, and on the orders of the Commandant General, he delivered to the enemy over the four guns which, from the beginning, were under his control and which he eventually had burrows due to lack of ammunition. During the war, in February 1901, he was made First Lieutenant and in March 1902 by Commandant General promoted to Captain.

Kapitein O. J. Hindon.

Captain O. Jack. Hindon

Captain O. J. Hindon, a Scotsman by birth, Jack Hindon arrived about twelve years ago, at the age of 18 old in the Transvaal and settled in Middelburg. With the Jameson raid he acquired full citizenship and subsequently joined the National Police Middelburg. When war broke out, he joined the commando as Commandant Piet Trichardt, and entered Natal under General Joubert. After the first having taken part in the battles for Ladysmith, he accompanied the reconnaissance troops to Estcourt, and joined Edward's on his return Reconnaissance Corps. He and Slegtkamp, his comrade throughout the war, were the two civilians who were decisive in victory at Spioenkop to hoist the Five Color to the highest point of the head. He then took a job share in the later war operations and, at the general withdrawal from Natal, was charged by General Joubert with making the coal mines in Elandslaagte unusable.

With the Natal branch of the Irish Brigade, he moved to the Free State and joined there under Captain Danie Theron. When General De Wet moved to Waterberg, this served corps as his rearguard, and first withdrew from the Krugersdorp district to Kliprivier to release about fifty political prisoners, all but a few of whom committed suicide connected. But it was mainly as a guerrilla leader that Hindon distinguished himself. Hii was nowadays when the first trains are taken by Theron's corps in the Free State, and then left with reports to General Botha, who was then in Roossekal. Here he was appointed captain of a corps whose main purpose was... to damage the enemy's railways and communications lines. One of his own men, Carel Cremor, later killed in Naboomspruit, discovered the Martini Henri gun lock device, through which Hindon managed to get several dozen trains into the to make air fly. On the Eastern Line he and his corps did it in one day hours, no less than three long, heavily loaded trains crashed, and he also often operated on the Zoutpansberg and other lines. After the peace, Hindon married a daughter of Mr J. H. Coetzee Bosjesspruit, Bethal, and left for Enropa in March 1903. After a while in Holland he went to America and will probably live in Mexico settle down to live. In a letter to "De Volksstem" before his departure Netherlands, Jack strongly advised all Afrikaners to leave their homeland.

Adjudant-General H. De Jager.

Adj.-General
Hilgaard de Jager

Among the Transvalers who lived through the last war, perhaps none are in it wider circle and more favorably known, than the subject of this sketch, Hilgaard de Jager, Chief Adjutant on the staff of the Commandant-General from the beginning hostilities in Natal, up to the end at Vereeniging. His military career is closely linked to that of his superior, General Louis Botha; de Jager followed him in Natal, fought at his side at Dundee and for a few days later also in the battle for Ladysmith, where General Botha served for the first time Acted as chief officer, took part in the tour of Estcourt, participated in the battle of Colenso service as adjutant and then to Pieters Heights, Biggarsberg, Brandfort, Kliprivier, Pretoria, Donkerhoek, Dalmanutha, Helvetia, Lijdenburg; accompanied the Commandant General from here to the western districts and back, served in Chrissies Meer, in the general attack on the Eastern Line, at Itala, Spieshoek, Onverwacht, Bakenlaagte. De Jager accompanied the General in Middelburg in May 1901 with Lord Kitchener was being negotiated and was also involved in the last negotiations today Vaal River.

Captain J. J. Naude.

Captain Jacobns Johannes Naudé

The subject of this sketch, Jacobns Johannes Naudé, was born in Smithfield, Orange Free State, January 28, 1876. When the last war broke out, he was in Pretoria sexton of the Dutch. Rev. or Geref. Church and joined Dr. Ambulance. vau der Merwe, who entered Natal with General Meyer's commando. Then Pretoria was taken, Naudé stayed there, but did not escape long afterwards and joined a division of Commandant Badenhorst's people, living west of Pretoria in the Schurvebergen lay. From here he undertook a series of successful plundering expeditions to the village, organized a complete search service there, in which he was faithful was helped by a large number of African women and others in the city. This information service extended so far that all kinds of official information were extracted from enemy sources managed to obtain sources, the military railway obtained time tables - with which the catching trains was advised - a regular exchange of letters with friends points in the city and the youngest couriers soon after arrival command read. After the conference in Branddrift - where Naudé told the Commandant - General presented a neatly embroidered flag, as a gift from some women from Pretoria - he was appointed Captain and determined himself, with the help of a number of civilians, mainly involved in investigative work until the end of the war. On the Hogeveld, he nevertheless also took part in various battles and smaller battles, and after the war, under the title "In Doodsgevaar," had a fascinating article in the Volksstem publish the story of his adventures.

Captain P. C. Paff.

Captain Paul Constant Paff

Captain Paul Constant Paff, born in Amsterdam, Holland, in 1863, was born at the age of 17 telegraphist. After passing his Telegraph and Postal examinations, he took up a convocation published in Holland in 1888 on behalf of the SA Republic, including in the same years to come to the Transvaal as a Telegraphist, with two years' leave. Before this leave was over, in January 1890, he was requested by General Joubert to serve in the Transvaal, in in connection with the Artillery Corps, to establish a Field Telegraph department, with which work he started in June of the same year, after his appointment as a Second Lieutenant Mobile Artillery and Police. In 1894 he served in that capacity in the war against Malaboch, as well as later in subjugating Magoeba, Majadje, Mamatolla and others riotous naturals heads in the north of Zoutpansberg. After the Jameson Raid he was taken as a witness for the English Parliamentary Inquiries Commission summoned to London, promoted to First Lieutenant the following year. In 1898 his department was part of the expedition against Magato (Mpefoe). The Paff went through the last war from start to finish and became Captain in 1901 promoted and most recently served as head of field telegraph and post at the Government in Velde.

Majoor J. L. Pretorius.

Major Johannes Lodewijk Pretorius

Major J. L. Pretorius was born May 24, 1872, was Johannes Lodewijk Pretorius, son of the late Koramandant Henning Pretorius, Some years before the Jaraeson raid, official on the office of the Koramandant General and started his military career as enlisted man (later officer) of the Pretoria Volunteer Cavalry Corps. In January 1896 he was appointed as a Second Lieutenant in the State Artillery and as such made the war against 'Mpefoe. On September 28, 1899, Pretorius left as chief of the third battery of the S.A. to Zandspruit, and then captured under generals Lukas Meyer and Botha took part in Talana Heuvel ahead of Dundee, Modderspruit, Colenso, Platrand and Peters Hoogten. After our commandos had been driven back from Natal, he left for the Free State with the Commandant General and a field artillery department, participated in the fighting that took place between Bloemfontein and Johannesburg for Pretoria and then the battles of Donkerhoek (Diamond Hill) and Dalmanutha. During the war he was promoted by the Commandant Goneraal to the rank of Major. In December 1901 Pretorius was captured on the Hogeveld and sent to prison St. Helena transported. In March after the peace he was consulted by General Botha the government in charge of a shipment to the unwilling prisoners of war in Bermuda, who however, remained without consequence.

Captain P. Scholtz.

Captain Piet Scholtz

Captain Piet Scholtz was a young man of 21 years old when the war broke out. He belonged to it Freedom commando and when they were in that district after the military operations in Natal When he returned, General Emmett appointed him Captain of a Reconnaissance Corps which was composed of four citizens from each district. This corps performed good services, and on one occasion Scholtz, with sixteen of his men, took a convoy of heavily loaded wagons and draft animals, 1 cannon with ammunition and 40 prisoners of war. He was seriously wounded in Graspan, but took part in many battles after that finally had to lose his life in the battle of Itala or Talene in Natal.

Captain H. P. Slegtkamp.

Henri P. Slegtkamp

Boer riflemen with their flag on Spion Kop.[419]

Captain Henri P. Slegtkamp, Dutch, born April 19, 1873, lived in Middelburg and performed his first military service against Jameson, under Veldkornet Matthijs Pretorius van the Middelburg district. He was among the first to arrive at the end of September 99, under Commandant Piet Trichardt, went to Zandspruit, and headed commando under the late Commandant General Joubert, Natal, took part in the battle in Modderspruit and took part in the reconnaissance tour around Estcourt. In Dec. of the same year, Slegtkamp joined Commandant Edward's Reconnaissance Corps, which was then served for the commandos along the Tugela.

On January 22, 1900, he and Captain Jack Hindon found and hoisted the only civilians on Spioenkop there, as is known, with such good results, the Five Color. Afterwards he served with the Irish Brigade under Blake in the Free State, as far as Brandfort, then joined under Capt. Danie Theron and stayed behind as the other commandos moved north. After After Theron was killed, Slegtkamp and Hindon left the Free State with reports General Botha, and when a corps was established under Hindon in Lydenburg, Slegtkamp became a member there, Lieutenant, and then Captain in May 1902. The work of this corps mainly limited itself to obstructing the enemy's communications lines, and it is known that as such it rendered excellent services, of which Slegtkamp, next to his commanding officer, always took a working role.

Captain D. Theron.

Captain Danie Theron

When the last war broke out, the not yet 30-year-old Danie Theron was born Cape Colony and intended to fulfill an important role on the military stage and a to make a deep and lasting impression on the African public mind - law officer too Krugersdorp, where he took a keen interest in the ever-changing national affairs. And already then the preservation of peace to our sharp-sighted foremen seemed hopeless, Theron had organized an information service, which the Government: the form became of great use and later, at the start of the Natal campaign adopted a bicycle corps. After the battle of Colenso this corps took the form of the T.V.K. (Theron's Scout Corps), under which it became famous is, and to which men belonged who later served as captains of Mears' Corps, Hindon's Korps, Williams' Korps and others have rendered great services to their people. Theron's career has been glorious from start to finish. Because of his naughty visit to Genl. Cronje, when he was surrounded by superior force near Paardeberg, he became generally known. With his T.V.K. he covered the rear of Genl. the Law's trait to Waterberg and then returned to the Free State, after which he traveled to both sides of Vaal river and in the western districts of Transvaal successfully operated against 's enemy lines of communication. In August 1901 this hero was killed near Buffelsdoorn, Gatsrand, hit by a bomb and killed, deeply regretted by his men, who carried him to his hands, and by the great Republican force. Danie Theron was as decisive as good-hearted, as determined as sincere and amiable. He was - which may not be known enough - the creator of it sung throughout South Africa war song: "Our country is in trouble again."

Captain H. Du Toit.

Captain Heinrich du Toit

Captain Heinrich du Toit originally came from the Cape Colony and settled as teacher in Pretoria. After passing the exam, he subsequently became an officer State Artillery, and when the latest war broke out, he moved with the First Battery to Ladysmith. At the battle of Modderspruit, October 30, 1899, he was in command it 7.5 cm. Creusot cannon ("Long Tom"), which was stationed on Pepworth's head, and was seriously injured here, so that he had to be hospitalized in Pretoria for several months lie. After the capture of the Transvaal capital, although still with his wound Suffering, du Toit moved with the artillery to the Hogeveld and took the blows Donkerhoek and Dalmanutha with him, and then went to Zoutpansberg, where he died General Beyers served with the remaining "Long Tom" until he finally surrendered this piece a fierce battle in Houtbosberg, in order to prevent it from happening would fall into the hands of the enemy. The artillery in this district gathered under him to a small corps that operated for quite some time north of Pietersburg, where it together with a section of Veldkornet M. W. Pretorius, in an attack by blacks led by English officers, suffered heavy losses, and was also plagued by fever became. Du Toit remained in the field until the end of the war.

Captain A. F. Du Toit.

Captain Andries F. du Toit

Captain Andries F. du Toit joined the Lijdenburg as a private citizen in Natal kommando, under Field Kornet M. P. Taute, and took part in the battles as a non-commissioned officer in Platrand and Spioenkop. In Belfast he was appointed Captain of a Reconnaissance corps and was currently at the battle of Dalmanutha and the smaller one fighting that followed, as well as the battle of Helvetia and the attack on the Eastern railway line. He took up a camp at Pisangkop, in collaboration with Commandant Mol from Australian Volunteers. In this battle Mol was seriously wounded and du Toit subsequently appointed in his place. Afterwards he operated with his commando, which numbered about 200 men, under General Ben Viljoen, on the Sabie and in the surrounding areas Pilgrim's peace. He was present on June 12, 1902, when the citizens of this section had to lay down their arms near Belfast.

Captain V. E. Von Wichmann.

Captain Friedrich Wilhelm von Wichmann.

Before he came to South Africa, Friedrich Wilhelm von Wichmann was a lieutenant the first Guards Infantry regiment in Potsdam, near Berlin, Germany. After working for some time at the Surveyor General's office in Pretoria, von Wichmann was appointed in 1897 as a Second Lieutenant at the Fortress Artillery in Johannesburg. On September 29, 1899, he left with the German commando joined under the command of the late General Jan Kock, to Kliprivier, but was from there recalled to the Hoofdlager in Zandspruit and posted to the Third State Artillery Battery, which entered Natal under General Lukas Meyer. In Natal he participated in the battles of Dundee, Moddersspruit, Colenso, the journey to Estcourt and most of the other battles were delivered south of Ladysmith. On the Hogeveld returned, he joined an artillery division under the late General Chris Botha and served with distinction until the end of the war.

Captain H. Williams.

Captain Henry Williams

Captain Henry Williams was born in Victoria West, Cape Colony, in the year 1874, moved in 89 to the Orange Free State, and came to the Transvaal after the Jameson Invasion. If As a citizen he joined the Witwatersrand-Pretoria commando, which went to Colesberg later served in the Afrikaner Kavallerie Korps, under Konmandant Abr. Malan, and later still with Theron's Volunteer Corps, until the death of the leader. Then he served as a lieutenant in Captain Mears' corps. He was born on June 6, 1901 Waterval Onder was appointed captain and was given the task of setting up a separate corps which was later known as Williams' Fighting Corps. In March before peace comes Williams captured and arrested in Pretoria on charges of having worn the English uniform and of the destruction of railways, whereupon he was subsequently sentenced to death; but peace was just then restored there became, he regained his freedom. He is killed several times during the war wounded and experienced many blows.

Captain B. G. V. De Witt Hamer.

Captain Boudewijn Gerrit Verselewel de Witt Hamer

Captain Boudewijn Gerrit Verselewel de Witt Hamer, born February 12, 1855 in Doesburg in Gelderland, Netherlands, was an officer in the Dutch army and came South in 1885 Africa. He was editor of the first Dutch news magazine in Natal, called "De Natal Farmers Friend," left for Transvaal in 1887, where he worked several has held government positions. He worked for many years as... civil servant of the Mining Department, was Mining Commissioner, first in Leydsdorp and afterwards at Barberton. He resigned from the latter position and was elected Member of the Union the Second Volksraad in 1897, to be active for the Cape Delverijen until the outbreak of war, when he was appointed Captain of the Dutch Corps, witnessed the battle of Elandslaagte, was captured there on October 21, 1899 and was taken to Simonstown. After a varied stay on transport ships in the False Bay and in the POW camp at Simonsstad, he was captured on April 1, 1900. transported to St. Helena, where he worked in various administrative areas for the benefit of his fellow prisoners of war until September 1902, when he gained his freedom regained.

Major J. F. Wolmarans

Major Jan Francois Wolmarans

Major Jan Francois Wolmarans, a son of the former Executive Councilor J. M. A. Wolmarans, received his military education in Breda, in the Netherlands. After returning to the Transvaal, some years before the Jameson raid, he was appointed Second Lieutenant in the State Artillery and gradually promoted to the rank of Major. When war broke out in 1899, he moved to Zandspruit with half the artillery. while the other, under General Cronjé, went to Mafeking and Kimberley. At Zandspruit the artiliry was again distributed, the Third Battery moved under General Lukas Meijer towards Dundee, while the First Battery via Charlestown and Newcastle crossed borders. Wolmarans was with the Third Battery and witnessed the battles which were delivered at Dundee and Moddersspruit. When he returned to the Transvaal, he left with the commandos to the Hogeveld and was presently in many battles there. Recently the remainder of the artillery operated under him in the vicinity of Bronkhorstspruit and Pretoria. He was captured on January 11, 1902, close to Ermelo, and taken to St. Helena.

Smuts' Commando

Smuts' Commando in April of 1902.[271]

The Smuts Commando. The group of officers posted here was taken near Okiep, then General Smuts, who sat in the middle of the group, had taken possession of this place. A page was devoted elsewhere in this Album to General Smuts himself.

[271] Jan Smuts is seated. Viewer left standing is General Jaap van Deventer wearing a Mauser C96 and standing viewer right with his leg up is General Manie Maritz. General Smuts Vol 1 by Millin Sarah Gertrude (1936) p.159 File:Smuts Boereoorlog.jpg - Wikimedia Commons Also at higher resolution: https://af.wikipedia.org/wiki/L%C3%AAer:Smuts_Boereoorlog.jpg

1. Gen. Jan Smuts	10. Field Cornet Willie Kotzé
2. Burger B.D. Bower S. "Oom Barend"	11. Corporal Hennie Vermaas
3. Gen. Jacob Louis van Deventer	12. Field Cornet Jan Van Brummelen
4. Field Cornet A. G. Boshoff	13. Commandant Stoffel Schoeman
5. Commandant Ben D. Bouwer	14. Jacob Brink, Military Commissioner
6. Field Cornet Cornelis Brink	15. Cornet A. Standers
7. Field Cornet Jan Van Der Berg	16. Field Cornet Ben Coetzee
8. Cornet Piet F. Visser	17. Marquis De Kersauson "Roberts"
9. Gen. S. G. Maritz	18. Carl T. Möller

About the other officers are reminded of the following: Starting with those on the right side of General Smuts in a standing position are located, the spectator encounters a bearded warrior, namely:

Burger B.D. Bower Sr., Known conveniently as "Oom Barend," 62 years old, he fought with the Free Staters 1865 against the Basoetoes, when he was seriously wounded. Next to this one:

General Van Deventer. to whom a separate sketch is dedicated elsewhere in this Album.

On his right side stands Field Cornet A. G. Boshoff, who fought under Commandant Bouwer and on 25 February 1902 in Windhoek, distr. Vanrhijnsdorp, was seriously injured during the capture of the Police camp.

In addition to these is located Commandant B. D. Bouwer, already mentioned elsewhere in this album. On the corner on this side of the group is:

Field Cornet Cornelis Brink who served under the aforementioned Commandant and came from Haaskraal, distr. Potchefstroom.

Rest at his feet Field Cornet Jan Van Der Berg, also from Potchefstroom, who in July 1901 during the storming of a railway fort in Brandfort was seriously injured, which did not prevent him from entering the Cape Colony insisted. There he was seriously injured for the second time in January 1902 of which he will remain disabled for the rest of his life. Rests on his left side

Field Cornet Piet F. Visser, a former Transvaal mine official, who became an officer in the K. Colony appointed. The group to the left of General Smuts is composed as follows: the heavy-set officer is next to him

General S. G. Maritz, years old, for a time a member of Theron's Reconnaissance Corps and present at the death of this beloved captain. He passed the Cape with General Wijnand Malan colony and, after fighting for a while, was sent to Calvinia district with three citizens where he managed to organize a commando, whose Commandant he was is appointed.

After some beautiful battles, "ou Manie" grew and became a general Maritz was called by his comrades - armed forces every day, so that on 7 August, 1901, took possession of Niewoudtsville and invaded Clanwilliam, where he settled with Captain Jan Theron and Captain Smit, and together into the district Piketberg, Moorreesburg, Darling and Malmesbury were extended. From there it went trip to Tulbach district, where the Town Guard of twenty-four rivers Wellington is disarmed. In November of this year General Smuts was met the Van Rhijnsdorp district; In January 1902 Maritz was appointed General. When the peace had become a fait accompli, "old Manie" left the Colony in dismay and visited German S.W. Africa and Europe. He was captured at Tontelboskolk, in Galvinia Dec. 1901 seriously injured. Standing behind him, next to Genl. Smuts,

Field Cornet Willie Kotzé, from Calvinia district, a known good marksman, and to the left of this

Corporal Hennie Vermaas, living in the Klerksdorp district, one of the brave four that died in January 1902 from Van Rhijnsdorp district on foot and armed the Bovenland, Paarl and Stellenbos spies for fourteen days with excellent results.

On his left side is Field Cornet Jan Van Brummelen, Dutch by birth, a determined and widely loved soldier.

In addition to these stands Commandant Stoffel Schoeman, a Free Stater, who succeeded General Maritz as Commandant, was previously a member of Theron's corps, and one of the three that entered Calvinia district with Maritz. He followed General Maritz to German South West Africa. Next to this one

Jacob Brink, Military Commissioner of Commandant Bouwer, from Johannesburg, competent and helpful. Finally, on this corner the group stands

Field Cornet A. Standers belonging to the command of Commandant Schoeman, and originating from the district Ventersburg, Orange Free State. He also used to belong to the late Danie's corps Theron and then followed Maritz to the old Colony.

At his feet sits the fearless Field Cornet Ben Coetzee, before the war an official at the Pretoria post office. He originally belonged to it Kommando van Bouwer, but later transferred to Schoeman.

On his right is seated Marquis De Kersauson, known to his fighters under the name of "Roberts," a Frenchman from nationality, who moved to Transvaal during the war out of sympathy for the Afrikaners had come. His bravery is especially evident as a depeche driver to separate. Twice he traveled through D. S. W. Africa to Europe and back with reports and has sacrificed a lot for the Boers.

At the latest: Carl T. Möller, from Graaff Reinet, at the outbreak of the war, student of medicine in Edinburgh, mot Sievwright's ambulance came to Trausvaal, under Dr. Neethling in Lijdenburg employed and then head of General Smut's ambulance. He nursed with success of General van Deventer, when he was seriously injured; as well Commandant Bouwer.

The Lijdenburg Kommando

Seated L to R.: Commander D. J. Schoem, Chris Muller and Wynand J Viljoen. Ben's younger brother. Christiaan H Muller. Unable to decipher who is in the back row.

The Lijdenburg Kommando. The adjacent group of six people represents the leaders of the Lijdenburg resistance during the latter part of the war, as can be seen from the signatures below the main characters were placed. General Muller and Commandants D. J. Schoeman and W. J. Viljoen were sent elsewhere dedicated separate pages in this album.

Civilian Authorities

Rev. E. C. Anderssen.

Rev. E. C. Anderssen

Rev. E. C. Anderssen, preacher of the Netherlands at the outbreak of the war. Rev. or Ger. Church in Vrijheid, was born on March 22, 1868 in Ceres, Cape Colony, of a German father and African mother. He was a student at the Z. A. Atheneum in Cape Town, and then from the Theological Training School in Stellenbos. After teaching at Umtata, Tembuland, and later at Mclean, distr. East London, Rev. Anderssen appealed to Freedom 9 months before the outbreak of the war. Rev. Anderssen has had the good fortune and distinction, thanks to many close ones escape, to experience the entire war until the end with his Vrijhrid Kommando. Only on June 16, 1902, after an absence of 21 months, he returned to his parsonage.

Military Commissioner B. H. V. K. Arnoldi.

Military Commissioner B. H. V. K. Arnoldi.

B. H. van Ketel-Arnoldi was mining commissioner of the Potchefstroom Gold Fields and was appointed in January 1900 appointed Chief Military Commissioner for the South Western borders headquarters in the large castle at Modderrivier, where General Cronjé commanded. He accompanied General Cronjé's force to Paardeberg and became a prisoner of war there made and then transferred to St. Helena

Military Commissioner W. J. H. Barter

Military Commissioner William Joseph Henry Barter

Military Commissioner William Joseph Henry Barter, of Irish descent and born in 1865, was born in Sept. 1900 by Komm. Genl. Botha was appointed Military Commissioner in Lijdenburg. As such he accompanied Genl. Gravett who sent him from the Highveld to Pilgrim's Rest at to gather provisions. Then I got the blank Government Notes in my hands had fallen to the enemy, Barter was appointed one of the members of the supervisory board State Printing House in the Field, and when the State Mint in the Field was created he was appointed Treasurer; afterwards he established a military tannery and tailoring

Dr. J. C Bierens De Haan.

Dr. J.C.J. Bierens De Haan

Dr. J.C.J. Bierens De Haan was born in the Netherlands in 1867 and studied medicine at Leiden University promoted, Dr. Bierens de Haan at the end of November with the first ambulance of the Dutch Red Cross the South African Republic and was sent to Moddersspruit - Natal - sent where he met Dr. Lingbeek organized and organized a large field hospital thus providing assistance to a large number of sick and injured people. He made the battles at Platrand, at Ladismith, and then those at the Upper Tugela, Spioenkop and Vaalkrans, personally, subsequently took over the management of the hospital in Pretoria, where mainly seriously injured people were nursed. After the fall of Bloemfontein, Dr. Bierens de Haan a field ambulance with which he operated in the Free State and witnessed several battles.. When Pretoria was captured, he remained with the large number of wounded, that surgeon those in need of help, behind; after recovery from this he obtained, not without long negotiations, a pass to return to the Boer lines. In Machadodorp Appointed a member of the Mediese Commission in August 1900, he remained in charge of the commandos accompany and set up with Dr. McLeod repeatedly set up large field hospitals, then withdrew with Commandant General Botha of Hektor Spruit through the Sabie area northwards and participated in several small battles in Lijdeuburg. In the late 1900's closed Dr. B. de Haan joined the Heidelberg commando, under General Spruyt, took part in all kinds of wanderings across the Hogeveld, in train attacks and in the battles at Boesmanskop, Greylingsstad, Bethal and Chrissiesmeer, where he captured but later released. In May 1901, Dr. B.de Haan sent to Europe to obtain medical aid for the Boers, but the numerous and vigorous attempts by the English Government to obtain the passage of doctors, remained without results.

Rev. M. P. A. Coetsee

Rev. M. P. A. Coetsee.

Rev. M. P. A. Coetsee, is one of the few remaining pioneers who took part on the great Trek from the Cape Colony in 1836. The latest war made ZEerw. proud his old age, to the end. Born on February 15, 1831, in Zitrivier, Colesberg district, he moved with his parents under Potgieter from the Cape Colony. Winburg was built and built by these pioneers then Potchefstroom. In those days he already took part in many small commandos against the native tribes, and was nowadays at Zwartkopjes and Boomplaats. Also in it In Lijdenburg, where he moved in 1852, he made many commandos against the blacks along. In 1864 he was called to study as a minister, and was ordained five years afterwards being, Rev. Coetseo in Middelburg K.K. professions. After 20 years here Having preached the Gospel, His Rev. in 1889 in Lijdenburg. In 1895 With the approval of the teacher and congregation, the church and parsonage were moved to Belfast; and when the enemy took possession of this place in August, 1900, His Rev. of the commandos to Machadodorp and from there to Komatiepoort. From this one instead it became a wandering on foot through the Bosveld, up to Ohrigstadsrivier, from where he traveled to the Steenkampsbergen and here always in the vicinity of General Viljoen's commando remained until finally on June 10, 1902, under General Chris. Muller, laid down his arms at the Monument in Belfast. Due to old age and due to physical infirmities, requested and obtained His Rev. retired since 1903.

Dr. H. J. Coster.

Dr. Hermanus Jacob Coster

Dr. Hermanus Jacob Coster was born in the Netherlands in 1866 and arrived shortly after had obtained his doctorate in law in Leiden, and moved to Pretoria, where he worked as a professor established a lawyer. After practicing for several years, he was appointed to the office of State Called a public prosecutor and performed his difficult task with commendable merit led the criminal proceedings against the Reformers, who launched the had threatened the government. At the outbreak of the last war, Dr. Coster, who had since returned to private practice, joined the Hollander Corps and was killed on 21 October 1899 in the battle of Elandslaagte, Natal, deeply regretted by a wide circle of citizens who suffered a heavy loss due to his untimely death caused, fully realized.

Adv. I. S. Ferreira.

Adv. I. S. Ferreira

Adv. I. S. Ferreira was born on January 25. 1875 at Zonnebloem, Ladybrand district, Orange Free State, Ignatius Stefanus Ferreira went to school there for 5 years, then from '90 – '95 at Gray College, Bloemfontein, then a year at ' S. African College in Cape Town and finally 3 years at the Middle Temple in London. Adv. Ferreira was military secretary to Genl during the war. de la Rey and occasionally acting general during his absence. He accompanied Genl. de la Rey to Europe after peace.

Rev. M. J. Goddefroy.

Rev. M. J. Goddefroy

Rev. M. J. Goddefroy completed his studies at the University of Utrecht, Netherlands, was pastor at several Dutch churches from February 11, 1877. Reformed municipalities and was called to the SA Republic in April 1887. This teacher stood until 1890 actually only for 13 different municipalities with 17 church places, as when one second pastor for Rustenburg and Marico was appointed. Rev. Goddefroy visited before the capture of Pretoria, where his station was, repeatedly the commandos and accompanied them when the capital was in the hands of the enemy fell. In June 1901 he and his family were infected with fever, 70 miles south captured east of Pietersburg and taken to Pietersburg's concentration camp fed. The following October he was sent to Trichonopoly, British India from there deported to Fort Govindhar, in the Pundjab, one of the unhealthiest provinces of India. In July 1902 ZEerw. first to leave this continent, and Because permission to return to Transvaal was first refused, Rev. Goddefroy pass return to Pretoria in mid-1903. The celebration of his 25th anniversary of Gospel ministry fourth SEerw. under great interest during his Indian captivity. Of the six sons of Rev. Goddefroy, four have distinguished themselves on the battlefield. The eldest, Joseph, served in Danie Theron's corps, where he became a lieutenant promoted; but died on October 27, 1900 in the Waterberg district of typhus. Two other sons, Anton and Marius, were taken prisoner of war shortly afterwards battle at Pienaars River and taken to St. Helena; there the latter lost on 9 February 1901 by a treacherous weather bullet. The fourth son, Edmond Goddefroy, was wounded at Colesberg in December 1899, then fought under General Beyers, but in May 1901, together with about thirty other loyal citizens, he was executed by attacked a large force, captured and sent to the Indies.

Dr. J. O. Hohls.

Dr. Johan Otto Hohls

Dr. Johan Otto Hohls was the son of a German missionary, and was born in Natal, about the year 1859. Sent to Germany for his education, he studied at Strasbourg and obtained his doctorate as a doctor at the University of Heidelberg. During it In 1889 he returned to South Africa and stayed in Natal for some time, but settled later practiced medicine in Pietersburg in Transvaal. In early 1899, Hohls was appointed by the government as a doctor at the State Artillery, with the rank of captain. When the war broke out he left with his ambulance with the main force, under General Joubert, to Ladysmith and was killed in the battle that was delivered to Pepworth's (Modderspruit) on October 30. There is a temporary one there there was a need for hospital orders, his pharmacist helped in this work and became wounded on top of Pepworth's head. Hohls, who was in the hospital working on the nursing the wounded, then went out in person with heroism exposing himself to a shower of shell-cracks: and, while he was giving his apothecary the first aid, he was fatally shot in the back. The corpse was taken to Pretoria brought and buried there in the German cemetery with full military honors. In Zoutpansberg the name of **Dr. Hohls in grateful memory held in connection with the well-known antipyretic drug he discovered.**

Adv. L. J. Jacobsz.

Adv. Louis Johannes Jacobsz

Adv. Louis Johannes Jacobsz was born in Bethlehem district, Orange Free State, November 6, 1863 born during the Basoetu war and first received his education at it Bloemfontein Gray College, later London University, where he practiced law Inner Temple was admitted in 1888. After practicing for some time in the Free State Jacobsz came to Transvaal and acted as Public Prosecutor in Johannesburg and subsequently as assistant State Attorney. After the capture of Pretoria he accompanied the Government in the field, as Wd. State Attorney and as such was currently with the peace negotiations in Klerksdorp and in Vereniging.

Dr. A. Jurriaanse

Dr. A. Jurriaanse,

Dr. A. Jurriaanse, born in 1872, completed his education in Leiden and Vienna, Austria, and left at her own expense at the outbreak of the last war complete equipment to Transvaal, from where the Government immediately appointed him as head of sent a field hospital to the south of the Free State. He was present at several battles, including the conquest, by Genl. B.C. de Wet - at the time still Commandant - of Lord Roberts' army; he subsequently joined the commandos of Genl. de la Rey, then van Genl. Philip Botha. After the battle at Zandrivier, Dr. Jurriaanse left without a command for the second time, went to Transvaal and connected joined the Ermelo commando, at the request of Genl. Tobias Smuts, lived the battles of Donkerhoek, Bronkhorstspruit and Dalmanutha and then shared it guerrilla life of said commando. For a long time he was the only physician between the two Machadodorp on Volksrust, from Swaziland to Standerton. In place of Mr W. Buhrmann, Ermelo district, he had established his central hospital, from where he and his ambulance crossed through 't Hogeveld. During the campaign to Zululand, especially at the battle of Italy, was Dr. Jurriaanse also nowadays. And besides injured civilians he had too the uncaptured women and children under his care. After the conclusion of peace, Dr. Jurriaanse moved to England, where he arrived in September took a medical examination in London in 1903 and was therefore entitled to practice medicine To practice Transvaal medicine. On Saint January 1904, Dr. A. Jurriaanse as physician based in Ermelo.

Rev. A. P. Kriel.

Rev. Rev. Abraham Paul Kriel

Rev. Rev. Abraham Paul Kriel was ordained as a minister in 1880 and had already served for several years the Witwatersrand was working when the last war broke out. Initially joined His Rev. joined the Krugersdorp commando, which moved to Dalmanutha after the battle Zoutpansberg and from there to the western districts. Remained under General Beyers However, a small part of this commando is in the Zoutpansbergse behind, and Among these was also Rev. Bantam. The battle of Nooitgedacht, the march around Pretoria and he attended the later battles on the Hogeveld and was known for his dedication and calm helpfulness in moments of danger. As a Field Chaplain and Patient Comforter he served General Beyers' command until the end of the war. Returning to Langlaagte, Rev. Kriel there, with the help of numerous friends, a blessed orphanage has been established, in which no fewer than 260 orphans and the children of needy parents from the last war are raised.

J. C. Krogh, L.U.R.

Johannes Christoffel Krogh, L.U.R.

Johannes Christoffel Krogh, L.U.R. born about 1846 in Uitenhage, Cape Colony, during the English intermediate government after 1877 in Transvaal state service, was then Landdrost van Standerton and later became part of the restored South African Republic, Landdrost van Wakkerstroom With the settlement of Swazieland affairs he was appointed representative of the Transvaal Government in Swazieland, and then in 1895 the Organizational Proclamation became valid, Mr. Krogh remained as Special Commissioner Bremersdorp. During the second part of the war he was Landdrost and then Member of the Executive Council, in which capacity he served until the end. Tans he is Magistrate in Belfast and member of the commission to investigate the South African natural situation.

Rev. J. M. Louw.

Rev. James Murray Louw

Rev. James Murray Louw, a son of Rev. A. A. Louw, was born in Murraysburg in the Cape Colony, and later went to live on the Paarl. He received his higher education Stellenbos, first at Victoria College and then at the Theological Training School. During the last war Rev. Louw continued to be a field preacher with the commandos and explained His Rev. together with the Heidelberg and some other commandos, on 5 June 1902 in Kraalstation, weapons removed. The adjacent portrait was cut from a group that met in Heidelberg the next day was taken when His Rev. was still in his commando clothes. Rev. Louw is preacher at the Ned. Rev. or Geref. municipality in Boksburg.

Rev. H. J. Neethling.

Rev. H. J. Neethling

Rev. H. J. Neethling was born in Prince Albert, K.K., and is the eldest son of the then teacher there, Rev. J. H. Neethling, who appealed to Stellenbos shortly afterwards became; The young Neethling spent his years of study here and in Cape Town. After After completing his studies he went to Europe, visited Palestine and America and was born in 1881 Lijdenburg ordained as teacher of the Dutch. Ref. Church. In the first General Meeting in which His Rev. took seat, negotiations were started with the Dutch government. Rev. Church that later led to the vereniging, and there the first-mentioned cork at that time had two teachers, he took an active part in the deliberations. Due to lack to teachers, the Dutch had to be served by him. Ger. municipalities of Lijdenburg, Middelburg, Pretoria, Zoutpansberg and Ermelo. In 1895 His Rev. delegated to the co-religionists at Humpata and accompanied by his brother, C. Neethling, currently a teacher Nieuwoudtsville, K.K., he traveled through not only Humpata, but also Damaraland. In the At the beginning of the last war Rev. Neethling our people at Ladysmith and along the Tugela, and later again in the Biggarsbergen and Allemansnek. Then finally the enemy reached Lijdenburg, he left his family and accompanied the commandos. Lator Mrs. Neethling and family were sent out, and remained out about six months, when they were captured. Rev. Neethling found a wide field of employment among the various commandos and people still outside who are in the eastern and northeastern districts ceased, and he therefore completely surrendered to them. Short for peace, H.Ew. the most sensitive blow, show his wife in the so-called Refugee camp at Balmoral died after they refused to send her to the family in the to let go of the old colony. The esteemed father also recently passed away in Stellenbos from Rev. Neethling. One of his brothers, Dr. A. M. Neethling, proved inestimable services to the commandos, and after the death of Dr. Hohls, an official appointment as a doctor in the artillery.

Staatssekretaris F. W. Reitz.

State Secretary
Francis William Reitz

State Secretary Francis William Reitz was born in Swellendam, October 5, 1844. He studied law and letters, in both of which he excelled. He completed his studies in Scotland and England and practiced as a lawyer in the Cape Colony, which he represented the district of Richmond in Parliament for a short time. In those days he worked diligently on the political organization of the Afrikaner party, and was also diamond digger. He was called to the Orange Free State for 18 years Chief judge and 8 years President of this Republic. In 1896 he came to Pretoria, where Reitz established himself as a lawyer until he was elected State Secretary in 1898. When the seat of the Transvaal Government was constantly moved after the capture of the capital had to be, and our administrators had to go through many tough days, State Secretary Reitz never lost his self-confidence, energy and cheerfulness. He even found many occasions of important military activities or other events to be displayed in verse time, to the cheer of his fellow citizens. Immediately after signing the Peace Treaty, Reitz went to Europe and North. America and has so far shown no intentions to go to his beloved to return to South Africa.

Rev. J. W. G. Strasheim.

Rev. Jacob Willem Gideon Strasheim

Rev. Jacob Willem Gideon Strasheim was born in Colesberg in the Cape Colony, where his father worked as a teacher for many years. He continued his higher education the Victoria College in Stellenbos and then at the Theologies Seminary. After exam and classification, became ZEerw. in 1899 in Klerksdorp, where he is still active. When the latest war broke out, Rev. Strasheim the commandos and he has it his fellow soldiers had the privilege of providing advice and assistance in the field until the end of the battle to support. When, after the war, the battle in the field of education was about to be waged, he was one of those who did everything possible to reach agreement with the authorities to reach an understanding, but has since worked with heart and soul for it Christian National Education. He is also an eloquent advocate of rapprochement between the three Dutch churches.

Dr. H. Tileman.

Dr. Hero Tileman

Dr. Hero Tileman was born March 17, 1873 in Norden in East Friesland, Germany. After having completed his preliminary studies in various universities, he did so Leipzig the state examination for medicine. In November after the outbreak of the war he joined the German section of the German-Belgian "Sanity Expedition" on board the "Herzog" to Delagoa Bay. Tileman first served in the Field Hospital in Ladysmith, then in Pretoria, Norvalspont and Springfontein. Then, recalled to Natal, he founded, together with Dr. Albrecht from Austria, entered a field hospital in Glencoe, and later in Volksrust and Standerton. From the latter place - when the Central Medieval Commission had ceased to exist - Tileman moved to Ermelo and set up an ambulance here, with which he subsequently worked under the Wakkerstroom, Utrecht, Vrijheid, Piet Retief, Swazieland commando and part of the State Artillery served. Tileman had to go often witness his ambulance being captured, his hospital burned down and his patients were taken prisoners of war. He himself became a prisoner of war three times was made, and was arrested and treated as such for the last time after six weeks are, north of the Delagoa bay, track to the Middelburg commando extradited. He traveled by rail again with the government in April 1901 and took orders upon request from the Commandant General in Ermelo took over the service of Dr. Bierens de Haan, that one when was about to travel to Europe. He subsequently served under the Heidelberg, Pretoria, Standerton, Bethal, Ermelo and Middelburg commandos, and part State Artillery; and then some of these commandos were driven over the Natal track his working area extended as far as the Wilge River in the Free State. In the name of the Boer Volk, it was given to him by the Commandant after the war General, presented with a written expression of thanks; on behalf of the ladies of Pretoria, Heidelberg, Bethal and Standerton he received several addresses of thanks. He returned to Europe together with the chief generals and received from the Germans Emperor's Red Cross medal and an appointment as "Kaiserlicher Marine assistant Arzt der Reserve," and currently works as an assistant doctor at the Eppendorfer Krankenhaus in Hamburg.

Reg. Secretary Dirk E. van Velden.

Reg. Secretary Dirk E. van Velden

Reg. Secretary Dirk E. van Velden, born November 22, 1869 in Pietermaritzburg, enjoyed his life education at the grammar school and Victoria College in Stellenbos; was before the war successively a civil servant in the office of the Auditor General, later a clerk at the secretary of the Second Volksraad, and became secretary after his death, which he remained until the war. He went on commando as an ordinary citizen and was appointed in September 1900 Secretary of the Government and as such was present at the peace negotiations. Van Velden finally accompanied General Botha on his trip to Europe as private secretary.

Adv. N. J. De Wet.

Adv. N. J. De Wet

Adv. N. J. De Wet was born in 1873 in the Aliwal Noord district, Cape Colony, son of N. J. de Wet, ex-member of the Legislative Council there, he first studied at the Public School, Aliwal North, later at Victoria College, Stellenbos, and finally at Cambridge University, England, where he obtained his doctorate in law. (L.L.B.) Returned to South Africa in January 1896 and was admitted as a lawyer in the K.K. and de Z. A. R., de Wet settled in Pretoria, where he began to practice. During the first two months of the war he was part of General de la's staff Rey, as Legal Advisor and Secretary. Then he came back to Pretoria as Censor and also for a time as Wd. Assistant State's Attorney. From early June 1900 to At the end of the war he was on the staff of Commandant General Botha as a Military Officer Secretary. He also acted as Wd. State Attorney for the Hogeveld until Feb. 1901, when the Act was passed by the Uitv. Council was appointed as Judge (to hear appeals from Landdrosten).

End Military Officer Album

Bibliography

Many works have been footnoted in the text, enthusiastically if also somewhat idiosyncratically as to form. Restating those references again here would extend this work unnecessarily. I would rather cite material directly in the location where it is most relevant than bury it at the end of the work or in the endnotes where I primarily include citations to images. The full and equally idiomatic bibliography for this series is over sixty pages long and will be included in an Appendix volume to be published at the end of the series. For this volume an abbreviated bibliography is as follows:

Viljoen References

Viljoen Ben, My Reminiscences of the Anglo-Boer war, London: Hood, Douglas, & Howard, 11, Clifford's Inn, E.C. 1902. Maps from Drawings by P. Van Breda. London: Hood, Douglas, & Howard, 11, Clifford's Inn, E.C. 1902. Cite as Viljoen Ben, My Reminiscences, 1902. Maps from Drawings by P. Van Breda.

General Ben J. Viljoen, My reminiscences of the Anglo-Boer war. St. Louis, Mo., Becktold printing & book mfg. co. 1905, Popular ed. Viljoen, Ben J. 1868–1917.

Catalogue Search Use: Internet Archive Search: Ben J. Viljoen

Library of Congress version: Internet Archive Location: https://archive.org/details/myreminiscenceso00viljiala

New York Public Library Better smaller copy from Internet Archive: My reminiscences of the Anglo-Boer war: Viljoen, Ben J. (Ben Johannis), 1868–1917: Free Download, Borrow, and Streaming: Internet Archive

University of Michigan Copy: https://archive.org/details/myreminiscences00viljgoog/mode/1up

Project Gutenberg Edition: My Reminiscences of the Anglo-Boer War by Ben J. Viljoen - Free Ebook (gutenberg.org) https://gutenberg.org/ebooks/25049

Hathi Trust Locations: Catalog Record: My reminiscences of the Anglo-Boer war | HathiTrust Digital Library

Hathi Trust Location: #7 - My reminiscences of the Anglo-Boer war, by General Ben J. Viljoen.... - Full View | HathiTrust Digital Library

My Reminiscences of the Anglo-Boer War (archive.org)

https://ia800201.us.archive.org/25/items/myreminiscences01viljgoog/myreminiscences01viljgoog.pdf

Times History

The Times' History of the War in South Africa 1899–1902 by Amery, L. S. (Leopold Stennett), 1873–1955; Childers, Erskine, 1870–1922; Tallboy, G. P; Williams, Basil, 1867–1950 Sampson Low. Marston and Company, Ltd, 1900–1902

Catalog Search: Internet Archive: Digital Library of Free & Borrowable Texts, Movies, Music & Wayback Machine And

Full Volumes: Times History Vol 1 (1901) https://archive.org/details/timeshistoryofwa01ameruoft

History up until October 19th, 1899

Times History Vol 2 (1902) https://archive.org/details/timeshistoryofwa02amer

Up until Colenso, December 15th 1899

Times History Vol 3 (1905) https://archive.org/details/timeshistoryofwa03amer December 1899 to March 1900

Times History Vol 4 (1906) https://archive.org/details/timeshistoryofwa04amer To Relief of Kimberley and Mafeking May 17th 1900

Times History Vol 5 (1907) https://archive.org/details/timeshistoryofwa05amer Concludes with the peace treaty in June 1902

Times History Vol 6 (1909) https://archive.org/details/timeshistoryofwa06ameruoft Up until 1909

Times History Vol 7 (1909) https://archive.org/details/timeshistoryofwa07ameruoft Appendices, Chronology and a thirty page bibliography followed by an Index (226 pages)

Most maps by Spottiswoode & Co. Ltd.[420]

Doyle, Arthur Conan. The Great Boer War. Vol 1 1900. Vol 1 ISBN: 9781535224338, 1535224339 Page count 314 Volume 1 concludes with Spion Kop. Tauchnitz, edition The Great Boer War - Google Books

Second Edition: https://archive.org/details/greatboerwar00doylgoog

Doyle, Arthur Conan. The Great Boer War: Seventh Impression 1900 (September) London, Smith, Elder & Co. 595 pages This edition concludes with "Halt at Pretoria" The great Boer war: Doyle, Arthur Conan, Sir, 1859-1930: Free Download, Borrow, and Streaming: Internet Archive

Our Military Officers. Album of portraits with life sketches.

Original Publication: Onze krijgs-officieren: album van portretten met levens-schetsen der. by Gustav Schoeman Preller, Frans Vredenrijk Engelenburg 1904 Publisher "Volksstem" kantore

Catalogue Record: https://archive.org/search?query=title%3A%28Onze%20Krijgs-Officieren%29

Direct Link: https://archive.org/details/onzekrijgsoffic00engegoog

Reproduction: Our Military Officers. Album of portraits with life sketches of the Transvaal Generals and Commandants), Volksstem, Pretoria 1904. In Dutch with a preface by Louis Botha. Transcribed and published by: Mr. Pieter van der Merwe.

Onze Krijgs-Officieren Album Van Portretten Met Levens-Schetsen Der Transvaalse Generaals En Kommandanten E. A. Uitgegeven Ten "Volksstem" Kantore, Pretoria. 1904.

File:Onze Krijgs-Officieren. Album van portretten met levens-schetsen, Pretoria, Volksstem, 1904.pdf

Download Access: https://en.wikipedia.org/wiki/File:Onze_Krijgs-Officieren._Album_van_portretten_met_levens-schetsen,_Pretoria,_Volksstem,_1904.pdf

Direct Link to File: https://upload.wikimedia.org/wikipedia/commons/5/52/Onze_Krijgs-Officieren._Album_van_portretten_met_levens-schetsen%2C_Pretoria%2C_Volksstem%2C_1904.pdf

Source Page: https://en.wikipedia.org/wiki/Annie_Botha#cite_note-4

License Links

My many thanks to all the authors, libraries, conservators, and scanners/uploaders who have allowed me to populate, illustrate and annotate this compilation. Responsible efforts have been made to ensure that material in this work public domain material and otherwise is governed by the following licenses.

Creative Commons 1.0: CC0 1.0 Deed | CC0 1.0 Universal | Creative Commons

Creative Commons 1.0: Creative Commons — Attribution 1.0 Generic — CC BY 1.0

Creative Commons 2.0: Creative Commons — Attribution 2.0 Generic — CC BY 2.0

Creative Commons 3.0: Creative Commons — Attribution 3.0 Unported — CC BY 3.0

Creative Commons 4.0: Creative Commons — Attribution 4.0 International — CC BY 4.0

And: CC BY-SA 4.0 Deed | Attribution-ShareAlike 4.0 International | Creative Commons

See Generally: Creative Commons license,[272] Please note that all images have been modified in one way or another during the course of publication.

Also included is material authorized under one or more of the following licenses:

UK Open Government License: This work contains public sector information licensed under the Open Government Licence v3.0. https://www.nationalarchives.gov.uk/doc/open-government-licence/version/3/

UK Open Parliament License: This work contains public sector information licensed under the Open Government Licence v3.0. https://www.parliament.uk/site-information/copyright-parliament/open-parliament-licence/

Some documents are reproduced pursuant to the UK Open Parliament License and this applies not only material in which either House owns the copyright but also material published before 1 August 1989 which is the operative Commencement date of the Copyright and Designs act of 1988; and 50 years having elapsed subsequent to such publication. The same terms applying to previously published material that at one time subsisted under Crown Copyright.[273]

[272] Creative Commons License: Creative Commons license - Wikipedia And About CC Licenses - Creative Commons

[273] If the work was commercially published within 75 years of the end of the calendar year in which it was made, Crown copyright will last for 50 years from the end of the calendar year in which it was published. https://copyrightservice.co.uk/copyright/p01_uk_copyright_law Flow Chart: Crown copyright flowchart 2015

Also, This work contains public sector information made available under the South African Promotion of Access to Information Act, 2000 ("PAIA"). https://www.gov.za/documents/promotion-access-information-act

Google Earth License Links

Google Earth License: https://www.google.com/intl/en-GB_ALL/permissions/geoguidelines/

Sales of these annotated, illustrated and supplemented historical narratives are anticipated to be less than the 5,000 copies permitted. If that is not the case the author will contact the Google Cloud Customer Team for additional permission(s). https://cloud.google.com/contact/

Copyright Discussion

This work is a comprehensive republication of the included works which have been compiled, annotated, illustrated and expanded in order to provide historical context and clarity.

Images and text used by permission or by law are clearly identified by citation and should not be reproduced; as to the remainder there is no copyright claimed by myself on any public domain images, text or any common property information contained in this work.

Future creators should continue to follow the permissions, guidelines and requirements from the original source and author/uploader who is cited. Links to Creative Commons, Google Earth and other licenses are noted in the license links section.

So far as can be determined, as of 2025, the text and images used to illustrate the first person accounts included in this volume reside in the public domain in the United States having all been first published prior to January 1st, 1930 as well as in all other jurisdictions which employ any of the following copyright regimens:

1. A copyright regimen that grants an author's works subsisting copyright for seventy (70) years pma[274] or less.
2. A copyright term subsisting for 120 years from the date of creation or first publication.
3. And such jurisdictions that apply the Rule of the Shorter term to an author's pma date or publication date.

This work will not be marketed or deliberately sold in any jurisdiction with a longer copyright term. See Also: 2025 in public domain - Wikipedia

Copyright for the assembled compilation of this material, or substantial portions thereof that would comprise separate articles or works unto themselves is retained. This work and the series in sum comprise a derivative work. Literary, dramatic, and adaptation rights of the saga demonstrated and explicated by this series is retained.

Otherwise, with respect to the underlying texts and images recast in a more permanently accessible form I choose to pay forward my creative energy to the people who lived through these experiences and it has been a pleasure to find and assemble their collective words and images in a new synoptic and synergistic arrangement.

Except where noted otherwise, every effort has been made to ensure that no material has been used without an attributed source. However, if any non-attributed source, incorrect source, or copyrighted material used without permission or justification, has been included then I would be grateful for information that would enable any errors or inappropriate inclusions to be corrected or omitted in subsequent impressions.

Starting from creation, to publication, through to purchase, archiving, and digitizing, this annotated, illustrated and expanded edition is the sixth step in the journey of these works. My thanks go to all who have helped preserve and perpetuate these important historical records.

[274] post mortem auctoris or "pma" [post mortem author].

Endnotes

[1] Vertical, black and white image of General Ben Viljoen, commander and president of the South African Boer War Exhibit Company General Ben Viljoen https://mohistory.org/collections/item/N35468 And: https://commons.wikimedia.org/wiki/File:General_Ben_Viljoen,_commander_and_president_of_the_South_African_Boer_War_Exhibit_Company.jpg

[2] Image source: britishbattles.com See Also: Ben Viljoen - Wikipedia

[3] General Ben Viljoen with armed Boer soldiers and a baboon (Papio ursinus) with a rifle. Gen. Ben Viljoen (4th from the right back row) and some of his commandos in the Pilgrim's Rest district in 1901. A medic with a Red Cross headband. https://commons.wikimedia.org/wiki/File:General_Ben_Viljoen_with_armed_Boer_soldiers_and_a_baboon_with_a_rifle,_1901.jpg

[4] Cunliffe, History of the Boer War Vol 1 (NY Library Scan).

[5] General Ben Viljoen, commander and president of the South African Boer War Exhibit Company. https://mohistory.org/collections/item/N35468

[6] Front of original Publication: https://commons.wikimedia.org/wiki/File:B.J._Viljoen_-_Mijne_herinneringen_uit_den_Anglo-Boeren-Oorlog,_Amsterdam,_W._Versluys,_1902.gif

[7] In general, look for libraries in northern climates when searching for the best copies of scans. The rate of paper oxidation, combustion, over time is slower.

[8] See the Bibliography, Infra for: The Times' History of the War in South Africa–1902 by Amery, L. S. (Leopold Stennett).

[9] They can be exported from the PDFs as PNGs and further modified to extract more on-screen definition. Most Times History maps were created by Spottiswoode & Co. Ltd.

[10] Crop of map from, My Early Life.

[11] Haldane, How We Escaped from Pretoria. Image credit: Neville Constantine, Neville C. at Angloboerwar.com Anglo Boer War - Armored Train Ambush 11-15-99 Chuchill & Haldane - Page 13 - Boer War Forum

[12] Haldane, How We Escaped from Pretoria. Image credit: Neville Constantine, Neville C. at Angloboerwar.com Anglo Boer War - Armored Train Ambush 11-15-99 Chuchill & Haldane - Page 13 - Boer War Forum

[13] Davitt, Boer Fight for Freedom, 3rd ed. 1902. (Duquesne Scan) Michael Davitt The Boer Fight For Freedom, Funk & Wagnalls Company 1902 Duquesne Scan https://archive.org/details/boerfightforfree00mich/page/n623/mode/2up

[14] Howard Clement Hillegas, The Boers in War. Published by, Appleton and company, New York 1900. Pages 320 to 322 of the US Edition: https://archive.org/details/boersinwarstoryo00howarich Hathi Trust: https://catalog.hathitrust.org/Record/009262432

[15] Library of Congress Maps and Charts # 2877 Copyright 1899 by Geo. H. Walker & Co. Boston. 12.5 inches by 5.25 Inches. Search Bing for Old Maps of South Africa. Link(s) to George H. Walker Map of SA. Link to Yellow Map. But it is a slightly sharper scan. Take out 70% of colour and up contrast ~70%. https://relicmaps.com/products/south-africa-catoosa-county-georgia-united-states-walker-county-whitfield-county-1899-poster-print-gift-33342?variant=47622134497587&msclkid=6d07220b96831debcf184a3085bf2c4d&utm_source=bing&utm_medium=cpc&utm_campaign=Shopping_Search_051025_142025&utm_term=4586337883546955&utm_content=Ad%20group%20%231

[16] In general, look for libraries in northern climates when searching for the best copies of scans. The rate of paper oxidation, combustion, over time is slower.

[17] See the Bibliography, Infra for: The Times' History of the War in South Africa–1902 by Amery, L. S. (Leopold Stennett).

[18] They can be exported from the PDFs as PNGs and further modified to extract more on-screen definition. Most Times History maps were created by Spottiswoode & Co. Ltd.

[19] Times History Vol 5 (1907) https://archive.org/details/timeshistoryofwa05amer Concludes with the peace treaty in June 1902.

[20] Eyre & Spottiswoode was the London-based printing firm established in 1739 that was the King's Printer, and subsequently, a publisher prior to being incorporated; it once went by the name of Spottiswoode, Ballantyne & co. ltd. https://en.wikipedia.org/wiki/Eyre_%26_Spottiswoode

[21] Times History Vol 4 (1906) To Relief of Kimberley and Mafeking May 17th 1900 https://archive.org/details/timeshistoryofwa04amer

[22] Times History Vol 5 (1907) https://archive.org/details/timeshistoryofwa05amer Concludes with the peace treaty in June 1902.

[23] Times History Vol 5 (1907) https://archive.org/details/timeshistoryofwa05amer Concludes with the peace treaty in June 1902.

[24] Times History Vol 5 (1907) https://archive.org/details/timeshistoryofwa05amer Concludes with the peace treaty in June 1902.
[25] Times History Vol 5 (1907) https://archive.org/details/timeshistoryofwa05amer Concludes with the peace treaty in June 1902.
[26] Blake, John Y. Fillmore, A West Pointer with the Boers: (Angel Guardian Press, 1903), U. Cal Scan: https://archive.org/details/westpointerwithb00blakrich
[27] Davitt, Boer Fight for Freedom, 3rd ed. 1902. (Duke Scan)
[28] General Ben Viljoen posing on a path. https://commons.wikimedia.org/wiki/File:General_Ben_Viljoen_posing_on_a_path.jpg
[29] https://commons.wikimedia.org/wiki/File:Hudson_Lowe_BNF_Gallica.jpg
[30] Rinderpest 1896-CN - Rinderpest - Wikipedia
[31] Arthur Mann The Boer in Peace and War, 1900.
[32] Davitt, Boer Fight for Freedom, 3rd ed. 1902. (Duke Scan)
[33] ABW Source.
[34] Anglo-Boer War Album, Cape Town, D. Edwards & co. [1902?]
[35] War pictures. v. 1 (no. 1-16) 1900 #1 - War pictures. v. 1 (no. 1-16) 1900. - Full View | HathiTrust Digital Library And: Catalog Record: War pictures | HathiTrust Digital Library
[36] Heroes of the Boer War. Review of Reviews, 1903.
[37] Davitt, Boer Fight for Freedom, 3rd ed. 1902. (Duke Scan)
[38] John Bull's crime; or, Assaults on republics. 1901 New York The Abbey Press. HathiTrust: Catalog Record: John Bull's crime | HathiTrust Digital Library Internet Archive Records: https://archive.org/search?query=title%3A%28John%20Bull%27s%20Crime%29 NYPL Copy: https://archive.org/details/johnbullscrimeo02davigoog/mode/2up
[39] After Pretoria: The Guerilla War Vol 3 Dec. 1900
[40] Times History, Vol 4.
[41] Heavily loaded ox wagon. - Atom site for DRISA PB1153_001.jpg (1400×922) (drisa.co.za)
[42] Hillegas, Boers in War, 1900.
[43] Hillegas, Boers in War, 1900.
[44] Black & White Weekly Vol 1–2 1899–1900.
[45] Credit Neville Constantine: https://www.angloboerwar.com/forum/21-uniforms-badges-and-equipment/31231-artillery-and-ammunition?start=252
[46] Webster Davis, John Bull's crime; or, Assaults on Republics. 1901 New York The Abbey Press.
[47] War with the Boers, H. Brown, London [1902?] Vol 2.
[48] Courtesy of Neville Constantine: Anglo Boer War - Artillery and Ammunition - Page 17 - Boer War Forum
[49] Courtesy of Neville Constantine: Anglo Boer War - Artillery and Ammunition - Page 17 - Boer War Forum
[50] Black & White Weekly Vol 1-2 1899-1900. "In Fort Pretoria: Boer artillery ready for action."
[51] Photograph of one of the Transvaal Staatsartillerie 75 mm QF Creusot guns (Thanks to MC Heunis) Anglo Boer War - Artillery and Ammunition - Page 17 - Boer War Forum
[52] Portrait photograph of Lieutenant and Adjutant Carl Roos (with thanks to MC Heunis). Anglo Boer War - Armored Train Ambush 11-15-99 Chuchill & Haldane - Page 21 - Boer War Forum
[53] Anglo-Boer War Album, Cape Town, D. Edwards & co. [1902]
[54] Cunliffe, History of the Boer War Vol 1 (NY Library Scan).
[55] Lieutenant-General Sir John French, K.C.B. Commanding Cavalry Div. 1899-1900 Commanding Cape Colony, 1901-1902 Times History War in S.A Vol 3 Photo from MacMillan & Co. from "With General French and the Cavalry in South Africa."
[56] BritishBattles.com https://www.britishbattles.com/
[57] Heroes of the Boer War. Review of Reviews, 1903.
[58] BritishBattles.com https://www.britishbattles.com/
[59] BritishBattles.com https://www.britishbattles.com/
[60] Imperial Light Horse at Elandslaagte by William Barnes Wolllen.
[61] Image by R. Canton Woodville File:Elandslaagte 1.jpg - Wikimedia Commons
[62] Cunliffe, History of the Boer War Vol 1 (NY Library Scan).
[63] Cunliffe, History of the Boer War Vol 1 (NY Library Scan)

[64] With Roberts to the Transvaal. C.N. Robinson [1900]
[65] File:Boer War; searching for the wounded from the battlefield at Wellcome V0015550.jpg - Wikimedia Commons Watercolour sketch by F. Craig, 1899.
[66] Please Note Indian Stretcher Bearer at battle site: File:Boer War; removing the wounded after the Battle of Elandslaa Wellcome V0015512.jpg - Wikimedia Commons
[67] Black & White Weekly Vol 1-2 1899-1900 Photo by D. Barnett. This must have been early in the battle or the next day since the battle ended at night.
[68] Image included to show R.A.M.C. personnel working with Indian bearers. "Removing the wounded after the battle of Elandslaagte. This operation had to be performed in a pouring rain and in almost total darkness." With the Flag to Pretoria, H. W. Wilson Vol 1 1900 Sketch by W. Dewdr.
[69] After Pretoria: The Guerilla War Vol 3 p. 51, 1900
[70] Hillegas, With the Boer forces, 1901 (Photograph by R. Steger.)
[71] File:Hollanderkorps Memorial at Elandslaagte.jpg - Wikimedia Commons Author: David Biggins
[72] Heroes of the Boer War. Review of Reviews, 1903.
[73] With the Flag to Pretoria, H. W. Wilson Vol 2 March 1900.
[74] File:Termite Magnetic DSC03613.jpg - Wikimedia Commons
[75] Heroes of the Boer War. Review of Reviews, 1903.
[76] Anglo-Boer War Album, Cape Town, D. Edwards & co. [1902?]
[77] Webster Davis, John Bull's crime; or, Assaults on Republics. 1901 New York The Abbey Press.
[78] Webster Davis, John Bull's crime; or, Assaults on Republics. 1901 New York The Abbey Press
[79]Cunliffe, History of the Boer War Vol 1 (NY Library Scan).
[80] Webster Davis, John Bull's crime; or, Assaults on Republics. 1901 New York The Abbey Press.
[81] Anglo-Boer War Album, Cape Town, D. Edwards & co. [1902?]
[82] Davitt, Boer Fight for Freedom, 3rd ed. 1902. (Duke Scan).
[83] Davitt, Boer Fight for Freedom, 3rd ed. 1902. (Duke Scan)
[84] Cunliffe, History of the Boer War Vol 1 (NY Library Scan).
[85] Image and data provided by Wellcome Collection https://www.jstor.org/stable/community.24840254
[86] File:The Second Boer War, 1899-1902 Q71973.jpg - Wikimedia Commons
[87] Black & White Weekly Vol 1-2 1899–1900.
[88] Hillegas, Boers in War, 1900.
[89] Haldane, How We Escaped from Pretoria.
[90] War with the Boers, H. Brown, London [1902?] Vol 1.
[91] War with the Boers, H. Brown, London [1902?] Vol 1.
[92] War with the Boers, H. Brown, London [1902?] Vol 1.
[93] Heroes of the Boer War. Review of Reviews, 1903.
[94] Heroes of the Boer War. Review of Reviews, 1903.
[95] Langsnek Tunnel, July 1900. Damaged northern portal after being blown up. - Atom site for DRISA
P1628_01.jpg (1400×814) (drisa.co.za)
[96] A train made up of artillery and ox-waggons. The latter are used in the making [of] laagers for defensive purposes and of course for transport between places where there is no railway communication. The man almost in the centre of the picture in a black uniform is a Transvaal artillery officer. Anglo Boer War - Images from 'On the Transvaal border' published by W H Gill, 1900 - Boer War Forum Credit Dr. David Biggins.
[97] Cunliffe, History of the Boer War Vol 1 (NY Library Scan).
[98] Cunliffe, History of the Boer War Vol 1 (NY Library Scan).
[99] Source:www.angloboerwar.com/forum/19-ephemera/3...l-1900?start=0#89968 Credit: Dr. David Biggins
[100] Source:www.angloboerwar.com/forum/19-ephemera/3...l-1900?start=0#89968 Credit: Dr. David Biggins
[101] Webster Davis, John Bull's crime; or, Assaults on Republics. 1901 New York The Abbey Press.
[102] Reports on military operations in South Africa and China. U.S. Adjutant-General's Office. 1901.
[103] Reports on military operations in South Africa and China. U.S. Adjutant-General's Office. 1901.
[104]Image source: britishbattles.com
[105] Image source: britishbattles.com

[106] Black & White Weekly Vol 1-2 1899–1900.
[107] Cunliffe, History of the Boer War Vol 1 (NY Library Scan).
[108] Image source: britishbattles.com
[109] 155 mm Creusot Long Tom - Wikipedia Image Source: 155 mm Creusot Long Tom - Wikipedia File:Long tom on platform at Mafeking.jpg - Wikimedia Commons
[110] Image source: britishbattles.com
[111] Image source: With the Flag to Pretoria, H. W. Wilson Vol 1 1900 and britishbattles.com
[112]Image Source: With the Flag to Pretoria, H. W. Wilson Vol 1 And: Buller's Campaign with the Natal Field Force Map by A.I. Ltd.
[113] Illustrated Sporting and Dramatic News - Saturday 17 March 1900 p. 21 https://www.britishnewspaperarchive.co.uk/viewer/bl/0001857/19000317/037/0021
[114] Black & White Weekly Vol 1-2 1899-1900 Sketch by Rene' Bull.
[115] Image source: With the Flag to Pretoria, H. W. Wilson Vol 1 1900 and britishbattles.com
[116] Image source: With the Flag to Pretoria, H. W. Wilson Vol 1 1900 and britishbattles.com
[117] Cunliffe, History of the Boer War Vol 1 (NY Library Scan).
[118] Image source: britishbattles.com
[119] Cunliffe, History of the Boer War Vol 1 (NY Library Scan).
[120] Image Source: With the Flag to Pretoria, H. W. Wilson Vol 1 1900 By O. Eckhardt after a sketch by E. Prater.
[121] Image source: With the Flag to Pretoria, H. W. Wilson Vol 1 1900 and britishbattles.com
[122] Image source: britishbattles.com Originally from Navy–Army Illustrated.
[123] Creswicke, Louis South Africa and the Transvaal War, Vol. 2 1900 Photo by Cribb, Southsea.
[124] Repository: Library of Congress Prints and Photographs Division Washington, D.C. 20540 USA https://www.loc.gov/pictures/item/2006684574/resource/
[125] Buller's Campaign with the Natal Field Force 1902
[126] The 155 mm Creusot Long Tom was a French siege gun (artillery piece) manufactured by Schneider et Cie in Le Creusot, France and used by the Boers in the Second Boer War as field guns. Four guns, along with 4,000 common shells, 4,000 shrapnel shells and 800 case shot were purchased by the South African Republic (informally known as the Transvaal) in 1897.
[127] 155 mm Creusot Long Tom - Wikipedia Image Source: 155 mm Creusot Long Tom - Wikipedia File:Long tom on platform at Mafeking.jpg - Wikimedia Commons
[128] Cunliffe, History of the Boer War Vol 1 (NY Library Scan).
[129] Ben Viljoen and his family: Helena Beatrix Els (1871-1950) and Wynand Eugene Viljoen (1891–1978). https://commons.wikimedia.org/wiki/File:Ben_Viljoen_en_gesin.jpg
[130] Image source: Navy–Army Illustrated & britishbattles.com
[131] File:4.7inchGunPietersHillBoerWar.jpeg - Wikimedia Commons Author: B.W. Caney.
[132] With Roberts to the Transvaal. C.N. Robinson [1900] Bacon's War Map.
[133] Anglo-Boer War Album, Cape Town, D. Edwards & co. [1902?]
[134]Black & White Weekly Vol 1–2 1899–1900.
[135] Black & White Weekly Vol 1–2 1899–1900.
[136] File:GiantsCastlePanoramaSmall.jpg - Wikimedia Commons
[137] The maps in WSC, My Early Life and Frontiers and Wars show identical features and labels as the London to Ladysmith version.
[138] A Handbook of the Boer War - Gale and Polden Limited, London 1910 Source: Anonymous ISBN 978-1-37-497455-5. https://babel.hathitrust.org/cgi/pt?id=uc1.l0061277489&seq=7
[139] Cunliffe, History of the Boer War Vol 1 (NY Library Scan).
[140] Davitt, Boer Fight for Freedom, 3rd ed. 1902. (Duke Scan)
[141] With the Flag to Pretoria, H. W. Wilson Vol 2 Dec., 1899. PF: The end of the barrel was damaged and it was cut down and returned to action.
[142] File:Photograph; 'Long Tom', RAMC; 801-22-3 Wellcome L0021295.jpg - Wikimedia Commons
[143] Hillegas, Boers in War, 1900.
[144] File:The Second Boer War, 1899-1902 Q72220.jpg - Wikimedia Commons Author: William Skeoch Cumming
[145] Twenty feet long by seven feet wide, drawn by 14, 16 or 18 trained bullocks "spanned" in pairs History of War SA, Birch, 1899.

[146] Cunliffe, History of the Boer War Vol 1 (NY Library Scan).
[147] Neville Constantine 3rd: Pretoria_Commando_Crop_3.jpg.
[148] File:The Second Boer War, 1899-1902 Q68471.jpg - Wikimedia Commons Author Unknown author
[149] Cunliffe, History of the Boer War Vol 1 (NY Library Scan).
[150] File:The Second Boer War, 1899-1902 Q72297.jpg - Wikimedia Commons Author: William Skeoch Cumming.
[151] Anglo-Boer War Album, Cape Town, D. Edwards & co. [1902?]
[152] Webster Davis, John Bull's crime; or, Assaults on Republics. 1901 New York The Abbey Press.
[153] Davitt, Boer Fight for Freedom, 3rd ed. 1902. (Duke Scan)
[154] Webster Davis, John Bull's crime; or, Assaults on Republics. 1901 New York The Abbey Press.
[155] Anglo-Boer War Album, Cape Town, D. Edwards & co. [1902?]
[156] File:Resa del bacino del Brandewater.jpg - Wikimedia Commons
[157] Benett Burleigh The Natal Campaign Chapman Hall Ltd. 1900 The Natal campaign: Burleigh, Bennet, d. 1914: Free Download, Borrow, and Streaming: Internet Archive
[158] History of War SA, Birch, 1899.
[159] Times History Vol 4.
[160] Anglo-Boer War Album, Cape Town, D. Edwards & co. [1902?]
[161] Anglo-Boer War Album, Cape Town, D. Edwards & co. [1902?]
[162] A Handbook of the Boer War With General Map of South Africa and 18 Sketch Maps and Plans Gale And Polden Limited London 1910 A Handbook of the Boer War - Free Ebook (gutenberg.org) Original by Stanford's Geography
[163] Times History Vol 4.
[164] Heroes of the Boer War. Review of Reviews, 1903. PF: Not sure about the identifications particularly of M. J. Prinsloo and Viljoen.
[165] Ian Hamilton's March 1900 The Copp, Clark Co. Toronto. And Frontiers & Wars.
[166] File:Doornkop Battle.jpg - Wikimedia Commons Author: AxelRavera.
[167] Richard Harding Davis With Both Armies in South Africa. Charles Scribner's Sons, New York 1900.
[168] Richard Harding Davis With Both Armies in South Africa. Charles Scribner's Sons, New York 1900.
[169] Anglo-Boer War Album, Cape Town, D. Edwards & co. [1902?]
[170] With the Flag to Pretoria, H. W. Wilson Vol 2 June, 1900.
[171] Richard Harding Davis With Both Armies in South Africa. Charles Scribner's Sons, New York 1900.
[172] Lord Roberts and staff cheering the Queen--Raising Union Jack on occupation of Pretoria, June 5th, S.A. Library of Congress Prints and Photographs Division Call Number: LOT 11960-8 [item] [P&P] https://www.loc.gov/pictures/item/00651051/resource/
[173] Times History, Vol. 5. E. & N.E. Transvaal.
[174] The Biograph in battle: its story in the South African War related with personal experiences. by William Kennedy Laurie Dickson, London: T. Fisher Unwin, 1901. Catalog Search: https://archive.org/search?query=title%3A%28The+Biograph+in+Battle%29
[175] The Biograph in battle: its story in the South African War related with personal experiences. by William Kennedy Laurie Dickson, London: T. Fisher Unwin, 1901. Catalog Search: https://archive.org/search?query=title%3A%28The+Biograph+in+Battle%29
[176] The Biograph in battle: its story in the South African War related with personal experiences. by William Kennedy Laurie Dickson, London: T. Fisher Unwin, 1901. Catalog Search: https://archive.org/search?query=title%3A%28The+Biograph+in+Battle%29
[177] Frederick Woods, Young Winston's Wars. Viking Press. 1972.
[178] Ian Hamilton's March 1900 The Copp, Clark Co. Toronto. And Frontiers & Wars.
[179] A Handbook of the Boer War - Gale And Polden Limited, London 1910 Source: Anonymous ISBN 978-1-37-497455-5. https://babel.hathitrust.org/cgi/pt?id=uc1.l0061277489&seq=7
[180] Black & White Weekly Vol 3, Sept, 1900. Photo by Biograph Co.
[181] The South African Republic Police (Dutch: Zuid-Afrikaansche Republiek Politie; ZARP). https://en.wikipedia.org/wiki/South_African_Republic_Police
[182] Hiley and Hassell, The Mobile Boer, Grafton Press, New York, (c. 1902).
[183] Hillegas, With the Boer forces, 1901. (No photo credit listed.)
[184] Davitt, Boer Fight for Freedom, 3rd ed. 1902. (Duke Scan)
[185] Hillegas, Boers in War, 1900.

[186] Hillegas, Boers in War, 1900.
[187] Battle of Witpoort, South African War Memorial (Halifax), Nova Scotia. https://commons.wikimedia.org/wiki/File:BattleAtWitpoortSouthAfrica1900.jpg Author: Hantsheroes.
[188] Navy & Army Illustrated Vol 11 p. 123 October 27th, 1900.
[189] Anglo-Boer War Album, Cape Town, D. Edwards & co. [1902?]
[190] Anglo-Boer War Album, Cape Town, D. Edwards & co. [1902?]
[191] File:Boer War Photograph Taken by B W Caney Q115367.jpg - Wikimedia Commons Photo by B W Caney.
[192] Credit Rob Davidson: Anglo Boer War - Artillery and Ammunition - Page 7 - Boer War Forum
[193] After Pretoria: The Guerilla War Vol 3 p. 86 1900.
[194] File:South Africa-Mpumalanga-Long Tom01.jpg - Wikimedia Commons Author: NJR ZA
[195] Map of the battle of Bergendal - final attack (Battles of the nineteenth century, 1901) https://en.wikipedia.org/wiki/Battle_of_Bergendal#/media/File:Map_of_the_battle_of_Bergendal_-_final_attack_(Battles_of_the_nineteenth_century,_1901).jpg
[196] Source: History of the War in South Africa. Author: Maurice & Grant. https://commons.wikimedia.org/wiki/File:Bergendalfarm.jpg
[197] https://commons.wikimedia.org/wiki/File:BergEnDalMemorial.JPG Author: Olivier.karin 2009.
[198] Hillegas, Boers in War, 1900.
[199] Waterval-Boven, circa 1901. Armoured train during Anglo-Boer War at railway station. - Atom site for DRISA N78689.jpg (1400×1045) (drisa.co.za)
[200] File:Termine cremagliera Saline-Volterra.jpg-Wikimedia Commons
[201] Rack railway-Wikipedia
[202] Courtesy of Neville Constantine: Anglo Boer War - Artillery and Ammunition - Page 20 - Boer War Forum
[203] Webster Davis, John Bull's crime; or, Assaults on Republics. 1901 New York The Abbey Press.
[204] Anglo Boer War - Images from 'On the Transvaal border' published by W H Gill, 1900 - Page 2 - Boer War Forum Credit Dr. David Biggins.
[205] File:StateLibQld 2 106408 Ambulance presented by the Women of Queensland for South African Service, 1900.jpg - Wikimedia Commons
[206] File:Photographs of General Hospitals etc. from the Boer War. Wellcome L0023815.jpg - Wikimedia Commons
[207] Times History War in S.A Vol 3 Photo by Steger.
[208] Heroes of the Boer War. Review of Reviews, 1903.
[209] Near Beaufort West, 1900. Train accident. - Atom site for DRISA P1559.jpg (1400×1090) (drisa.co.za)
[210] Anglo-Boer War Album, Cape Town, D. Edwards & co. [1902?]
[211] Haldane, How We Escaped from Pretoria. Image credit: Neville Constantine, Neville C. at Angloboerwar.com Anglo Boer War - Armored Train Ambush 11-15-99 Chuchill & Haldane - Page 13 - Boer War Forum
[212] With Roberts to the Transvaal. C.N. Robinson [1900] This from the Modder River Battlefield in OFS.
[213] Merebank station after arrival of a refugee train. Author LSE Library File:Refugees at Merebank Station, c.1901. (22955486916).jpg - Wikimedia Commons
[214] Heroes of the Boer War. Review of Reviews, 1903.
[215] Anglo-Boer War Album, Cape Town, D. Edwards & co. [1902?]
[216] Anglo-Boer War Album, Cape Town, D. Edwards & co. [1902?]
[217] Anglo-Boer War Album, Cape Town, D. Edwards & co. [1902?]
[218] Courtesy of Neville Constantine: Anglo Boer War - Artillery and Ammunition - Boer War Forum
[219] Coke's Hartebeest (Alcelaphus buselaphus cokii), https://commons.wikimedia.org/wiki/File:Coke%27s_Hartebeest.jpg Authored. Gordon E. Robertson
[220] Richard Harding Davis With Both Armies in South Africa. Charles Scribner's Sons, New York 1900.
[221] Image Source: With the Flag to Pretoria, H. W. Wilson Vol 1.
[222] Johnston, W. and A.K. - South African Republic. Orange Free State, Natal, Basuto Land, Etc. [1897] File:Johnston, W. and A.K. - South African Republic. Orange Free State, Natal, Basuto Land, Etc..jpg - Wikimedia Commons
[223] Webster Davis, John Bull's crime; or, Assaults on Republics. 1901 New York The Abbey Press.
[224] Webster Davis, John Bull's crime; or, Assaults on Republics. 1901 New York The Abbey Press.
[225] Heroes of the Boer War. Review of Reviews, 1903.
[226] Hillegas, Boers in War, 1900.

[227]https://commons.wikimedia.org/wiki/File:StateLibQld_2_236658_Bubbles_the_horse_harnessed_into_a_pony_trap_outside_%27Warwillah%27,_Indooroopilly,_Brisbane,_1900.jpg
[228] These are from Hillegas 1900 Pages 320 to 322 of the US Edition titled The Boers in War.
[229] Heroes of the Boer War. Review of Reviews, 1903.
[230] The Sphere - Saturday 09 June 1900 p. 16.
https://www.britishnewspaperarchive.co.uk/viewer/bl/0001861/19000609/015/0016
[231] Haldane, How We Escaped from Pretoria. p. 191.
[232] Haldane, How We Escaped from Pretoria. p. 182.
[233] Haldane, How We Escaped from Pretoria. Image credit: Neville Constantine, Neville C. at Angloboerwar.com
[234] Courtesy of Neville Constantine: Anglo Boer War - Artillery and Ammunition - Page 4 - Boer War Forum
[235] File:The Second Boer War, 1899-1902 Q72291.jpg - Wikimedia Commons Author: William Skeoch Cumming\
[236] Credit Rob Davidson: Anglo Boer War - Artillery and Ammunition - Page 7 - Boer War Forum
[237] Credit Rob Davidson: Anglo Boer War - Artillery and Ammunition - Page 7 - Boer War Forum
[238] Credit Rob Davidson: Anglo Boer War - Artillery and Ammunition - Page 7 - Boer War Forum
[239] War pictures. v. 1 (no. 1–16) 1900 #1 - War pictures. v. 1 (no. 1-16) 1900. - Full View | HathiTrust Digital Library And: Catalog Record: War pictures | HathiTrust Digital Library
[240] File:State artillery corps, Orange Free State - Korps van die Staatsartillerie, Oranje-Vrystaat, near Colesberg.jpg - Wikimedia Commons
[241] https://commons.wikimedia.org/wiki/File:Heliograph-2.jpg
[242] https://en.wikipedia.org/wiki/Heliograph
[243] October 21 Folder Anglo Boer War - Small Arms and Ammunition - Page 4 - Boer War Forum
[244] Per Neville Constantine: Anglo Boer War - Artillery and Ammunition - Page 10 - Boer War Forum
[245] Navy & Army Illustrated Vol 10.
[246] https://commons.wikimedia.org/wiki/File:Steenkampsbergpas1.JPG Author: Michaelphillipr.
[247] Blake, John Y. Fillmore, A West Pointer with the Boers: (Angel Guardian Press, 1903), U. Cal Scan:
[248] PF: I did not record a 'source' for this picture when it was found and processed. Probably Times History.
[249] With Roberts to the Transvaal. C.N. Robinson [1900].
[250] The reminiscences of Lady Randolph Churchill, by Mrs. George Cornwallis-West. New York: Century 1908.
[251]. War pictures. v. 1 (no. 1-16) 1900 #1 - War pictures. v. 1 (no. 1-16) 1900. - Full View | HathiTrust Digital Library And: Catalog Record: War pictures | HathiTrust Digital Library
[252] The Reverse at Helvetia - The Argus (Melbourne), 2 January 1901.
https://commons.wikimedia.org/wiki/Category:Battle_of_Helvetia#/media/File:The_Boer_War._The_Reverse_at_Helvetia_-_The_Argus_(Melbourne),_2_January_1901.jpg
[253] With Roberts to the Transvaal. C.N. Robinson [1900]
[254] Image source: With the Flag to Pretoria, H. W. Wilson Vol 1 1900 and britishbattles.com
[255] Davitt, Boer Fight for Freedom, 3rd ed. 1902. (Duke Scan)
[256] Heroes of the Boer War. Review of Reviews, 1903.
[257] Boer War: picket on the guard at night. Reproduction of a watercolour. Image and data provided by Wellcome Collection https://www.jstor.org/stable/community.24840267
[258] File:The Second Boer War, 1899-1902 Q72521.jpg - Wikimedia Commons Author: William Skeoch Cumming.
[259] File:The Second Boer War, 1899-1902 Q72085.jpg - Wikimedia Commons Author: William Skeoch Cumming.
[260] Pattern 2 train. Circa 1901. New pattern bogie on No 12 armoured train. (Publication on armoured trains in the Anglo Boer War) - Atom site for DRISA
PB3032_007.jpg (1237×969) (drisa.co.za)
[261] File:The Second Boer War, 1899-1902 Q72436.jpg - Wikimedia Commons Author: William Skeoch Cumming
[262] Anglo-Boer War Album, Cape Town, D. Edwards & co. [1902?]
[263] Creswicke Louis, South Africa and the Transvaal War Vol 7 p. 88 Photo: London Stereoscopic Co.
[264] Navy & Army Illustrated Vol 9.
[265] Black & White Weekly Vol 1–2 1899–1900.
[266] Original Source: Edwin Sharpe Grew: Field-Marshal Lord Kitchener. His Life and Work for the Empire. London 1916. Volume 2. File:Kitchener at the Peace Conference that ended the Second Boer War.png - Wikimedia Commons On 28 February 1901 British commander Lord Kitchener met Boer commander Louis Botha at Middelburg, Transvaal.

Category:Middelburg Peace Conference - Wikimedia Commons Also: File:The Second Boer War, 1899-1902 Q68497.jpg - Wikimedia Commons

[267] File:Kitchener at the Peace Conference that ended the Second Boer War.png - Wikimedia Commons

[268] Image: Hillegas, With the Boer forces, 1901 (Photograph by Leo Weinthal, Pretoria.) and quote: With the Flag, After Pretoria: The Guerilla War Vol 3 p. 91 1900.

[269] Page 13 https://www.britishnewspaperarchive.co.uk/viewer/bl/0001684/19010309/091/0013

[270] Southport Guardian - Saturday 23 March 1901 page 5. Col 7 https://www.britishnewspaperarchive.co.uk/viewer/bl/0001783/19010323/121/0005

[271] Southport Guardian - Saturday 23 March 1901 page 5. Col 7 https://www.britishnewspaperarchive.co.uk/viewer/bl/0001783/19010323/121/0005

[272] Southport Guardian - Saturday 23 March 1901 page 5. Col 7. https://www.britishnewspaperarchive.co.uk/viewer/bl/0001783/19010323/121/0005

[273] File:Philip Rudolph Botha (1851-1901), Boer War general.jpg - Wikimedia Commons

[274] Map of Sekhukhuneland or Sekukuniland. Author: Xufanc https://commons.wikimedia.org/wiki/File:Sekukuniland-South_Africa.png

[275] Map showing the location of the former bantustan of Lebowa in South Africa. Author: Htonl https://commons.wikimedia.org/wiki/File:Lebowa_in_South_Africa.svg

[276] Bechuanaland Protectorate: https://en.wikipedia.org/wiki/Bechuanaland_Protectorate

[277] File:Chief-khama-III.jpg - Wikimedia Commons See: Botswana - Wikipedia

[278] War in SA, Harding 1899.

[279] Portrait of Sebele in his twenties taken by German anthropologist Gustav Fritsch at Ntsweng (nowadays, Old Molepolole) in 1865 https://commons.wikimedia.org/wiki/File:Sebele_I_by_Fritsch_1865.jpg

[280] https://commons.wikimedia.org/wiki/File:Bathoen_I_in_london1895.jpg

[281] South Africa; its history, heroes and wars by Mackenzie, William Douglas (William Douglas), 1859-1936; Assisted by Alfred Stead, 1877-1933, joint author. Published 1899 by The Co-Operative Publishing Company, Chicago, Ill. Page 160 South Africa; its history, heroes and wars: Mackenzie, W. Douglas (William Douglas), 1859-1936: Free Download, Borrow, and Streaming: Internet Archive

[282] SA National Archives: http://www.nationalarchives.gov.za/node/1236

[283]To The Bitter End: a photographic history of the Boer War, 1899–1902, by Lee, Emanoel C. G Harmondsworth: Penguin, 1986. https://archive.org/details/tobitterend00eman_0

[284] Hillegas, Boers in War, 1900.

[285] Davitt, Boer Fight for Freedom, 3rd ed. 1902. (Duke Scan)

[286] Anglo-Boer War Album, Cape Town, D. Edwards & co. [1902?]

[287] To The Bitter End: a photographic history of the Boer War, 1899–1902, by Lee, Emanoel C. G. Harmondsworth: Penguin, 1986. https://archive.org/details/tobitterend00eman_0

[288] War with the Boers, H. Brown, London [1902?] Vol 1.

[289] Hiley and Hassell, The Mobile Boer, Grafton Press, New York, (c. 1902)

[290] Heroes of the Boer War. Review of Reviews, 1903.

[291] War with the Boers, H. Brown, London [1902?] Vol 2.

[292] Hillegas, Boers in War, 1900.

[293] BoerWarBlockHouse SouthAfrica - Second Boer War - Wikipedia Credit: Danie van der Merwe.

[294] With Roberts to the Transvaal. C.N. Robinson [1900].

[295] A gun carrying carriage named "Cock of the North" of No. 3 Artillery Train at the railway station at Bronkhorstspruit, Gauteng Province. File:The Second Boer War, 1899-1902 Q72437.jpg - Wikimedia Commons Author: William Skeoch Cumming.

[296] Circa 1901. No 18 armoured train. (Publication on armoured trains in the Anglo Boer War) - Atom site for DRISA PB3032_008.jpg (1257×1001) (drisa.co.za)

[297] With the Flag to Pretoria, H. W. Wilson Vol 2 June, 1900. Photo by Sergt. Haslam, R.E.

[298] Anglo-Boer War Album, Cape Town, D. Edwards & co. [1902?]

[299] Haldane, How We Escaped from Pretoria. Image credit: Neville Constantine, Neville C. at Angloboerwar.com Anglo Boer War - Armored Train Ambush 11-15-99 Chuchill & Haldane - Page 13 - Boer War Forum

[300] Navy & Army Illustrated Vol 11 p. 123 October 27th, 1900.

[301] File:The Second Boer War, 1899-1902 Q72248.jpg - Wikimedia Commons Author: William Skeoch Cumming.

[302] File:The Second Boer War, 1899-1902 Q72102.jpg - Wikimedia Commons
[303] Navy & Army Vol 14 p. 285, June 7th 1902. By the time of publication the hostilities had ended.
[304] File:The Second Boer War, 1899-1902 Q72204.jpg - Wikimedia Commons
[305] File:The Second Boer War, 1899-1902 Q72160.jpg - Wikimedia Commons Author: William Skeoch Cumming.
[306] File:The Second Boer War, 1899-1902 Q72284.jpg - Wikimedia Commons Author: William Skeoch Cumming
[307] File:Andries Petrus Johannes Cronjé (1849-1923) with ammo.jpg - Wikimedia Commons
[308] Hillegas, With the Boer forces, 1901. (Photograph by the Author.)
[309] War with the Boers, H. Brown, London [1902?] Vol 2.
[310] Hillegas, With the Boer forces, 1901. (Photograph by Reginald Sheppard, Pretoria.)
[311] Blake, John Y. Fillmore, A West Pointer with the Boers: (Angel Guardian Press, 1903), U. Cal Scan:
[312] After Pretoria: The Guerilla War Vol 3 April, 1901.
[313] Ganna, circa 1902. Damage to rolling stock after a train was captured by Boer forces at Ganna siding near Nelspoort. - Atom site for DRISA
[314] To The Bitter End: a photographic history of the Boer War, 1899–1902, by Lee, Emanoel C. G Harmondsworth: Penguin, 1986. https://archive.org/details/tobitterend00eman_0
[315] Anglo-Boer War Album, Cape Town, D. Edwards & co. [1902?]
[316] General Ben Viljoen posing on a path. https://commons.wikimedia.org/wiki/File:General_Ben_Viljoen_posing_on_a_path.jpg
[317] File:The Second Boer War, 1899-1902 Q72256.jpg - Wikimedia Commons Author: William Skeoch Cumming.
[318] File:The Second Boer War, 1899-1902 Q71973.jpg - Wikimedia Commons
[319] File:The Second Boer War, 1899-1902 Q72196.jpg - Wikimedia Commons Author: William Skeoch Cumming.
[320] Cunliffe, History of the Boer War Vol 2. (Composite of Mich. St. & U. Cal. Scans.)
[321] File:The Second Boer War, 1899-1902 Q72139.jpg - Wikimedia Commons Author: William Skeoch Cumming.
[322] File:The Second Boer War, 1899-1902 Q72137.jpg - Wikimedia Commons Author: William Skeoch Cumming.
[323] Anglo Boer War - Images from 'On the Transvaal border' published by W H Gill, 1900 - Page 3 - Boer War Forum Credit Dr. David Biggins.
[324] Anglo Boer War - Images from 'On the Transvaal border' published by W H Gill, 1900 - Page 3 - Boer War Forum Credit Dr. David Biggins.
[325] File:The Second Boer War, 1899-1902 Q72301.jpg - Wikimedia Commons Author: William Skeoch Cumming
[326] Eight heads of Cornish stamps powered by a waterwheel. https://commons.wikimedia.org/wiki/File:Geevor_waterwheel_stamps.jpg
[327] Bisbee, Arizona ca. 1900. https://westernmininghistory.com/state/arizona/
[328] Stamp Mill. https://commons.wikimedia.org/wiki/File:FiveStampMill5629.JPG The original uploader was Leonard G.
[329] After Pretoria: The Guerilla War Vol 3 July, 1900.
[330] War pictures. v. 1 (no. 1–16) 1900 #1 - War pictures. v. 1 (no. 1-16) 1900. - Full View | HathiTrust Digital Library And: Catalog Record: War pictures | HathiTrust Digital Library
[331] Blake, John Y. Fillmore, A West Pointer with the Boers: (Angel Guardian Press, 1903), U. Cal Scan:
[332] Blake, John Y. Fillmore, A West Pointer with the Boers: (Angel Guardian Press, 1903), U. Cal Scan:
[333] Heroes of The Boer War.
[334] Blake, John Y. Fillmore, A West Pointer with the Boers: (Angel Guardian Press, 1903), U. Cal Scan:
[335] File:Phoenix Stereopticon.png - Wikimedia Commons
[336] File:TradeCardMasonAndHamlinOrganCoCirca1880s.jpg - Wikimedia Commons
[337] File:Canadian troops in armoured train South Africa Jan 1902 LAC 3407086.jpg - Wikimedia Commons Author Roberts, George F.
[338] Credit Neville Constantine. Armoured Train Manned by Terrible's Marines.
[339] Circa 1901. Six-inch gun. (Publication on armoured trains in the Anglo Boer War) - Atom site for DRISA P3032_04.jpg (1400×1085) (drisa.co.za)
[340] Waterval-Boven, circa 1901. Armoured train during Anglo-Boer War at railway station. - Atom site for DRISA N78689.jpg (1400×1045) (drisa.co.za)
[341] General Ben Viljoen and his secretary after their capture by the British in front of a tent, January 1902. Handwritten caption: "Gen. Ben Viljeon & Secretary being / brought into line after capture. Jan./02." https://commons.wikimedia.org/wiki/File:General_Viljoen_and_his_secretary_after_their_capture,_January_1902.jpg
[342] Image Source: Boer PoWs | Saint Helena Island Info: All about St Helena, in the South Atlantic Ocean

[343] https://commons.wikimedia.org/wiki/File:Longwood_House.jpg

[344] https://commons.wikimedia.org/wiki/File:Longwood_House_le_12_janvier_2008.jpg Author: Michel Dancoisne-Martineau.

[345] Image Source: Boer PoWs | Saint Helena Island Info: All about St Helena, in the South Atlantic Ocean

[346] The first shipment of 514 Pow's arrived on the 10th April 1900, including General Cronjé and his wife, Colonel Schiel and 21 other officers. Image Source: Exiles | Saint Helena Island Info: All about St Helena, in the South Atlantic Ocean

[347] Image Source: Boer PoWs | Saint Helena Island Info: All about St Helena, in the South Atlantic Ocean Credit: Domaines Francais de Sainte Helene boercampdeadwood2_ishe.jpg

[348] Image Source: Exiles | Saint Helena Island Info: All about St Helena, in the South Atlantic Ocean

[349] Image Source: Exiles | Saint Helena Island Info: All about St Helena, in the South Atlantic Ocean

[350] Pocket tinderbox with firesteel and flint. This type was used during the Boer War due to a scarcity of matches. Author: Theo V. Bresler, File:Tinderbox.jpg - Wikimedia Commons

[351] File:The Second Boer War, 1899-1902 Q72037.jpg - Wikimedia Commons Author: William Skeoch Cumming

[352] File:The Second Boer War, 1899-1902 Q72296.jpg - Wikimedia Commons Author: William Skeoch Cumming

[353] Anglo-Boer War Album, Cape Town, D. Edwards & co. [1902?]

[354] Anglo-Boer War Album, Cape Town, D. Edwards & co. [1902?]

[355] Basuto-Pony - Basuto pony - Wikipedia

[356] File:Howard C. Hillegas.jpg - Wikimedia Commons Source: With the Boer Forces Howard C. Hillegas, Methuen & Co. 36 Essex Street W.C., London, 1900. Author: Thomas Franklin Fairfax Millard for Howard C. Hillegas.

[357] War with the Boers, H. Brown, London [1902?] Vol 1.

[358] War with the Boers, H. Brown, London [1902?] Vol 1.

[359] The file is most likely misnamed. It is in use in the SAR (Transvaal) page: File:Seal of the Orange Free State.png - Wikimedia Commons

[360] General Ben Viljoen delivering his lecture on the Boer War at Queen's Hall, October 16 1902 - G, Montbard. Illustrated London News - Saturday 25 October 1902 p 602 https://commons.wikimedia.org/wiki/File:General_Ben_Viljoen_delivering_his_lecture_on_the_Boer_War_at_Queen%27s_Hall,_October_16_1902_-_G,_Montbard.jpg

[361] Boer PoWs | Saint Helena Island Info: All about St Helena, in the South Atlantic Ocean License: You are welcome to quote from this site. We are happy for you to reproduce content found on Saint Helena Island Info BUT with a few constraints: You must credit this site with a link to at least http://sainthelenaisland.info/, or ideally to the URL of the page where the content was found. Failure to do so will constitute an infringement of our Copyright. Not all content on this site is actually owned by us. If you see a credit marker thus: {a} beside or superimposed on content, click on the marker to see the item's owner. You will need to contact that person/organisation for permission to use the content. If permission is granted you must still credit this site (perhaps with Found on http://sainthelenaisland.info/, as 1: (above) You do not need to ask our explicit permission as long as these requirements are followed. If you cannot meet these requirements but still wish to reproduce our content please contact us - we may be prepared to make other arrangements.

[362] The Governor of St Helena | Saint Helena Island Info: All about St Helena, in the South Atlantic Ocean

[363] Robert Armitage Sterndale - Wikipedia Was Governor June 7th, 1897 to July 25th July 1902.

[364] Black & White Weekly Vol 3 1900.

[365] The first shipment of 514 Pow's arrived on the 10th April 1900, including General Cronjé and his wife, Colonel Schiel and 21 other officers. Image Source: Exiles | Saint Helena Island Info: All about St Helena, in the South Atlantic Ocean

[366] Image Source: Boer PoWs | Saint Helena Island Info: All about St Helena, in the South Atlantic Ocean

[367] Image Source: Boer PoWs | Saint Helena Island Info: All about St Helena, in the South Atlantic Ocean

[368] Image Source: Boer PoWs | Saint Helena Island Info: All about St Helena, in the South Atlantic Ocean

[369] Image Source: Boer PoWs | Saint Helena Island Info: All about St Helena, in the South Atlantic Ocean

[370] Image Source: Boer PoWs | Saint Helena Island Info: All about St Helena, in the South Atlantic Ocean

[371] Image Source: Boer PoWs | Saint Helena Island Info: All about St Helena, in the South Atlantic Ocean

[372] Image Source: Boer PoWs | Saint Helena Island Info: All about St Helena, in the South Atlantic Ocean

[373] Image Source: Boer PoWs | Saint Helena Island Info: All about St Helena, in the South Atlantic Ocean

[374] Image Source: Boer PoWs | Saint Helena Island Info: All about St Helena, in the South Atlantic Ocean

[375] Image Source: Boer PoWs | Saint Helena Island Info: All about St Helena, in the South Atlantic Ocean

[376] Image Source: Boer PoWs | Saint Helena Island Info: All about St Helena, in the South Atlantic Ocean

[377] Image Source: Boer PoWs | Saint Helena Island Info: All about St Helena, in the South Atlantic Ocean

[378] Image Source: Boer PoWs | Saint Helena Island Info: All about St Helena, in the South Atlantic Ocean

[379] Image Source: Boer PoWs | Saint Helena Island Info: All about St Helena, in the South Atlantic Ocean
[380] Image Source: Boer PoWs | Saint Helena Island Info: All about St Helena, in the South Atlantic Ocean
[381] Image Source: Boer PoWs | Saint Helena Island Info: All about St Helena, in the South Atlantic Ocean
[382] Image Source: Boer PoWs | Saint Helena Island Info: All about St Helena, in the South Atlantic Ocean
[383] Black & White Weekly Vol 3 1900 Photo by Innes, Jamestown, St. Helena.
[384] Anglo-Boer war prisoners in St Helena showing old men and young boys with toys made at the camp. Source: Parliament Archive. Women and Children in White Concentration Camps during the Anglo-Boer War, 1900-1902 | South African History Online (sahistory.org.za)
[385] Image Source: Boer PoWs | Saint Helena Island Info: All about St Helena, in the South Atlantic Ocean
[386] File:German prisoners of war on the Island of St Helena.jpg - Wikimedia Commons
[387] Smorenburg's crate is apparently on display at the Museum Africa in Johannesburg. Image Source: Boer PoWs | Saint Helena Island Info: All about St Helena, in the South Atlantic Ocean
[388] Image Source: Boer PoWs | Saint Helena Island Info: All about St Helena, in the South Atlantic Ocean
[389] Image Source: Boer PoWs | Saint Helena Island Info: All about St Helena, in the South Atlantic Ocean
[390] Image Source: Boer PoWs | Saint Helena Island Info: All about St Helena, in the South Atlantic Ocean
[391] Image Source: Boer PoWs | Saint Helena Island Info: All about St Helena, in the South Atlantic Ocean
[392] File:1904 worlds fair boer war program.jpg - Wikimedia Commons
[393] Boer War parade, World's Fair, St. Louis, Mo. | Library of Congress (loc.gov)
[394] Arrival of the President and Mrs. Roosevelt for luncheon at West Pavilion, World's Fair, St. Louis, U. S. A. - digital file from original stereograph | Library of Congress (loc.gov)
[395] Inspiring view NW from German Building over Festival Hill, World's Fair, St. Louis, 1904 | Library of Congress (loc.gov)
[396] Festival Hall, celebrating the world's advance since the Louisiana Purchase (1803-1904), St. Louis, U. S. A. | Library of Congress (loc.gov)
[397] West pavilion and floral garden, Louisiana Purchase Exposition, St. Louis, Mo., U. S. A. | Library of Congress (loc.gov)
[398] Grand Basin and splendid Electricity Bldg., from east pavilion of Festival Hall, World's Fair, St. Louis, U. S. A. | Library of Congress (loc.gov)
[399] Night thoughts of the Great World's Fair - Palace of Electricity and Machinery from "East Cascade," St. Louis, U. S. A. | Library of Congress (loc.gov)
[400] Pharus-map World's Fair St. Louis, 1904. | Library of Congress (loc.gov)
[401] Gen. Cronje, Viljoen and Boshoff, Boer War, Louisiana Purchase Exposition, St. Louis, Mo., U.S.A. | Library of Congress (loc.gov) Keystone View Co.
[402] General Ben Viljoen, commander and president of the South African Boer War Exhibit Company. https://mohistory.org/collections/item/N35468
[403] https://mohistory.org/collections/item/N15932
[404] A daring leap, illustrating Gen. DeWet's escape, Boer War spectacle, World's Fair, St. Louis, Mo., | Library of Congress (loc.gov)
[405] File:Re-enactment of a Boer War battle scene on the Pike at the 1904 World's Fair.jpg - Wikimedia Commons
[406] An Exciting skirmish in the Battle of Colenso, Boer War, World's Fair, St. Louis, U.S.A. | Library of Congress (loc.gov)
[407] File:"Boer Cavalry ready for Mimic War." Boer War Exhibit at the 1904 World's Fair.jpg - Wikimedia Commons
[408] End of the first battle, the Boers victorious, Boer War, World's Fair, St. Louis, U.S.A. | Library of Congress (loc.gov)
[409] After the Battle of Colenso, Boer War exhibit, Louisiana Purchase Exposition | Library of Congress (loc.gov)
[410] Explosion of an ammunition wagon https://mohistory.org/collections/item/N35467
[411] Explosion of an ammunition wagon during the Battle of Paardeberg, Boer War, World's Fair, St. Louis, U.S.A. | Library of Congress (loc.gov)
[412] https://mohistory.org/collections/item/N35465
[413] https://mohistory.org/collections/item/N35466
[414] Hand-to-hand fight, Highland Brigade exterminated, Battle of Colenso, Boer War, Louisiana Purchase Exposition, St. Louis, Mo., U.S.A. | Library of Congress (loc.gov)
[415] Gordon Highlanders leaving the arena, Boer War, World's Fair, St. Louis, U.S.A. | Library of Congress (loc.gov)
[416] Ben Viljoen in front of his tent in St. Louis in 1904. https://commons.wikimedia.org/wiki/File:Ben_Viljoen_in_front_of_his_tent,_St._Louis_1904.jpg

[417] Commander_Boshoff,_Generals_Cronjé_and_Viljoen,_St._Louis_1904.
https://commons.wikimedia.org/wiki/File:Commander_Boshoff,_Generals_Cronj%C3%A9_and_Viljoen,_St._Louis_1904.jpg

[418] https://commons.wikimedia.org/wiki/File:Genl_Christiaan_Frederik_Beyers.jpg

[419] Image source: With the Flag to Pretoria, H. W. Wilson Vol 1 1900 and britishbattles.com

[420] https://en.wikipedia.org/wiki/Eyre_%26_Spottiswoode

Made in United States
Orlando, FL
05 November 2026

90230216R00201